MULTIPLE COMPARISONS, SELECTION, AND APPLICATIONS IN BIOMETRY

STATISTICS: Textbooks and Monographs

A Series Edited by

D. B. Owen, Founding Editor, 1972–1991

W. R. Schucany, Coordinating Editor
Department of Statistics
Southern Methodist University
Dallas, Texas

R. G. Cornell, Associate Editor
for Biostatistics
University of Michigan

W. J. Kennedy, Associate Editor
for Statistical Computing
Iowa State University

A. M. Kshirsagar, Associate Editor
for Multivariate Analysis and
Experimental Design
University of Michigan

E. G. Schilling, Associate Editor
for Statistical Quality Control
Rochester Institute of Technology

1. The Generalized Jackknife Statistic, *H. L. Gray and W. R. Schucany*
2. Multivariate Analysis, *Anant M. Kshirsagar*
3. Statistics and Society, *Walter T. Federer*
4. Multivariate Analysis: A Selected and Abstracted Bibliography, 1957–1972, *Kocherlakota Subrahmaniam and Kathleen Subrahmaniam*
5. Design of Experiments: A Realistic Approach, *Virgil L. Anderson and Robert A. McLean*
6. Statistical and Mathematical Aspects of Pollution Problems, *John W. Pratt*
7. Introduction to Probability and Statistics (in two parts), Part I: Probability; Part II: Statistics, *Narayan C. Giri*
8. Statistical Theory of the Analysis of Experimental Designs, *J. Ogawa*
9. Statistical Techniques in Simulation (in two parts), *Jack P. C. Kleijnen*
10. Data Quality Control and Editing, *Joseph I. Naus*
11. Cost of Living Index Numbers: Practice, Precision, and Theory, *Kali S. Banerjee*
12. Weighing Designs: For Chemistry, Medicine, Economics, Operations Research, Statistics, *Kali S. Banerjee*
13. The Search for Oil: Some Statistical Methods and Techniques, *edited by D. B. Owen*
14. Sample Size Choice: Charts for Experiments with Linear Models, *Robert E. Odeh and Martin Fox*
15. Statistical Methods for Engineers and Scientists, *Robert M. Bethea, Benjamin S. Duran, and Thomas L. Boullion*
16. Statistical Quality Control Methods, *Irving W. Burr*
17. On the History of Statistics and Probability, *edited by D. B. Owen*
18. Econometrics, *Peter Schmidt*

Additional Volumes in Preparation

MULTIPLE COMPARISONS, SELECTION, AND APPLICATIONS IN BIOMETRY

A Festschrift in Honor of Charles W. Dunnett

edited by

FRED M. HOPPE

McMaster University
Hamilton, Ontario, Canada

Marcel Dekker, Inc.   New York • Basel • Hong Kong

Library of Congress Cataloging-in-Publication Data

Multiple comparisons, selection, and applications in biometry : a
    festschrift in honor of Charles W. Dunnett / edited by Fred M.
    Hoppe.
        p.   cm. -- (Statistics, textbooks and monographs)
        Includes bibliographical references and index.
        ISBN 0-8247-8895-8 (alk. paper)
        1. Multiple comparisons (Statistics) 2. Ranking and selection
    (Statistics) 3. Biometry I. Hoppe, Fred M.
    II. Dunnett, Charles W.   III. Series.
    QA278.4.M85 1993
    519.5--dc20   92-31061
                                                                   CIP

This book is printed on acid-free paper.

MARCEL DEKKER, INC.
270 Madison Avenue, New York, New York 10016

Current printing (last digit):
10 9 8 7 6 5 4 3 2 1

PRINTED IN THE UNITED STATES OF AMERICA

Preface

This collection of research papers has been prepared for the seventieth birthday of Charles W. Dunnett. Many describe current developments in *multiple comparison methods* and *selection procedures*, to which he has made important contributions. There are close affinities between the two topics, discussed in a number of these papers, and readily apparent in the context of multiple comparisons of treatments with a control and subset selection in single factor experiments, where both methodologies use the same tables developed by Dunnett for multiple comparisons with a control. The combination of these two topics in one book therefore provides a unique fusion of papers to stimulate further work. The authors are not only affiliated with universities, but also come from industry, government, and private practice. Consequently the papers reflect an unusual blend of theory and applications. It is my hope that this volume will be an up-to-date reference to the literature for professional statisticians working in the areas represented, a useful resource for end users in the fields of application, and a readable monograph for graduate students seeking out research topics.

These papers were presented at a Symposium on Biostatistics held at McMaster University, Hamilton, Canada to honour Charles Dunnett. His stature has been recognized not only in Canada, through his Presidency of the Statistical Society of Canada and receipt of the Society's Gold Medal, but also internationally, by his election as a Fellow of the American Statistical Association and his Presidency of the Biometric Society (ENAR). During his career, spanning the dual research pillars of industry and academia, he has influenced the understanding of statistical methodology in the medical and pharmaceutical fields. As Peter Armitage points out in his paper, the most common requests he received as Editor of *Biometrics* for permission to reproduce tables and other material, were for Dunnett's "New tables for multiple comparisons with a control" in *Biometrics* **20** 482-491.

Charles Dunnett's papers reveal clarity of thought and presentation. Some are directed at health and pharmaceutical professionals, such as his expository surveys "Biostatistics in pharmacological testing" in *Medicinal Research: Biology and Chemistry, III Pharmacological Testing Methods,* or "Drug screening: The never-ending search for new and better drugs" in *Statistics, A Guide to the Unknown*. Others are highly mathematical, though motivated by substantive applications. Mathematics is used because it provides the right tools, but the ultimate aim has always been to develop methods for better understanding data; this theme is recurrent throughout his work. Many students of the subject of statistics, seduced by the austere beauty and the refined elegance of its mathematical component as persuasively presented in polished tomes, develop neither an interest in nor skill at analyzing data. They would benefit much were they encouraged to get down into the trenches with the experimenters.

It is my great pleasure to thank the contributors for their enthusiasm and co-operation in helping to bring this entire project, the conference and the publication of their research, to happy fruition. From all of us, and many other well-wishers who could not be included here, this volume is affectionately dedicated to Charlie.

I am grateful for the financial backing of the Natural Sciences and Engineering Research Council of Canada, whose initial support seeded this undertaking. Generous funding was subsequently provided by the Faculty of Science and the Department of Clinical Epidemiology and Biostatistics at McMaster University, the Medical Research Council of Canada, and the Clinical Trials Methodology Group of the Hamilton Civic Hospitals Research Centre. Support was also provided by The Upjohn Company and by Merck Sharp & Dohme Research Laboratories. I thank all these organizations for their splendid foresight, and I wish to express appreciation to George Browman, Charles Goldsmith, Bob McNutt, and especially Michael Gent for their efforts to help secure funding.

I gladly acknowledge the assistance of Patsy Chan whose administrative skills were invaluable. Communications and transmission of papers were facilitated by the magical world of e-mail and fax, and many referees gave freely of their time carefully reading the material. Typesetting was accomplished by the equally wondrous TeX and LaTeX systems and I thank Deborah Iscoe for her speed and precision in technical typing. I thank Maria Allegra, Associate Aquisitions Editor, Marcel Dekker, Inc., for her interest in this book and it has been a pleasure working with Walter Brownfield, Production Editor, Marcel Dekker, Inc., who supplied excellent pagesetting recommendations.

Finally, I thank my wife Marla for her support and encouragement once I thrust myself into this project. She proofread the entire manuscript and, together with Daniel and Tamara, withstood an intense three weeks at the end when the final camera-ready version was being composed.

Fred M. Hoppe

Contents

PART B   SELECTION

PART C   BIOMETRY AND DESIGN

Contributors

Peter Armitage   Department of Statistics, University of Oxford, Oxford OX1 3TG, United Kingdom

Robert E. Bechhofer   School of Operations Research and Industrial Engineering, Cornell University, Ithaca, NY 14853-3801

David F. Bray   Health Policy Research and Evaluation Unit, Department of Community Health and Epidemiology, Queen's University, Kingston, Ontario K7L 3N6, Canada

David R. Bristol   Biostatistics Department, Schering-Plough Research Institute, Kenilworth, NJ 07033

Pinyuen Chen   Department of Mathematics, Syracuse University, Syracuse, NY 13244-1150

Randall A. Coates   Department of Preventive Medicine and Biostatistics, University of Toronto, Toronto, Ontario M5S 1A8, Canada

Robert N. Curnow   Department of Applied Statistics, University of Reading, Reading RG6 2AN, United Kingdom

H.A. David   Department of Statistics, Iowa State University, Ames, Iowa 50011

Stefan Driessen   Medical Research and Development Unit, Organon International B.V., 5340 BH, Oss, The Netherlands

Edward J. Dudewicz   Department of Mathematics, Syracuse University, Syracuse, NY 13244-1150

Don Edwards   Department of Statistics, University of South Carolina, Columbia, SC 29208

Vern T. Farewell   Departments of Health Studies and Statistics and Actuarial Science, University of Waterloo, Waterloo, Ontario N2L 3G1, Canada

Walter T. Federer   Biometrics Unit, Cornell University, Ithaca, NY 14853-7801

David J. Finney   Department of Statistics, Edinburgh University, Edinburgh EH9 3JZ, Scotland

Spencer M. Free Jr.   Private Consultant, 302 Belpaire Court, Newtown Square, PA 19073

Joseph Glaz   Department of Statistics, University of Connecticut, Storrs, CT 06269-3120

David Goldsman   School of Industrial and Systems Engineering, Georgia Institute of Technology, Atlanta, GA 30332-0205

G.V.S. Gopal   Z.S. Associates, 1800 Sherman Ave., Evanston, IL 60201

Shanti S. Gupta   Department of Statistics, Purdue University, West Lafayette, IN 47907

Mark Hartmann   Department of Operations Research, The University of North Carolina, Chapel Hill, NC 27599

Anthony J. Hayter   School of Industrial and Systems Engineering, Georgia Institute of Technology, Atlanta, GA 30332-0205

Manfred Horn   Biometrical Unit, Hans Knöll Institute of Natural Product Research, D-6900 Jena, Germany

Satish Iyengar   Department of Mathematics and Statistics, University of Pittsburgh, Pittsburgh, PA 15260

Christopher Jennison   School of Mathematical Sciences, University of Bath, Bath BA2 7AY, United Kingdom

Peter W.M. John   Department of Mathematics, University of Texas, Austin, TX 78712

H.J. Keselman   Department of Psychology, The University of Manitoba, Winnipeg, Manitoba R3T 2N2, Canada

Jerry F. Lawless   Department of Statistics and Actuarial Science, University of Waterloo, Waterloo, Ontario N2L 3G1, Canada

TaChen Liang   Department of Mathematics, Wayne State University, Detroit, MI 48202

Yuning Liao   Department of Statistics, Purdue University, West Lafayette, IN 47907

Wei Liu   Department of Mathematics, University of Southampton, Southampton SO9 5NH, United Kingdom

Robert Lussier   Canadian Centre for Health Information, Statistics Canada, Ottawa, Ontario K1A 0T6, Canada

Charles E. McCulloch   Biometrics Unit, Cornell University, Ithaca NY 14853-7801

Nitis Mukhopadhyay   Department of Statistics, University of Connecticutt, Storrs, CT 06269-3120

S. Panchapakesan   Department of Mathematics, Southern Illinois University, Carbondale, IL 62901-4408

Janet M. Raboud   Department of Health Care and Epidemiology, University of British Columbia, Vancouver, British Columbia V6T 1W5, Canada

Ping Sa   Department of Mathematics and Statistics, University of North Florida, Jacksonville, FL 32216

Thomas J. Santner   Department of Statistics, The Ohio State University, Columbus, OH 43210-1247

Eugene Seneta   School of Mathematics and Statistics, University of Sydney, Sydney, NSW 2006, Australia

Milton Sobel   Department of Statistics, University of California, Santa Barbara, CA 93106

John D. Spurrier   Department of Statistics, University of South Carolina, Columbia, SC 29208

Ajit C. Tamhane   Departments of Statistics and Industrial Engineering and Management Sciences, Northwestern University, Evanston, IL 60208-4070

Baldeo K. Taneja   Statistical Evaluation and Research Branch, Division of Biometrics, Center for Drug Evaluation and Research, U.S. Food and Drug Administration, Rockville, MD 20857

John W. Tukey   Department of Mathematics, Princeton University, Princeton, NJ 08544-1000

Bruce W. Turnbull   School of Operations Research and Industrial Engineering, Cornell University, Ithaca, NY 14853-3801

Rüdiger Vollandt   Department of Medical Informatics and Biomathematics, Friedrich Schiller University, D-6900 Jena, Germany

Anne Whitehead   Department of Applied Statistics, University of Reading, Reading RG6 2AN, United Kingdom

Ping Yan   Department of Statistics and Actuarial Science, University of Waterloo, Waterloo, Ontario N2L 3G1, Canada

Chapter 1

A Conversation with Charles W. Dunnett

Charles W. Dunnett was born in 1921 in Windsor, Ont. He graduated in 1942 with a BA in Mathematics and Physics from McMaster University. He served in the Royal Navy during World War II, and was awarded an MBE for work on radar. Returning to Canada, he obtained an MA in Mathematics at the University of Toronto in 1946. Following two years at Columbia University and a year spent teaching at the New York State Maritime College, in 1949 he joined the Food and Drug Laboratories of the Department of National Health and Welfare in Ottawa as a biometrician.

He spent 1952-53 on leave at Cornell University. In 1953 he accepted a position as statistician at Lederle Laboratories, a division of American Cyanamid Company, and remained with them until 1974, when he was appointed Professor of Clinical Epidemiology and Biostatistics in the Health Sciences Faculty at McMaster University. He was chairman of the Applied Mathematics Department at McMaster from 1977 to 1979, and subsequently a member of the Department of Mathematical Sciences. He was awarded the title of Professor Emeritus in 1987 in the Departments of Clinical Epidemiology and Biostatistics and Mathematics and Statistics.

Professor Dunnett obtained his doctorate in 1960 working with Professor D.J. Finney at Aberdeen University. He is a Fellow of the American Statistical Association and a Member of the International Statistical Institute. He served as President of the Statistical Society of Canada in 1982. In 1986 the Society awarded him its Gold Medal for his contributions to Statistics.

1

The following are excerpts from a conversation recorded at McMaster University and is reprinted from Liaison, The Bulletin of the Statistical Society of Canada.

L= *Liaison*,   **D**= Dunnett.

L. *Would you start by telling us about your radar work, for which you were awarded the MBE?*
D. When I graduated from here [McMaster University] with a Bachelor's degree I went right into the service, and used the electronics background from my physics training to enlist as a radar officer. I was on loan to the Royal Navy. I went to England and took some training courses, and then went to a research establishment where they were developing radar equipment for aircraft in the R.A.F. and the Fleet Air Arm of the Navy. I spent about six months working with the research people on a new type of radar for detecting submarines and other vessels on the surface of the sea, and then went out into the field, so to speak, with the radar equipment. First to a training squadron, where they trained the air crew to operate the radar. Then I spent a year and a half with the first operational Fleet Air Arm squadron to use the radar. We were on board a carrier for some months, patrolling in the Atlantic and on the Murmansk convoy run, and then before D-Day we were disembarked from the carrier and stationed on an RAF airfield in the south of England, where the squadron supported the D-Day landings by searching for and attacking enemy shipping in the English Channel at night. We moved over to Belgium when Northern Europe was liberated and operated from airfields there until the end of the war in Europe. So that was my naval career. I was in charge of the maintenance of the radar, and there were lots of new problems with getting the radar in service. We were the first ones to use it, so the first ones to discover the problems. My job was to find solutions to the technical problems if I could, and consult with the research people in England when their help was needed.

L. *Did you go back to university right after the war?*
D. Yes. I was scheduled to transfer to the Pacific area, but when the war suddenly ended I decided to do a Master's degree in mathematics at the University of Toronto. One of the courses I took happened to be a statistics course, which fascinated me, and I decided that statistics was what I was interested in doing. There was another student at Toronto that year, Colin Blyth, who was also a major influence on my going into statistics, besides the course I was taking. His plan after finishing his master's degree at Toronto was to go to Iowa State University, which had of course Snedecor, Cochran and Gertrude Cox at that time. He had applied to go to Iowa State

so I did also. We were both accepted, planning to go there, when we learned that Cochran and Gertrude Cox were leaving to go to North Carolina. For all we knew there was no one left at Iowa State! Actually that was not the case, but Colin transferred to the University of North Carolina and I applied to go to Columbia University was Wald and Wolfowitz were.

I stayed at Columbia in the Department of Mathematical Statistics for two years. I was planning to do my doctorate there, but didn't do the thesis work. Took all the course work. Passed the qualifying examination. Then decided I had studied enough theory and wanted some practical experience. After one year teaching mathematics and physics at the New York State Maritime College in the Bronx, I went to Ottawa and joined the Food and Drug Laboratories as a biometrician.

That was an interesting period for me because I became exposed to lots of applied problems in statistics. We had problems in sampling, for instance. I had a couple of trips to Halifax where they took samples of food products coming off the ships. I worked with Dr. J.W. Hopkins who was a biometrician at the National Research Council. He had developed an interesting theory for sampling these types of products. The physical characteristics of the problem required a two-stage sampling design, because the food products – primarily nuts, dates or figs – were packed in containers such as boxes or bags. The primary sampling unit had to be the container and a sample of containers was selected, and then within each container a secondary sample of items was selected for examination. Each individual item was defective if it was contaminated, for example with insect infestation or mould, and the problem was to determine the proportion defective in the shipment. The theory was based on assuming that within a container the defectives were randomly distributed, in an infinite population, so that if you had a sample of a certain size from only one container the binomial distribution was applicable. But then the binomial parameter could vary from container to container in the shipment, and they had data to show that this was the case. In Dr. Hopkins' model the binomial parameter varied in a beta distribution and that produced what he called the negative hypergeometric distribution but has since become known as beta-binomial.

L. *Did the use of this distribution involve much computation?*
D. Yes, there were heavy computations involved in estimating the parameters of this negative hypergeometric distribution. We used moment estimates of the parameters and then developed an acceptance sampling procedure for the shipments (Dunnett and Hopkins 1951).

Another major problem was biological assay. They did a lot of experiments to determine potencies of drugs relative to a standard drug. Probit analysis was used, and again the computations were pretty heavy. The standard procedure was that the laboratory person would spend one day in the laboratory doing the experiment, taking the observations on the animals. The next day he'd spend at his desk calculator, doing the calculations needed to determine the potency estimates.

L. *When did you first become involved in academic research?*
D. I left the Food and Drug Laboratories on a one-year leave of absence to go to Cornell University, where I was invited to work with Bob Bechhofer and Milton Sobel, who had been graduate students at Columbia at the same time I was there. Bob Bechhofer had a research grant to work on selection procedures. I had a very productive year there. We were officially in the Mathematics Department, but actually our offices were in another building, in Warren Hall, which was in the agricultural area. Walt Federer and Doug Robson had their offices there and we really saw as much of them as we did of Jack Kiefer and others in the mathematics faculty.

L. *What were the selection procedures you were working on?*
D. The main application was selecting the best treatment: for example, selecting the population with the largest population mean. The idea was that if in an experiment several treatments are being compared, you could do an analysis of variance and test the hypothesis that all the means are equals with the F test and so on, but in many applications the null hypothesis of no treatment differences doesn't answer the real problem. The real problem is to determine which is the best of the treatments. We wanted a procedure that would guarantee a specified probability of correctly selecting the best population mean. Of course this probability depends upon the configuration of the population means, which is unknown. If the true population means are very close to each other, you can't guarantee a very high probability of correctly selecting the highest one. But if the true population means are very close together, it's not so important that the best one be correctly selected. It's only important if the best one is more than a specified amount better. You specify the difference that's important to be able to detect. Then you can determine what size of experiment is required. Bob Bechhofer had formulated the selection problem in this way, which has become known as the "indifference zone" approach, and developed the necessary theory for the case of normal populations with a common, known variance. This was described in his 1954 *Annals* paper (Bechhofer 1954).

In extending his methods to the unknown variance case we encountered or formulated the multivariate Student's t distribution. I suppose that was the main theoretical contribution made during the year I spent at Cornell. There were two papers published in *Biometrika* (Bechhofer et al. 1954), (Dunnett and Sobel 1954). The multivariate distribution was defined there, though mostly we dealt with the two-dimensional case because again, all we had were desk calculators to work with. Also, we considered that two dimensions were a major step forward from one!

L. *The Cornell experience seems to have led to a decision to leave the Food and Drug Laboratories permanently.*
D. I decided that I wanted to move to the United States, only because there were more statisticians around in the U.S. I had an offer of a position at Lederle Laboratories, a pharmaceutical company in Pearl River, New York. There was a small statistical group there already. Dr. Frank Wilcoxon was head of the group. The interesting thing is that immediately on joining Lederle I found another application for the multivariate Student's t distribution that I think may have turned out to be even more important than the selection applications, namely comparing treatments with a standard treatment. In fact, multiple comparison methods had just come into use at that time. Scheffé's paper (Scheffé 1953), which would be published in *Biometrika*, was known to quite a few people. Tukey's work on multiple comparisons was known, and Duncan's work was known. The day I started work at Lederle, Dr. Wilcoxon went through with me some problems they were working on, and he mentioned this problem of multiple comparisons between treatments and a control treatment as one which Tukey and Scheffé had not dealt with. He showed me a paper by Roessler in the *Journal of Horticultural Science*, predating Tukey and Scheffé, on this very problem. The method Roessler had developed was not rigorous, in that it ignored the correlations between the comparisons of treatment vs. control that are present if the same control issued in all the comparisons. I took Roessler's method and saw that if the correlations were taken into account you were led to exactly the same multivariate Student's t distribution that we had developed the year before at Cornell University. If all the sample sizes are equal for the treatments and the control, and if the variances are homogeneous, then the correlations are all 0.5, because of the presence of the same control mean in all of the comparisons. In the selection problem the same correlation 0.5 comes in because all the comparisons are with the best treatment.

L. *The 1955 JASA paper (Dunnett 1955) which gave critical values for the most extreme difference between treatment and control means, and the 1964 Biometrics paper (Dunnett 1964) which refined and extended the method, are still very widely cited in the scientific literature.*
D. Yes, they've been cited a lot.
   The second major problem that I became interested in at Lederle Laboratories was drug screening, the problem of testing chemicals for biological activity. A drug company will have a large file of chemical compounds that they or others have developed, and the question is, do any of these

compounds have any biological activity that might eventually make them useful as drugs for treating disease? The first step is to do a quick test in animals of a large number (20 or 30) of the compounds, then select a subset for further testing, accepting or rejecting the others. I had taken a course from Abraham Wald at Columbia University on sequential analysis, and was interested in finding applications for sequential methods. In drug screening, for a particular drug, the typical experiment would be completed in a time frame of a week or two, where the animal is treated with the drug, the results are observed, and then the experimenter can decide whether to repeat the test. It is actually feasible to repeat the test several times, in successive experiments, before coming to a decision on activity or non-activity. There were a number of applications that I worked on at Lederle, setting up sequential procedures for deciding: accept the drug, reject the drug, or test it again.

In 1957 I learned that Dr. O.L. Davies of I.C.I. Pharmaceuticals was scheduled to present an invited paper (Davies 1958) on drug screening at the ISI meetings in Stockholm. Lederle's management considered this to be an important enough reason for me to travel to Sweden to attend the conference. For me, it was doubly fortunate that Professor D.J. Finney also attended and presented a paper (Finney 1958) on similar problems in the agricultural area: selection of new plant varieties. This led to an exchange of correspondence between us, resulting in my going to Aberdeen the following year with my family on a two-year educational leave from Lederle to work with Professor Finney on a D.Sc. thesis.

L. *What was the department like at Aberdeen?*
D. Professor Finney was head of the Department of Statistics. He was also head of the Agricultural Research Council's Unit of Statistics, which was based at Aberdeen. He had the two groups, but actually they were pretty well merged: faculty members worked on A.R.C. problems, and A.R.C. members taught courses in the Department of Statistics. Michael Sampford was there, as well as Bill Brass, and Peter Fisk. Richard Cormack was working on his doctorate at that time. So was Robert Curnow, and he and I worked very closely together because he was working on plant selection problems which involved a similar theory to those I was working on.

L. *Was the paper (Dunnett 1960) which was read before the Royal Statistical Society based on your thesis?*
D. Well, that was part of it. I went there to work on a thesis on drug screening, and at the same time I continued some work on selection. That *R.S.S.* paper was the selection part of it. I worked on a different model from what we had worked on at Cornell. It was a sort of Bayesian model

where prior information was assumed about the population means. So it turned out that the thesis was divided between the two problems of drug screening and selection. The screening part (Dunnett 1961) was published in the proceedings of a conference on pharmacological statistics held at the University of Leiden in the Netherlands.

L. *Again there was a good deal of computation involved. I have the impression you must enjoy working out numerical methods.*
D. Yes, I enjoy computing. We got our first computer at Lederle in 1957. It was a Royal McBee LGP-30, and had a magnetic drum memory. We eventually upgraded it to 8000 words, but I think it was about half that to start with. All computations had to be programmed into machine language. It was actually fairly easy to program. There were sixteen different operations, like add, subtract, store... each operation followed by a four-digit address, to indicate where the number was stored. Programs and data had to be punched on a long roll of paper tape. That was fun. You had to make sure the paper tape didn't get tangled up. Our offices were on the sixth floor, in the tallest building at Lederle, so to unravel the tapes we just tossed them out the window. We had six floors to let them unroll down.

L. *Things have come a long way since then.*
D. I'm afraid so.

L. *Recently using more modern devices you published a couple of papers in JASA (Dunnett 1980a,b) on pairwise multiple comparisons which were very interesting.*
D. I guess the main point of those papers was that they showed by computer simulation that the studentized range method developed by Tukey in the early 1950's for doing all pairwise treatment comparisons could be extended to cases where the treatment groups had unequal sample sizes. This had been proposed years earlier, proposed by Tukey himself, but there was no rigorous justification for it. In Tukey's method with equal sample sizes you'd use the studentized range point times $s/\sqrt{n}$ as the allowance or the critical value for comparing any one treatment with another treatment in the group. If the treatments have unequal sample sizes, one method which was proposed was to replace $s/\sqrt{n}$ with $\frac{s}{\sqrt{n}}\sqrt{\frac{1}{n_1} - \frac{1}{n_2}}$, the standard error of the difference between two means, divided by $\sqrt{2}$ to make it comparable. This was an intuitive method only: it wasn't known whether the joint confidence coefficient was $1 - \alpha$.

There were a number of other methods proposed that did have the mathematical justification that it could be shown their joint confidence coefficient became greater than the nominal value as the sample sizes became unequal. But my computer simulation experiments showed that the confidence intervals based on the natural extension of the studentized range, which of course were narrower, had this property also. And that was since mathematically proved for the equal variances case in a paper by Hayter in the *Annals of Statistics* (Hayter 1984). In the unequal variance case I think computer simulation is the only possible way of providing justification for methods analogous to the studentized range. Another paper (Dunnett 1982) dealt with robust methods for multiple comparisons.

L. *Was it the medical program which made you decide to come back to McMaster?*
D. Yes, primarily. From my pharmaceutical work I was interested in medical applications.
L. *Your 1977 Biometrics paper with Gent (Dunnett and Gent 1977) on equivalence testing was motivated by medical applications.*
D. Yes, some medical studies are carried out with the aim of showing that the usual null hypothesis of equal treatment effects is true rather than false. It was Mike Gent's idea that this required a change in the usual philosophy of hypothesis testing. There have been other papers published on a similar theme, but ours was one of the first, I believe.

L. *What are you working on these days?*
D. I'm just getting back to drug screening now. I have some ideas for further work in that area I hope to work on next year. Still working on multiple comparisons, and have a paper with Bob Bechhofer at Cornell and Ajit Tamhane at Northwestern University that I'll be giving at the Biometric Conference in Belgium later this month. This is on a method for doing multiple comparisons between treatments and a control, and we're interested in joint confidence interval estimation for each treatment vs. the control treatment. But the aspect that's different is that not all of the treatments have a sufficiently high mean to be of interest, so a preliminary experiment is done to try to weed out the inferior treatments, and then the main experiment is done on those that survive the first phase. We examine under which situations it's more efficient to use both phases to determine joint confidence intervals for the comparisons of interest, rather than using the first phase merely to eliminate inferior treatments and the second phase to determine the confidence intervals.

L. *You also have some connection with the University of Swansea in Wales, haven't you?*
D. No official connection now, but I spent two research leaves there from McMaster and had honorary appointments as a visiting professor. My wife and I frequently visit the Swansea area, and the Department of Management

Science and Statistics at the university has been very generous in providing me with facilities such as computer time and work space when I am there.

L. *Thank you, Professor Dunnett.*

References

Bechhofer, R.E. (1954). *Ann. Math. Statist.* **25** 16 - 39.

Bechhofer, R.E., Dunnett, C.W. and Sobel, M. (1954). *Biometrika* **41** 170 - 176.

Davies, O.L. (1958). *Bull. I.S.I.* **36** 226 - 241.

Dunnett, C.W. (1955). *Jour. Amer. Statist. Assoc.* **50** 1096 - 1121.

Dunnett, C.W. (1960). *J. Roy. Statist. Soc. B* **22** 1 - 40.

Dunnett, C.W. (1961). In *Quantitative Methods in Pharmacology* (H. de Jonge, ed.), 212 - 231, North-Holland.

Dunnett, C.W. (1964). *Biometrika* **20** 482 - 491.

Dunnett, C.W. (1980a). *Jour. Amer. Statist. Assoc.* **75** 789 - 795.

Dunnett, C.W. (1980b). *Jour. Amer. Statist. Assoc.* **75** 796 - 800.

Dunnett, C.W. (1982). *Comm. Statist. A* **11** 2611 - 2629.

Dunnett, C.W. and Gent, M. (1977). *Biometrics* **33** 593 - 602.

Dunnett, C.W. and Hopkins, J.W. (1951). In *Symposium on Bulk Sampling* Spec. Tech. Pub. No. 114, Amer. Soc. for Testing Materials 13 - 18.

Dunnett, C.W. and Sobel, M. (1954). *Biometrika* **41** 153 - 159.

Finney, D.J. (1958). *Bull. I.S.I.* **36** 242 - 268.

Hayter, A.J. (1984). *Ann. Statist.* **12** 61 - 75.

Scheffé, H. (1953). *Biometrika* **40** 87 - 104.

Chapter 2

Whither Biometry?

DAVID J. FINNEY Former Head, Department of Statistics, Edinburgh University, Edinburgh, Scotland

1 Introduction

I am grateful for the privilege of being the first speaker at this Symposium, so pleasingly planned to honour CHARLES DUNNETT, a colleague whom we all respect for his many contributions to our profession during the past forty-five years, and who is a dear friend to many of us here. I think I should be wrong to spend my whole time in eulogy of CHARLES, richly though such tribute would be deserved, but I begin by briefly reviewing his career.

In 1958, Charles came to the University of Aberdeen to work as a doctoral student under my guidance. I did not then know him personally, but he was already becoming known for researches relating to the development and testing of new drugs. To take an unknown student, however well recommended he may be, in the hope that he will undertake research of doctoral quality, always has an element of gambling! In such speculation, I have been more successful than I deserve, but never more so than with CHARLES. I believe that I may have been the first to arouse his interest in problems of screening new chemical compounds with intent to identify those that seem promising as therapeutic agents. He rapidly mastered the logical strands of the problem, related them to his own experience in the pharmaceutical industry, and employed his great mathematical and practical good sense to develop, and to describe clearly, procedures comprising

11

optimal rules for selection. Subsequently, he has published many important papers in this field.

That same practical relevance has characterized his career. He has contributed to an impressive range of medical and health problems, often as a member of a multi-disciplinary research team. His publications, like his outlook on teaching, and his collaboration in clinical and epidemiological studies, have always manifested awareness of statistical method as an aid to better understanding of our world rather than as a branch of pure mathematics. This attitude of mind has been very evident in the attention he has given to *multiple comparison techniques.* Were I ever to encounter a situation needing such a procedure, my first thought would be to consult CHARLES DUNNETT.

Even if I had the time, evaluation of his whole corpus of research would be beyond my competence. I propose to speak to you, I hope a little provocatively, on "Whither Biometry?" If I could, I would delight in telling you what vital new results will be achieved between now and AD 2000. I cannot claim such prescience: instead I shall present personal comments on things that I believe we need to do in order to put our biometric house in order!

2   Biometry

What do we mean by biometry, and wherein is it distinct from statistics? I think that today we mean much more than the purely etymological "measurement of life", yet let us never forget that good measurement and reliable numerical recording are prerequisites for *"Application of statistical inference to quantitative properties of living material"*, which phrase I suggest as an adequate definition of the sense in which the word biometry has been used for almost a century. "Biostatistics" ought to be synonymous, although some people appear to restrict that word to medical applications. Note that I do not say "biometrics", for I see no need for that ugly word, even though, regrettably, it was chosen as the name for a journal! We never write of "geometrics" of "calorimetrics" or of similar word-formations (with the exception of the equally ugly "econometrics")!

When we think of biometry, whether as practitioners or as teachers, we must never forget the role of measurement: under that head, I include the process of counting — a process often less easy than perhaps we thought it at the confident age of five. In 1938, I was fortunate in becoming a student with R.A. FISHER, then at the height of his powers. Recognizing in me a young man who knew a little about statistical theory but who was totally ignorant on biology, Fisher assigned to me the task of recording tail growth in a strain of short-tailed mice. With the aid of an instrument constructed by that ingenious geneticist Alexander Fabergé (whose grandfather had made elaborate jewelry for the Tsars), I measured assiduously for many weeks. No remarkable truths emerged, but I acquired a proper concern for care in measuring and recording, not forgetting correct identification of the individuals measured. Two or three years later, when I had become an assistant to FRANK YATES at Rothamsted Experimental Station, the lesson was reinforced by the Rothamsted practice of regarding its junior statisticians as available labour for field work that involved cutting or digging of crop samples. At the time, this probably seemed to me a waste of my valuable time, but it did much to inculcate appreciation of the realities of field research. Many years later, I was shocked to hear an eminent chemist, recently appointed as an administrator of agricultural research, state that he could not understand how valid research could be conducted outside laboratories!

3   Inference

In looking to the future of biometry, let us keep in mind that sound quantitative inference is always conditioned by the nature, origin, and correctness of the available data, as well as by the exact form of questions that are asked. An elementary textbook of statistical method may do much harm to our students, even perhaps to ourselves, if it follows a sensible introductory chapter on analysis of variance by exercises for the reader of the type: "Test the significance of differences between the means of these three groups", after which appear three columns of ten numbers without mention of what they represent, of how they were measured and in what units, or of anything relevant to their meaning.

In several of his early papers on human genetics, Fisher emphasized how the method by which individuals are ascertained affects unbiased estimation of genetic parameters. Similarly, all who practice sample survey must be on guard against biases that may enter as a result of the manner in which subjects or other sampling units are selected from the population. There is ample documentation of the dangers of apparently haphazard selection, reliance on volunteers, and other subjective processes. Randomization and random selection, of course within legitimate stratificatory or like constraints, have long impressed me as one of Fisher's deepest contributions to scientific method. Random selection, indeed, may introduce ethical anxieties: these must never be ignored, but commonly they can be overcome, as often they are today in medical research. I suggest to you that regard for professional standards should make those of us who practise statistics or biometry hesitate before we analyze data from a source where needed randomization was neglected.

I have worries today about the mechanics of randomization. Like many, I was brought up to use tables of so-called random numbers, yet the history of these may disclose that they were constructed by rule, maybe from remote digits of tables of logarithms. I suspect that almost all who randomize today use computer-generated pseudo-random numbers, produced deterministically from a recommended function. System software may enable new numbers to be called successively by a simple command such as RAND, yet sequences of numbers may repeat, albeit with very long period. Can this procedure properly be claimed as random in the sense that Fisher intended? We may make comforting noises, saying "These numbers are random enough for practical purposes", but what does that really mean?

I am no philosopher able to clarify these matters. I hope that others will study, I have no idea how, the effects of randomization as currently practised on the quality of inference. How many of us in fact know by what rule our computer will generate 256 random numbers for the experiment that we design tomorrow? We could imagine having a small radioactive source within a computer, particle emissions from which would generate digits complying with philosophical definitions of randomness. Doubtless most scientific research will continue as at present, but I wish I could be sure that my worries are without foundation. Ought editors to insist that every published paper shall include specification of any pseudo-random generator used? I have no idea what use could be made of that information!

One still too often encounters the notion that a list of data from an experiment, or from a purely observational study, classified according to treatments or other factors, suffices to determine the appropriate statistical analysis. This can lead to nonsense. Insistence on explicit statement of all assumptions implicit in a chosen form of analysis provides some safeguard. Misleadingly naive conclusions are less likely if the statistician forces himself to face all consequences of assuming independence, randomness, distributional Normality, and the like. Those who adopt distribution-free or non-parametric procedures do not thereby banish all dangers: uncritical assumption of Normality can be a lesser danger than unthinking assumption that observational units are statistically independent or have been randomly chosen, questions about which are unanswerable except when real data are presented with full specification of origins, manner of acquisition, etc.

If we agree that information in greater detail than mere listing of numbers is needed before we can properly start a process of statistical inference, should we not try to specify more formally what information is essential and the manner of its use? If data relate to an experiment, with little fear of dispute I assert that a full account of experimental design, nature of treatments (including quantitative values and logical interrelations), randomization procedures, and units of measurement is essential. Analogous requirements apply to a sample survey and its frame. Greater difficulties arise over specifying the mathematical formulation, commonly termed the *model*, on which statistical analysis is to be based. This may involve questions of Normality, linearity, additivity, logical independence, and functional forms of regression equations, possibly augmented by matters concerning the relevance of any concomitant observations as also by necessity of checking and verifying any questionable values.

Will an expert system ever satisfactorily replace the experienced and collaborative consultant biometrician? If a proposal to write a system forces some who think themselves moderately good consultants to focus thought on the information they need before they either design an experiment or advise on how to analyze a set of data, the outcome should be healthy. There can be no rigid rules: the attempt to specify guiding principles or personal practices may not lead to an ideal expert system, but it can help in clarifying what we mean by professional experience and so stimulate improved teaching of statistical science.

4   Choice of Analytical Techniques

Any thought about expert systems should persuade you of the urgent need for biometricians to give greater attention to the task of choosing the form of analysis for each new data set. My generation of statisticians appears to have failed to transmit the message that, although modern software facilitates the consequential arithmetic, wise choice still requires statistical expertise. The computer revolution, and the proper wish of a good scientist to handle his own data, encourage the use of attractively presented and seductively documented software packages, even though the scientist himself may lack sufficient understanding of statistical principles to guide his choice of method from a package menu. Current economic stresses may force him into self-dependence because he lacks the professional statistical support that, thirty years ago, was becoming customary in research institutes.

I shall illustrate the danger. In a recent biological journal, I saw a paper on a specialized problem in physiology about which I knew nothing. The author was unknown to me and I saw no indication that he had any knowledge of biometry. Possibly he was qualified to select his biometric techniques or had had professional advice, but he made no mention of either. In his text, I noted the statement that curve fitting and statistical analysis had been performed with the help of a software package PlotIT. Had the reference been to a standard package, such as GENSTAT or BMDP, any reader could easily discover what regression and other techniques were available and could begin to guess at those likely to have been used. With a package unfamiliar to most readers, no one can judge what facilities it offered to the research physiologist. Was this author even certain that methods offered by PlotIT are correctly programmed (see Section 6 below)? Nowhere did he tell what methods he used, or what was meant by interval estimates shown in his diagrams. Can we be sure that he never opted for analyses based upon totally inappropriate assumptions? Too few editors seek advice from a biometrician on every submitted manuscript that mentions biometric techniques and terminology. In this instance, ought they not to have asked the author to make his biometric methods as clear as he did his experimental?

Our failure to ensure critical and well-balanced use of statistics in the biological sciences is apparent when one scans almost any range of journals. Faults, alas, are not limited to those who lack formal training in our discipline. I recently saw with horror a paper by two professional biometricians that related to a large sample of animals divided among geographically distinct sub-groups. For the most interesting variable, the authors took each sub-group in turn, and sought to determine which of Pearson's probability density functions gave the best fit. Without mentioning efficiency or possible biases, they estimated parameters by the method of moments; moreover, they gave no hint of why, in 1990, examination of all these functions might be scientifically relevant! A few months ago, I had occasion to survey the contents of current issues of several Indian journals of agricultural research. Many papers reported conclusions from statistical analyses, sometimes even with a biometrician as a co-author. It was often evident that computer output had been uncritically inserted into the text. One table relating to a regression study displayed a column of regression coefficients without any thought for the number of digits that accuracy of measurement and realistic meaning of conclusions could justify: successive entries looked something like $2.39758632 \pm 0.032455019$. That far worse can occur in print is exemplified by two extracts from a recent issue of a British journal of high repute:

"For the purpose of these experiments we defined a significant persistent change in the response to test stimulation in the following way. If $(B \pm b)$ represents the mean and standard deviation of the 12 responses during the last 2 min. of the first control period, and $(A \pm a)$ are the mean and standard deviation of the 12 responses during the ninth and tenth min. of test stimulation, after the second period of conditioning stimulation, then significant persistent changes are defined as those in which $(A-B) > (a+b)$, and the percentage change is expressed as $100 \times [(A/B) - 1]$." (Bradley et al. 1991).

"The data were analyzed on a Macintosh Plus or SE. For each absorbance curve, all the raw data, including baseline measurements, were first smoothed using a five-point Golay-Savitzky (sliding average) routine applied before baseline subtraction. An estimated λ_{max} was obtained by visual inspection of the baseline-corrected absorbance curve. A fifth order polynomial at 95% confidence was fit (sic) to the data for the 25 nm shorter to 100 nm longer than this estimated λ_{max}." (Silman et al. 1991).

The way out is not easy. Further dangers may lie in the corruption or re-definition of language by successful packages. I recently heard an intrinsically excellent and interesting lecture that was almost incomprehensible to me because it was presented entirely in GLIMSPEAK. I respect GLIM as a valuable and powerful package. For reasons that I cannot regard as totally shameful, I have never myself had occasion to use it: I was therefore denied understanding of methods used in parameter estimation and in testing hypotheses because, despite being essentially maximum likelihood procedures, they were presented in a totally unfamiliar jargon. I simply did not know that a symbol such as "%f" implied a scalar quantity! Of course, every scholarly discipline creates its own special vocabulary. Are we content that communication among biometricians shall be restricted to those who are regular users of a particular package?

5   Trends in the Literature of Biometry

I urge that, in our own research and in our influencing of students, we who are concerned for the future of biometry keep before us all needs of biology in respect of quantitative phenomena that are not fully deterministic but have a stochastic component. Even if reading ostensibly biometric journals has caused him no worries, a biometrician who talks with biologists will hear complaints about the tendency for statistical method to become mathematically more complicated and increasingly distant from practical application. The charge is not entirely just: there are many examples of theoretical results in pure mathematics later becoming vital to practical problems in the physical world. Yet I believe that we should repeatedly remind ourselves, and our students, that the explosive growth of biometric science in this century has resulted from the great pioneers, Francis Galton, Karl Pearson, and above all R.A. Fisher, being stimulated by contemporary problems in biological science. The concepts of regression, families of frequency distributions, experimental design, and many more are readily traced to these origins.

Are we today neglecting the importance of such stimuli to new research?

I urge you to examine recent volumes of journals established with the intention of fostering biometric understanding. The 1300 pages of Volume 45 of *Biometrics* (1989) contained 108 papers, some of which undoubtedly, were important contributions to the development of biometric practice. On a rapid subjective assessment, I classified 80 papers as explicitly directed at specific problems of biology and 28 as mathematical theory motivated solely by the urge to generalize other theory; even of the 80, as many as 12 lacked any genuine numerical example, although simulations were often used to verify, to demonstrate, or to assess some aspect of efficiency. Despite my subjectivity and my limitation to one volume, I am disturbed that, in a journal whose stated objectives stress the *application* of methods and their *exemplification* for the benefit of experimental science, so many papers lack an example of when and how a newly proposed method should be used, and one paper in four is so theoretical as not to mention motivation or applicability. In its early years, *Biometrika* provided a vehicle for publication of new statistical work bearing upon biological problems. I have looked at Volume 74 (1987). I classified 56 of the 109 papers as wholly theoretical, without obvious applicability; of the 53 that seemed motivated by practical problems, 24 lacked any examples other than simulations, and few of the other 29 showed any concern for the first three letters of the journal's name!

I neither censure editorial policies nor condemn authors. A good research mind needs freedom to pursue a new topic purely on account of its personal appeal, and without limitation to potential applicability. Nevertheless, I question the health of our science if so great a part of the research that we choose to publish shows little relation to the questions that concern modern biologists. Let us hope to follow our illustrious predecessors in being stimulated by challenges from new fields of biological inquiry, perhaps especially those where new techniques of observation, manipulation of material, and measurement demand innovation in the techniques of statistical biometry. Sometimes no more may be needed than a small adaptation of familiar procedures, but sometimes a novel line of biometric research may develop; I do not believe that such an approach to our discipline will ever fail to give us intellectual satisfaction and a sense of achievement.

6   Software for Biometry

We can surely be confident that techniques in today's better known statistical packages have been correctly programmed, with due regard to the numerical accuracy of computations and output. The user is responsible for deciding which facilities and options he will employ in his current analysis;

wisely or not, he may even try a sequence of different choices on the same data. Yet he may reach grossly erroneous (even absurd) conclusions if, for example, he implicitly uses irrational assumptions about frequency distributions and error structures, confuses independent and dependent variables, omits proper weighting of observations, or ignores the constraints of design that governed an experiment or sample survey.

In most countries, manufacturers of electrical equipment are required to safeguard users against misuse or accidents. Misuse of statistical software is rarely life-threatening, but I think we should expect its authors to incorporate safeguards and warnings against he most likely abuses, these to include information on distributional assumptions implicit in each procedure provided, clear description of the consequences that flow from the options on offer, and advice on seeking professional statistical help if difficulties or apparent absurdities of output are encountered.

For the first twenty years of my career, I practised biometry with the aid of hand-driven or electro-mechanical calculators. Are any of you old enough to remember the coming of those remarkable gadgets that gave automatic division and even automatic cumulation of sums of squares? I believe these machines forced on us an intimate understanding of data and of the processes of arithmetic: we now benefit tremendously from doing almost all statistical analysis by programmable computers, but let us not forget what we have lost.

I have already indicated my concern that good software be produced and that it be used wisely. Any biometrician who becomes involved in software writing should do everything possible to ensure that it is well tested for its internal correctness and numerical accuracy, as also, of course, its conformity to the theory of whatever techniques it offers the user. There is a long-established tradition that the author of a paper on biological research shall, in a section entitled Methods, describe apparatus, instruments, and materials or animals that he has used. I suggest that we should inquire into the reasons for a practice that I personally believe to be excellent. We may then ask whether, perhaps, journal editors ought to regard identification of any statistical package used as equally essential to proper presentation of a publishable paper.

When we engage in consultative and collaborative activities, we should discourage use of any package of unknown quality and reliability, however attractively simple its use may seem. Unless we make this our practice, our profession may be brought into disrepute among scientists who discover too late that software available to them, or used by statisticians on their problems, has produced absurdities. Related opinions that apply particularly to the teaching of statistics have been well expressed by Searle (1989) and Dallal (1990).

.7 Biometry and Ethics

We who work as biometricians can range over a broad spectrum of interactions with other persons and with society as a whole. The general ethical standards of scholarly decency, intellectual truth and honesty, allied to fair dealing with the work and reputations of others, apply to us as to other professionals such as doctors, lawyers, and bankers. Is this enough? When we advise on the planning of quantitative investigations, analyze data that others have collected, prepare reports that may influence society at large, and teach these skills to our students, new ethical considerations can enter. I hope that I have never departed seriously from proper standards, but I have certainly failed to discuss them in my teaching. In the 1990's, we may need to make more explicit the principles of ethics under which we try to operate professionally (Finney 1991).

In 1985, the International Statistical Institute published a *Declaration On Professional Ethics* (ANON. 1985). I think none of us would dissent from its contents, but is it more than a call for standards that an earlier age would have considered *natural to a gentleman*? It specifies the obligations of a statistician in respect of objective approach to information, expertise in the choice of whatever analytical methods he uses, refusal to accept contractual limitations on the outcome of an inquiry, care for the confidentiality of all data that may be shown to him, and especial care for the human subjects of inquiries by minimizing invasion of privacy, obtaining informed consent to any experimental study, and preserving confidentiality of personal information.

During recent years, there have been well-documented reports of apparently reputable scientists publishing papers based upon data that were grossly distorted, or even forged, so as to support a preconceived conclusion. Surely the practice of research in any discipline ought to inculcate a dedication to truth, both in the reporting of facts and in relations with fellow scientists. In the past, I may have complacently believed that general standards of honesty, and of decent behaviour towards working associates would suffice to ensure maintenance of scientific ethics. Regretfully, I now accept that pious optimism is not enough, and that the training of a scientist must contain explicit guidance on ethics. I understand that students of medicine and of law already receive something of this kind.

I have no evidence that biometricians are notorious sinners! Why then do I raise the issue here? I believe that various aspects of our own professional ethics deserve more explicit discussion than they have had in the past.

Before he analyzes data, a biometrician ought to scrutinize them with an informed eye, endeavoring to confirm that they accord with what he has been told about their character and to detect any gross anomalies (Finney· 1990). In my opinion, to omit such scrutiny is irresponsible. The eye may notice much in the simple act of reading a list of observations. Simple software can aid the task; more sophisticated software may one day do much more.

Individually and as a profession, we should seek to earn respect for our honesty and objectivity in analysis and interpretation of all data we handle. We cannot avoid all risk of human misjudgment and mistake, but we should not give explicit or implicit approval to anything emerging from our work that does not confirm to the highest standards of propriety. I therefore assert that accurate and truthful recording of facts must always be in the forefront of our minds. If we cannot trust the correctness of data set before us, whether numerical measurements and counts or factual descriptions of the nature and sources of these data, our own science becomes a nonsense. One consequence of this should be that we ourselves exercise great care over verbal and numerical accuracy. Carelessness in the spelling of names in a list of references may seem no worse than a discourtesy, although it may cause confusion. On the other hand wrong spelling of the name of a chemical compound or biological organism could cause misunderstanding or even serious harm to someone affected by practical application of research in which we have collaborated.

If this principle is to guide all that we undertake, we must expect to obtain from anyone who seeks our collaboration or consultative help full disclosure of all relevant information: often, only we ourselves can assess the relevance of particular information to an analysis that we are about to make. Hence the ethics of statistical practice also place demands upon our employer, scientific colleague, or client. Is it ever ethical of a scientist from another discipline to submit percentages for statistical analysis without making available to the biometrician the numerators and denominators from which they were calculated? Quite apart from the question of where lies the responsibility for arithmetical correctness, in deciding his form of analysis, the biometrician may need to consider special aspects of discrete frequencies, transformations, and Normality. He should not be asked to analyze any quantities, such as percentage increases or other ratios derived from raw data by some preliminary arithmetic, without first seeing the calculations from the original records. Yet I have known instances of a young biometrician who asks about such matters being regarded as impertinent.

A statistician engaged in his professional capacity to aid any organization, whether as full-time employee or as temporary consultant, obviously has duties towards his current employer. His relations with that employer are analogous to those of other professional experts such as lawyers and physicians. He must bring to his employment, to consultation, and to analysis, all his professional skill and experience, supporting his counsel by specifying the methods that he has adopted. Commonly these will be standard methods, well-documented in methodological literature. He is not responsible for any difficulty his client may have in understanding technical details, but the confidentiality that can apply to data has no place in relation to methods and techniques of statistical analysis.

I think I can properly draw attention to one more duty that I believe to lie with any research biologist whose work benefits from our profession. In an age obsessed with priority of publication, he should not unreasonably prevent his biometric colleagues from using data, in teaching or in methodological publication, where new ideas developed during collaboration may add to the general pool of statistical knowledge. Safeguards may be needed, but not to the extent of delay until all novelty has gone from the biometrician's contribution to the research.

8   Ethics and Statistical Software

The close connection between ethical conduct and what I have said about statistical software should now be evident. If a biometrician reports to his client that data have been analyzed by the *WONDERSTAT* package, he implicitly takes responsibility for the quality of the package and its appropriateness to the problem in hand. That a research scientist should undertake his own statistical analyses, using standard packages on a PC, can have many merits, but has dangers that I have already discussed. The ease with which a micro-chip, programmed to undertake statistical computations, can be incorporated into sophisticated laboratory instruments such as auto-analyzers can aggravate the risks. The manufacturer of such an instrument, perhaps destined for use as a hospital diagnostic aid by means of immunoassay, may regard the program as a commercial secret. In my view such secrecy should be legally prohibited as no more acceptable than the marketing of drugs without declaration of their constituents; the risk that the software may contain mistakes, or that the instrument may use it uncritically on data for which an experienced biometrician would have judged it unsuitable, may have ethical implications for a clinician or his diagnostic laboratory.

A new ethical danger has arisen through the power of micro-chip technology. A sophisticated laboratory instrument may produce and record data that will subsequently require some form of statistical analysis. How easy, then, to incorporate into the instrument a chip containing the appropriate program, so that the data can *b* analyzed and summarized at the time of collection, and without need for any subsequent processing. This has certainly happened with some commercial auto-analyzers used in clinical biochemistry laboratories for radioimmunoassay, which is today a widely practised diagnostic aid for patients who may be sufferers from hormonal imbalance. The form of computation needed for estimating hormone content of serum samples, by use of the counts of particle emissions produced during radioimmunoassay, is well known, but the procedure is complicated for a non-statistician, and sound programs are not widely available. An instrument manufacturer who inserts a chip programmed for this purpose will not willingly disclose details that he regards as a commercial secret, probably not even the source of the program, its author, or publications on which it was based. Consequently, a physician may be expected to diagnose and treat a patient on the basis of a result produced by a program that is concealed from him and that may even contain gross mistakes. As a biometrician with some experience of bioassay theory and method, I am very conscious of the faults that could easily enter into this micro-chip if it has been programmed by someone unfamiliar with bioassay practice: for example, it may involve untested assumptions about the form of a response curve, it may employ arbitrary rules for rejection of outliers, the estimation of parameters may take no account of unequal weighting or any criteria of optimality, and indeed its arithmetical logic may not have been adequately checked. What can we think of the ethics of permitting patient care to depend upon a black-box procedure, the methods of which are regarded as secrets that cannot be exposed to informed scientific criticism? As biometricians, there is little that we can do directly; we may, however, have opportunities of urging on medical colleagues that proper analysis of these assays is not a purely mechanical process, and of ourselves taking interest in developing further understanding of types of anomalous behaviour that radioimmunoassay data may show. I further believe that a manufacturer should be expected to make public all statistical principles and algorithms used in his software, even though the precise manner of their implementation remains commercially confidential.

9 Rejection of Data

In the years ahead, we may need to face many issues of principle concerning rejection of observations, especially in relation to use of standard software. Those of us who have been brought up by way of designed experiments and their analyses have long been familiar with so-called *missing plot techniques*, methods of modifying a standard analysis of variance so as to give the correct least squares estimation when some observations that belong to the complete design are absent. I have often urged that these methods be used only when the reason for an observation being missing is known to have had no association with the particular experimental treatment. An assumption to that effect is questionable if the treatment might possibly have encouraged attack by a predator or occurrence of a disease, but it may be justifiable if a page has been carelessly lost from a laboratory notebook, or if an experimental animal has been accidentally killed.

In any experimental situation, we can easily envisage the problem caused by one datum that is patently absurd: the 15 kg chicken, the rat with birth weight 3 mg, the human blood pressure or hemoglobin level far outside any plausible range. Plausibility is inevitably judged subjectively. A program can be so written as to draw attention to all data that fall outside stated limits, and, if desired, to analyze the reduced data set after rejecting such values. This is so temptingly easy that a software author may choose to make rejection automatic, or to incorporate rules for rejection within what is on its way to becoming an expert system for a class of experiments. Ought we not to give serious thought to the principles that should guide the choice of limits? My purpose in mentioning this is that I would like to learn how to formulate principles that should guide the conscientious biometrician, and also guide the many scientists who, without reliance upon a professional biometrician, operate their own computer for analyzing their own data.

Let me give you a very simple illustration. A body falling freely under gravity from a height h, under conditions devised to make air resistance negligible, will reach the ground in t seconds, where $t = (2/g)^{0.5} \times h$, g being the gravitational acceleration. Thus g might be estimated by making a series of observations on t for different h. One could then estimate $(2/g)^{0.5}$ by computing a linear regression of t on h constrained to pass through the origin. What is to be done if the regression departs significantly from linearity, if some observations are clearly far from the fitted line, or if the fitting process shows conclusively a departure from the origin? Obviously one first looks for explanation in terms of inaccurate measurements, biases in the dropping procedure, uncontrolled air resistance, and so on. Having failed to reach satisfaction on these, in what circumstances should any discrepancies be ignored or observations discarded?

My example is not as silly as you may think. It has similarities to bioassay problems that interest me. In the process of standardizing therapeutic drugs, WHO organizes large collaborative studies. There may be bulk supplies of two materials, S and T, for which chemical stability can be assumed during a study. Small samples of S and T are taken for sending to perhaps 30 collaborating laboratories, in many different countries. All may be of good repute for conscientious, accurate, and trustworthy work, although possibly few will have an experienced biometrician on hand. At each laboratory, a standard type of experiment will be conducted, each having 8K rats randomly assigned in fourfold replication among K differ-ent doses of S and K of T (where K will commonly be chosen as a small integer between 2 and 5). At a fixed time after each animal received its dose, a response y will be measured, perhaps some biochemical property of the rat's blood. For T, the regression function of y on logarithm of dose is known to be identical with that for S except for a horizontal displacement of amount μ on the log-dose scale; this statement is the logical analogue of the physical law connecting time and distance of fall in my gravity ex-ample. Although experience suggests that the regression of y on log(dose) is approximately linear, no theory requires this to be so: unknown, but undoubtedly finite, lower and upper limits to possible values of y, make it likely that, if very low or very high doses have been chosen, curvature may appear even though the theoretical μ property remains. The statistical procedure is to fit two regression equations subject to the condition of con-stancy of horizontal separation, and from them to calculate M, an estimate of μ. If linearity and constant error variance obtain, this is standard least squares regression procedure. It is easily modified to take account of some alternative algebraic formulation of the regression function. On the other hand, even with linearity and variance homogeneity in no doubt, the two regression lines may depart significantly from parallelism, possibly just be-yond the 5% level but possibly with $t = 6.18$. The reason could be chance, a bad flaw in laboratory technique, or contamination of the sample of T by an unwanted substance that ought never to have been near the experiment (dish washing fluid?). Before subsequent compilation of evidence in order to obtain a composite estimate of μ, one would like to be able to reject any experiment flawed by technical errors but to retain for their proper contribution to assessment of precision those where chance alone produced apparent discrepancy.

Each experiment is likely to be analyzed locally in its own laboratory, by a program that we can but hope is free from algebraic error. The results, and also the raw data on every animal, will be sent to a centre where values of M can be studied with a view to preparing a report that will state a compounded estimate for $\rho = antilog(\mu)$, the ratio that converts any dose of T into the equivalent dose of S.

Further questions now arise. Even if most laboratories show regressions deviating little from linearity, perhaps two experiments have individual responses that deviate by more than 4 standard errors from lines fitted to the remainder. The variance among replicate animals at a dose may usually be fairly homogeneous, but perhaps in one or two laboratories the estimate of variance at one dose is much smaller than at others. There will be a tendency to bias, by focusing attention on the greatest departure from the norm in any of these respects, but I worry most about evidence that shouts so loudly as to forbid simply ignoring it. Once again, how does one decide what indications to disregard, so allowing the associated M to be included in a final summarizing process? What indications should cause a particular M to be rejected from all subsequent work? Are there situations in which one can justify rejecting two apparently anomalous values of y, or rejecting all from one dose at which results seem odd?

My concern is with situations where conscientious inquiry fails to disclose gross mistakes in data entry, or the anomaly of a rat that lost its tail in a fight before the experiment began! Non-constancy of horizontal displacement, (or non-parallelism of regressions), as mentioned above, makes that experiment invalid as a contributor to a composite estimate of μ, but one wishes not to reject an experiment if chance alone has caused apparent deviation from parallelism. Presumably we can never tolerate a practice of rejecting or disregarding portions of data solely because they do not confirm to preconceived ideas or solely because deviations transgress some arbitrary level of statistical significance. Even if we reject certain data because they are considered to be inconceivably remote from the truly, rejection followed by uncritical averaging of the remainder is far from ideal. Whatever decisions are taken, any responsible scientist needs to have a care for the ethical basis of a potency estimation that will later be used as a basis for treating sick persons.

10   Whither?

The difficulties and cautions that I have been discussing are not to be interpreted as advising biometricians to protect the purity of their consciences by avoiding problems that present ethical difficulties. On the contrary, I firmly believe that, if we are asked to participate (as consultants or as committee members) in inquiries of importance to public welfare where policies and decisions may turn upon essentially biometric considerations, we have a professional duty, if at all possible, to make our services available and to handle any ethical difficulties in accordance with the dictates of conscience. That in no way conflicts with the desirability now, and in the near future, of seeking to distil general principles from our collective experience.

When I finally try to answer the question of my title, I must give priority to the importance of effective evangelism for the better use of sound statistical and biometric methods throughout biological research. This implies encouragement to biologists to perform many of their own analyses, but help and guidance to ensure choice of appropriate methods. When we have done all that we can to ensure availability of software of the highest quality, we can properly discourage development, distribution, or use of unnecessary and inadequately validated software. My concern for greater attention to biometric and general scientific ethics relates closely to these matters, but is of wider significance.

I confidently hope that some of our best and most experienced colleagues will participate in attempts to develop expert systems for various aspects of biometric practice. Although I think them unlikely to succeed in producing systems that can properly replace the human mind, I believe that this activity will contribute much to the expression, discussion, understanding, and acceptance of important general principles. Among the consequences, I trust, will be exposure of the logical folly of the so-called *Bayesian Methods* that infect us today, and that may gravely mislead the many biologists whose researches have true need of biometric science.

References

Anon. (1985). A Declaration on Professional Ethics. *Bulletin of the International Statistical Institute* **51** (5) 319-345.

Bradley, P.M., Burns, B.D., and Webb, A.C. (1991). Potentiation of synaptic responses in slices from the chick forebrain. *Proceedings of the Royal Society* **B243** 19-24.

Dallal, G.E. (1990). Statistical computing packages: Dare we abandon their teaching to others? *American Statistician* **44** 265-266.

Finney, D.J. (1990). Statistical data — Their care and maintenance. *Indian Society of Agricultural Statistics.* ISAS Bulletin No. 1.

Finney, D.J. (1991). Ethical aspects of statistical practice. *Biometrics* **47** 331-339.

Searle, S.R. (1989). Statistical computing packages: Some words of caution. *American Statistician* **43** 189-190.

Silman, A.J., Ronan, S.J., and Loew, E.R. (1991). Histology and microspectrophotometry of the photoreceptors of a crocodilian, *Alligator mississippiensis. Proceedings of the Royal Society* **B243** 93-98.

Tufte, E.R. (1983). *The Visual Display of Quantitative Information.* Cheshire, Connecticut USA, The Graphics Press.

Chapter 3

Probability Inequalities and Dunnett's Test

EUGENE SENETA   School of Mathematics and Statistics,
University of Sydney, Sydney, Australia

Abstract   Two kinds of lower bounds, of first and second degree, are considered for $P(\cap_{i=1}^{k} A_i)$ where A_i, $i = 1, \cdots, k$ are events. The first kind of inequality applies to an arbitrary number of arbitrary events in an arbitrary probability space; these inequalities are Boole-type, or additive. The second kind is of multiplicative type, and may be thought of as originating from Chebyshev's Covariance Inequality; these inequalities hold in somewhat restrictive situations. Work on multiplicative inequalities, which has flourished recently, has its roots in the note of Dunnett and Sobel (1955) the prototype application being the multiple comparisons procedure known as Dunnett's test, in the context of a multivariate t-distribution. The purpose of the present paper is to give some coherence and insight into existing theory, and to pay tribute to the work of C.W. Dunnett. There are no new technical results.

1   Simultaneous Inference and Dunnett's Test

Let $A_1, A_2, \cdots, A_k$ be an arbitrary number of arbitrary events in an arbitrary probability space. The best-known lower bound for the probability of intersection is given by

$$1 - \sum_{i=1}^{k} P(\overline{A}_i) \leq P(\bigcap_{i=1}^{k} A_i). \qquad (1.1)$$

In a simultaneous inference (characteristically multiple comparisons) setting each A_i has the form

$$A_i = \{|Y_i| < c_i\} \qquad \text{or} \qquad \{Y_i < c_i\}) \qquad (1.2)$$

for dependent random variables Y_i, $i = 1, \cdots, k$ and the A_i's (that is, c_i's) are chosen so that

$$1 - \sum_{i=1}^{k} P(\overline{A}_i) = 1 - \alpha$$

for specified α. Then, if this is done under H_0, the significance level of the test is at most α; or the confidence level of the corresponding simultaneous confidence intervals is at least $1 - \alpha$.

The inference procedure is thus a conservative one, but has the advantage of involving only marginal probabilities $P(\overline{A}_i)$. The inequality (1.1) is Boole's (1854); in simultaneous inference it is inevitably misnamed "Bonferroni's Inequality" and the corresponding simultaneous confidence bounds Bonferroni bounds.

Dunnett's (1955) test is concerned with k treatments of corresponding sample sizes n_i, $i = 1, \cdots, k$, and a control treatment of sample size n_0, and focusses on comparing the treatments with the control. The samples are assumed drawn from $\mathcal{N}(\mu_i, \sigma^2)$, $i = 0, 1, \cdots, k$ distributions; the sample means $\overline{X}_i$, $i = 0, 1, \cdots, k$ are therefore independent of the residual mean square S^2. Here we take in (1.2)

$$Y_i = \left(\overline{X}_i - \mu_i - (\overline{X}_0 - \mu_0)\right)\Big/S\sqrt{\frac{1}{n_i} + \frac{1}{n_0}} \qquad (1.3)$$

$i = 1, \cdots, k$. These random variables are dependent (having $\overline{X}_0$ and S in common) and have jointly a multivariate t-distribution, introduced in a more general setting by Dunnett and Sobel (1954) as the joint distribution of

$$T_i = Z_i/S, \qquad i = 1, \cdots, k \qquad (1.4)$$

where Z_i's are individually $\mathcal{N}(0, \sigma^2)$ but have correlation matrix $R = \{\rho_{ij}\}$ and $\nu S^2/\sigma^2 \sim \chi_\nu^2$, distributed independently of the Z_i's. The multivariate t-distribution has undergone a number of tabulations (beginning with Dunnett and Sobel 1954, Dunnett 1955, through Hahn and Hendrikson 1971, for example). However, because of the great number of parameters involved (specifically the possibly differing ρ_{ij}'s), simultaneous inference via bounds such as (1.1) has retained its relevance to the present day.

Indeed in this setting Dunnett's test holds pride of place as the archtypical example on which to illustrate bound methodology. We mention Šidák (1962, 1968); Hunter (1976b); Stoline (1983); Glaz and Johnson (1984); Bauer and Hackl (1985); Block et al. (1988).

Further, we shall see in the sequel also that work on multiplicative inequalities has its roots in the note of Dunnett and Sobel (1955) which first addressed, in this setting, the question of lower bounds for $P(\cap A_i)$.

2 Additive Inequalities

Boole's Inequality (1.1) is, by visual form, an additive inequality, of degree 1 (intersections of at most one event occur in the bound. An additive degree 2 lower bound, which holds in the same totally general setting as (1.1) is Hunter's Inequality

$$1 - \sum_{i=1}^{k} P(\overline{A}_i) + \sum_{i=2}^{k} \max_{1 \leq s \leq i-1} P(\overline{A}_s \cap \overline{A}_i) \leq P(\bigcap_{i=1}^{k} A_i) \qquad (2.1)$$

which sharpens (1.1) at the cost of an increase of 1 in degree. It should be noted that the degree 2 component of the left-hand side is not invariant (for given events $A_1, \cdots, A_k$) under permutation of labels $1, \cdots, k$ of the events, and thus the left-hand side needs to be maximized over the set $\tilde{\prod}$ of all permutations. (In the situation of exchangeable A_i, $i = 1, \cdots, k$, (2.1) reduces to the well known degree 2 Sobel-Uppuluri (1972) bound.)

Providing the degree 2 quantities $P(\overline{A}_s \cap \overline{A}_i)$ can be evaluated on the left of (2.1), it will provide a better test (smaller c_i's will do for a given α) in (1.2).

The inequality (2.1), in graph-theoretic form, was discovered by Hunter (1976a) and rediscovered by Worsley (1982). The awkward graph-theoretic form has been used till as recently as the work of Stoline (1983), Glaz (1987), and Block et al. (1988). The inequality, in the analytical form (2.1) was rediscovered by Margaritescu (1986), and a simple proof based on an identity of Hoppe (1985) and Boole's (1854) degree 1 upper bound $P(\cap A_i) \leq 1 - \max P(\overline{A}_i)$ was given in Seneta (1988). The equivalence of the graph theoretical and analytical form is completed in Hoppe and Seneta (1990).

A simple algorithm for determining a permutation of indices which maximizes

$$\sum_{i=2}^{k} \max_{1 \le s \le i-1} P(\overline{A}_s \cap \overline{A}_i) \tag{2.2}$$

is that of Jarník (1930), which is described in English by Graham and Hell (1985, p. 47, column 1) and (in the multivariate t setting with graph-theoretic emphasis) by Hunter (1976b) and Stoline (1983, p. 368). We give it here in its general form for the reader's information.

Put $r_{ij} = P(\overline{A}_i \cap \overline{A}_j)$, $i, j = 1, \cdots, k$ (in fact, in the algorithm the r_{ij}'s need not be probabilities). Initially let $C = \phi$, $U = \{1, \cdots, k\}$ (ϕ is the empty set). Take any subscript i_0 from U and place it in C so now $C = \{i_0\}$, $U = \{1, 2, \cdots, k\} - \{i_0\}$. Then

1. Find the largest r_{ij} for $i \in C$, $j \in U$, and denote a corresponding pair of (i, j) by (i^*, j^*).

2. Redefine U and C by setting $U = U - \{j^*\}$, $C = C + \{j^*\}$. If $U \ne \phi$ go to 1; otherwise stop.

The final C, taking the initial i and the j^*s added in sequence, is an optimizing permutation. That this is so is easily proved by using the graph-theoretic ("tree") formulation of the problem, and adapting the simple proof of the often-cited minimal spanning-tree algorithm of Kruskal (1956).

However, it is clear that the algorithm as described above, just as (2.2) itself, needs no graph-theoretic concepts; and their continuing presence in bound theory may not be necessary.

3   Multiplicative Inequalities and Relation to Additive

The following inequalities which hold in certain special structural situations, but not in general (in contrast to (1.1) and (2.1)), are of multiplicative form and respectively of degree 1 and 2:

$$\prod_{i=1}^{k} P(A_i) \le P(\bigcap_{i=1}^{k} A_i) \qquad (\text{"product bound"}) \tag{3.1}$$

$$P(A_1)P(A_2|A_1) \cdots P(A_k|A_{k-1}) \le P(\bigcap_{i=1}^{k} A_i) \ (\text{"sub-Markov bound"})$$

$$\tag{3.2}$$

Work on (3.1) goes back to Dunnett and Sobel (1955), though most authors date its origins later. The same note also produced a degree 2 bound and ideas for much subsequent work, though credit for them has rarely been given.

Focus on the left-hand sides of (3.1) and (3.2). First note that the latter clearly is at least as high as the former if $P(A_i \cap A_j) \geq P(A_i)P(A_j)$ for all i, j [*positive pairwise dependence*]. Next, the algebraic identity:

$$1 - \sum_{i=1}^{k} q_i \leq \prod_{i=1}^{k}(1 - q_i), \qquad 0 \leq q_i \leq 1 \qquad (3.3)$$

shows by putting $q_i = P(\overline{A_i})$, as noted in Dunnett and Sobel (1955), that the bound in (3.1) is higher than the Boole bound (1.1). To see, however, that (3.1) does not hold in general it is necessary only to take an example where $k = 2$ and $P(A_1 \cap A_2) < P(A_1)P(A_2)$, to see the bound 'overshoot'.

Analogously, although for arbitrary events A_i, $i = 1, \cdots, k$, Glaz (1987) has shown

$$1 - \sum_{i=1}^{k} P(\overline{A_i}) \; + \; \sum_{i=2}^{k} P(\overline{A}_{i-1} \cap \overline{A_i})$$
$$\leq P(A_1)P(A_2|A_1) \cdots P(A_k|A_{k-1}) \qquad (3.4)$$

(the left-hand side is not greater than the left in (2.1), but is a popular bound), we use an example to show that even though positive dependence obtains the right hand side may overshoot $P(\cap_{i=1}^{k} A_i)$.

Example   Take $k = 3$ and the sample space Ω the unit interval $[0, 1)$, with probability measure Lebesgue. Let $\overline{A}_1 = (0.2, \ 0.8]$, $\overline{A}_2 = (0.1, \ 0.6]$, $\overline{A}_3 = [0, \ 0.5)$. Clearly

$$1 - P(\overline{A}_1) - P(\overline{A}_2) - P(\overline{A}_3) + P(\overline{A}_1 \cap \overline{A}_2) + P(\overline{A}_2 \cap \overline{A}_3) = 0.2 = P(\cap A_i)$$

so the Hunter bound (2.1) here can be taken as of form (3.4) and is optimal (without any permutation of indices). On the other hand

$$P(A_1 A_2)P(A_2 A_3)/P(A_2) = 0.24$$

so (3.2) breaks down. In this example we have, further, positive pairwise dependence.

We have also constructed an example for $k = 4$ (by specifying all 16 elementary disjunctions of form $B_1 \cap B_2 \cap B_3 \cap B_4$ where each $B_i = A_i$ or $\overline{A_i}$) where: (1) the (optimalized) Hunter bound (2.1) which has value 0.43 cannot be put into the form on the left of (3.4); (2) $P(A_1 \cap A_2 \cap A_3 \cap A_4) =$

0.48; (3) positive pairwise dependence obtains; (4) (3.2) holds under all permutations of the index set $(1, 2, \cdots, k)$ but the highest value it attains under these permutation is 0.394, not as good as the Hunter bound.

These notes reveal that little can be said, under positive pairwise dependence, about the relative merit of the sub-Markov bound compared to the Hunter bound; while we have not yet examined conditions for validity of (3.1), (3.2) (see Section 6).

The reason for the use of the term sub-Markov in (3.2) is seen by comparison with the equality

$$P(A_1)P(A_2|A_1)P(A_3|A_2, A_1) \cdots P(A_k|A_{k-1}, \cdots, A_1) = P(\bigcap_{i=1}^{k} A_i).$$

Positive pairwise dependence is the only situation relevant to considering the bounds (3.1) and (3.2), although, as we have seen, it is not sufficient for them to hold.

Pairwise *negative* dependence does occur in some simultaneous inference contexts.

Example

$$A_i = \{\frac{X_i - \overline{X}}{\sigma} < c\}, \qquad i = 1, \cdots, k$$

where the X_i's are independent $\mathcal{N}(\mu, \sigma^2)$ is a situation of interest in outlier detection. Here the A_i's are exchangeable and $P(A_i \cap A_j) \leq P^2(A_i)$ (Doornbos 1976).

However, an elegant argument in Dykstra et al. (1973) shows that if $\alpha = \sum_{i=1}^{k} q_i$ (with $q_i = P(\overline{A}_i)$ as before) then

$$0 \leq P(\bigcap_{i=1}^{k} A_i) - (1 - \alpha) \leq \alpha^2/2.$$

Thus for a small α ($\alpha = 0.05$ say), the bound provided by Boole's Inequality is within $(.05)^2/2 = 0.00125$ of $P(\cap A_i)$, so essentially nothing will improve much on Boole.

4   Chebyshev's Inequality. The Dunnett and Sobel Bounds

Chebyshev's covariance inequality (which dates to Chebyshev 1882) states that if X is any random variable and g, f are monotone functions such that $E\ g^2(X) < \infty$, $E\ f^2(X) < \infty$, then if f and g are concordantly monotone

$$\mathrm{Cov}\,(f(X),\ g(X)) \geq 0 \qquad (4.1)$$

(while if f and g are discordantly monotone the inequality in (4.1) is reversed). An elegant proof is given by Kingman (1978, p. 184). Clearly, for concordantly monotone functions f_i, $i = 1, \cdots, n$, satisfying $E\ f_i^2(X) < \infty$, it follows that

$$E(\prod_{i=1}^{n} f_i(X)) \geq \prod_{i=1}^{n} E\ f_i(X). \qquad (4.2)$$

Thus taking

$$A_i = \{Y_i < c_i\},$$

with $c_i > 0$, and Y_i defined by (1.3), we find

$$P(\bigcap_{i=1}^{k} A_i) \;=\; E\{P(Y_i < c_i, i = 1, \cdots, k | \overline{X}_0, S)\}$$

$$\;=\; E\{\prod_{i=1}^{k} P(Y_i < c_i | \overline{X}_0, S)\}$$

by joint independence of $\overline{X}_i$, $i = 0, \cdots, k$ and S^2;

$$= E_S E_{\overline{X}_0}\{\prod_{i=1}^{k} P(\overline{X}_i - \mu_i \leq \overline{X}_0 - \mu_0 + c_i S \sqrt{\frac{1}{n_i} + \frac{1}{n_0}} | \overline{X}_0, S)\}$$

and since for fixed $\overline{X}_0$, S, $P(Y_i < c_i)$ is increasing with $\overline{X}_0$, by (4.2)

$$\geq\; E_S \prod_{i=1}^{k} E_{\overline{X}_0}(P(\overline{X}_i - \mu_i \leq \overline{X}_0 - \mu_0 + c_i S \sqrt{\frac{1}{n_i} + \frac{1}{n_0}} | \overline{X}_0, S))$$

$$=\; E_S \prod_{i=1}^{k} P(\overline{X}_i - \mu_i \leq \overline{X}_0 - \mu_0 + c_i S \sqrt{\frac{1}{n_i} + \frac{1}{n_0}} | S)$$

and using (4.2) again, since again all the P's are monotone with S

$$P(\bigcap_{i=1}^{k} A_i) \geq \prod_{i=1}^{k} P(A_i). \qquad (4.3)$$

The steps are those carried out by Dunnett and Sobel (1955) to obtain (3.1).

To get a degree 2 multiplicative bound, they take the A_i's two at a time and non-overlapping to give (in this special setting)

$$P(\bigcap_{i=1}^{k} A_i) \geq \left[\begin{array}{ll} \prod_{i=1}^{k/2} P(A_{2i-1} \cap A_{2i}), & k \text{ even} \\ P(A_1) \prod_{i=1}^{(k-1)/2} P(A_{2i} \cap A_{2i+1}), & k \text{ odd.} \end{array} \right. \tag{4.4}$$

Since it is clear from the previous paragraph that we have *positive pairwise dependence*, the bounds in (4.4) are better than in (4.3).

However in the case of $n_1 = n_2 = \cdots = n_k$, $c_1 = c_2 = \cdots = c_k$ the events A_i, $i = 1, \cdots, k$ are exchangeable, and one can avoid Chebyshev's inequality altogether, by using the fact that the A_i's are conditionally independent, and Liapunov's inequality. Thus $P(A_1|\overline{X}_0, S) = P^{1/2}(A_1 \cap A_2|\overline{X}_0, S)$ and

$$\begin{aligned} P(\bigcap_{i=1}^{k} A_i) &= E(\prod_{i=1}^{k} P(A_i|\overline{X}_0, S)) = E(P^{k/2}(A_1 \cap A_2|\overline{X}_0, S)) \\ &\geq E^{k/2} P(A_1 \cap A_2|\overline{X}_0, S) = P^{k/2}(A_1 \cap A_2) \end{aligned} \tag{4.5}$$

which is sharper (for odd k) than (4.4) because of *positive pairwise dependence*. Further the sub-Markov inequality holds in this setting (Glaz and Johnson 1984), and since

$$P^{k/2}(A_1 \cap A_2) \leq P(A_1)P^{k-1}(A_2|A_1)$$

again by positive pairwise dependence, the sub-Markov bound is sharper. Finally note that (3.4) predicts a superiority of the sub-Markov bound over the Hunter bound (2.1), which in this exchangeable can becomes the Sobel-Uppuluri inequality.

We illustrate numerically in the exchangeable case $n_i = n$, $i = 0, \cdots, k$ which gives $\rho_{ij} = 1/2$, $i \neq j$; $c_i = C$, $i = 1, \cdots, k$ by obtaining first C such that

$$P(\bigcap_{i=1}^{k} A_i) = 0.95$$

(which can be determined for various k and ν from Dunnett (1955, Table 1). We then go to Dunnett and Sobel's (1954) tables of the bivariate t with this C to look up $P(A_1 \cap A_2)$ and hence calculate the bounds. See Table 1.

Notice that the degree 2 bounds are substantially better than degree 1 bounds generally; the effectiveness of all bounds declines with increasing k, and the Dunnett-Sobel bound is substantially better, for large k, than the Sobel-Uppuluri-Hunter bound.

The Chebyshev argument cannot be applied to produce a bound in the case of a two-sided interval: $P\{|Y_1| < c_1, |Y_2| < c_2, \cdots, |Y_k| < c_k\}$, in the

Table 1: Numerical comparison of bounds for one-sided Dunnett intervals.

	ν	C	Boole (1.1)	Product (3.1)	Dunnett Sobel (4.5)	Sobel Uppuluri Hunter (2.1)	Sub- Markov (3.2)
k=3	10	2.34	0.938	0.939	0.945	0.946	0.946
	15	2.24	0.939	0.940	0.946	0.947	0.947
	20	2.19	0.939	0.940	0.945	0.946	0.946
	60	2.10	0.940	0.941	0.944	0.945	0.945
k=9	10	2.81	0.917	0.920	0.921	0.919	0.921
	15	2.67	0.921	0.924	0.924	0.921	0.924
	20	2.60	0.923	0.926	0.926	0.923	0.927
	60	2.48	0.928	0.930	0.934	0.929	0.936

case of n_i's and c_i's different (since monotonicity is lost), but in the special case of exchangeability, there is no problem with the Liapunov argument and (4.5) holds.

In this two-sided setting, Dunnett (1955) finds C^* so that

$$P^{k/2}\{|Y_1| < C^*,\ |Y_2| < C^*\} = \gamma$$

where γ (e.g. 0.95) is fixed and C is the value sought satisfying

$$P\{|Y_1| < C,\ |Y_2| < C, \cdots,\ |Y_k| < C\} = \gamma. \qquad (4.6)$$

Then $C \leq C^*$, by (4.5), and C^* would give, for example, a conservative simultaneous $100\,\gamma\%$ level confidence interval.

If we turn to the more general setting of the exchangeable multivariate t-distribution with correlation matrix with $\rho_{ij} = \rho,\ i \neq j$, we can compare the Boole and Sobel-Uppuluri-Hunter bounds in this way by seeking C_B and C_H such that

$$\begin{aligned} 1 - kP\{|T_1| > C_B\} = \gamma \ &=\ 1 - k\,P\{|T_1| > C_H\} \\ &+\ (k-1)P\{|T_1| > C_H, |T_2| > C_H\}. \end{aligned} \qquad (4.7)$$

Clearly

$$C_B \geq C_H \geq C.$$

Table 2: C/C_H when $\gamma = 0.95$.

ρ	0.2	0.5	0.8	0.9
$k = 3$	0.997	0.992	0.985	0.983
$k = 9$	0.986	0.964	0.924	0.907

While C_B is the $\alpha = (1 - \gamma)/(2k)$ point of a univariate t-distribution, C_H requires an algorithmic procedure, including calculation of bivariate t-probabilities. Such a numerical comparison has been carried out by Stoline (1983). Below is a portion of his results (the values of C were obtained in part from Hahn and Hendrikson 1971). The value $\rho = 0.5$ corresponds to the Dunnett (1955) setting. A comparison with $C_M (\geq C)$ arising from the sub-Markov bound is clearly of interest where

$$P(|T_1| \leq C_M)P^{k-1}(|T_2| \leq C_M \mid |T_1| \leq C_M) = \gamma.$$

5 Monotonicity and the Exchangeable Case of Multivariate t

Dunnett and Sobel's (1955) framework was a slightly more general setting than that of Dunnett's test, specifically (1.4) where

$$Z_i = (1 - b_i^2)^{\frac{1}{2}} W_i - b_i W_0, \qquad i = 1, \cdots, k$$

where $0 \leq b_i \leq 1$, and W_i, $i = 0, 1, \cdots, k$ are $NID(0,1)$. This gives $\rho_{ij} = b_i b_j$, $i \neq j$, $i, j = 1, \cdots, k$. (When specialized to $b_i = \sqrt{\rho}$, $i = 1, \cdots, k$, this gives the exchangeable case of the last section). One of the first papers to take up Dunnett and Sobel's work in precisely this setting with Šidák (1962). The title of this Czech-language paper already suggests a motivation by the case of unequal n_i's, $i = 1, \cdots, k$ in Dunnett's test, for which situation, as Šidák (1962, p. 299) noted, the Chebyshev inequality does not work in the two-sided case. Šidák was led to prove the concordant monotonicity with each b_i (where $\rho_{ij} = b_i b_j$, $i \neq j$) of each of:

$$P(\bigcap_{i=1}^{k} \{|T_i| \leq c_i\}), \ P(\bigcap_{i=1}^{k} \{T_i \leq c_i\}). \tag{5.1}$$

His purpose was to obtain two-sided bounds on the quantities (5.1) using the monotonicity and the resulting exchangeable case $\rho_{ij} = \rho$, $i \neq j$. He later extended these results to a more general structure of the ρ_{ij}'s in the better known English-language paper Šidák (1968). Slepian (1962) had already obtained the monotonicity with the ρ_{ij}'s of the one-sided quantity in (5.1). Using obvious notation where $\rho = \min(i > j)\rho_{ij}$, we have

$$P(\bigcap_{i=1}^{k} A_i|\{\rho_{ij}\}) \geq P(\bigcap_{i=1}^{k} A_i|\rho). \tag{5.2}$$

By taking $\rho = 0$, Šidák (1962) (1968) shows that the degree 1 product bound (3.1) obtains, even in the two-sided case of (5.1).

By taking $c_1 = c_2 = \ldots = c_k$ we can apply the various lower bounds developed in the exchangeable case to the right of (5.2). Šidák does not go beyond Dunnett and Sobel's (1955) degree 2 bound (4.5); as we have seen this can be sharpened to

$$P(\bigcap_{i=1}^{k} A_i|\rho) \geq P^{k-1}(A_1 \cap A_2)/P^{k-2}(A_1). \tag{5.3}$$

There seems little point in applying the reduction to exchangeability when one is considering the Hunter lower bound (which *always obtains*), because its general form does not go beyond use of bivariate t-values.

6   Validity of the Sub-Markov Inequality

The key roles in regard to multiplicative inequalities in the previous two sections, albeit in the setting of multivariate t, have been played by: 1) Chebyshev's covariance inequality in the presence of monotonicity in establishing multiplicative bounds; 2) the fact that exchangeability, via Liapunov's inequality, enables bounds to be established without monotonicity; 3) positive pairwise dependence in showing that degree two lower bounds are sharper than the degree 1 bound.

We have also shown by example in Section 3 that pairwise positive dependence alone does not guarantee the validity of the sub-Markov bound (3.2).

We now examine these issues in a more general setting.

If the underlying probability space is $\mathbf{R}^k$ and $A_i = \{X_i \in (-\infty, a_i]\}, i = 1, \cdots, k$; or $A_i = \{X_i \in [b_i, \infty)\}, i = 1, \cdots, k$ and the joint density of $X_1, \cdots, X_k$, satisfies a MTP_2 condition, then

$$\prod_{i=1}^{k} P(A_i) \leq P(A_1) \prod_{i=2}^{k} P(A_i|A_{i-1}) \leq P(\bigcap_{i=1}^{k} A_i)$$

(and in fact a third order bound improves on the second order bound etc.).

This result, relevant to the characteristically statistical situation (1.2), is due to Glaz and Johnson (1984, Theorem 2.3), whose paper (we refer to it as G.J. henceforth) gave recent impetus to the study of multiplicative bounds of order greater than 1.

It is relevant to note in passing that Esary et al. (1967) have proved that the (degree 1) product inequality (3.1) holds if for every pair of coordinatewise increasing real valued functions f and g, $\mathrm{Cov}\,(f(\mathbf{X}), g(\mathbf{X})) \geq 0$ where $\mathbf{X} = (X_1, \cdots, X_n)$. That is, if Chebyshev's covariance inequality holds in a multivariate setting.

The basis of G.J.'s result is a sequence of results, culminating in Corollary 4.1, of Karlin and Rinnott (1980) from which we take the following definitions.

A real valued function $f(x, y)$ of two variables is *totally positive* of order two, TP_2, if

$$f(x_1, y_1)f(x_2, y_2) - f(x_1, y_2)f(x_2, y_1) \geq 0$$

for all $x_1 < x_2$ and $y_1 < y_2$. A real valued function of k variables $f(x_1, \cdots, x_k)$ is said to be *multivariate totally positive* of order 2, MTP_2, if for any pair of arguments x_1 and x_j keeping the others fixed, the function is TP_2.

In the context of the result cited above of G.J., we note that since the MTP_2 condition is invariant under permutation of indices we may deduce that

$$\max_{\Pi} P(A_1) \prod_{i=2}^{k} P(A_i | A_{i-1}) \leq P(\bigcap_{i=1}^{k} A_i). \qquad (6.1)$$

Further the condition continues to hold for marginal densities, so, taking $k = 2$, we see that it implies pairwise positive dependence.

An obvious question to investigate in view of our Section 2, and (3.4), is the validity of the inequality, sharper than (6.1):

$$\max_{\Pi} P(A_1) \prod_{i=2}^{k} \max_{1 \leq s < i} P(A_i | A_s) \leq P(\cap_{i=1}^{k} A_i). \qquad (6.2)$$

The value of the left hand side may be constructed as follows. Take i_0 such that $P(A_{i_0}) = \max_i P(A_i)$, and let $C = \{i_0\}$, $U = \{1, 2, \cdots, k\} - \{i_0\}$. Then proceed as in Steps 1 and 2 of the Jarník algorithm described at the end of Section 2, taking $r_{ij} = P(A_i | A_j)$. For an investigation somewhat along these lines but in a graph-theoretic setting, see Block et al. (1988).

We return now to an examination of the relevant sequence of Karlin and Rinnott's (1980) results, and G.J.'s main results (Theorems 2.3 and 2.5).

The essence of Karlin and Rinnott's (1980, Section 2) is to prove that if f is a probability density with respect to product measure σ on $\mathbf{R}^k$, and f satisfies an MTP_2 condition then for any pair of coordinatewise increasing (or decreasing) functions φ and ψ on $\mathbf{R}^k$

$$\int \varphi(\mathbf{x})\psi(\mathbf{x})f(\mathbf{x})d\sigma(\mathbf{x}) \geq \int \varphi(\mathbf{x})f(\mathbf{x})d\sigma(\mathbf{x}) \int \psi(\mathbf{x})f(\mathbf{x})d\sigma(\mathbf{x}) \quad (6.3)$$

which is a *multivariate generalization of Chebyshev's covariance inequality* (4.1).

Now let $S_1 = \{X_s \in A_s\}$, $S_2 = \{X_j \in A_j, \ j = s+1, \cdots, i-1\}$, $S_3 = \{X_i \in A_i\}$, $h(\mathbf{x})$ the joint MTP_2 density of $X_1 \cdots, X_k$, and I_j, $j = 1, 2, 3$ are the corresponding indicator functions of S_1, S_2, S_3 in $\mathbf{R}^k$. If the A_i are all semi-infinite intervals of the same kind (e.g. all of form $(-\infty, a]$), then the I_j are all concordantly monotone on $\mathbf{R}^k$; further $I_2 h$ is an MTP_2 function, and $I_2 h / \int I_2 h d\mathbf{x}$ is an MTP_2 probability density. Taking $d\sigma(\mathbf{x}) = d\mathbf{x}$, $f(\mathbf{x})$ this density, $\varphi = I_3$, $\psi = I_1$ we obtain from (6.2)

$$E(I_1 I_2 I_3)E(I_2) \geq E(I_3 I_2)E(I_1 I_2)$$

which is tantamount to

$$P(\mathbf{X} \in S_3 | \mathbf{X} \in S_1 \cap S_2) \geq P(\mathbf{X} \in S_3 | \mathbf{X} \in S_2),$$

the cornerstone of G.J.'s proof of their Theorem 2.3.

Theorem 2.5 of G.J. is concerned with the situation where random variables $T_1, \cdots, T_k$ are conditionally independent relative to a σ-field G. Let $A_i = \{T_i \in (a_i b_i]\}$, say, $i = 1, \cdots, k$, and $g_i = E(I_{A_i}|G) = P(A_i|G)$. Then

$$P(\bigcap_{i=1}^{k} A_i) = \int \prod_{i=1}^{k} g_i d\mu$$

where μ is a probability measure on the G sets. The joint density of X_i, $i = 1, \cdots, k$, where $X_i = I_{A_i}$ is

$$f(x_1, \cdots, x_k) = \int \prod_{r=1}^{k} g_r^{x_r}(1 - g_r)^{1-x_r} d\mu$$

where $x_r = 0$ or 1, for each $r = 1, \cdots, k$. This function will be MTP_2 if we take arbitrary i, j, $i \neq j$, and setting $x_i = x_j = 0$ keeping the other x_r's arbitrary

$$f(\ldots, 0, \ldots, 0, \ldots)f(\ldots, 1, \ldots, 1\ldots) \geq f(\ldots, 0, \ldots, 1, \ldots)f(\ldots, 1, \ldots, 0, \ldots) \quad (6.4)$$

i.e.

$$\int g_i g_j d\nu \int (1-g_i)(1-g_j)d\nu \geq \int (1-g_j)g_i d\nu \int (1-g_i)g_j d\nu \qquad (6.5)$$

where

$$d\nu = K \prod_{\substack{r=1 \\ r\neq i,j}}^{k} g_r^{z_r}(1-g_r)^{1-z_r} d\mu$$

where K is a normalizing constant to make $\int d\nu = 1$. But (6.5) is tanta-mount to

$$\int g_i g_j d\nu \geq \int g_i d\nu \int g_j d\nu \qquad (6.6)$$

which is *Chebyshev's covariance inequality*. Thus if (6.6) holds, and if the intervals are of the same infinite kind, G.J.'s Theorem 2.3 stated earlier in this section can be applied to the X_i's to give (6.1).

Clearly (6.6) will hold (Liapunov's inequality) if the g_r's are *identical* for all r, and this occurs in the situation of multivariate t with exchangeable random variables, and each interval the same.

More generally, the above discussion reveals that the sub-Markov bound (3.2) will hold for k specific semi-infinite intervals of the same kind $A_1, ..., A_k$, if (6.4) holds where f is the joint density of the indicator functions $I(A_i)$, $i = 1, \cdots, k$. These conditions on the probabilities of elementary disjunctions (each of degree k) of the sets A_i, $i = 1, \cdots, k$, from which positive pairwise dependence is an elementary consequence, are in themselves quite stringent. Since the sub-Markov bound (3.2) is of degree 2, one might have hoped that degree 2 conditions would suffice. However, it becomes clearer that degree k conditions may need to be involved in general since we are attempting to assert that

$$P(\bigcap_{i=1}^{k} A_i) = P(A_1) \prod_{i=2}^{k} P(A_i | A_{i-1} \cap \cdots \cap A_1) \geq P(A_1) \prod_{i=2}^{k} P(A_i | A_{i-1}).$$

In order to check the condition (6.4) in general we need a quantity of information which will not be available. Thus the validity of the sub-Markov bound will be known in special situation only, such as the exchangeable case of the multivariate t above.

Acknowledgments

I wish to thank Fred Hoppe for bringing several references to my attention, and for his hospitality.

References

Barnett, V. and Lewis, T. (1984). *Outliers in Statistical Data* (Second edition). John Wiley and Sons, Inc., Chichester.

Bauer, P. and Hackl, P. (1985). The application of Hunter's inequality in simultaneous testing. *Biom. Jour.* **27** 25 - 38.

Block, H.W., Costigan, T. and Sampson, A.R. (1988). Optimal product-type probability bounds. Technical Report No. 88-07. Series in Reliability and Statistics, Department of Mathematics and Statistics, University of Pittsburgh, Pittsburgh, PA.

Boole, G. (1854). *An Investigation of the Laws of Thought on Which are Founded the Mathematical Theories of Logic and Probabilities.* London: Macmillan. [Reprinted Dover, New York, 1958.]

Chebyshev, P.L. (1882). On approximate expressions of integrals through others between the same limits [in Russian]. *Soobsch. Mat. Obsch. Kharkovsk. Imp. Univ.* **2** 93 - 98. [More recently in his collected works: Chebyshev, P.L. (1948). *Polnoe Sobranie Sochinenii* **3** 128 - 131. Izd. AN SSSR. Moscow - Leningrad.]

Cornish, A.E. (1954). The multivariate t-distribution associated with a set of normal sample deviates. *Austral. Jour. Phys.* **7** 531 - 542.

Doornbos, R. (1976). *Slippage Tests* (Second edition). Mathematical Centre Tracts, No. 15. Mathematisch Centrum, Amsterdam.

Dunnett, C.W. (1955). A multiple comparison procedure for comparing several treatments with a control. *Jour. Amer. Statist. Assoc.* **50** 1096 - 1121.

Dunnett, C.W. and Sobel, M. (1954). A bivariate generalization of Student's t-distribution with tables for certain special cases. *Biometrika* **41** 153 - 169.

Dunnett, C.W. and Sobel, M. (1955). Approximations to the probability integral and certain percentage points of a multivariate analogue of Student's t-distribution. *Biometrika* **42** 258 - 260.

Dykstra, R.L., Hewett, J.E. and Thompson Jr., W.A. (1973). Events which are almost independent. *Ann. Statist.* **1** 674 - 681.

Esary, J.D., Proschan, F. and Walkup, D. (1967). Association of random variables with applications. *Ann. Math. Statist.* **38** 1466 - 1474.

Glaz, J. (1987). A comparison of Bonferroni-type and product-type inequalities in presence of dependence. *Symposium on Dependence in Statistics and Probability*. Hidden Valley Conference Center, Pennsylvania. [Tech. Report No. 87 - 27, Department of Statistics, University of Connecticut, Storrs CT.]

Glaz, J. and Johnson, B. McK. (1984). Probability inequalities for multivariate distributions with dependence structures. *Jour. Amer. Statist. Assoc.* **79** 436 - 440.

Graham, R.L. and Hell, P. (1985). On the history of the minimal spanning tree problem. *Ann. Hist. Comput.* **7** 43- 57.

Hahn, G.J. and Hendrickson, R.W. (1971). A table of percentage points of the distribution of the largest absolute value of k Student t variates and its application. *Biometrika* **58** 323 - 332.

Hoover, D.R. (1988). Comparisons of improved Bonferroni and Sidak/Slepian bounds with application to normal Markov processes. Technical Report, Department of Statistics, University of South Carolina, Columbia, SC.

Hoppe, F.M. (1985). Iterating Bonferroni bounds. *Statist. Probab. Lett.* **3** 121 - 125.

Hoppe, F.M. and Seneta, E. (1990). A Bonferroni-type identity and permutation bounds. *Intern. Statist. Rev.* **58** 253 - 261.

Hunter, D. (1976a). An upper bound for the probability of a union. *Jour. Appl. Prob.* **13** 597 - 603.

Hunter, D. (1976b). Simultaneous t-tests in normal models. IBM Thomas J. Watson Research Center, IBM Research Report RJ 1757 (25565), Yorktown Heights, New York.

Jarník, V. (1930). O jistém problému minimálním. [On a certain minimal problem.] *Práce Moravské Přírodovedecké Spolecnosti v Brně (Acta Societ. Scient. Natur. Moravicae)* **6** 57 - 63.

Jogdeo, K. (1977). Association and probability inequalities. *Ann. Statist.* **5** 495 - 504.

Karlin, S. and Rinott, Y. (1980). Classes of orderings of measures and related correlation inequalities - I. Multivariate totally positive distributions. *Jour. Multivar. Anal.* **10** 467 - 498.

Kingman, J.F.C. (1978). Uses of exchangeability. *Ann. Prob.* **6** 183 - 197.

Kruskal, J.B. (1956). On the shortest spanning tree of a graph and the travelling salesman problem. *Proc. Amer. Math. Soc.* **7** 48 - 50.

Margaritescu, E. (1986) A note on Bonferroni's inequalities. *Biom. Jour.* **28** 937 - 943.

Seneta, E. (1988). Degree, iteration and permutation in improving Bonferroni-type bounds. *Austral. Jour. Statist.* **30A** 27 - 38.

Šidák, Z. (1962). Nestejne počty pozorování při srovnávání několika skupin s jednou kontrolní. [Unequal numbers of observations in comparing several treatments with one control.] *Aplikace Matematiky* **7** 292 - 314.

Šidák, Z. (1968). On multivariate normal probabilities of rectangles: their dependence on correlations. *Ann. Math. Statist.* **39** 1425 - 1434.

Slepian, D. (1962). The one-sided barrier problem for Gaussian noise. *Bell System Tech. Jour.* **41** 463 - 501.

Sobel, M. and Uppuluri, V.R.R. (1972). On Bonferroni-type inequalities of the same degree for the probability of unions and intersections. *Ann. Math. Statist.* **43** 1549 - 1558.

Stoline, M.R. (1983). The Hunter method of simultaneous inference and its recommended use for applications having large known correlation structures. *Jour. Amer. Statist. Assoc.* **78** 366 - 370.

Worsley, K.J. (1982). An improved Bonferroni inequality and applications. *Biometrika* **69** 297 - 302.

Chapter 4

Multiple Comparisons in Split Block and Split-Split Plot Designs

WALTER T. FEDERER Biometrics Unit, Cornell University, Ithaca, New York

CHARLES E. MCCULLOCH Biometrics Unit, Cornell University, Ithaca, New York

Abstract Multiple comparisons in split block and split-split plot designs encounter statistical, conceptual, and philosophical difficulties. Multiple error terms and distributional properties of some of the error terms account for the statistical problems. Structure or lack of it in treatment design is what causes the conceptual and philosophical difficulties. When there is structure in the treatment design, specific contrasts in the form of orthogonal single degree of freedom, or sets of degrees of freedom, comparisons are required. When the different entries of a factor have a nominal scale of measurement and no structure, multiple comparisons procedures usually will be indicated. The error rate base will need to be addressed regardless of the nature and structure of the treatment design. Selection of the error rate base is the important consideration. Then, the selection of the particular multiple comparisons procedure to be used follows. The term multiple comparisons needs to be precisely defined. The above situations are addressed for these more complex designs.

1   Introduction

This topic was selected for presentation because of the contributions of Charles W. Dunnett in this area, especially his papers dealing with experimentwise error rates for comparisons with a control. The paper is a sequel to the ones by Federer (1975), Federer and McCulloch (1984) and Federer and Meredith (1992) on analyses for these designs. In this paper, we shall consider the complete set of all possible comparisons or some subset thereof as falling in the realm of multiple comparisons. Some authors appear to consider multiple comparisons to be confined to all possible pairwise comparisons of means, some as all comparisons with a control, some as all possible comparisons, and some as a selected subset of all possible comparisons. We consider all these situations to be in the realm of multiple comparisons.

In many situations involving the use of split block and split plot designs, there is considerable structure associated with the treatment design. Hence the problem of considering all possible comparisons does not arise even though the problem of selecting the error rate will remain. As Tukey (1991) has so aptly pointed out in his thoughtful article, multiple comparisons are not always appropriate, especially when there is structure in the treatment design. When structure is absent, multiple comparisons are appropriate. One particular case in which multiple comparisons will be appropriate for a split-split plot design is when the main or whole plot treatments are populations (grandparents, species, hospitals, e.g.), the split plot treatments are subpopulations (parents, crosses or families, procedures,), and the split-split plot treatments are individuals within a subpopulation (children, lines or selections, technicians). Many such situations can be found in practice for the simple split plot design. A variety of multiple comparisons may be desired among the whole plot treatments, among split plot treatments, and/or among the split-split plot treatments. In any event, every experiment involves a set of comparisons and the need for selecting an error rate base.

In the next section, a particular member of the family of split block designs is selected to illustrate the procedures for multiple comparisons. The first situation considered is for the case when there is considerable structure among the levels of the two factors which form the two-way whole plots. Then, the situation for which all possible comparisons among pairs of means is desired, is presented. Multiple error terms causes complications for both cases and the differing sums of squares of coefficients for the different contrasts necessitates the computation of a different range for each contrast.

In the third section a third situation is considered for a split block design where there is no structure among the levels of the two factors. A simulation-based approach is illustrated.

The subject of multiple comparisons for a split-split plot design is discussed in the fourth section. The particular split-split plot design selected to illustrate multiple comparisons procedures is the one which has populations as the whole or main plots, subpopulations as the split plots, and individuals within subpopulations as the split-split plots.

Three particular and different error rate bases are to be used in the following. These are comparisonwise or per comparison, per experiment, and experimentwise. There are more bases (e.g., Tukey 1953, 1991; Hartley 1955; Federer 1961; Chew 1977; Saville 1990, 1991; Holland 1991; and Lea 1991, for a discussion of error rates and multiple comparisons procedures), but our attention is confined to these three. An error rate base needs to be selected prior to selecting a multiple comparisons procedure. This means that it will not be possible to use only one multiple comparisons procedure for all situations as advocated by some (e.g., Saville 1990, 1991). A per comparison or comparisonwise error rate is defined to be

Error rate per comparison

$$= \frac{E[\ \sharp \text{ of erroneous inferences of comparisons}]}{[\sharp \text{ of inferences on comparisons attempted}]}$$

$= $ proportion of all comparisons expected to be erroneous when the null hypothesis is true.

A per experiment error rate is said to be

Error rate per experiment

$$= \frac{E[\ \sharp \text{ of erroneous inferences}]}{[\sharp \text{ of experiments}]}$$

$= $ expected number of erroneous statements per experiment when the null hypothesis is true.

An experimentwise error rate is defined to be

Experimentwise error rate

$$= \frac{E[\ \sharp \text{ of experiments with one or more erroneous statements}]}{[\sharp \text{ of experiments}]}$$

$= $ expected proportion of experiments with one or more erroneous inferences when the null hypothesis is true.

Following are four multiple comparisons procedures we consider which, for illustrative purposes, are described only for comparing means in the one-way layout.

lsd: The comparisonwise confidence interval for a pair of means is computed as $\bar{y}_{i\cdot} - \bar{y}_{i'\cdot} \pm t_{\alpha,f}(2s^2/r)^{1/2}$, where s^2 is an estimate of the experimental error mean square with f degrees of freedom, $\bar{y}_{i\cdot}$ is the sample treatment mean $i \neq i' = 1, 2, \cdots v$, r is the number of replicates for the ith treatment mean, and $t_{\alpha,f}$ is the tabulated value for the two-tailed Student's t at the α percent level for f degrees of freedom. This is the least significant difference procedure.

esd: The per experiment confidence interval for m pairs of means is computed as $\bar{y}_{i\cdot} - \bar{y}_{i'\cdot} \pm t_{\alpha/m,f}(2s^2/r)^{1/2}$. This is sometimes called the Bonferroni procedure.

hsd: The experimentwise confidence interval on pairs of means is computed as $\bar{y}_{i\cdot} - \bar{y}_{i'} \pm q_{\alpha,f,v}(s^2/r)^{1/2}$, where $q_{\alpha,f,v}$ is the tabulated value of the Studentized-range statistic at the α percent level for f degrees of freedom and v treatments in the experiment. This is commonly called the honestly significant difference or Tukey's range procedure.

multiple comparisons with the best and subset selection: Choose the ith population to be in the selected subset if

$$\bar{y}_{i\cdot} \geq \max_{i' \neq i}\{\bar{y}_{i'\cdot}\} - d_{v,f,\alpha}(2s^2/r)^{1/2},$$

where $d_{v,f,\alpha}$ is the one-sided α point of a $(v-1)$-variate t-distribution with f degrees of freedom and common correlation $\rho = 1/2$. These values have been tabulated by Dunnett (1955), for example, where the tables are entered using $k = v - 1$. Simultaneous intervals which correspond to the subset selection procedure confidence intervals for $\mu_{[v]} - \mu_i$ are given by $[0, D_i]$, where

$$D_i = \max\{0, \ (\max_{i' \neq i}\{\bar{y}_{i'\cdot}\} - \bar{y}_{i\cdot} + d_{v,f,\alpha}(2s^2/r)^{1/2})\}$$

and $\mu_{[i]}$, $i = 1, 2, \cdots, v$, represent the ordered population means ($\mu_{[1]} \leq \mu_{[2]} \leq \cdots \leq \mu_{[v]}$) (Hsu 1981). Dunnett's one-sided comparisons with a control are related to simultaneous confidence intervals with the best. If the control treatment has the largest sample mean, the upper one-sided confidence intervals for the differences between the control mean and the other treatment means will be the same as those for $\mu_{[v]} - \mu_i$. The error rate is experimentwise.

2   Split Block Design

The particular split block, or two-way whole plot, design we use as an example will have the levels of one factor, say A, laid out in a latin square design with the rows corresponding to complete blocks and the columns corresponding to orders within the complete blocks. The experimental units for the levels of the second factor, say B, are laid out across the experimental units for the levels of factor A and are designed as randomized complete block design. Let r represent the number of blocks for A and let a be the number of levels of A, then for $r = 4 = a$, a schematic plan would appear like this:

Block (row)		Order (column)			
		1	2	3	4
1	B_1				
	B_2	A_1	A_2	A_3	A_4
	B_3				
2	B_3				
	B_1	A_2	A_3	A_4	A_1
	B_2				
3	B_1				
	B_2	A_3	A_4	A_1	A_2
	B_3				
4	B_2				
	B_1	A_4	A_1	A_2	A_3
	B_3				

There are $a = 4$ experimental units for the levels of factor A and $b = 3$ experimental units for levels of factor B in each block. There will be r randomizations for the a levels of factor A, restricted to form a latin square, and another r randomizations (unrestricted) for the b levels of factor B. The experimental unit for the combination of level i of factor A and level j of factor B is $1/ab$ of the block size. The fact that there are three levels of randomization and three different sizes of experimental units leads to the consideration of three different error mean squares. A linear model for the above designed experiment would be

$$Y_{ghij} = \mu + \rho_g + \gamma_h + \tau_i + \pi_j + \delta_{ghi} + (\tau\pi)_{ij} + \lambda_{gj} + \epsilon_{ghij}, \qquad (2.1)$$

where Y_{ghij} is the response for the $ghij^{th}$ observation, μ is a mean effect common to all observations, ρ_g is the gth row effect, γ_h is the hth column effect, τ_i is the effect of the ith level of factor A, π_j is the effect of the jth level of factor B, δ_{ghi} is a random error effect for the subdivision within row g and column h associated with the levels of factor A, γ_{gj} is a random error effect for the subdivision within row g associated with the levels of factor B, $(\tau\pi)_{ij}$ is an interaction effect of the ith level of factor A and the jth level of factor B, and ϵ_{ghij} is a random error effect associated with the $ghij^{th}$

Table 1: Analysis of variance for a split block design.

Source of Variation	Degrees of Freedom	Mean Square
Total	rab	
Correction for mean	1	
Block (row)	$r - 1$	
Column	$r - 1$	
Factor A	$a - 1 = r - 1$	
Contrast A_1	1	
Contrast A_2	1	
$\vdots$	$\vdots$	
Contrast A_{a-1}	1	
Error (a)	$f_a = (r-1)(r-2)$	E_a
Factor B	$b - 1$	
Contrast B_1	1	
Contrast B_2	1	
$\vdots$	$\vdots$	
Contrast B_{b-1}	1	
Error (b)	$f_b = (r-1)(b-1)$	E_b
A $\times$ B	$(a-1)(b-1)$	
$A_1 \times B_1$	1	
$A_1 \times B_2$	1	
$\vdots$	$\vdots$	
$A_2 \times B_1$	1	
$A_2 \times B_2$	1	
$\vdots$	$\vdots$	
$A_{a-1} \times B_{b-1}$	1	
Error (ab)	$f_{ab} = (r-1)(a-1)(b-1)$	E_{ab}

observation. Given that there is sufficient structure in the treatment design to construct orthogonal single degree of freedom contrasts, a partitioning of the factor and interaction degrees of freedom may be made as given in Table 1. The multiple error mean squares is a feature of the experiment design and not of multiple comparisons procedures.

Given the $ab - 1$ single degree of freedom contrasts in Table 1, the error rate base for constructing confidence intervals (or making tests of significance) could be per contrast (comparisonwise), per the $ab - 1$ set of contrasts (per experiment error rate base), or some other base. The problem of determining an error rate is a feature of multiple comparisons and not of the statistical design of either the treatments or the experiment.

For the $a - 1$ contrasts A_1 to A_{a-1} for factor A, we use E_a to construct the $a - 1$ confidence intervals. For the $b - 1$ contrast B_1 to B_{b-1} for factor B, use is made of E_b to construct the $b - 1$ confidence intervals. For the $(a - 1)(b - 1)$ contrasts $A_1 \times B_1$ to $A_{a-1} \times B_{b-1}$ for the interaction, we make use of E_{ab} to construct the $(a - 1)(b - 1)$ confidence intervals. For a comparisonwise error rate, use

$$t_{f_a, \alpha} (E_a \sum_i c_i^2 / rb)^{1/2},$$

$$t_{f_b, \alpha} (E_b \sum_j d_j^2 / ra)^{1/2}, \text{ and}$$

$$t_{f_{ab}, \alpha} (E_{ab} \sum_i \sum_j \{c_i d_j\}^2 / r)^{1/2},$$

for the A, B, and A $\times$ B contrasts, respectively. In the above, c_i represent the coefficients in one of the contrasts among the means for levels of factor A, i.e.,

$$\sum_i c_i \bar{y}_{i\cdot}.$$

d_j are the coefficients in a contrast among levels of factor B. For a per experiment error rate, we replace α with $\alpha/(ab - 1)$. According to Ghosh (1955), this is what would be done for an experimentwise error rate for this set of contrasts. A more efficient approach would be to use percentage points of the multivariate t-distribution, though tables of these percentage points are less accessible.

If multiple comparisons among pairs of the ab means $\bar{y}_{\cdot\cdot ij}$, five different variances of a difference between two means would be involved (See Federer and Meredith 1992). These are

$$V(\bar{y}_{\cdot\cdot i\cdot} - \bar{y}_{\cdot\cdot i'\cdot}) = 2E_a / rb,$$

$$\begin{aligned}
V(\bar{y}_{\cdot\cdot\cdot j} - \bar{y}_{\cdot\cdot\cdot j'}) &= 2E_b/ra, \\
V(\bar{y}_{\cdot\cdot ij} - \bar{y}_{\cdot\cdot i'j}) &= 2[E_a + (b-1)E_{ab}]/rb, \\
V(\bar{y}_{\cdot\cdot ij} - \bar{y}_{\cdot\cdot ij'}) &= 2[E_b + (a-1)E_{ab}]/ra, \text{ and} \\
V(\bar{y}_{\cdot\cdot ij} - \bar{y}_{\cdot\cdot i'j'}) &= 2[aE_a + bE_b + (ab-a-b)E_{ab}]/rab.
\end{aligned}$$

Note that the five different standard errors is a feature of the experiment design and not of multiple comparisons procedures. These multiple variances will result in five different ranges for comparing pairs of means. The number of degrees of freedom, f^*, for the last three mean squares must be approximated. (See Grimes and Federer 1984, and also the next section for an alternate approach.) The resulting comparisonwise error rate and the per experiment error rate $(1 - \alpha)\%$ confidence intervals are computed as

comparisonwise:

A mean: $\bar{y}_{\cdot\cdot i\cdot} - \bar{y}_{\cdot\cdot i'\cdot} \pm t_{f_a,\alpha}(2E_a/rb)^{1/2}$

B mean: $\bar{y}_{\cdot\cdot\cdot j} - \bar{y}_{\cdot\cdot\cdot j'} \pm t_{f_b,\alpha}(2E_b/ra)^{1/2}$

AB mean: $\bar{y}_{\cdot\cdot ij} - \bar{y}_{\cdot\cdot i'j} \pm t_{f^*,\alpha}(2[E_a + \{b-1\}E_{ab}]/rb)^{1/2}$

                  $\bar{y}_{\cdot\cdot ij} - \bar{y}_{\cdot\cdot ij'} \pm t_{f^*,\alpha}(2[E_b + \{a-1\}E_{ab}]/ra)^{1/2}$

                  $\bar{y}_{\cdot\cdot ij} - \bar{y}_{\cdot\cdot i'j'} \pm t_{f^*,\alpha}(2[aE_a + bE_b + \{ab-a-b\}E_{ab}]/rab)^{1/2}$

per experiment: In the above confidence interval formulae, replace the value for α by

$$\alpha/[a(a-1)/2 + b(b-1)/2 + ab(ab-1)/2],$$

where the denominator is the total number of pairwise comparisons of means made in comparing the A, the B, and the $A \times B$ means.

3   Simulation-based Methods for the Split Block Design

In complicated experimental designs, such as the split-block or split-split plot experiments illustrated here, it is quite likely that the experimenter will wish to control the error rate for a subset of all the possible comparisons. In such cases the lsd procedure is not applicable, since it only controls the comparisonwise error rate. The esd procedure may be easily adapted by appropriately adjusting the α level, but can be overly conservative (Edwards and Berry 1987). Given the variety of error terms and correlations between contrast that are possible, a simulation approach similar to that used by

Edwards and Berry (1987) seems the most feasible. We sketch such an approach for the split-block experiment and model (2.1) using the hsd and subset selection approaches. We assume that the experimenter is interested in simultaneously controlling the error rate for both comparisons among A and B. This would be appropriate, for example, when A and B are applied together and our inferences depend on the combination of the two.

For the hsd approach let S_A be the set of indices, (i, i'), representing all of the comparisons of interest among levels of A, that is, $(i, i') \in S_A$ implies we wish to make inferences about $\tau_i - \tau_{i'}$. Let S_B have a similar definition for factor B. Simultaneous confidence intervals for all the comparisons in S_A and S_B would be of the form:

$$\bar{y}_{\cdot i \cdot} - \bar{y}_{\cdot i' \cdot} \pm d\sqrt{\frac{2E_a}{rb}} \text{ for all } (i, i') \in S_A$$

and

$$\bar{y}_{\cdots j} - \bar{y}_{\cdots j'} \pm d\sqrt{\frac{2E_b}{ra}} \text{ for all } (j, j') \in S_B.$$

The problem is to determine d. This can be achieved by a simulation approach. Straightforward calculations show that $\bar{y}_{\cdot i \cdot} - \bar{y}_{\cdot i' \cdot}$ and $\bar{y}_{\cdots j} - \bar{y}_{\cdots j'}$ are uncorrelated. An algorithm follows.

1. Generate a set of normal variates, $Z_{ii'}$, with the same variance-covariance structure as

$$\frac{\bar{y}_{\cdot i \cdot} - \bar{y}_{\cdot i' \cdot}}{\sqrt{\frac{2\sigma_a^2}{rb}}} \quad (i, i') \in S_A.$$

2. Generate an independent (of $Z_{ii'}$) set of normal variates, $X_{jj'}$, with the same variance-covariance structure as

$$\frac{\bar{y}_{\cdots j} - \bar{y}_{\cdots j'}}{\sqrt{\frac{2\sigma_b^2}{ra}}} \quad (j, j') \in S_B.$$

3. Generate a χ^2 variate, U, with f_a degrees of freedom.

4. Generate a χ^2 variate, V, with f_b degrees of freedom.

5. Calculate $W = \max \left\{ \max_{(i,i') \in S_A} \frac{Z_{ii'}}{\sqrt{U/f_a}}, \max_{(j,j') \in S_B} \frac{X_{jj'}}{\sqrt{V/f_b}} \right\}$

6. Repeat steps 1 - 5 N times (a large number).

7. Order the values in 6 from smallest to largest and use the αN largest one as d.

This approach is fully efficient in that it incorporates the exact number of comparisons used and the exact correlation structure.

For the subset approach suppose we wish to derive simultaneous subsets and multiple comparisons with the best for both τ_i (factor A) and π_j (factor B). This would be appropriate if we simultaneously needed the level of A and the level of B to be within a certain distance of the "best". Since contrasts among the factor A means are independent of the factor B means (assuming normality) inferences can straightforwardly be made by using simple independence arguments. We will, however, illustrate the simulation approach to show its flexibility. The ith level of factor A is in the first selected subset if

$$\bar{y}_{..i.} \geq \max_{i' \neq i}\{\bar{y}_{..i'.}\} - d(\frac{2E_a}{rb})^{\frac{1}{2}}.$$

the jth level of factor B is in the second selected subset if

$$\bar{y}_{...j} \geq \max_{j' \neq j}\{\bar{y}_{...j'}\} - d(\frac{2E_b}{ra})^{\frac{1}{2}}.$$

The corresponding simultaneous confidence intervals with the best are given by

$$[0, D_i] \text{ for } \tau_{[a]} - \tau_i \text{ and}$$
$$[0, G_j] \text{ for } \pi_{[b]} - \pi_j$$

where

$$D_i = \max\{0, \max_{i' \neq i}\{\bar{y}_{..i'.}\} - \bar{y}_{..i.} + d(\frac{2E_a}{rb})^{\frac{1}{2}}\}$$

and

$$G_j = \max\{0, \max_{j' \neq j}\{\bar{y}_{...j'}\} - \bar{y}_{...j} + d(\frac{2E_b}{ra})^{\frac{1}{2}}\}.$$

Again the problem is to determine d. An algorithm follows:

1. Generate a set of standard normal variates, Z_i, with correlation $\frac{1}{2}$ ($i = 1, 2, \cdots a - 1$).

2. Generate an independent (of Z_i) set of standard normal variates, X_j, with correlation $\frac{1}{2}$ ($j = 1, 2, \cdots b - 1$).

3. Generate a χ^2 variate, U, with f_a degrees of freedom.

4. Generate a χ^2 variate, V, with f_b degrees of freedom.

5. Calculate $W = \max \left\{ \max_i \dfrac{z_i}{\sqrt{U/f_a}}, \max_j \dfrac{x_j}{\sqrt{V/f_b}} \right\}$

6. Repeat steps 1 - 5 N times (a large number).

7. Order the values in 6 from smallest to largest and use the αN largest one as d.

For an error rate of $\alpha = .05$ Edwards and Berry (1987) recommend choosing N in the range 3,200 to 320,000. Even the upper value is easily implemented on a fast personal computer. We used an 80486 IBM-PC compatible computer running at 33MHz and the matrix language GAUSS. As an example, the subset-selection problem with $a = 4$, $b = 5$ and $N = 320,000$ required 22 minutes. Note this is 3 times faster than the mainframe results reported by Edwards and Berry (1987)! Problems with complicated correlation structures do not require significantly more time since the Cholesky decomposition required to simulate the correlation structure needs to be computed only once.

4   Split-Split Plot Design

To illustrate multiple comparisons procedures for a split-split plot design, we select a particular member of the family of these designs. Our design will have individuals nested within subpopulations and subpopulations nested within populations. The set of a populations will be the whole or main plots which will be arranged in a randomized complete block design of r blocks. Each population has b subpopulations which are nested within populations. These b subpopulations of a population will be randomly assigned to the split plot experimental units within each whole plot or population. Note that subpopulation 1 from population 1 has nothing in common with subpopulation 1 from population 2 except the number 1. Each subpopulation has c individuals which are randomly allocated to the split-split plot experimental units within each split plot experimental unit or a subpopulation. There are r randomizations performed for the populations in the whole plots, ra randomizations on the subpopulations in the split plots, and rab randomizations on the individuals in the split-split plots within a subpopulation. A linear model for the response from the $ghij^{th}$ split-split plot experimental unit is

$$Y_{ghij} = \mu + \rho_g + \alpha_h + \delta_{gh} + \beta_{hi} + \pi_{ghi} + \gamma_{hij} + \epsilon_{ghij}, \qquad (4.1)$$

where μ is an effect common to every observation, ρ_g is the gth block effect, α_h is the hth population effect, δ_{gh} is a random error effect associated with whole plot experimental units, β_{hi} is the effect of subpopulation i from population h, π_{ghi} is a random error effect associated with split plot experimental units, γ_{hij} is the effect for individual j from subpopulation i from population h, and ϵ_{ghij} is a random error effect associated with split-split plot experimental units. An analysis of variance for this response model is outlined in Table 2. In the table, additional partitioning of various sets of degrees of freedom are made. A comparison of the error terms making up E_b and E_c may be made if desired. Likewise the relative sizes of the mean squares making up the pooled mean square for subpopulations within populations and of the individuals within subpopulations are of interest in certain situations.

The various variances of a difference between two means that are of interest in this design are

$$
\begin{aligned}
V(\bar{y}_{.h..} - \bar{y}_{.h'..}) &= 2E_a/rbc \\
V(\bar{y}_{.hi.} - \bar{y}_{.hi'.}) &= 2E_b/rc \\
V(\bar{y}_{.hi.} - \bar{y}_{.h'i.}) &= V(\bar{y}_{.hi.} - \bar{y}_{.h'i'.}) \\
&= 2[E_a + (b-1)E_b]/rbc \\
V(\bar{y}_{.hij} - \bar{y}_{.hij'}) &= 2[E_c]/r \\
V(\bar{y}_{.hij} - \bar{y}_{.hi'j}) &= V(\bar{y}_{.hij} - \bar{y}_{.hi'j'}) \\
&= 2[E_b + (c-1)E_c]/rc \\
V(\bar{y}_{.hij} - \bar{y}_{.h'ij}) &= V(\bar{y}_{.hij} - \bar{y}_{.h'i'j}) \\
&= V(\bar{y}_{.hij} - \bar{y}_{.h'ij'}) = V(\bar{y}_{.hij} - \bar{y}_{.h'i'j'}) \\
&= 2[E_a + (b-1)E_b + b(c-1)E_c]/rbc.
\end{aligned}
$$

Here again, the multiple error mean squares is a feature of the experiment design and not of the fact that multiple comparisons procedures are being considered.

For each of the above variances of a difference between two means, a batchwise error rate could be set. For an experimentwise error for the batch of comparisons pertaining to pairwise comparisons among the a population means, v would be set equal to a, the number of populations, in $q_{\alpha,f,v}$. The corresponding confidence interval would be

$$
\bar{y}_{.h..} - \bar{y}_{.h'..} \pm q_{\alpha,f,a}(E_a/rbc)^{1/2}.
$$

The value for v using the second variance and the resulting set of pairwise comparisons above, would be $v = ab$, and the corresponding confidence interval would be

$$
\bar{y}_{.hi.} - \bar{y}_{.hi'.} \pm q_{\alpha,f,ab}(E_b/rc)^{1/2}, \quad i \neq i'.
$$

Table 2: Analysis of variance for a split-split plot design.

Source of Variation	Degrees of Freedom	Mean Square
Total	$rabc$	
Correction for mean	1	
Block	$r - 1$	
Whole Plot	$a - 1$	
(populations = Factor A)		
Error (a)	$f_a = (r - 1)(a - 1)$	E_a
Split plots	$a(b - 1)$	
(subpopulations within populations)		
Subpop'ns within pop'n 1	$b - 1$	
Subpop'ns within pop'n 2	$b - 1$	
$\vdots$	$\vdots$	
Subpop'ns within pop'n a	$b - 1$	
Error (b)	$f_b = a(r - 1)(b - 1)$	E_b
Subpop'n 1 × block	$(r - 1)(b - 1)$	
$\vdots$	$\vdots$	
Subpop'n b × block	$(r - 1)(b - 1)$	
Individuals within subpopulations	$ab(c - 1)$	
Ind. wn subpop 1, pop 1	$c - 1$	
Ind. wn subpop 2, pop 1	$c - 1$	
$\vdots$	$\vdots$	
Ind. wn subpop b, pop 1	$c - 1$	
Ind. wn subpop 1, pop 2	$c - 1$	
$\vdots$	$\vdots$	
Ind. wn subpop b, pop a	$c - 1$	
Error (c)	$f_c = ab(r - 1)(c - 1)$	E_c
Ind. wn subpop 1, pop 1× block	$(r - 1)(c - 1)$	
Ind. wn subpop 2, pop 1× block	$(r - 1)(c - 1)$	
$\vdots$	$\vdots$	
Ind. wn subpop b, pop a × block	$(r - 1)(c - 1)$	

For the third variance, $v = ab$, and the corresponding confidence intervals are

$$\bar{y}_{.hi.} - \bar{y}_{.h'i'.} \quad \pm \quad q_{\alpha,f*,ab}\{[E_a + (b-1)E_b]/rbc\}^{1/2},$$
$$h \neq h' \text{ and } i \neq i',$$

where the degrees of freedom, f^*, for this variance may be approximated via a Cochran type (Cochran and Cox 1957, p. 101) approximation (but see the previous section for an alternate approach). For the fourth variance, $v = abc$ and the corresponding confidence interval is

$$\bar{y}_{.hij} - \bar{y}_{.hij'} \pm q_{\alpha,f,abc}(E_c/r)^{1/2}, \quad j \neq j'.$$

For the last two variances, $v = abc$ and f^* needs to be approximated for each variance. The corresponding confidence intervals are

$$\bar{y}_{.hij} - \bar{y}_{.hi'j'} \quad \pm \quad q_{\alpha,f*,abc}\{[E_b + (c-1)E_c]/rc\}^{1/2},$$
$$i \neq i' \text{ and } j \neq j'$$

and

$$\bar{y}_{.hij} - \bar{y}_{.h'i'j'} \quad \pm \quad q_{\alpha,f*,abc}\{[E_a + (b-1)E_b + b(c-1)E_c]/rbc\}^{1/2},$$
$$h \neq h', i \neq i' \text{ and } j \neq j'.$$

For an approximately experimentwise error rate, the v used for all six variances would be $v = abc$.

For the subset selection procedure, the rules in the preceding paragraph would apply except that in each case, $v - 1 = k$ would be used in place of v when entering Dunnett's tables. The appropriate degrees of freedom would be used in each case.

5   Discussion

We have considered both traditional and simulation-based approaches to multiple comparisons in complicated experimental designs with multiple error terms. Procedures such as the lsd or esd generalize in a straightforward manner. However, because of the multiple error terms and multiple degrees of freedom, tables for the hsd and subset selection are not available. In such cases, simulation-based approaches are straightforward to implement, though computationally intensive.

Acknowledgments

This research was partially supported by the Mathematical Sciences Research Institute at Cornell University and Hatch grants 151-401 and 151-406. We wish to thank the two referees for constructive comments.

References

Chew, V. (1977). Comparisons among treatment means in an analysis of variance. Agricultural Research Service, USDA, ARS/H/6, iv + 64 pp.*

Cochran, W.G. and G.M. Cox (1957). *Experimental Designs* (Second edition). John Wiley and Sons, Inc., New York.

Dunnett, C.W. (1955). A multiple comparisons procedure for comparing several treatments with a control. *Journal of the American Statistical Association* **50** 1096 - 1121.

Edwards, D. and Berry J.J. (1987). The efficiency of simulation-based multiple comparisons. *Biometrics* **43** 913 - 928.

Federer, W.T. (1975). The misunderstood split plot. In *Applied Statistics* (R. P. Gupta, ed.), North Holland Publishing Company, Amsterdam, 9 - 39.

Federer, W.T. (1961). Experimental error rates. *American Society for Horticultural Science* **78** 605 - 615.

Federer, W.T. and C.E. McCulloch (1984). Multiple comparisons procedures for some split plot and split block designs. In *Design of Experiments - Ranking and Selection* (T. J. Santner and A. C. Tamhane, eds.), Marcel Dekker, Inc., New York and Basel.*

Federer, W.T. and M.P. Meredith (1992). Covariance analysis for split plot and split block designs. *The American Statistician* **46** 155 - 162.

Ghosh, M.N. (1955). Simultaneous tests of linear hypotheses. *Biometrika* **42** 441 - 449.

Grimes, B.A. and W.T. Federer (1984). Comparison of means from populations with unequal variances. In *W.G. Cochran's Impact on Statistics* (P.S.R S. Rao and J. Sedransk, eds.), John Wiley and Sons, Inc., New York, pp. 353 - 374.*

Hartley, H.O. (1955). Some recent developments in analysis of variance. *Communications in Pure and Applied Mathematics* **8** 47 - 72.∗

Holland, B. (1991). Comment on Saville. *The American Statistician* **45** 165.

Hsu, J.C. (1981). Simultaneous confidence intervals for all distances from the "best". *Annals of Statistics* **9** 1026 - 1034.

Lea, P. (1991). Multiple confusions. *The American Statistician* **45** 165 - 166.

Saville, D.J. (1990). Multiple comparisons procedures: The practical solution. *The American Statistician* **44** 174 - 180.

Saville, D.J. (1991). Reply to Holland and Lea. *The American Statistician* **45** 166 - 168.

Tukey, J.W. (1953). The problem of multiple comparisons. Unpublished manuscript, Princeton University.

Tukey, J.W. (1991). The philosophy of multiple comparisons. *Statistical Science* **6** 100 - 116.∗

∗ See these papers for additional references.

Chapter 5

Multiple Comparisons with a Control in Response Surface Methodology: The Problem, Examples, and Relative Efficiency Results

DON EDWARDS Department of Statistics, University of South Carolina, Columbia, South Carolina

PING SA Department of Mathematics and Statistics, University of North Florida, Jacksonville, Florida

Abstract Quadratic response surface methodology often focuses on finding the levels of some (coded) predictor variables $x = (x_1, x_2, ..., x_k)$ that optimize the expected value of a response variable y. Typically the experimenter starts from some best guess for the predictors (usually coded to $x = 0$) and performs an experiment varying them in a region about this "control", and then models $E(y)$ as a second-order polynomial in x. The question of interest is whether any x in the experimental region provides a better $E(y)$ than the control, and if so, how much better? This paper reviews recent results of Sa and Edwards (1992) on the construction of simultaneous confidence intervals for $E(Y|x) - E(Y|0)$ for all x within a specified distance of 0. New examples are provided, as well as efficiency results relative to the natural adaptation of the Scheffé method.

1   A Gift for an Inspirational Researcher

We are pleased and proud to contribute to this symposium in honor of Professor Dunnett, but the first author has "a bone to pick" with him. Whenever a research idea related to multiple comparisons with a control (MCC) comes up, it is necessary to search the literature for papers related to the new idea. A logically sound strategy for this search is to simply look for papers referencing the pioneering papers Dunnett (1955) and Dunnett (1964). Though any MCC-related paper would surely reference these, the strategy is not very practical because these papers are referenced too often. Looking up "Dunnett" in the Science Citation Index is like looking up "Johnson" in an American phone book. Figure 1, our gift to Professor Dunnett, shows citations to these papers versus time. The articles counted range across a remarkably broad spectrum of fields of study.

As data analysts, with an exponential growth rate evident in Figure 1, we feel morally compelled to make an environmental impact assessment. By the year 2005, when the first of these two papers is fifty years old, it will be referenced approximately 1166 times per year. By 2055, the figure will be 172,544 citations per year: an entire volume of the Science Citation Index will be devoted solely to citations to Professor Dunnett. By the year 2155, when the paper is 200 years old, it will be referenced approximately 3,774,852,617 times per year. The hardcopy necessary to record so many citations will consume most remaining plant life, creating an ecosystem imbalance resulting in depletion of all breathable oxygen and extinction of the human race (we have used robust nonparametric regression in these calculations, and are hence protected against any faulty assumptions).

2   Response Surfaces: Approaches to Formal Inference

In Response Surface Methodology (RSM) the experimenter faces the task of investigating the relationship between some response variable y and a number of predictors $\xi = (\xi_1, ..., \xi_k)'$, often for the purpose of optimizing $E(y)$. In what follows we assume that optimization is in fact the goal, and

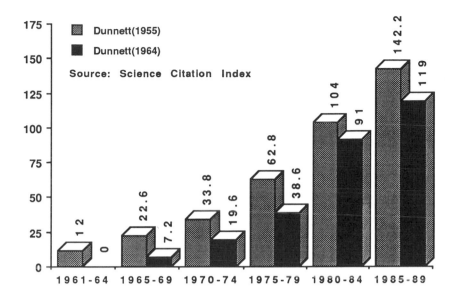

Figure 1: References/year to Charles Dunnett's comparisons with control papers.

without loss of generality that large responses are desirable, so that "optimizing" means "maximizing". Two fine texts on RSM have recently been published: Box and Draper (1987) and Khuri and Cornell (1987). Myers et al. (1989) give a review of methodological advances and applications of RSM in the years 1966-1988.

Usually the ξ_i will be converted to coded variables x_i by

$$x_i = \frac{\xi_i - \xi_{i0}}{(sc)_i},$$

where ξ_{i0} is a centering constant and $(sc)_i > 0$ is a scaling constant, $i = 1, ..., k$. The centering constants ξ_{i0} often represent the experimenter's best guess for optimizing values, or the values currently in use; hence the point $\xi = \xi_0 (x = 0)$ can be considered a "control" or "standard" combination of the predictors. In this case the somewhat ambitious experimental goal of optimizing $E(y)$ might be rephrased to the more achievable one of finding combinations of ξ that improve upon ξ_0, and quantifying the potential amount of improvement.

The form of the mean response $E(y|x)$ as a function of $x = (x_1, x_2, .., x_k)$ is often approximated by a quadratic polynomial. In this article, we assume the model

$$y = \beta_0 + \delta(x) + (\text{other terms}) + \varepsilon \tag{2.1}$$

where

$$\delta(x) = \sum \beta_i x_i + \sum \beta_{ii} x_i^2 + \sum \sum_{i<j} \beta_{ij}(\sqrt{2} x_i x_j) \tag{2.2}$$

for β_0, β_i, β_{ij} unknown constants $i, j = 1, ..., k$ and error $\varepsilon \sim N(0, \sigma^2)$. "Other terms" may include block effects, covariates, and so on but must be fixed effects or regressors not interacting with the regressors in $\delta(x)$. The $\sqrt{2}$ factor at right in (2.2) is somewhat unconventional in the RSM literature, but is useful for certain mathematical results here.

As a backdrop for discussion, consider a data set taken from Schmidt et al. (1979) concerning (we assume) the maximization of $y = $ cohesiveness of whey protein gel in a process controlled in part by $\xi_1 = $ cysteine and $\xi_2 = $ Cacl$_2$. The heavy dots in Figure 2 show the combinations of ξ_1 and ξ_2 used in the experiment. Also shown are contours of the fitted surface, which is of the form (2.1) with no "other terms". The symbol $\hat{x}_S$ identifies the maximizer of the fitted surface, and if we suppose that the center point represents the process combination currently in use for production, the natural question, requiring a careful answer, is:

(*) "should we trust this fitted surface so much as to move the process to the location $'\hat{x}'_S$, and if so how much improvement in $E(y)$ would this change be likely to produce?"

Myers et al. (1989) give a review of some available methodology for formal inference relevant to optimization of $Q(x) = \beta_0 + \delta(x)$. One approach has been to construct a confidence region for the true stationary point x_S of Q, the point giving $\partial Q/\partial x \equiv 0$. If Q has a maximizer, it is x_S. These confidence regions can be very useful, but not if x_S lies far from the experimental region, or if it is a saddle point. Also, such a region can be large either if the surface is not well estimated, or if it is well estimated but fairly flat; a confidence region for x_S seems unable (by itself) to distinguish between these two very different outcomes. Finally, a confidence region for x_S will also not shed light on the issue of how much improvement is likely to be obtained by moving away from $x = 0$.

Other approaches to formal inference in RSM use the matrix expression $Q(x) = \beta_0 + x'\beta + x'Bx$ where $\beta = (\beta_1, ..\beta_k)'$ and B is a square matrix of order k with diagonals β_{ii} and off-diagonals $\beta_{ij}/\sqrt{2}$. The eigenvalues and eigenvectors of B determine the shape of Q in the vicinity of x_S. Confidence

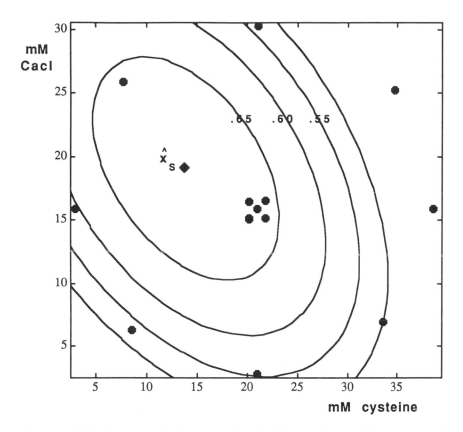

Figure 2: Cohesiveness of whey protein gel design points and fitted second-order surface.

intervals for eigenvalues can shed useful information as to whether x_S is a maximizer versus a saddle point. Again, however, these will not be useful if x_S is far from the experimental region, and what information they do provide will again not directly answer the question "*" above.

A third approach has been to obtain confidence bounds for $Q(x_S)$ or simultaneous confidence bounds for $Q(x)$ for all x, or for all x in some region. These can provide very useful information if there are no block effects, and if it is expected that β_0 will remain constant in the future. Often however β_0 represents "background effects" like batch, machine, or operator effects that have been fixed for the purpose of experimentation but will change in the future. Finally, even without these problems, such interval(s) will not provide careful *comparisons* between the mean response at the predictor combination currently in use and the mean response at alternative combinations.

The quantity $\delta(x) = E(y|x) - E(y|0)$ measures the improvement in mean response at a given x relative to the mean response at the control (in a particular block and for fixed values of any covariates). The MCC in RSM problem is then defined as follows: obtain simultaneous confidence intervals for $\delta(x)$ for all x in a specified region C_x about the origin. The present paper provides a summary of progress made to date, with new examples and a small efficiency study. For more details, the reader is referred to Sa and Edwards (1990).

3   An Exact Solution for the Case of One Predictor

When $k = 1$, (2.2) reduces to

$$\delta(x) = E(y|x) - E(y|0) = \beta_1 x + \beta_{11} x^2$$

and we seek simultaneous confidence intervals for all x in a set $C_x = [a, b]$, where $a \le b$ are prespecified constants. Uusipaikka (1983) generalized and expanded results of Halperin and Gurian (1968) and Wynn and Bloomfield (1971) for obtaining confidence bands for a linear function $\beta_1 + \beta_{11} x$ for all $x \in [a, b]$. These bands can simply be multiplied by x to obtain confidence bands for $\delta(x)$ in our setting.

Formally, suppose $(\hat{\beta}_1, \hat{\beta}_{11})$ are estimators distributed as bivariate Normal with mean $(\beta_1, \beta_{11})'$ and covariance matrix $\sigma^2 V$ for

$$V = \begin{pmatrix} v_L & v_{LQ} \\ v_{LQ} & v_Q \end{pmatrix}$$

and independent of an error mean square s^2 satisfying $\nu s^2/\sigma^2 \sim X^2(\nu)$ for integer $\nu > 0$. These assumptions will of course hold in the case of quadratic regression, with V the submatrix of $(X'X)^{-1}$ corresponding to $(\hat{\beta}_1, \hat{\beta}_{11})$. The $(1 - \alpha) \times 100\%$ exact confidence bands for $\delta(x)$ are:

$$\hat{\beta}_1 x + \hat{\beta}_{11} x^2 \pm \sqrt{2} m_\alpha s (x^2 v_L + 2x^3 v_{LQ} + x^4 v_Q)^{1/2} \qquad (3.1)$$

for all $x \in [a, b]$, where m_α is the upper $\alpha \times 100\%$ point from Uusipaikka (1983, Table 3 ($\alpha = .05$) or Table 4 ($\alpha = .01$)). Sa and Edwards (1992) provide tables for $\alpha = .10$ and $\alpha = .20$. The critical points m_α depend on α, ν and on the design and choice of $[a, b]$ through the quantity

$$\bar{p} = \frac{[v_L + (a + b)v_{LQ} + abv_Q]}{\{[(v_L + 2av_{LQ} + a^2 v_Q)(v_L + 2bv_{LQ} + b^2 v_Q)]^{1/2}\}}.$$

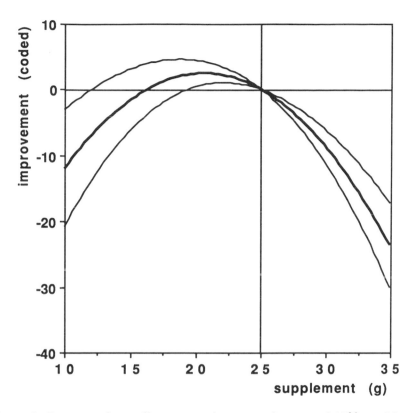

Figure 3: Rat growth vs. dietary supplement estimate and 95% confidence bands for weight gain improvement over control dose 25 gm.

We typically choose the constants a, b to be at or near the boundaries of the experimental region. Choosing these closer to 0 will typically yield tighter intervals, but on a smaller range of x.

Example Box et al. (1978, p. 480) consider the modeling of $y =$ growth rate of rats (coded units) vs. $\xi =$ amount of dietary supplement, in the range $10 - 35$ grams. The data are readily available and so will not be reprinted here. Suppose that the current wisdom in the rat-growing literature is that $\xi_0 = 25$ maximizes mean growth; let $x = \xi - 25$. A quadratic regression of y_i on $x_i (i = 1, ..., 10)$ fits well, giving $\hat{\beta}_1 = -1.121$, $\hat{\beta}_{11} = -0.1277$, and $s^2 = 6.456$ with $\nu = 7$ df. Also, $v_L = 0.002977$, $v_{LQ} = 0.0001830$, $v_Q = 0.00002542$; choosing $a = -15$, $b = 10$ gives $\bar{p} = -0.3227$. From Uusipaikka (1983, Table 3) $m_{0.05} \simeq 2.1271$, interpolating linearly on $\bar{p}$. Simultaneous 95% confidence bands for $\delta(x)$ are given by appropriate evaluation of (2.1) and are shown with an estimate $\hat{\delta}(x)$ in Figure 3. Note that the lower bound is positive for $-5.4 \leq x \leq 0$. The experimenter can then be 95% confident that dietary supplements in the range $[19.6, 25)$ grams will give greater mean growth than the 25 gram supplement. Using the lower and upper bounds, he/she can also make a judgement as to how much larger the mean growth might be, in case there are other reasons to avoid using $\xi < 25\ g$.

Obviously the width of these simultaneous intervals will reduce to 0 as $x \to 0$; this is as it should be, since $\delta(x) \to 0$ also. It is quite possible that the bounds always include 0, though; a sufficient condition for this to occur is that the F ratio for the hypothesis $\beta_1 = \beta_{11} = 0$ is less than $2m_\alpha^2$.

4   Some MCC Approaches for Two or More Predictors

For the case $k > 1$, the MCC problem is much more difficult. This section considers the following goal for the model (2.1): obtain simultaneous confidence intervals for $\delta(x)$ for all x within a specified distance R_I of the origin. That is, for all x in $C_x = \{x : x'x = \sum x_i^2 \leq R_I^2\}$. R_I is called the "radius of inference". Some specialized notation will be needed. Define:

$$
\begin{aligned}
w &= w(x) \\
&= (x_1, ..., x_k, x_1^2, ..., x_k^2, \sqrt{2}x_1 x_2, ..., \sqrt{2}x_{k-1}x_k),
\end{aligned}
$$

so $w : p \times 1$ for $p = 2k + k(k-1)/2$. Also, let

$$
\beta_w = (\beta_1, ..., \beta_k, \beta_{11}, ..., \beta_{kk}, \beta_{12}, ..., \beta_{k-1,k})'.
$$

Let $\hat{\beta}_w$ be an estimator with multivariate normal distribution, mean β_w and covariance matrix $\sigma^2 V$, V known. Let s^2 be an independent error mean square satisfying $\nu s^2/\sigma^2 \sim X^2(\nu)$ for $\nu > 0$. The estimator of $\delta(x)$ is $\hat{\delta}(x) = w'\hat{\beta}_{w'}$ and its estimated standard error is $S(x) = s(w'Vw)^{1/2}$.

A conservative method which can always be applied to the MCC problem is an adaptation of Scheffé simultaneous confidence intervals:

$$
\hat{\delta}(x) \pm (pF_{\alpha,p,\nu})^{1/2}S(x) \text{ for all } x \tag{4.1}
$$

where $F_{\alpha,p,\nu}$ is the upper$-\alpha$ percent point of the F distribution with p and ν degrees of freedom. These confidence bands are probably very conservative, since they hold for all w, not just those which satisfy $w_{k+1} = w_1^2$, $w_{k+2} = w_2^2$, ..., etc., and also for $w(x)$ for x far from the experimental region.

When the design is rotatable, the bands (4.1) can be improved using a result of Casella and Strawderman (1980). Let I_k and J_k be the $k \times k$ identity matrix and a $k \times k$ matrix of ones, respectively. For our purposes, the design is rotatable if V is a block-diagonal matrix of the following form:

$$V = \begin{pmatrix} V_L & 0 & 0 \\ 0 & V_Q & 0 \\ 0 & 0 & V_C P \end{pmatrix} \qquad (4.2)$$

where $V_L = aI_k$, $V_Q = bI_k + cJ_k$, $V_{CP} = bI_{k(k-1)/2}$ for known constants a, b, c. Rotatable designs provide $S(x)$ of very simple form,

$$S(x) = [a(x'x) + (b + c)(x'x)^2]^{1/2}. \qquad (4.3)$$

If the model is purely quadratic in $x_1, ..., x_k$, as in (2.1) with "other terms" orthogonal to w-columns, then it is well known that the design is rotatable if the design points $(x_{1u}, ..., x_{ku})$, $u = 1, ..., n$, satisfy the following constraints (all summations are for $u = 1, ..., n$):

1. $\sum x_{iu}^2 = n\lambda_2$, for each $i = 1, ..., k$

2. $\sum x_{iu}^4 = 3 \sum x_{iu}^2 x_{ju}^2 = 3n\lambda_4$, for each $1 \leq i \neq j \leq k$

3. $\sum x_{hu}^s x_{iu}^r x_{ju}^m x_{tu}^f = 0$, for all other $s, r, f, m \leq 2$, $s + r + f + m \leq 4$,

for some constants λ_2, λ_4. In this case, the V-constants $a = 1/n\lambda_2$, $b = 1/2n\lambda_4$, $c = (\lambda_2^2 - \lambda_4)/[2n\lambda_4((k + 2)\lambda_4 - k\lambda_2^2)]$.

Denote the first m components of w by w_1 and the remainder by w_2. Casella and Strawderman (1980) found that (in our notation) if $V = I$, the Scheffé critical point can be replaced by a smaller value c_α if the confidence bounds need only apply on the set $\{w : w_1' w_1 \geq q^2 w_2' w_2\}$ for some $q^2 > 0$. The value c_α depends on ν, q^2, and the length m of the first part of the vector. The result can be applied in our case with $m = k$ since x in C_x means

$$x'x \leq R_I^2$$
$$\Longleftrightarrow (w_1' w_1) \leq R_I^2$$
$$\Longleftrightarrow (w_1' w_1)^2 \leq R_I^2 (w_1' w_1)$$
$$\Longleftrightarrow w_2' w_2 \leq R_I^2 (w_1' w_1)$$

since $(w_1' w_1)^2 = w_2' w_2$. Our $V \neq I$, but it is of a simple enough form that adaptation of the Casella and Strawderman result is tractable.

Theorem (Sa and Edwards 1992)

If the design is rotatable,

$$\hat{\delta}(x) \pm c_\alpha S(x)$$

are conservative simultaneous $(1 - \alpha) \times 100\%$ confidence intervals for $\delta(x)$ for all x such that $x'x \leq R_I^2$, where c_α is the Casella and Strawderman (1980) critical point with ν degrees of freedom, $m = k$ and

$$
\begin{aligned}
q^2 &= a/[R_I^2(b+c)] \\
&= \{2\lambda_4[(k+2)\lambda_4 - k\lambda_2^2]\}/\{R_I^2\lambda_2[(k+1)\lambda_4 - (k-1)\lambda_2^2]\}
\end{aligned}
$$

Casella and Strawderman provide a limited table of values c_α^2 for $\alpha = .05$ and $\nu = \infty$. We have written a Fortran program calling International Mathematical and Statistical Library routines (IMSL 1987) to numerically compute these critical points for any given α, m, p, ν, and q^2.

Example   Returning to the Schmidt et al. (1979) data on maximization of mean cohesiveness of Whey protein gel with respect to $\xi_1 = $ cysteine (mM) and $\xi_2 = CaCl_2$ (mM), define $x_1 = (\xi_1 - 21)/13$ and $x_2 = (\xi_2 - 16.2)/9.7$, hence $\xi_{10} = 21$ and $\xi_{20} = 16.2$ is the control combination. Table 1 lists the design points (x_{1u}, x_{2u}), $u = 1, ..., 13$ and the responses Y_i. The design is a uniform precision rotatable central composite design. The design constants $\lambda_2 = 8/13$ and $\lambda_4 = 4/13$; V is of the form (4.2) with $a = b = .125$, $c = .01875$.

The fitted quadratic surface contoured in Figure 1 is

$$\hat{y} = .666 - .092x_1 - .010x_2 - .099x_1^2 - .061x_2^2 - 0.49(\sqrt{2})x_1x_2.$$

We can construct simultaneous 95% confidence surfaces for $\delta(x)$ for all x within a radius of $R_1 = \sqrt{2}$ as follows: $q^2 = .43478$ and using the Fortran program we obtain $c_{.05} = 4.4021$. Figure 4 at center shows contours for estimated improvement $\hat{\delta}(x) = \hat{y} - .666$ with positive improvement areas shaded. At top and bottom in Figure 4 are the simultaneous 95% confidence upper bounds $U(x) = \hat{\delta}(x) + c_{.05}S(x)$ and lower bounds $L(x) = \hat{\delta}(x) - c_{.05}S(x)$. Since there is a region of x such that the lower bound is positive, we can be confident that some improvement in mean cohesiveness is possible over the control choice of 21 mM cysteine and 16.2 mM $CaCl_2$.

5   A Small Relative Efficiency Study

The Casella-Strawderman adaptation given by Theorem 1, when applicable, yields uniformly shorter intervals than the Scheffé adaptation (4.1).

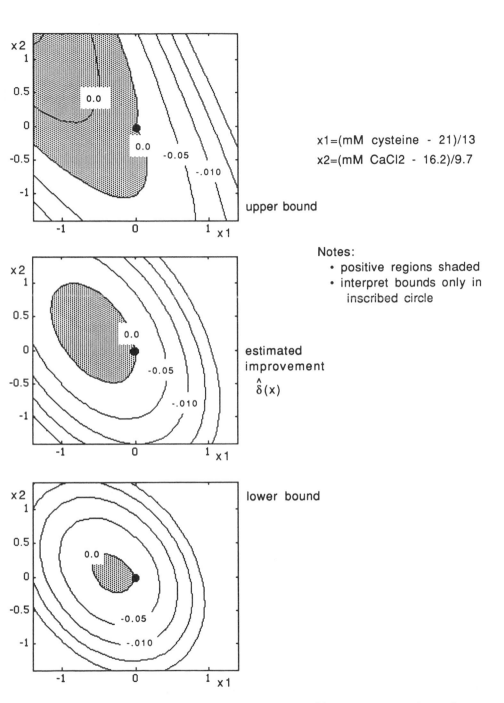

Figure 4: Cohesiveness of whey protein gel 95% confidence surfaces for improvement over control.

Table 1: Experimental results: Cohesiveness y of whey protein gel vs. $x_1 = (mMcysteine - 21)/13$ and $x_2 = mMCaCl_2 - 16.2)/9.7$ (Source: Schmidt et al. 1979).

run	1	2	3	4	5	6	7
x_1	-1	1	-1	1	$-\sqrt{2}$	$\sqrt{2}$	0
x_2	-1	-1	1	1	0	0	$-\sqrt{2}$
y	.55	.52	.67	.36	.59	.31	.54

run	8	9	10	11	12	13
x_1	0	0	0	0	0	0
x_2	$\sqrt{2}$	0	0	0	0	0
y	.51	.66	.66	.66	.66	.66

The asymptotic relative efficiency of the Casella-Strawderman adaptation to the Scheffé adaptation is the limiting ratio of sample sizes needed to achieve equal confidence interval widths for all x; this is just the ratio of squared $\nu = \infty$ critical points $[pF^2_{\alpha,p,\infty}]/c^2$. For finite ν, we will here define relative efficiency as the ratio of these squared critical points at the value ν determined by some number of replicates of a basic design. The relative efficiency then depends on k, α, the design, the number of replicates, and the choice of R_I. We choose $\alpha = .05$ and consider rotatable uniform-precision central composite designs for $k = 2$ and $k = 3$ (Khuri and Cornell 1987 pp. 116 - 120). A single replicate of the basic $k = 2$ design is the design given in Table 1. A single replicate of the basic $k = 3$ design consists of 8 cube points $(\pm 1, \pm 1, \pm 1)$, 6 axial points with axial value $8^{1/4}$, and 6 center points.

Table 2 shows the relative efficiencies. Considerable improvement (at least 20 - 34% of sample size) over the Scheffé adaptation is possible by choosing the radius of inference R_I small. However, this value will usually be chosen at the limits of the experimental region, $R_I \simeq \sqrt{k}$. Sample size savings here are considerably more modest, 2 - 3%.

Table 2: Relative efficiencies of the Casella-Strawderman vs. Scheffé adaptations (rotatable uniform-precision central composite designs, $\alpha = 0.05$).

number of predictors k	radius of inference R_1	relative efficiency 1 rep.	relative efficiency 2 rep.	relative efficiency ∞ rep.
2	0.5	1.246	1.222	1.203
	1.0	1.071	1.067	1.063
	$\sqrt{2}$	1.032	1.030	1.029
3	.05	1.338	1.307	1.285
	1.0	1.092	1.086	1.082
	$\sqrt{3}$	1.021	1.020	1.020

6 Conclusions

The problem of multiple comparisons with a control in response surface methodology has been defined and motivated. For quadratic regression ($k = 1$), an exact solution has been provided, and an improvement over Scheffé-type bounds has been offered when the design is rotatable. A substantial amount of work remains to be done. An exact solution in the rotatable case when $k > 1$ has been elusive, yet the problem seems a tractable one. Extensions to less restrictive designs, perhaps designs orthogonal in the linear terms only, are needed. The authors are pursuing these and other related research avenues.

References

Box, G.E.P. and Draper, N. (1987). *Empirical Model-Building and Response Surfaces*. John Wiley and Sons, Inc., New York.

Box, G.E.P., Hunter, W.G. and Hunter, J.S. (1978). *Statistics for Experimenters*. John Wiley and Sons, Inc., New York.

Casella, G. and Strawderman, W.E. (1980). Confidence bands for linear regression with restricted predictor variables. *Jour. Amer. Statist.*

Assoc. **75** 862 - 868.

Dunnett, C.W. (1955). A multiple comparison procedure for comparing several treatments with a control. *Jour. Amer. Statist. Assoc.* **50** 1096 - 1121.

Dunnett, C.W. (1964). New tables for multiple comparisons with a control. *Biometrics* **20** 482 - 491.

Halperin, M., and Gurian, J. (1968). Confidence bands in linear regression with constraints on independent variables. *Jour. Amer. Statist. Assoc.* **63** 1020-1027.

IMSL(1987). *Fortran routines for mathematical applications.* IMSL Inc., Houston, Texas.

Khuri, A I., and Cornell, J.A. (1987). *Response Surfaces.* Marcel Dekker, New York.

Myers, R.H., Khuri, A.I. and Carter, W.H., Jr. (1989). Response surface methodology: 1966-1988. *Technometrics* **31** 137 - 157.

Sa, P. and Edwards, D. (1992). Multiple comparisons with a control in response surface methodology. In review.

Schmidt, R.H., Illingworth, B.L., Deng, J.C., and Cornell, J.A. (1979). Multiple regression and response surface analysis of the effects of calcium chloride and cysteine on heat-induced whey protein gelatin. *Jour. Agricult. Food Chem.* **27** 529 - 532.

Uusipaikka, E. (1983). Exact confidence bands for linear regression over intervals. *Jour. Amer. Statist. Assoc.* **78** 638 - 644.

Wynn, H.P., and Bloomfield, P. (1971). Simultaneous confidence bands in regression analysis. *Jour. Royal Statist. Soc.* (Ser. B) **33** 202 - 217.

Chapter 6

One-Sided Multiple Comparisons of Response Rates with a Control

DAVID R. BRISTOL   Biostatistics Department, Schering-Plough Research Institute, Kenilworth, New Jersey

Abstract   Many experiments are designed and performed to simultaneously compare the response rates of a binary ("success-failure") variable for k (≥ 1) experimental treatments to that of a control treatment with the goal of determining if any of the experimental treatments have a response rate larger than that of the control treatment. Four asymptotic single-stage testing procedures for performing such comparisons with $k = 2$ and $k = 3$ are examined with respect to the approximate power for specified sample sizes.

1   Introduction

Many experiments are designed and performed to simultaneously compare k (≥ 1) experimental treatments to a control treatment with respect to a binary ("success-failure") variable with the goal of determining if any of the experimental treatments have a response rate larger than that of the control treatment. Let p_i denote the response rate (probability of success) for the i-th experimental treatment, $i = 1, \cdots, k$, and let p_0 denote the response rate for the control treatment. Interest lies in determining if

$p_i > p_0$ for any i. Here this problem is solved by testing the null hypothesis $H_0 : p_0 = p_1 = \ldots = p_k$ against the alternative hypothesis $H_1 : p_i > p_0$ for at least one i. The proposed testing procedures are based on a random sample of size n from each experimental treatment and a random sample of size n_0 from the control treatment. Let x_i denote the number of success from the i-th experimental treatment with $\hat{p}_i = x_i/n$ denoting the corresponding proportion of successes and x_0 denotes the number of successes from the control treatment with $\hat{p}_0 = x_0/n_0$ denoting the corresponding proportion of successes.

The asymptotic normality of the binomial distribution is used to construct testing procedures similar to, and motivated by, the testing procedure considered by Dunnett (1955, 1964) for comparing the means of several normal distributions to the mean of a normal distribution associated with a control when the common variance is unknown. Testing procedures for comparison of independent binomial distributions to a specified binomial distribution have not been adequately examined.

Four testing procedures $(R_1, R_2, R_3$ and $R_4)$ are examined as various ways to compare $\hat{p}_i$ and $\hat{p}_0$, and thus to test H_0 against H_1, using the asymptotic normality of the binomial distribution. The power of these test procedures is evaluated for $n_0 = 20, 30,$ and 40, and $n = 50 - n_0, 100 - n_0$, and n_0, and specified values of $p_0, p_1, \cdots, p_k$ satisfying $p_0 \leq p_1 \leq \ldots \leq p_k$ for $k = 2$ and $k = 3$.

2   Testing Procedure R_1

Let $T_{1i} = (\hat{p}_i - \hat{p}_0)/\{\hat{p}_i(1 - \hat{p}_i)/ n + \hat{p}_0(1 - \hat{p}_0)/ n_0\}^{1/2}, 1 \leq i \leq k$. Testing procedure R_1 is:

Reject H_0 in favor of H_1 if $\max\{T_{1i}; i = 1, \cdots, k\} \geq C_1$,

where C_1 is chosen so that the test is an asymptotically α-level test. Since $\hat{p}_i$ is asymptotically normal with mean p_i and variance $\sigma_i^2 = p_i(1 - p_i)/ n$ and $\hat{p}_0$ is asymptotically normal with mean p_0 and variance $\sigma_0^2 = p_0(1 - p_0)/ n_0$, the asymptotic distribution of $(T_{11}, \ldots, T_{1k})'$ is k-variate normal, with means $E(T_{1i}) = p_i - p_0$, unit variances, and correlations $\rho(T_{1i}, T_{1j}) = \sigma_0^2/\{(\sigma_0^2 + \sigma_i^2)(\sigma_0^2 + \sigma_j^2)\}^{1/2}, i \neq j$. Thus, under H_0, the asymptotic distribution of $(T_{11}, \ldots, T_{1k})'$ is k-variate normal, with zero means, unit variances, and common correlations equal to $n/(n + n_0), i \neq j$. The critical value C_1 is determined from the asymptotic distribution of $\max\{T_{1i}; i = 1, \cdots, k\}$ under H_0, and thus values of C_1 are available from many sources, including Dunnett (1955), Gupta et al. (1985), and Bechhofer and Dunnett (1988). A procedure similar to R_1 was proposed by Piegorsch (1991) to construct simultaneous $(1 - \alpha)100\%$ confidence intervals for all $p_i - p_0$, $i = 1, \cdots, k$. Let $\Delta_i = p_i - p_0$ and let Φ and Φ^{-1} denote the standard normal c.d.f. and its inverse, respectively. The approximate power using testing procedure R_1 is given by

$$\Pi(R_1) = 1 - \int_0^1 \prod_{i=1}^k \Phi((\sigma_0\Phi^{-1}(t) + C_1(\sigma_i^2 + \sigma_0^2)^{1/2} - \Delta_i)/\sigma_i)dt.$$

3   Testing Procedure R_2

Testing procedure R_2 uses a different estimate of the asymptotic standard deviation than that used for testing procedure R_1. Let $\hat{p}_{0i} = (n\hat{p}_i + n_0\hat{p}_0)/(n + n_0)$ denote the pooled estimate of p_0 and p_i. Also, let $T_{2i} = (\hat{p}_i - \hat{p}_0)/\{\hat{p}_{0i}(1 - \hat{p}_{0i})(1/n + 1/n_0)\}^{1/2}$, $i = 1, \cdots, k$. Testing procedure R_2 is:

Reject H_0 in favor of H_1 of $\max\{T_{2i}; i = 1, \cdots, k\} \geq C_2$,

where C_2 is chosen so that the test is an asymptotically α-level test. Let $p_{0i} = (np_i + n_0p_0)/(n + n_0)$ and $\sigma_{0i}^2 = p_{0i}(1 - p_{0i})(1/n + 1/n_0)$. The asymptotic distribution of $(T_{21}, \ldots, T_{2k})'$ is k-variate normal, with means $E(T_{2i}) = (p_i - p_0)/\sigma_{0i}$, variances $V(T_{2i}) = (\sigma_i^2 + \sigma_0^2)/\sigma_{0i}^2$, and correlations $\rho(T_{2i}, T_{2j}) = \sigma_0^2/\{(\sigma_0^2 + \sigma_i^2)(\sigma_0^2 + \sigma_j^2)\}^{1/2}$, $i \neq j$. Thus under H_0, the asymptotic distribution of $(T_{21}, \ldots, T_{2k})'$ is k-variate normal, with zero means, unit variances, and common correlations equal to $n/(n + n_0)$, $i \neq j$. Since $(T_{21}, \ldots, T_{2k})'$ and $(T_{11}, \ldots, T_{1k})'$ have the same asymptotic distribution under H_0, $C_2 = C_1$. The approximate power using R_2 is

$$\Pi(R_2) = 1 - \int_0^1 \prod_{i=1}^k \Phi\left((\sigma_0\Phi^{-1}(t) + C_2\sigma_{0i} - \Delta_i)/\sigma_i\right) dt.$$

Each term of the integrand of $\Pi(R_2)$ is less than or equal to the corresponding term of the integrand of $\Pi(R_1)$ if and only if $\sigma_{0i}^2 \leq \sigma_0^2 + \sigma_i^2$. Comparison of the power of the corresponding procedures for the two-sample problem was presented by Robbins (1977), and Eberhardt and Fligner (1977). They showed that, for $k = 1$, R_1 is more powerful than R_2 when $n = n_0$. The question regarding the comparison when $n \neq n_0$ was raised in the former, whereas a detailed comparison of the two procedures when $n \neq n_0$ was presented in the latter. Their results may be useful for comparison of $\Pi(R_1)$ and $\Pi(R_2)$.

4   Testing Procedure R_3

Testing procedure R_3 is a generalization of R_2. Let $\hat{p}_* = (n \sum_{i=1}^{k} \hat{p}_i + n_0\hat{p}_0)/(kn + n_0)$ denote the pooled estimate of $p_0, p_1, \cdots, p_k$. Furthermore, let $T_{3i} = (\hat{p}_i - \hat{p}_0)/\{\hat{p}_*(1 - \hat{p}_*)(1/n + 1/n_0)\}^{1/2}$, $i = 1, \cdots, k$. Testing procedure R_3 is:

Reject H_0 in favor of H_1 if $\max\{T_{3i}; i = 1, cdots, k\} \geq C_3$,

where C_3 is chosen so that the test is an asymptotically α-level test. Let $p_* = (n \sum_{i=1}^{k} p_i + n_0p_0)/(kn + n_0)$ and $\sigma_*^2 = p_*(1 - p_*)(1/n + 1/n_0)$. The asymptotic distribution of $(T_{31}, \ldots, T_{3k})'$ is k-variate normal, with means $E(T_{3i}) = (p_i - p_0)/\sigma_*$, variances $V(T_{3i}) = (\sigma_i^2 + \sigma_0^2)/\sigma_*^2$, and correlations $\rho(T_{3i}, T_{3j}) = \sigma_0^2/\{(\sigma_0^2 + \sigma_i^2)(\sigma_0^2 + \sigma_j^2)\}^{1/2}$, $i \neq j$ Thus under H_0, the asymptotic distribution of $(T_{31}, \ldots, T_{3k})'$ is k-variate normal, with zero means, unit variances, and common correlations $n/(n + n_0)$, $i \neq j$. Since $(T_{31}, ..., T_{3k})'$, $(T_{21}, \ldots, T_{2k})'$ and $(T_{11}, \ldots, T_{1k})'$ have the same asymptotic distribution under H_0, $C_3 = C_2 = C_1$. The approximate power using R_3 is

$$\Pi(R_3) = 1 - \int_0^1 \prod_{i=1}^{k} \Phi\left((\sigma_0\Phi^{-1}(t) + C_3\sigma_* - \Delta_i)/\sigma_i\right) dt.$$

Testing procedure R_3 has the undesirable property that misleading pairwise comparisons may result because the weighted average of the sample proportions is used for estimation of the asymptotic variance. For example, for $p_0 = p_1 = 0.5$ and $p_2 = 0.9$, one may incorrectly conclude that $p_1 > p_0$ only because the pooled estimate is used.

The small-sample behavior of the exact distribution of

$$((N - 1)/N)^{1/2} \max\{T_{3i}; i = 1, \ldots, k\}$$

under H_0, where $N = kn + n_0$, was examined by Williams (1988) using the conditional distribution of $(n_0\hat{p}_0, n\hat{p}_1, \ldots, n\hat{p}_k)$ given $N\hat{p}_*$.

5   Testing Procedure R_4

Testing procedure R_4 is based on the arc-sine transformation. As this is the variance-stabilizing transformation, it has the desirable property that the asymptotic variance is independent of any unknown parameters. Let $\theta_i = \sin^{-1}(p_i^{1/2})$ and $\theta_0 = \sin^{-1}(p_0^{1/2})$. Furthermore, let $\hat{\theta}_i = \sin^{-1}(\hat{p}_i^{1/2})$ and $\hat{\theta}_0 = \sin^{-1}(\hat{p}_0^{1/2})$. Then $\hat{\theta}_i$ is asymptotically normal with mean θ_i and variance $1/4n$, and $\hat{\theta}_0$ is asymptotically normal with mean θ_0 and variance $1/4n_0$. Let $T_{4i} = 2(\hat{\theta}_i - \hat{\theta}_0)/(1/n + 1/n_0)^{1/2}$. Testing procedure R_4 is:

Reject H_0 in favor of H_1 if $\max\{T_{4i}; i = 1, \cdots, k\} \geq C_4$,

where C_4 is chosen so that the test is an asymptotically α-level test. The asymptotic distribution of $(T_{41}, \ldots, T_{4k})'$ is k-variate normal, with means $E(T_{4i}) = 2(\theta_i - \theta_0)/(1/n + 1/n_0)^{1/2}$, unit variances, and common correlations equal to $\rho(T_{4i}, T_{4j}) = n/(n + n_0), i \neq j$. Thus, under H_0, the asymptotic distribution of $(T_{41}, \ldots, T_{4k})'$ is k-variate normal, with zero means, unit variances, and common correlations equal to $n/(n + n_0), i \neq j$. Since $(T_{41}, \ldots, T_{4k})'$, $(T_{31}, \ldots, T_{3k})'$, $(T_{21}, \ldots, T_{2k})'$, and $(T_{11}, \ldots, T_{1k})'$ have the same asymptotic distributions under H_0, $C_4 = C_3 = C_2 = C_1$. The approximate power using R_4 is

$$\Pi(R_4) = 1 - \int_0^1 \prod_{i=1}^k \Phi((n/n_0)^{1/2}\Phi^{-1}(t) - 2n^{1/2}(\theta_i - \theta_0) + n^{1/2}(1/n + 1/n_0)^{1/2}C_4)dt.$$

Although R_4 has the desirable property that the asymptotic covariance matrix is independent of any unknown parameters, it has the undesirable property that it is only applicable when inference regarding $p_i - p_0$ can be expressed in terms of $\theta_i - \theta_0$. A two-sided analogue of R_4 was discussed by Bristol (1989) for sample size determination when $n_0 = n$.

6 Example

Suppose that $\hat{p}_0 = 0.5$ was observed based on a random sample of size $n_0 = 40$ from the control treatment, and $\hat{p}_1 = 0.8$ and $\hat{p}_2 = 0.9$ were observed based on random samples of size $n = 10$ from $k = 2$ experimental treatments. The four testing procedures reduce to comparison of $\max\{T_{mi}; i = 1, 2\}$, $m = 1, \ldots, 4$, to 1.94560, the critical value obtained from page 138 of Bechhofer and Dunnett (1988), with $p = 2$, $\rho = 0.2$, and $\nu = \infty$. The observed pairwise statistics are:

	T_{1i}	T_{2i}	T_{3i}	T_{4i}
i=1	2.0112	1.7094	1.7452	1.8201
i=2	3.2391	2.2923	2.3270	2.6228

Using R_1, it is concluded that both of the experimental treatments have larger response rates than the control treatment; using either of the other three testing procedures, it is concluded that only one experimental treatment has a larger response rate than the control treatment.

7   Approximate Power of the Four Testing Procedures

The approximate power of the test for the four testing procedures for $\alpha = 0.05$, $n_0 = 20$, 30 and 40, and $n = 50 - n_0$, $100 - n_0$, and n_0 is given in Table 1 for $k = 2$ and $p_0 = 0.1(0.1)0.6$, $p_0 \leq p_1 \leq p_2$, and in Table 2 for $k = 3$ and $p_0 = 0.1(0.1)0.7$, $p_0 \leq p_1 \leq p_2 \leq p_3$. Values of $(p_0, p_1, \ldots, p_k)$ presented here result in moderate to high power. An expanded version of this table is available from the author. Details regarding the computations are given in the next section.

To compare the testing procedures and make some general guidelines, a test procedure is said to dominate the other test procedures for given sample sizes and configuration $(p_0, p_1, \cdots, p_k)$ if it is more powerful than the others for the specified sample sizes and configuration. Note that no test procedure dominates for all sample sizes and configurations. Because of this lack of a test procedure that dominates for all sample sizes and configurations, some general observations are given, although some exceptions to these general observations occur. Most of these exceptions occur due to the equality of the approximate power (to three decimal places) for two or more procedures, which usually occurs for configurations with very high or very low power. For most sample sizes and configurations of probabilities, R_1 dominates. For $n_0 = n$, R_2 does not dominate. A general discussion of Tables 1 and 2 follows.

For $k = 2$ and $n_0 = n = 20$, R_1 dominates for all entries in the table, except for two configurations where R_4 dominates ($p_0 = 0.5, p_1 = p_2 = 0.9$ and $p_0 = 0.6$, $p_1 = 0.8$, $p_2 = 0.9$). For $n_0 = 20$ and $n = 30$, R_1 dominates for all configurations with $p_0 \leq 0.4$ and R_4 dominates for most configurations with $p_0 \geq 0.5$. For $n_0 = 20$ and $n = 80$, R_1 dominates for all configurations with $p_0 \leq 0.3$, and R_2 dominates for all configurations with $p_0 \geq 0.4$, except those configurations with $p_0 \geq 0.4$, $p_1 = p_2$, where R_3 dominates.

For $k = 2$, $n_0 = 30$, and $n = 20$, R_1 dominates for all configurations with $p_0 \geq 0.3$; R_2 and R_3 each dominate for few configurations with $p_0 \leq 0.2$; and R_4 dominates for no configurations. For $n_0 = n = 30$, R_1 dominates for all configurations, except one where R_3 dominates ($p_0 = p_1 = 0.1$, $p_2 = 0.4$), and two where R_4 dominates ($p_0 = 0.5$, $p_1 = p_2 = 0.9$ and $p_0 = 0.6$, $p_1 = 0.8$, $p_2 = 0.90$. For $n_0 = 30$ and $n = 70$, R_1 dominates for all configurations with $p_0 \leq 0.3$ and most configurations with $p_0 = 0.4$; and each of the other procedures dominates for some configurations with $p_0 \geq 0.5$.

For $k = 2$, $n_0 = 40$, and $n = 10$, R_1 dominates for all configurations with $p_0 \geq 0.4$; R_2 dominates for most configurations with $p_0 \leq 0.3$; R_3

dominates for most configurations with $p_0 = p_1 \leq 0.3$; and R_4 does not dominate. For $n_0 = n = 40$, R_1 dominates for most configurations. For $n_0 = 40$ and $n = 60$, R_1 dominates for all configurations with $p_0 \leq 0.4$; R_3 and R_4 each dominate for various configurations with $p_0 \geq 0.5$.

For $k = 3$ and $n_0 = n = 20$, R_1 dominates for all configurations with $p_0 \leq 0.5$, except for three configurations where R_3 dominates ($p_0 = p_1 = p_2 = 0.1$, $p_3 = 0.4$; $p_0 = p_1 = p_2 = 0.1$, $p_3 = 0.5$; and $p_0 = p_1 = p_2 = 0.2$, $p_3 = 0.5$); and R_4 dominates for all configurations with $p_0 \geq 0.6$, except for one configuration where R_3 dominates ($p_0 = 0.7$, $p_1 = p_2 = p_3 = 0.9$). For $n_0 = 20$ and $n = 30$, R_1 dominates for all configurations with $p_0 \leq 0.5$, except for two configurations where R_3 dominates ($p_0 = p_1 = p_2 = 0.1$, $p_3 = 0.4$ and $p_0 = 0.5$, $p_1 = p_2 = 0.7$, $p_3 = 0.8$) and one configuration where R_4 dominates ($p_0 = 0.5$, $p_1 = 0.6$, $p_2 = 0.7$, $p_3 = 0.8$); R_4 dominates for all configurations with $p_0 \geq 0.6$, except for two configurations where R_3 dominates ($p_0 = 0.6$, $p_1 = p_2 = p_3 = 0.9$ and $p_0 = 0.7$, $p_1 = p_2 = p_3 = 0.9$). For $n_0 = 20$ and $n = 80$, R_1 dominates for configurations with $p_0 \leq 0.3$; R_2 dominates for all configurations with $p_0 \geq 0.4$, except for three configurations where R_3 dominates ($p_0 = 0.6$, $p_1 = 0.8$, $p_2 = p_3 = 0.9$; $p_0 = 0.6$, $p_1 = p_2 = p_3 = 0.9$; and $p_0 = 0.7$, $p_1 = p_2 = p_3 = 0.9$) and two configurations where R_4 dominates ($p_0 = 0.6$, $p_1 = 0.6$, $p_2 = 0.8$, $p_3 = 0.9$ and $p_0 = 0.6$, $p_1 = 0.7$, $p_2 = p_3 = 0.9$).

For $k = 3$, $n_0 = 30$, and $n = 20$, R_1 dominates for most configurations with $0.2 \leq p_0 \leq 0.6$; R_2 dominates for most configurations with $p_0 = 0.1$; R_3 dominates for several configurations with $p_0 = 0.1$; and R_4 dominates for $p_0 = 0.7$. For $n_0 = n = 30$, R_1 dominates for most configurations with $p_0 \leq 0.5$, except for configurations with $p_0 = p_1 = p_2 = 0.1$ for which R_3 dominates; and R_4 dominates for most configurations with $p_0 \geq 0.6$ and no configurations with $p_0 \leq 0.5$. For $n_0 = 30$ and $n = 70$, R_1 dominates for all configurations with $p_0 \leq 0.4$ and no configurations with $p_0 \geq 0.5$; R_2 dominates for most configurations with $p_0 = 0.5$; R_3 dominates for few configurations; and R_4 dominates for most configurations with $p_0 \geq 0.6$.

For $k = 3$, $n_0 = 40$, and $n = 10$, R_1 dominates for no configurations with $p_0 \leq 0.3$ and for all configurations with $p_0 \geq 0.4$; R_2 dominates for most configurations with $p_0 \leq 0.3$; R_3 dominates for $p_0 = p_1 = p_2 \leq 0.3$; and R_4 dominates for no configurations. For $n_0 = n = 40$, R_1 dominates for most configurations with $p_0 \leq 0.5$, except for configurations with $p_0 = p_1 = p_2 = 0.1$ for which R_3 dominates; and R_4 dominates for most configurations with $p_0 \geq 0.6$. For $n_0 = 40$ and $n = 60$, R_1 dominates for most configurations with $p_0 \leq 0.4$ and none with $p_0 \geq 0.6$; R_2 dominates for no configurations; R_3 dominates for few configurations especially with $p_0 = p_1 = p_2 = 0.1$; and R_4 dominates for most with $p_0 \geq 0.6$.

8   Evaluation of Approximate Power

The approximate power was determined for specified values of n_0, n, and $(p_0, p_1, \cdots, p_k)$ using IMSL subroutine QDAGS to evaluate the integral expression for the power for each of the four testing procedures, with an absolute error (ERRABS) of 0 and a relative error (ERRREL) of 0.0001. This gives accuracy of at least three decimal places. For values resulting in large power, the accuracy is even greater since the integral evaluated is the probability of not rejecting the null hypothesis. The critical values used in evaluation of the power were obtained from Table A.3 of Bechhofer and Dunnett (1988), except for some calculations for R_4 which used critical values from Gupta et al. (1985). The critical values given in the latter references are the same as the former (reported to five decimal places) for all sample sizes considered here, except for $k = 2$ with $n_0 = 30$ and $n = 70$ and with $n_0 = 20$ and $n = 80$, where the critical value from the latter is greater by 0.00001 than that from the former. The difference is of little, if any, consequence.

9   Discussion

Four popular procedures for one-sided multiple comparisons of $k = 2$ or $k = 3$ response rates with a control were examined with respect to asymptotic approximations to the power for various sample sizes and configurations of probabilities. It is seen that no one procedure has the largest power for all cases. This limited tabulation of the power is not intended to solve the problem, but is definitive evidence that the choice of procedure should be made according to the power to be achieved. Many other questions, such as detailed analytic guidelines, $\alpha \neq 0.05$, more than three experimental treatments, two-sided alternatives, and sample size determination, remain unanswered. In some cases, the choice of procedure will have a tremendous impact on the choice of sample size. Furthermore, other procedures could be developed as competitors to these four procedures.

Based on this limited tabulation, and results not presented here, some general recommendations can be made. None of these recommendations is always correct, but they can be useful for determining a single testing procedure under rather general considerations. For $k = 2$ and $k = 3$, a procedure is recommended based on (1) the values of n_0 and $n : n_0 > n$, $n_0 = n$, or $n_0 < n$; (2) the value of $p_0 : p_0 \leq 0.3$, $p_0 = 0.4$ or 0.5, $p_0 \geq 0.6$; (3) the configuration: the slippage configuration with all values equal the largest, the slippage configuration with all values equal except the smallest (p_0), or any other configuration.

General Recommendations

$k = 2$		$n_0 > n$	$n_0 = n$	$n_0 < n$
$p_0 \leq 0.3$	$p_0 = p_1$	R_3	R_3	R_1
	$p_1 = p_2$	R_2	R_1	R_1
	other	R_2	R_1	R_1
$p_0 = 0.4, 0.5$	$p_0 = p_1$	R_1	R_1	R_1
	$p_1 = p_2$	R_1	R_1	R_1
	other	R_1	R_1	R_1
$p_0 \geq 0.6$	$p_0 = p_1$	R_1	R_1	R_1
	$p_1 = p_2$	R_1	R_3	R_3
	other	R_1	R_1	R_4

$k = 3$		$n_0 > n$	$n_0 = n$	$n_0 < n$
$p_0 \leq 0.3$	$p_0 = p_1 = p_2$	R_3	R_3	R_3
	$p_1 = p_2 = p_3$	R_2	R_1	R_1
	other	R_2	R_1	R_1
$p_0 = 0.4, 0.5$	$p_0 = p_1 = p_2$	R_1	R_1	R_1
	$p_1 = p_2 = p_3$	R_1	R_1	R_1
	other	R_1	R_1	R_1
$p_0 \geq 0.6$	$p_0 = p_1 = p_2$	R_1	R_1	R_2
	$p_1 = p_2 = p_3$	R_1	R_3	R_3
	other	R_1	R_4	R_4

Table 1: Power(%) for R_1, R_2, R_3, R_4; $\alpha = 0.05$, $k = 2$.

| | $n_0=20$ | | | | | | | | | | | |
| | n=20 | | | | n=30 | | | | n=80 | | | |
(p_0,p_1,p_2)	R1	R2	R3	R4	R1	R2	R3	R4	R1	R2	R3	R4
(.1,.1,.4)	66.5	61.8	67.3	64.9	78.4	68.6	75.1	73.1	94.7	82.6	88.7	85.5
(.1,.1,.5)	87.6	82.7	86.5	84.5	94.9	89.2	92.4	90.5	99.7	97.6	98.7	96.9
(.1,.2,.4)	68.9	64.6	64.7	66.3	79.7	70.3	71.9	73.8	94.8	82.8	86.0	85.5
(.1,.2,.5)	88.2	83.6	84.7	84.9	95.0	89.6	91.1	90.6	99.7	97.6	98.3	96.9
(.1,.3,.4)	75.4	71.5	67.7	71.3	84.5	76.0	73.6	77.4	95.7	84.9	85.0	86.4
(.1,.3,.5)	90.3	86.3	85.0	86.4	95.9	91.1	90.8	91.3	99.7	97.7	97.9	96.9
(.1,.4,.4)	84.8	81.1	77.1	80.2	91.8	85.2	82.6	85.1	98.3	91.8	91.1	91.2
(.2,.2,.6)	82.0	76.8	79.1	77.5	89.7	84.4	86.3	84.8	97.5	94.7	95.4	93.7
(.2,.3,.6)	82.4	77.4	78.1	77.9	89.8	84.6	85.4	84.9	97.5	94.7	95.0	93.7
(.2,.4,.5)	64.9	61.3	59.4	61.4	72.9	66.6	65.5	67.6	85.4	76.7	76.7	77.9
(.2,.4,.6)	84.1	79.6	79.0	79.6	90.6	85.8	85.6	85.7	97.6	94.7	94.8	93.8
(.2,.5,.5)	75.4	71.6	69.5	71.3	82.4	76.6	75.3	76.8	91.3	84.6	84.3	84.5
(.2,.5,.6)	88.0	84.1	83.0	83.6	93.0	88.9	88.5	88.4	98.0	95.4	95.4	94.4
(.3,.3,.6)	53.6	49.9	51.0	51.1	62.1	57.7	58.8	58.9	76.9	72.7	73.0	72.8
(.3,.3,.7)	80.1	74.9	75.5	75.4	87.0	83.1	83.2	82.9	95.0	93.8	93.0	92.6
(.3,.4,.6)	55.1	51.6	51.5	52.5	62.9	58.6	58.7	59.7	77.0	72.7	72.5	72.8
(.3,.4,.7)	80.4	75.3	75.4	75.8	87.0	83.2	83.0	83.0	95.0	93.8	93.0	92.6
(.3,.5,.6)	60.0	56.7	56.1	57.4	66.8	62.7	62.4	63.5	78.1	74.0	73.8	74.0
(.3,.5,.7)	81.7	77.0	76.8	77.4	87.6	84.0	83.8	83.8	95.0	93.8	93.3	92.6
(.3,.6,.6)	70.5	66.9	66.5	67.5	76.7	72.7	72.7	73.2	85.1	81.6	81.9	81.2
(.3,.6,.7)	85.2	81.1	81.1	81.4	89.9	86.7	86.9	86.6	95.5	94.3	94.2	93.2
(.3,.7,.7)	91.3	87.9	88.3	88.2	94.4	92.1	92.6	92.0	97.5	96.8	97.1	96.1
(.4,.4,.7)	53.5	49.9	49.4	51.1	61.0	58.5	57.5	58.9	73.7	74.2	71.8	72.8
(.4,.4,.8)	81.9	76.7	75.5	77.5	87.3	85.1	83.3	84.8	93.8	94.8	92.7	93.7
(.4,.5,.7)	54.7	51.2	50.9	52.4	61.6	59.2	58.6	59.6	73.7	74.2	72.5	72.8
(.4,.5,.8)	82.1	77.0	76.2	77.8	87.3	85.1	83.9	84.9	93.8	94.8	93.2	93.7
(.4,.6,.7)	58.9	55.6	56.0	57.0	64.8	62.5	63.0	63.3	74.7	75.0	74.6	73.9
(.4,.6,.8)	82.8	78.0	78.1	79.1	87.6	85.5	85.2	85.4	93.8	94.8	93.9	93.8
(.4,.7,.7)	69.3	65.7	66.9	67.5	74.5	72.2	73.5	73.2	81.6	82.0	82.8	81.2
(.4,.7,.8)	85.5	81.2	82.4	82.8	89.3	87.4	88.2	87.9	94.1	95.0	95.0	94.2
(.5,.5,.8)	57.2	53.2	51.1	54.9	63.6	63.0	59.8	63.0	74.1	79.1	74.0	76.7
(.5,.5,.9)	87.5	82.6	79.6	84.5	90.5	90.0	86.7	90.5	94.3	97.2	94.3	96.9
(.5,.6,.8)	58.0	54.1	53.5	56.0	64.0	63.4	61.9	63.5	74.1	79.1	75.8	76.7
(.5,.6,.9)	87.6	82.7	81.1	84.7	90.5	90.0	87.9	90.6	94.3	97.2	95.1	96.9

Table 1: Power(%) for R_1, R_2, R_3, R_4; $\alpha = 0.05$, $k = 2$ (continued).

		$n_0=20$										
		$n=20$				$n=30$				$n=80$		
(p_0,p_1,p_2)	R_1	R_2	R_3	R_4	R_1	R_2	R_3	R_4	R_1	R_2	R_3	R_4
(.5,.7,.8)	61.4	57.8	59.2	60.3	66.4	65.9	66.9	66.7	74.7	79.5	78.7	77.5
(.5,.7,.9)	87.8	83.1	83.3	85.5	90.6	90.1	89.6	90.8	94.3	97.2	96.0	96.9
(.5,.8,.8)	71.4	67.6	70.6	71.3	75.4	74.9	77.6	76.8	80.9	85.2	86.5	84.5
(.5,.8,.9)	89.0	84.8	86.9	88.1	91.3	90.8	92.0	92.3	94.4	97.2	97.1	97.0
(.5,.9,.9)	93.4	90.1	92.7	94.0	94.7	94.3	96.1	96.4	96.1	98.2	98.7	98.6
(.6,.6,.9)	66.3	61.6	57.0	64.9	70.9	72.4	66.4	73.1	78.0	87.3	79.9	85.5
(.6,.7,.9)	66.6	62.0	60.7	65.6	71.0	72.5	69.8	73.4	78.0	87.3	82.7	85.5
(.6,.8,.9)	68.6	64.3	66.8	69.1	72.2	73.7	75.1	75.7	78.2	87.4	86.2	85.8

		$n_0=30$										
		$n=20$				$n=30$				$n=70$		
(p_0,p_1,p_2)	R_1	R_2	R_3	R_4	R_1	R_2	R_3	R_4	R_1	R_2	R_3	R_4
(.1,.1,.4)	70.2	71.7	75.3	72.2	82.8	79.4	83.5	81.5	96.9	91.6	94.7	92.6
(.1,.1,.5)	90.1	89.5	91.6	90.1	96.7	94.8	96.3	95.3	99.9	99.3	99.7	99.1
(.1,.2,.4)	72.9	74.8	73.8	73.9	84.3	81.2	81.5	82.3	97.0	91.8	93.2	92.7
(.1,.2,.5)	90.8	90.4	90.6	90.4	96.9	95.1	95.7	95.4	99.9	99.4	99.5	99.1
(.1,.3,.4)	79.6	81.5	77.8	79.2	89.0	86.6	84.1	85.9	97.9	93.8	93.4	93.6
(.1,.3,.5)	92.8	92.8	91.4	91.8	97.7	96.3	95.8	96.0	99.9	99.4	99.5	99.1
(.1,.4,.4)	88.4	89.4	86.1	87.3	95.0	93.3	91.3	92.2	99.4	97.6	97.2	96.8
(.2,.2,.5)	62.8	62.2	64.3	62.1	74.3	70.9	73.5	71.8	90.4	85.2	87.5	85.8
(.2,.2,.6)	86.7	84.4	85.9	84.1	93.9	91.4	92.6	91.2	99.2	98.2	98.5	97.7
(.2,.3,.5)	64.9	64.6	64.1	63.8	75.5	72.3	72.6	72.8	90.5	85.3	86.3	85.9
(.2,.3,.6)	87.2	85.1	85.3	84.6	94.1	91.6	92.0	91.4	99.2	98.2	98.3	97.7
(.2,.4,.5)	71.1	71.0	68.8	69.4	80.2	77.5	76.1	77.0	91.9	87.4	87.1	87.3
(.2,.4,.6)	88.9	87.3	86.5	86.3	94.8	92.7	92.4	92.2	99.2	98.2	98.3	97.7
(.2,.5,.5)	81.5	81.0	78.7	79.3	88.9	86.5	85.1	85.7	96.3	93.4	93.1	92.7
(.2,.5,.6)	92.4	91.1	89.9	90.0	96.6	95.1	94.6	94.4	99.5	98.7	98.7	98.2
(.3,.3,.6)	60.1	57.6	58.5	58.0	70.4	67.1	68.2	67.7	85.6	82.4	83.0	82.4
(.3,.3,.7)	86.4	82.3	82.9	82.2	92.8	90.2	90.5	89.8	98.3	97.6	97.5	97.1
(.3,.4,.6)	61.8	59.6	59.4	59.7	71.4	68.2	68.3	68.7	85.6	82.6	82.7	82.5
(.3,.4,.7)	86.7	82.8	82.9	82.7	92.9	90.3	90.3	90.0	98.3	97.6	97.5	97.1
(.3,.5,.6)	67.3	65.4	64.5	65.2	75.7	72.9	72.4	73.0	87.1	84.2	84.2	84.1
(.3,.5,.7)	88.0	84.6	84.3	84.3	93.5	91.1	91.0	90.7	98.3	97.6	97.5	97.1
(.3,.6,.6)	78.1	75.9	75.1	75.8	85.1	82.5	82.2	82.5	92.9	90.8	91.0	90.4
(.3,.6,.7)	91.1	88.4	88.1	88.2	95.2	93.3	93.4	93.1	98.6	98.1	98.1	97.6
(.3,.7,.7)	95.7	93.7	93.6	93.6	98.0	96.8	97.0	96.7	99.5	99.2	99.3	99.0

Table 1: Power(%) for R_1, R_2, R_3, R_4; $\alpha = 0.05$, $k = 2$ (continued).

							$n_0=30$					
		$n=20$				$n=30$				$n=70$		
(p_0,p_1,p_2)	R_1	R_2	R_3	R_4	R_1	R_2	R_3	R_4	R_1	R_2	R_3	R_4
(.4,.4,.7)	61.1	56.7	56.6	58.0	70.4	67.1	66.7	67.7	83.7	83.0	81.8	82.4
(.4,.4,.8)	89.1	83.6	83.2	84.1	93.9	91.4	90.8	91.2	98.1	98.0	97.4	97.7
(.4,.5,.7)	62.5	58.3	58.2	59.6	71.1	67.9	67.7	68.6	83.8	83.1	82.3	82.5
(.4,.5,.8)	89.3	83.9	83.7	84.5	93.9	91.4	91.1	91.3	98.1	98.0	97.6	97.7
(.4,.6,.7)	67.3	63.4	63.7	64.9	74.7	71.8	72.2	72.7	85.0	84.3	84.3	83.9
(.4,.6,.8)	90.0	85.1	85.3	85.8	94.2	91.9	91.9	91.9	98.1	98.0	97.8	97.7
(.4,.7,.7)	78.0	74.0	74.7	75.8	84.0	81.4	82.3	82.5	91.0	90.5	91.2	90.4
(.4,.7,.8)	92.2	88.1	88.8	89.3	95.4	93.5	94.0	93.9	98.3	98.2	98.3	98.1
(.5,.5,.8)	66.4	59.4	58.2	62.1	74.2	70.8	69.0	71.8	84.9	86.6	83.8	85.8
(.5,.5,.9)	94.5	88.6	87.3	90.1	96.7	94.8	93.5	95.3	98.6	99.1	98.2	99.1
(.5,.6,.8)	67.2	60.6	60.5	63.4	74.7	71.4	70.8	72.5	85.0	86.6	85.0	85.9
(.5,.6,.9)	94.6	88.6	88.1	90.2	96.7	94.8	94.2	95.3	98.6	99.1	98.5	99.1
(.5,.7,.8)	71.0	64.9	66.2	68.3	77.3	74.3	75.4	76.1	85.7	87.2	87.3	86.9
(.5,.7,.9)	94.7	89.1	89.6	91.0	96.8	94.9	95.0	95.6	98.6	99.1	98.8	99.1
(.5,.8,.8)	81.2	75.2	77.5	79.3	85.8	83.2	85.2	85.7	91.0	92.2	93.3	92.7
(.5,.8,.9)	95.6	90.8	92.2	93.4	97.2	95.6	96.4	96.7	98.7	99.1	99.2	99.2
(.5,.9,.9)	98.1	94.9	96.3	97.3	98.7	97.8	98.5	98.9	99.3	99.6	99.7	99.8
(.6,.6,.9)	77.5	67.3	64.6	72.2	82.7	79.3	75.8	81.5	89.2	92.6	88.7	92.6
(.6,.7,.9)	77.8	67.9	67.7	73.1	82.9	79.5	78.4	81.8	89.2	92.6	90.5	92.6
(.6,.8,.9)	79.8	70.7	73.3	77.0	84.0	80.9	82.6	84.3	89.4	92.7	92.6	93.1

							$n_0=40$					
		$n=10$				$n=40$				$n=60$		
(p_0,p_1,p_2)	R_1	R_2	R_3	R_4	R_1	R_2	R_3	R_4	R_1	R_2	R_3	R_4
(.1,.1,.4)	47.3	62.3	63.4	55.0	91.7	89.7	92.2	90.8	97.1	94.4	96.3	95.1
(.1,.1,.5)	68.9	79.8	80.9	75.4	99.2	98.6	99.1	98.7	99.9	99.7	99.8	99.6
(.1,.2,.4)	51.6	67.5	64.7	58.4	92.5	90.7	90.9	91.2	97.3	94.8	95.4	95.2
(.1,.2,.5)	71.3	82.4	80.9	77.0	99.3	98.7	98.9	98.7	99.9	99.7	99.7	99.6
(.1,.3,.4)	59.2	75.6	70.8	65.8	95.4	94.1	92.8	93.4	98.4	96.7	96.1	96.2
(.1,.3,.5)	75.7	86.7	83.6	80.7	99.5	99.1	99.0	98.9	99.9	99.7	99.7	99.6
(.1,.4,.4)	69.8	84.0	79.4	75.9	98.5	97.8	97.0	97.1	99.6	99.0	98.7	98.5
(.2,.2,.5)	43.8	50.5	51.1	46.0	85.3	82.8	84.7	83.2	92.2	89.2	90.8	89.5
(.2,.2,.6)	67.7	71.1	71.9	67.6	98.1	97.1	97.6	96.9	99.5	99.0	99.2	98.7
(.2,.3,.5)	47.1	54.5	53.0	49.0	86.0	83.4	83.9	83.8	92.4	89.5	90.0	89.6
(.2,.3,.6)	69.4	73.3	72.4	69.2	98.2	97.2	97.4	96.9	99.5	99.0	99.1	98.7
(.2,.4,.5)	54.0	62.2	59.2	55.9	89.4	87.6	86.6	86.9	94.2	91.8	91.3	91.4
(.2,.4,.6)	73.1	77.5	75.4	72.9	98.5	97.6	97.5	97.2	99.5	99.1	99.1	98.8

Table 1: Power(%) for R_1, R_2, R_3, R_4; $\alpha = 0.05$, $k = 2$ (continued).

						$n_0=40$						
		$n=10$				$n=40$				$n=60$		
(p_0,p_1,p_2)	R_1	R_2	R_3	R_4	R_1	R_2	R_3	R_4	R_1	R_2	R_3	R_4
(.2,.5,.5)	65.0	72.4	69.1	66.6	95.3	94.0	93.2	93.2	97.8	96.5	96.1	95.8
(.2,.5,.6)	79.2	83.3	80.7	79.1	99.1	98.6	98.4	98.2	99.7	99.5	99.4	99.2
(.3,.3,.6)	43.4	44.0	44.3	42.6	81.9	79.4	80.2	79.6	88.9	86.5	87.2	86.5
(.3,.3,.7)	70.4	66.3	66.7	65.4	97.7	96.5	96.6	96.1	99.1	98.6	98.7	98.4
(.3,.4,.6)	46.3	47.4	46.6	45.5	82.5	80.1	80.2	80.1	89.0	86.8	86.9	86.7
(.3,.4,.7)	71.6	68.1	67.6	66.9	97.7	96.5	96.6	96.2	99.1	98.6	98.6	98.4
(.3,.5,.6)	52.6	54.2	52.7	51.9	85.8	83.7	83.4	83.5	90.8	88.9	88.7	88.6
(.3,.5,.7)	74.5	71.8	70.7	70.4	97.9	96.9	96.8	96.5	99.2	98.7	98.7	98.4
(.3,.6,.6)	64.1	64.7	63.0	62.8	92.8	91.3	91.1	91.0	95.7	94.5	94.5	94.2
(.3,.6,.7)	80.1	77.8	76.4	76.6	98.6	97.9	97.9	97.6	99.4	99.1	99.1	98.9
(.3,.7,.7)	88.5	85.6	84.4	84.9	99.6	99.3	99.3	99.1	99.8	99.7	99.8	99.6
(.4,.4,.7)	46.1	40.8	40.8	42.6	81.9	79.4	79.1	79.6	88.1	86.7	86.2	86.5
(.4,.4,.8)	77.6	65.2	65.3	67.6	98.1	97.1	96.9	96.9	99.2	98.9	98.7	98.7
(.4,.5,.7)	48.6	43.6	43.4	45.3	82.3	79.9	79.7	80.1	88.2	86.9	86.6	86.7
(.4,.5,.8)	78.3	66.6	66.5	68.9	98.1	97.1	97.0	96.9	99.2	98.9	98.8	98.7
(.4,.6,.7)	54.5	49.8	49.4	51.6	85.0	82.9	83.2	83.3	89.6	88.5	88.7	88.4
(.4,.6,.8)	80.3	69.7	69.5	72.0	98.2	97.3	97.3	97.1	99.2	98.9	98.9	98.8
(.4,.7,.7)	66.5	60.4	60.0	62.8	92.1	90.5	91.0	91.0	94.8	94.0	94.5	94.2
(.4,.7,.8)	84.7	75.3	75.2	78.0	98.7	98.0	98.2	98.0	99.3	99.1	99.2	99.1
(.5,.5,.8)	53.3	40.1	39.9	46.0	85.3	82.8	81.3	83.2	89.9	89.6	88.0	89.5
(.5,.5,.9)	90.3	69.1	68.7	75.4	99.2	98.6	98.2	98.7	99.6	99.5	99.3	99.6
(.5,.6,.8)	55.2	42.5	42.8	48.4	85.5	83.0	82.6	83.6	90.0	89.7	88.9	89.6
(.5,.6,.9)	90.5	69.9	70.1	76.3	99.2	98.6	98.4	98.7	99.6	99.5	99.4	99.6
(.5,.7,.8)	60.5	48.1	48.9	54.6	87.2	85.1	85.9	86.2	90.8	90.6	91.0	90.9
(.5,.7,.9)	91.1	72.0	72.9	78.7	99.2	98.7	98.7	98.8	99.6	99.6	99.5	99.6
(.5,.8,.8)	73.0	58.6	60.0	66.6	93.3	91.8	93.0	93.2	95.3	95.1	95.9	95.8
(.5,.8,.9)	93.0	76.4	77.9	83.8	99.3	98.9	99.1	99.2	99.6	99.6	99.7	99.7
(.5,.9,.9)	97.4	84.8	86.4	91.7	99.8	99.6	99.7	99.8	99.9	99.8	99.9	99.9
(.6,.6,.9)	69.6	42.5	41.9	55.0	91.7	89.6	87.3	90.8	94.1	94.6	92.5	95.1
(.6,.7,.9)	70.5	44.4	45.2	57.1	91.8	89.7	88.9	91.0	94.1	94.6	93.7	95.1
(.6,.8,.9)	74.0	49.1	51.5	62.9	92.3	90.5	91.5	92.4	94.4	94.8	95.2	95.7

Table 2: Power(%) for R_1, R_2, R_3, R_4; $\alpha = 0.05$, $k = 3$.

	\multicolumn{12}{c}{$n_0=20$}										
	\multicolumn{4}{c}{n=20}				\multicolumn{4}{c}{n=30}				\multicolumn{4}{c}{n=80}			
(p_0,p_1,p_2,p_3)	R1	R2	R3	R4	R1	R2	R3	R4	R1	R2	R3	R4
(.1,.1,.1,.4)	61.2	56.0	65.8	59.4	74.2	62.9	74.4	68.4	93.5	78.7	89.3	82.9
(.1,.1,.1,.5)	84.4	78.3	85.7	80.8	93.3	85.9	92.2	88.0	99.6	99.6	98.8	96.1
(.1,.1,.2,.4)	63.4	58.5	63.3	60.7	75.5	64.5	71.4	69.1	93.5	78.8	86.9	83.0
(.1,.1,.2,.5)	85.0	79.2	83.9	81.1	93.4	86.2	90.9	88.1	99.6	96.6	98.4	96.1
(.1,.1,.3,.4)	70.0	65.4	66.4	65.8	80.6	70.3	73.2	72.8	94.6	81.1	85.9	83.9
(.1,.1,.3,.5)	87.2	82.1	84.1	82.7	94.4	87.9	90.4	88.8	99.6	96.7	98.1	96.1
(.1,.1,.4,.4)	80.5	75.9	75.8	75.4	89.2	80.7	82.1	81.4	97.8	89.2	91.5	89.4
(.1,.2,.2,.4)	65.3	60.7	60.8	61.9	76.5	65.8	68.5	69.6	93.6	79.0	84.5	83.0
(.1,.2,.3,.3)	54.4	51.5	45.8	50.3	63.4	53.4	49.7	55.6	79.3	57.9	58.3	64.8
(.1,.2,.3,.4)	71.3	67.0	63.7	66.5	81.2	71.2	70.4	73.2	94.6	81.2	83.7	83.9
(.1,.2,.4,.4)	81.2	76.8	73.6	75.8	89.4	81.1	79.9	81.6	97.8	89.2	90.2	89.4
(.1,.3,.3,.3)	62.7	59.5	51.1	57.2	71.4	61.4	54.7	62.2	84.6	64.7	61.7	69.6
(.1,.3,.3,.4)	75.9	71.8	65.5	70.1	84.5	75.3	71.3	75.8	95.3	82.8	82.8	84.6
(.1,.3,.3,.5)	89.3	85.0	82.4	84.2	95.2	89.4	88.8	89.5	99.6	96.8	97.3	96.1
(.1,.3,.4,.4)	83.9	79.9	74.0	77.8	90.9	83.3	79.7	82.8	97.9	89.7	89.2	89.5
(.1,.4,.4,.4)	88.8	85.2	79.4	82.7	94.3	88.2	84.6	86.9	98.9	93.0	92.0	91.8
(.2,.2,.2,.5)	51.6	47.4	52.6	49.3	62.4	54.8	61.0	57.8	81.4	70.8	77.0	73.3
(.2,.2,.2,.6)	77.9	71.7	76.3	73.0	87.0	80.4	84.5	81.4	96.9	93.2	94.9	92.3
(.2,.2,.3,.5)	53.4	49.3	51.8	50.6	63.3	55.9	59.6	58.5	81.4	70.9	75.2	73.4
(.2,.2,.3,.6)	78.3	72.3	75.2	73.3	87.1	80.6	83.4	81.5	96.9	93.2	94.4	92.3
(.2,.2,.4,.5)	58.9	54.9	55.4	55.4	67.8	60.7	62.2	62.4	82.8	72.6	75.1	74.5
(.2,.2,.4,.6)	80.1	74.6	75.8	75.0	88.0	81.9	83.4	82.3	96.9	93.2	94.0	92.4
(.2,.2,.5,.5)	70.0	65.6	65.5	65.6	78.2	71.3	72.1	72.1	89.4	81.3	82.5	81.7
(.2,.2,.5,.6)	84.5	79.5	79.8	79.4	90.8	85.8	86.1	85.4	97.4	94.0	94.4	93.0
(.2,.3,.3,.5)	54.8	50.9	51.0	51.7	64.1	56.9	58.3	59.1	81.5	71.0	73.7	73.4
(.2,.3,.3,.6)	78.7	72.9	74.2	73.7	87.2	80.8	82.5	81.6	96.9	93.2	94.0	92.3
(.2,.3,.5,.5)	70.5	66.3	64.6	66.1	78.4	71.7	71.0	72.3	89.4	81.3	81.7	81.7
(.2,.4,.4,.5)	64.0	60.3	57.2	59.8	71.6	64.8	62.9	65.6	83.8	74.0	74.0	75.5
(.2,.4,.5,.5)	73.0	69.0	66.0	68.2	80.1	73.6	71.8	73.8	89.7	81.7	81.5	82.0
(.2,.5,.5,.5)	79.1	75.1	72.0	74.0	85.1	79.3	77.5	78.9	92.4	85.7	85.3	85.3
(.3,.3,.5,.6)	53.9	50.3	50.4	51.3	61.5	56.9	57.3	58.1	74.8	70.1	70.2	70.4
(.3,.3,.6,.6)	64.8	60.6	60.6	61.6	71.9	67.3	67.4	68.2	82.4	78.3	78.1	78.1
(.3,.4,.5,.6)	54.8	51.3	50.7	52.1	61.9	57.4	57.2	58.5	74.9	70.1	69.9	70.4
(.3,.4,.6,.6)	65.3	61.2	60.7	62.1	72.1	67.5	67.4	68.4	82.4	78.3	78.1	78.1
(.3,.5,.5,.6)	58.4	55.0	54.0	55.6	64.8	60.4	60.0	61.3	75.8	71.1	70.9	71.3
(.3,.5,.6,.6)	67.5	63.6	62.9	64.3	73.6	69.2	69.0	69.8	82.7	78.6	78.7	78.4
(.3,.6,.6,.6)	73.6	69.7	69.3	70.3	79.0	74.8	75.1	75.3	86.1	82.5	83.0	82.1

Table 2: Power(%) for R_1, R_2, R_3, R_4; $\alpha = 0.05$, $k = 3$ (continued).

	$n_0=20$											
	n=20				n=30				n=80			
(p_0,p_1,p_2,p_3)	R1	R2	R3	R4	R1	R2	R3	R4	R1	R2	R3	R4
(.4,.4,.4,.7)	47.9	43.9	43.6	45.5	55.7	53.0	51.9	53.6	70.2	70.7	67.9	69.1
(.4,.5,.5,.7)	49.8	46.1	45.8	47.7	56.7	54.2	53.3	54.8	70.2	70.7	68.3	69.2
(.4,.6,.6,.7)	56.7	53.3	54.1	55.1	62.3	59.8	60.5	60.8	71.9	72.2	71.7	71.1
(.5,.5,.6,.8)	52.2	48.0	45.9	50.2	58.7	58.1	54.5	58.3	70.5	76.2	70.1	73.3
(.5,.6,.6,.8)	52.9	48.8	47.9	51.1	59.0	58.4	56.2	58.7	70.6	76.3	71.4	73.4
(.5,.6,.7,.8)	56.0	52.1	52.8	55.0	61.4	60.8	60.4	61.8	71.1	76.6	73.6	74.1
(.5,.7,.7,.8)	58.5	54.8	57.3	58.1	63.3	62.7	64.4	64.1	71.6	76.9	75.9	74.8
(.6,.6,.6,.9)	60.8	55.6	45.8	59.4	66.1	67.8	58.5	68.4	74.8	85.5	74.3	82.9
(.6,.6,.7,.9)	61.1	56.0	51.3	60.1	66.2	67.9	60.9	68.7	74.8	85.5	76.2	83.0
(.6,.6,.8,.9)	63.1	58.2	56.4	63.5	67.4	69.1	65.2	71.0	75.0	85.5	78.6	83.3
(.6,.6,.9,.9)	72.4	67.4	67.4	75.4	75.4	77.0	75.3	81.4	80.1	89.2	85.3	89.4
(.6,.7,.7,.9)	61.4	56.3	54.3	60.7	66.3	68.0	63.7	69.0	74.8	85.5	78.4	83.0
(.6,.7,.8,.9)	63.3	58.5	59.6	64.0	67.5	69.2	68.2	71.2	75.0	85.5	81.0	83.3
(.6,.7,.9,.9)	72.4	67.5	70.6	75.6	75.4	77.0	78.2	81.5	80.1	89.2	87.4	89.4
(.6,.8,.8,.9)	64.9	60.3	64.8	66.6	68.5	70.2	72.6	73.0	75.1	85.6	83.7	83.6
(.6,.8,.9,.9)	73.1	68.4	74.8	76.8	75.8	77.3	81.7	82.1	80.1	89.3	89.7	89.4
(.6,.9,.9,.9)	77.7	73.1	81.7	82.7	79.7	81.0	87.4	86.9	82.4	90.8	93.3	91.8
(.7,.7,.9,.9)	45.0	42.2	42.3	48.5	48.0	51.8	50.0	54.5	53.2	68.0	62.1	64.6
(.7,.9,.9,.9)	51.4	48.5	58.2	57.2	53.5	57.3	65.7	62.2	56.7	71.2	75.4	69.6

	$n_0=30$											
	n=20				n=30				n=70			
(p_0,p_1,p_2,p_3)	R1	R2	R3	R4	R1	R2	R3	R4	R1	R2	R3	R4
(.1,.1,.1,.3)	37.2	40.6	46.5	39.2	48.2	45.6	54.1	47.8	72.0	57.1	70.0	64.2
(.1,.1,.1,.4)	64.8	66.6	73.3	67.0	78.9	74.8	82.4	77.4	95.9	88.9	94.7	90.8
(.1,.1,.1,.5)	87.2	86.5	90.7	87.1	95.5	92.9	96.1	93.7	99.8	99.0	99.7	98.8
(.1,.1,.2,.3)	42.6	46.7	46.8	43.5	53.2	50.7	53.0	51.3	74.1	59.7	66.4	65.4
(.1,.1,.2,.4)	67.4	69.5	71.8	68.5	80.4	76.6	80.5	78.1	96.0	89.2	93.4	90.8
(.1,.1,.2,.5)	87.9	87.4	89.7	87.5	95.7	93.2	95.3	93.8	99.8	99.0	99.6	98.8
(.1,.1,.3,.3)	56.0	60.3	58.4	56.1	67.7	64.9	64.8	64.1	85.8	73.3	76.3	76.0
(.1,.1,.3,.4)	74.3	76.7	75.9	74.1	85.6	82.4	83.2	82.1	97.2	91.5	93.5	91.9
(.1,.1,.3,.5)	90.2	90.1	90.4	89.1	96.6	94.6	95.4	94.4	99.9	99.1	99.5	98.8
(.1,.1,.4,.4)	84.5	85.9	84.6	83.4	93.0	90.6	90.6	89.6	99.2	96.5	97.2	95.8

Table 2: Power(%) for R_1, R_2, R_3, R_4; $\alpha = 0.05$, $k = 3$ (continued).

						$n_0=30$					
		$n=20$				$n=30$				$n=70$	
(p_0,p_1,p_2,p_3)	R_1	R_2	R_3	R_4	R_1	R_2	R_3	R_4	R_1	R_2	R_3	R_4
(.1,.2,.2,.3)	47.3	51.7	46.2	47.1	57.2	54.9	51.3	54.1	75.7	61.7	62.8	66.4
(.1,.2,.2,.4)	69.6	72.0	70.1	69.9	81.6	78.1	78.5	78.8	96.1	89.4	92.0	90.8
(.1,.2,.3,.3)	59.2	63.6	56.5	58.3	69.9	67.3	62.1	65.5	86.2	74.1	72.9	76.3
(.1,.2,.3,.4)	75.9	78.3	74.0	74.9	86.3	83.3	81.2	82.5	97.2	91.6	92.3	91.9
(.1,.2,.4,.4)	85.3	86.7	83.1	83.8	93.2	90.9	89.3	89.7	99.2	96.5	96.6	95.8
(.1,.3,.3,.3)	67.8	72.0	62.9	66.0	78.0	75.5	68.2	72.6	91.0	80.9	77.3	81.2
(.1,.3,.3,.4)	80.6	82.9	76.5	78.7	89.5	87.0	82.8	85.0	97.8	93.1	92.3	92.6
(.1,.3,.3,.5)	92.3	92.4	89.8	90.5	97.3	95.7	94.8	95.0	99.9	99.2	99.3	98.8
(.1,.3,.4,.4)	87.9	89.3	84.0	85.7	94.6	92.7	89.6	90.8	99.3	96.9	96.3	95.9
(.1,.4,.4,.4)	92.2	93.1	88.5	89.9	97.0	95.6	93.1	93.8	99.7	98.4	97.8	97.3
(.2,.2,.2,.5)	57.1	56.4	60.3	56.3	69.4	65.5	70.3	66.8	88.1	81.7	86.1	82.9
(.2,.2,.2,.6)	83.2	80.3	83.4	80.2	92.0	88.6	91.2	88.7	98.8	97.4	98.2	96.9
(.2,.2,.3,.5)	59.0	58.6	60.0	57.8	70.6	66.9	69.4	67.7	88.2	81.9	84.9	82.9
(.2,.2,.3,.6)	83.6	81.0	82.7	80.6	92.1	88.8	90.5	88.8	98.8	97.4	98.0	96.9
(.2,.2,.4,.5)	65.1	65.0	64.6	63.4	75.6	72.2	72.8	72.1	89.8	84.1	85.6	84.5
(.2,.2,.4,.6)	85.5	83.3	83.7	82.4	93.0	90.1	90.8	89.7	98.9	97.5	97.9	97.0
(.2,.2,.5,.5)	76.4	75.8	75.0	74.1	85.4	82.4	82.4	81.7	95.1	91.2	91.8	90.7
(.2,.2,.5,.6)	89.6	87.8	87.5	86.7	95.2	93.0	93.1	92.3	99.2	98.1	98.3	97.5
(.2,.3,.3,.5)	60.7	60.5	59.6	59.3	71.6	68.0	68.5	68.4	88.3	82.0	83.8	83.0
(.2,.3,.3,.6)	84.1	81.6	82.0	81.0	92.2	89.1	89.9	88.9	98.9	97.4	97.8	96.9
(.2,.3,.4,.5)	66.3	66.4	64.0	64.4	76.2	73.0	71.8	72.6	89.9	84.2	84.6	84.5
(.2,.3,.5,.5)	77.1	76.6	74.3	74.6	85.7	82.7	81.6	81.9	95.1	91.2	91.3	90.7
(.2,.4,.4,.4)	56.2	57.7	52.8	54.5	64.5	62.2	58.3	60.8	76.7	69.1	67.2	70.0
(.2,.4,.4,.5)	70.8	71.0	67.2	68.4	79.6	76.7	74.3	75.6	91.0	85.7	85.2	85.6
(.2,.4,.5,.5)	79.7	79.4	76.0	77.0	93.7	91.1	90.6	83.4	95.4	91.7	91.3	91.1
(.2,.5,.5,.5)	85.4	84.9	81.7	82.5	91.5	89.3	87.4	87.9	97.1	94.5	94.1	93.6
(.3,.3,.3,.6)	54.3	51.5	53.4	52.1	65.2	61.5	63.6	62.4	82.6	78.8	80.4	79.0
(.3,.3,.5,.6)	61.2	59.0	58.7	59.0	70.5	67.2	67.4	67.6	84.3	80.7	81.0	80.8
(.3,.3,.6,.6)	72.7	70.0	69.7	70.1	81.0	77.8	77.8	78.0	90.9	88.3	88.3	88.0
(.3,.4,.5,.6)	62.2	60.2	59.1	60.0	71.1	67.9	67.4	68.2	84.3	80.8	80.8	80.8
(.3,.4,.6,.6)	73.3	70.7	69.8	70.7	81.2	78.1	77.7	78.3	90.9	88.3	88.3	88.0
(.3,.5,.5,.6)	66.2	64.4	62.9	61.9	74.2	71.3	70.5	71.3	85.5	82.1	82.0	82.0
(.3,.5,.6,.6)	75.6	73.3	72.1	73.1	82.7	79.8	79.4	79.8	91.2	88.7	88.8	88.4
(.3,.6,.6,.6)	81.7	79.3	78.4	79.1	87.6	85.0	84.8	84.9	93.9	91.8	92.1	91.4

Table 2: Power(%) for R_1, R_2, R_3, R_4; $\alpha = 0.05$, $k = 3$ (continued).

	\multicolumn — $n_0=30$										
	$n=20$				$n=30$				$n=70$			
(p_0,p_1,p_2,p_3)	R_1	R_2	R_3	R_4	R_1	R_2	R_3	R_4	R_1	R_2	R_3	R_4
$(.4,.4,.4,.7)$	55.3	50.4	50.8	52.1	65.2	61.4	61.1	62.4	80.5	79.7	78.2	79.0
$(.4,.4,.5,.7)$	56.5	51.9	51.7	53.5	65.9	62.2	61.8	63.2	80.6	79.7	78.3	79.1
$(.4,.4,.6,.7)$	61.2	56.8	56.8	58.6	69.5	66.2	66.0	67.3	81.8	81.0	80.0	80.6
$(.4,.5,.5,.7)$	57.6	53.2	53.1	54.8	66.5	63.0	62.6	63.9	80.6	79.8	78.6	79.2
$(.4,.5,.6,.7)$	62.0	57.8	57.9	59.6	69.9	66.7	66.8	67.8	81.9	81.0	80.4	80.7
$(.4,.6,.6,.7)$	65.5	61.6	62.1	63.4	72.6	69.6	70.3	70.8	82.8	82.0	82.1	81.8
$(.5,.5,.5,.8)$	60.7	52.9	51.1	56.3	69.4	65.4	62.5	66.8	81.9	83.9	78.0	82.9
$(.5,.5,.6,.8)$	61.5	54.0	53.0	57.4	69.7	65.9	63.8	67.4	81.9	83.9	80.3	82.9
$(.5,.6,.6,.8)$	62.2	55.0	54.9	58.5	70.1	66.3	65.4	68.0	81.9	83.9	81.3	82.9
$(.5,.6,.7,.8)$	65.8	58.9	60.1	63.0	72.7	69.2	69.7	71.4	82.7	84.5	83.3	84.0
$(.5,.7,.7,.8)$	68.6	62.2	64.7	66.6	74.7	71.5	73.5	74.1	83.3	85.1	85.2	84.9
$(.6,.6,.6,.9)$	72.7	60.8	56.2	67.0	78.8	74.6	68.6	77.4	86.7	90.9	84.5	90.8
$(.6,.6,.7,.9)$	73.0	61.3	58.8	67.8	78.9	74.8	70.7	77.7	86.7	90.9	85.8	90.8
$(.6,.6,.8,.9)$	75.0	64.1	64.0	71.7	80.1	76.3	74.7	80.3	86.9	91.1	87.6	91.2
$(.6,.6,.9,.9)$	84.1	74.1	75.3	83.4	87.4	84.2	84.2	89.6	91.2	94.3	92.9	95.8
$(.6,.7,.7,.9)$	73.3	61.9	61.6	68.5	79.0	74.9	73.0	78.0	86.7	90.9	87.3	90.8
$(.6,.7,.8,.9)$	75.2	64.6	66.8	72.2	80.2	76.4	77.1	80.5	86.9	91.1	89.2	91.3
$(.6,.7,.9,.9)$	84.2	74.2	77.8	83.6	87.4	84.2	86.1	89.6	91.2	94.3	94.1	95.8
$(.6,.8,.8,.9)$	76.8	66.8	71.6	75.1	81.2	77.7	80.8	82.3	87.2	91.2	91.1	91.6
$(.6,.8,.9,.9)$	84.8	75.3	81.2	84.8	87.7	84.7	88.6	90.2	91.2	94.3	95.3	95.8
$(.6,.9,.9,.9)$	88.7	80.1	87.2	89.9	90.7	88.1	92.9	93.8	93.0	95.6	97.3	97.3
$(.7,.7,.9,.9)$	55.7	45.6	47.2	56.1	60.2	57.4	57.5	64.1	66.6	75.1	71.8	76.0
$(.7,.8,.9,.9)$	56.6	46.7	52.7	57.5	60.7	58.0	62.7	64.9	66.7	75.1	76.1	76.1
$(.7,.9,.9,.9)$	63.3	53.0	63.0	66.0	66.6	63.9	72.6	72.6	70.8	78.8	83.6	81.2

	\multicolumn — $n_0=40$										
	$n=10$				$n=40$				$n=60$			
(p_0,p_1,p_2,p_3)	R_1	R_2	R_3	R_4	R_1	R_2	R_3	R_4	R_1	R_2	R_3	R_4
$(.1,.1,.1,.3)$	23.6	36.8	38.2	27.7	60.1	57.6	65.9	59.8	72.7	64.5	74.4	68.9
$(.1,.1,.1,.4)$	41.5	57.6	59.9	48.9	89.3	86.6	91.5	88.2	96.1	92.4	96.1	93.6
$(.1,.1,.1,.5)$	63.3	76.0	78.4	70.2	98.9	98.0	99.0	98.1	99.8	99.4	99.8	99.4
$(.1,.1,.2,.3)$	28.5	44.3	41.6	32.6	64.8	62.5	64.9	63.0	75.8	68.0	72.2	70.7
$(.1,.1,.2,.4)$	45.1	62.4	60.9	52.0	90.2	87.7	90.3	88.6	96.3	92.8	95.2	93.7
$(.1,.1,.2,.5)$	65.4	78.6	78.2	71.7	98.9	98.1	98.8	98.2	99.8	99.5	99.7	99.4

Table 2: Power(%) for R_1, R_2, R_3, R_4; $\alpha = 0.05$, $k = 3$ (continued).

							$n_0=40$					
		$n=10$				$n=40$				$n=60$		
(p_0,p_1,p_2,p_3)	R_1	R_2	R_3	R_4	R_1	R_2	R_3	R_4	R_1	R_2	R_3	R_4
(.1,.1,.3,.3)	38.0	56.9	52.7	43.4	79.2	76.9	76.9	75.8	87.8	81.6	82.8	81.8
(.1,.1,.3,.4)	52.2	70.8	67.0	59.1	93.6	91.7	92.2	91.2	97.7	95.1	96.0	94.9
(.1,.1,.3,.5)	69.7	83.2	80.9	75.5	99.2	98.6	98.9	98.4	99.9	99.6	99.7	94.4
(.1,.1,.4,.4)	63.0	80.0	76.1	69.8	97.7	96.7	96.7	95.9	99.4	98.4	98.6	97.8
(.1,.2,.2,.3)	33.0	50.6	43.7	36.8	68.5	66.3	63.1	65.4	78.0	70.6	69.6	72.1
(.1,.2,.2,.4)	48.4	66.5	61.2	54.7	90.8	88.6	89.0	88.9	96.5	93.1	94.3	93.8
(.1,.2,.3,.3)	41.8	61.7	53.2	46.7	80.9	78.8	74.6	76.8	88.5	82.6	80.3	82.2
(.1,.2,.3,.4)	55.0	73.8	66.7	61.2	93.9	92.2	91.0	91.4	97.8	95.3	95.2	94.9
(.1,.2,.4,.4)	65.1	82.0	75.4	71.2	97.8	96.8	96.1	95.9	99.4	98.4	98.3	97.8
(.1,.3,.3,.3)	49.3	70.0	60.0	54.6	87.6	85.8	80.4	83.0	93.2	88.7	85.0	87.1
(.1,.3,.3,.4)	60.7	79.4	70.7	66.5	95.7	94.4	92.1	92.9	98.5	96.6	95.6	95.7
(.1,.3,.3,.5)	74.9	88.0	82.1	79.5	99.4	98.9	98.7	98.6	99.9	99.7	99.6	99.5
(.1,.3,.4,.4)	69.4	85.7	77.8	74.9	98.3	97.6	96.3	96.4	99.6	98.7	98.2	98.0
(.1,.4,.4,.4)	76.0	90.1	82.8	80.8	99.3	98.9	98.0	97.9	99.8	99.5	99.1	98.9
(.2,.2,.2,.5)	38.1	45.1	46.3	40.1	81.7	78.6	82.4	79.3	90.0	86.2	89.5	86.8
(.2,.2,.2,.6)	61.9	65.9	67.5	61.8	97.4	95.9	97.0	95.7	99.3	98.5	99.0	98.2
(.2,.2,.3,.5)	40.9	48.6	47.8	42.7	82.4	79.5	81.5	79.9	90.3	86.5	88.6	87.0
(.2,.2,.3,.6)	63.5	67.9	67.8	63.3	97.4	96.0	96.7	95.7	99.3	98.5	98.9	98.2
(.2,.2,.4,.4)	32.7	41.5	39.2	34.4	66.0	63.9	63.8	63.3	74.3	69.4	70.1	69.8
(.2,.2,.4,.5)	47.1	55.9	53.7	49.0	86.2	83.8	84.2	83.3	92.3	89.1	89.9	88.9
(.2,.2,.4,.6)	67.0	72.2	70.7	66.9	97.7	96.5	96.9	96.1	99.3	98.6	98.8	98.3
(.2,.2,.5,.5)	58.1	66.5	63.8	59.8	93.4	91.6	91.6	90.8	96.8	94.9	95.2	94.3
(.2,.2,.5,.6)	73.5	78.5	76.4	73.4	98.7	97.9	97.9	97.4	99.6	99.1	99.2	98.8
(.2,.3,.3,.5)	43.4	51.7	49.0	45.1	83.0	80.3	80.7	80.3	90.4	86.8	87.8	87.1
(.2,.3,.3,.6)	64.9	69.7	68.0	64.6	97.4	96.1	96.5	95.8	99.3	98.5	98.7	98.2
(.2,.3,.4,.4)	35.7	45.1	40.8	37.3	67.6	65.6	63.1	64.5	75.1	70.3	69.0	70.5
(.2,.3,.4,.5)	49.3	58.5	54.3	51.0	86.6	84.3	83.5	83.6	92.4	89.2	89.2	89.0
(.2,.3,.5,.5)	59.7	68.3	63.9	68.0	93.5	91.8	91.1	90.9	96.9	95.0	94.8	94.4
(.2,.4,.4,.4)	42.5	52.9	47.3	44.2	75.4	73.5	70.0	71.8	81.7	77.4	75.0	76.8
(.2,.4,.4,.5)	54.5	64.1	58.7	56.0	89.0	87.0	85.3	85.9	93.7	90.9	90.1	90.2
(.2,.4,.5,.5)	63.6	72.4	66.9	64.9	94.4	92.9	91.7	91.8	97.2	95.5	95.0	94.7
(.2,.5,.5,.5)	70.7	78.6	73.0	71.6	96.8	95.7	94.8	94.7	98.5	97.4	97.0	96.7
(.3,.3,.3,.6)	37.7	38.3	38.9	37.0	77.9	74.8	76.6	75.2	86.1	83.2	84.6	83.3
(.3,.3,.5,.6)	45.8	47.5	46.3	45.2	82.0	79.4	79.6	79.4	88.2	85.7	86.0	85.6
(.3,.3,.6,.6)	57.1	57.8	56.4	55.9	90.3	88.2	88.2	88.1	94.2	92.5	92.6	92.2

Table 2: Power(%) for R_1, R_2, R_3, R_4; $\alpha = 0.05$, $k = 3$ (continued).

	$n=10$				$n=40$				$n=60$			
(p_0,p_1,p_2,p_3)	R_1	R_2	R_3	R_4	R_1	R_2	R_3	R_4	R_1	R_2	R_3	R_4
(.3,.4,.5,.5)	33.3	36.8	34.7	33.6	60.8	59.0	57.9	58.8	67.4	64.4	63.8	64.7
(.3,.4,.5,.6)	47.7	49.7	47.6	47.0	82.3	79.8	79.5	79.7	88.3	85.8	85.8	85.7
(.3,.4,.6,.6)	58.6	59.5	57.2	57.2	90.4	88.4	88.1	88.2	94.2	92.5	92.5	92.3
(.3,.5,.5,.5)	39.8	43.6	40.9	40.0	68.6	66.7	65.3	66.3	74.3	71.4	70.6	71.4
(.3,.5,.5,.6)	52.5	54.8	52.0	51.8	84.7	82.5	81.9	82.2	89.7	87.4	87.2	87.2
(.3,.5,.6,.6)	62.0	63.3	60.5	60.8	91.3	89.5	89.2	89.2	94.6	93.0	93.0	92.7
(.3,.6,.6,.6)	69.3	70.0	67.0	67.8	94.5	93.0	92.9	92.7	96.7	95.5	95.6	95.2
(.4,.4,.4,.7)	40.3	34.8	34.9	37.0	77.9	74.8	74.5	75.2	85.1	83.5	82.8	83.3
(.4,.4,.5,.7)	42.3	37.2	37.0	39.2	78.3	75.3	74.9	75.7	85.3	83.6	83.0	83.5
(.4,.4,.6,.7)	47.7	42.7	42.4	44.8	81.1	78.5	78.3	79.1	86.8	85.4	85.0	85.4
(.4,.5,.5,.7)	44.2	39.4	39.0	41.2	78.6	75.7	75.5	76.2	85.4	83.8	83.2	83.6
(.4,.5,.6,.7)	49.3	44.7	44.1	46.6	81.4	78.8	78.8	79.4	86.9	85.5	85.3	85.5
(.4,.6,.6,.6)	39.6	38.3	37.6	38.8	65.7	63.9	64.4	64.7	70.8	69.2	69.8	69.8
(.4,.6,.6,.7)	53.8	49.2	48.7	51.2	83.4	81.1	81.6	81.8	88.1	86.7	87.1	86.9
(.5,.5,.5,.8)	47.2	33.5	33.2	40.1	81.6	78.6	76.2	79.3	87.3	87.0	84.3	86.8
(.5,.5,.6,.8)	48.8	35.6	35.6	42.2	81.8	78.8	77.2	79.7	87.4	87.0	84.9	86.9
(.5,.6,.6,.8)	50.3	37.5	38.0	44.1	82.0	79.1	78.3	80.0	87.4	87.1	85.7	87.0
(.5,.6,.7,.8)	55.1	42.2	43.2	49.4	83.9	81.2	81.6	82.7	88.4	88.0	87.7	88.4
(.5,.7,.7,.8)	59.1	46.3	48.0	54.0	85.3	82.9	84.3	84.7	89.1	88.8	89.5	89.5
(.6,.6,.6,.9)	63.9	34.6	33.6	48.9	89.3	86.5	82.3	88.2	92.3	93.0	89.1	93.6
(.6,.6,.7,.9)	64.7	36.3	36.5	50.7	89.3	86.6	83.7	88.4	92.3	93.0	90.1	93.6
(.6,.6,.8,.9)	68.2	40.5	42.0	56.2	90.0	87.5	86.3	90.0	92.6	93.3	91.6	94.3
(.6,.6,.9,.9)	80.7	50.2	52.9	69.8	94.6	92.9	92.9	95.9	95.9	96.3	95.9	97.8
(.6,.7,.7,.9)	65.6	37.9	39.4	52.4	89.3	86.7	85.2	88.5	92.4	93.0	91.2	93.6
(.6,.7,.8,.9)	68.9	41.8	44.9	57.6	90.0	87.5	87.9	90.1	92.6	93.3	92.7	94.3
(.6,.7,.9,.9)	81.0	51.2	55.6	70.6	94.6	92.9	94.0	95.9	95.9	96.3	96.6	97.8
(.6,.8,.8,.8)	47.5	33.4	36.9	44.2	69.8	67.8	72.3	71.8	73.4	74.5	78.1	76.8
(.6,.8,.8,.9)	71.7	45.4	50.1	62.0	90.5	88.2	90.2	91.2	92.9	93.5	94.2	94.7
(.6,.8,.9,.9)	82.3	53.7	59.9	73.4	94.8	93.1	95.2	96.2	95.9	96.4	97.4	97.9
(.6,.9,.9,.9)	87.9	59.9	67.6	80.8	96.4	95.1	97.4	97.9	97.1	97.4	98.7	98.9
(.7,.7,.9,.9)	50.7	25.2	27.7	43.4	72.0	69.6	69.6	75.8	75.4	78.4	77.0	81.8
(.7,.8,.9,.9)	52.3	27.1	32.0	45.7	72.3	69.9	74.0	76.3	75.6	78.5	80.7	82.0
(.7,.9,.9,.9)	60.7	31.9	39.6	54.6	77.7	75.5	82.5	83.0	79.9	82.6	87.6	87.1

References

Bechhofer, R.E. and Dunnett, C.W. (1988). Tables of percentage points of multivariate Student t distributions. In *Selected Tables in Mathematical Statistics, No. 11*, American Mathematical Society, Providence, RI.

Bristol, D.R. (1989). Designing clinical trials for two-sided multiple comparisons with a control. *Control. Clin. Trials* **10** 142 - 152.

Dunnett, C.W. (1955). A multiple comparison procedure for comparing several treatments with a control. *Jour. Amer. Statist. Assoc.* **50** 1096 - 1121.

Dunnett, C.W. (1964). New tables for multiple comparisons with a control. *Biometrics* **20** 482 - 491.

Eberhardt, K.R. and Fligner, M.A. (1977). A comparison of two tests for equality of two proportions. *The American Statistician* **31** 151 - 155.

Gupta, S.S., Panchapakesan, S., and Sohn, J.K. (1985), On the distribution of the Studentized maximum of equally correlated normal random variables. *Commun. Statist. — Simul. Comput.* **14(1)** 103 - 135.

Piegorsch, W.W. (1991). Multiple comparisons for analyzing dichotomous response. *Biometrics* **47** 45 - 52.

Robbins, H. (1977). A fundamental question of practical statistics (Letter to the Editor). *The American Statistician* **31** 97.

Williams, D.A. (1988). Tests for differences between several small proportions. *Appl. Statist.* **37(3)** 421 - 434.

Chapter 7

Distribution-Free and Asymptotically Distribution-Free Comparisons with a Control in Blocked Experiments

JOHN D. SPURRIER Department of Statistics, University of South Carolina, Columbia, South Carolina

Abstract   It is desired to compare the effects of $p \geq 2$ experimental treatments with that of a control or standard treatment in a randomized block setting without assuming normality. Suppose b blocks of size k are to be observed. Steel (1959b) proposed a distribution-free many-one sign test for the case where each of the $p + 1$ treatments are observed exactly once in each block. While distribution-free, the test suffers from requiring that $k = p + 1$, from using a design which is generally inefficient and from not making use of between block information. Hollander (1966) proposed an asymptotically distribution-free many-one signed rank procedure for the same design. It makes use of between block information but has the same design limitations. Spurrier (1988) generalized Steel's procedure to allow for the possibility that the control appears more than once in each block and that not all experimental treatments appear in each block. He also addressed the question of finding asymptotically optimal designs within a broad class of designs. The present manuscript generalizes Hollander's procedure to allow the use of a more general set of designs. The generalized procedure is shown to be asymptotically distribution-free. The question of finding asymptotically optimal designs for use with this generalized procedure is addressed. The generalized Steel, generalized Hollander and normal theory procedures are compared based on a type of asymptotic relative efficiency.

1   Introduction

A common experiment in evaluating proposed drug or medical treatments is to compare $p \geq 2$ experimental treatments with a control or standard treatment. It is desired to determine which, if any, of the experimental treatments perform significantly better than the control while controlling the experimentwise error rate at α. The dual estimation problem is to make simultaneous $100(1 - \alpha)\%$ confidence statements about the amount of improvement that each experimental treatment offers relative to the control. Denote the control by treatment 0 and the experimental treatments by treatment numbers $1, \ldots, p$.

Dunnett (1955) presented the first discussion of comparison with control experiments. He considered the one-way layout under the assumption of i.i.d. normal random errors. Dunnett (1964), Bechhofer (1969), Bechhofer and Tamhane (1983), Spurrier and Nizam (1990) and others have presented additional results in this setting. Steel (1959a) presented a distribution-free comparison with control procedure based on Wilcoxon (1945) rank sum statistics for the one-way layout. Spurrier (1991a) provided an efficient way for computing probability points for Steel's (1959a) method and also generalized the method to allow for the use of any Chernoff-Savage (1958) rank statistic. For normal theory and nonparametric analyses with both α and total sample size fixed, it is desirable in the one-way layout to observe the control more often than the individual experimental treatments. For small values of α, the optimal ratio of sample sizes is approximately $(p)^{1/2}$.

Frequently, comparison with control experiments are done using b blocks of size k. Let Y_{ijh} denote observation h from treatment i in block j. The two-way model is

$$Y_{ijh} = \tau + \alpha_i + \beta_j + \varepsilon_{ijh} \qquad (1.1)$$

where τ is a location parameter, α_i is the ith treatment effect, β_j is the jth block effect and ε_{ijh} is a random error. It is assumed that the error terms are i.i.d. random variables with continuous distribution function F. Denote the distribution function of the difference of two error terms by G. Assume that large values of Y are desirable. It is desired to declare which, if any, of the α_i's are larger than α_0 for $i = 1, \ldots, p$.

Dunnett's (1955) procedure based on normal errors can be easily modified for designs where each experimental treatment appears once and the control appears $c > 0$ times in each block. Setting $c = 1$ yields a randomized complete block design. Bechhofer and Tamhane (1981) presented a class of incomplete block designs known as balanced treatment incomplete blocks designs and developed normal theory comparison with control procedures for these designs. Hedayat et al. (1988) survey results on optimal designs of comparison with control experiments with normal errors. In the two-way model with normal theory analysis, it is desirable to observe the control more often than the individual experimental treatments.

Steel (1959b) presented a distribution-free many-one sign procedure for randomized complete block designs. The procedure is based on counting, for each experimental treatment, the number of blocks in which the observation from the experimental treatment exceeds the observation from the control. The use of randomized complete block designs in comparison with control experiments can be undesirable in that it forces the control to be observed the same number of times as each experimental treatment. It can also be difficult in some experiments to have blocks of size $p+1$ for large values of p. Spurrier (1988) generalized Steel's (1959b) procedure to a class of designs analogous to balanced treatment incomplete block designs and investigated asymptotically optimal designs. In this generalization, the control is allowed to appear more than once in each block and the restriction that all test treatments appear in each block is removed. This procedure will be referred to as the *generalized sign procedure*. It has the disadvantage that it does not make use of between block comparisons.

Nemenyi (1963) presented a many-one signed rank procedure for randomized complete block designs. The procedure is based on the p paired-sample Wilcoxon (1945) signed rank statistics comparing each experimental treatment with the control. The use of the signed rank statistic allows for comparisons between blocks. While the two-sample signed rank statistic is distribution-free, Hollander (1966) showed that the joint distribution of the p signed rank statistics depends on F. Specifically, the null correlation between two signed rank statistics is a linear function of

$$\begin{aligned} \mu(F) &= \Pr(\varepsilon_1 < \varepsilon_2; \varepsilon_1 + \varepsilon_7 < \varepsilon_5 + \varepsilon_6) \text{ and} \qquad (1.2) \\ \lambda(F) &= \Pr(\varepsilon_1 + \varepsilon_4 < \varepsilon_2 + \varepsilon_3; \varepsilon_1 + \varepsilon_7 < \varepsilon_5 + \varepsilon_6), \end{aligned}$$

where $\varepsilon_1, \ldots, \varepsilon_7$ i.i.d. according to F. As $b \to \infty$, this correlation converges to $12\lambda(F) - 3$. Thus, even the limiting distribution is not distribution-free.

For continuous F,

$$3/10 \quad \leq \quad \mu(F) < [(2)^{(1/2)} + 6]/24 \cong 0.3089 \text{ and} \qquad (1.3)$$
$$0.2825 \quad \cong \quad 89/315 \leq \lambda(F) \leq 7/24 \cong 0.2917.$$

The lower bounds are due to Spurrier (1991b) and the upper bounds for for $\mu(F)$ and $\lambda(F)$ are due to Hollander (1967) and Lehmann (1964), respectively. Bounds for the limiting correlation are $123/315 \equiv 0.3905$ and $1/2$. The limiting correlation has a small effect on the large sample simultaneous inference probability point. Hollander (1966) modifies Nemenyi's (1963) procedure to make it asymptotically distribution-free by replacing $\lambda(F)$ in the limiting correlation by a consistent estimate.

The first purpose of this paper is to generalize Hollander's (1966) asymptotically distribution-free procedure so that it can be used with a wider class of designs. The generalized procedure will be referred to as the *generalized signed rank procedure*. The first class of designs, *extended complete block designs*, has $k = p + c$ with each block containing c observations from the control and one from each experimental treatment. The second class of designs, *balanced control incomplete block designs*, has $k < p + c$ with each block containing c observations from the control and the remaining $k - c$ observations are assigned to the experimental treatments according to a balanced incomplete block design in the experimental treatments. Such designs have been proposed by Cox (1958, p. 238) and Pearce (1960). These designs allow the control to be observed more often that the individual experimental treatments and the second type allows the blocks to have fewer than $p + 1$ observations. Extended complete block designs are considered in Section 2. In Section 3, balanced control incomplete block designs are studied.

The second purpose of this paper is to compare the generalized sign, generalized signed rank and the normal theory procedures in terms of a type of asymptotic efficiency. Asymptotically optimal designs of both classes are compared for the generalized sign, generalized signed rank and normal theory procedures in Section 4.

2   Generalized Signed Rank Tests: Extended Complete Block Designs

In extended complete block designs, the control appears $c > 0$ times and each experimental treatment appears once in each block. Hence, $k = p + c$. Let $\psi(a)$ denote the indicator function for $a < 0$.

Denote the difference between observation h from the control and the observation from treatment i in block j by

$$Z_{ijh} = Y_{0jh} - Y_{ij1}$$

for $i = 1, \ldots, p$; $j = 1, \ldots, b$; $h = 1, \ldots, c$. Rank the bc absolute differences $|Z_{ijh}|$ involving treatment i in ascending order. Let R_{ijh} denote the rank of $|Z_{ijh}|$ in this ranking. The generalized signed rank statistic for treatment i and the control is

$$T_i = \sum_{j=1}^{b} \sum_{h=1}^{c} R_{ijh} \psi(Z_{ijh}), \tag{2.1}$$

for $i = 1, \ldots, p$. Thus, T_i is the sum of the ranks assigned to the differences with the treatment i observation exceeding the control observation. Large values of T_i suggest that $\alpha_i > \alpha_0$. The data in Table 1 serves as an example with $b = 15$, $p = 2$ and $c = 2$. Note that $T_1 = 372$ and $T_2 = 423$.

To develop distribution theory for $(T_1, \ldots, T_p)$ it is convenient to use an equivalent formulation of the generalized signed rank statistic

$$T_i = S_{i1} + S_{i2} + S_{i3}, \tag{2.2}$$

where

$$S_{i1} = \sum_{j=1}^{b} \sum_{h=1}^{c} \psi(Z_{ijh}), \tag{2.3}$$

$$S_{i2} = \sum_{j=1}^{b} \sum_{h<h'}^{c} \psi(Z_{ijh} + Z_{ijh'}),$$

$$S_{i3} = \sum_{j<j'}^{b} \sum_{h=1}^{c} \sum_{h'=1}^{c} \psi(Z_{ijh} + Z_{ij'h'}).$$

Hollander (1966) shows for $c = 1$ and $\alpha_0 = \alpha_i = \alpha_{i'}$, for $i \neq i'$ that the marginal distribution of T_i is independent of F but $\mathrm{Cov}(T_i, T_{i'})$ depends on F through $\mu(F)$ and $\lambda(F)$ in (1.2).

If $c > 1$, both the marginal distribution and the covariance depend on F. Let $\varepsilon_1, \ldots, \varepsilon_6$ be i.i.d. according to F. Define

$$
\begin{aligned}
\Delta(F) &= \Pr(2\varepsilon_1 > \varepsilon_2 + \varepsilon_3), \tag{2.4}\\
\omega(F) &= \Pr(2\varepsilon_1 > \varepsilon_2 + \varepsilon_3, \varepsilon_1 + \varepsilon_5 > \varepsilon_2 + \varepsilon_4),\\
\eta(F) &= \Pr(2\varepsilon_1 > \varepsilon_2 + \varepsilon_3, 2\varepsilon_1 > \varepsilon_4 + \varepsilon_5),
\end{aligned}
$$

Table 1: Extended complete block example with $b = 15$, $p = 2$, and $c = 2$.

Block	Observations		
	Control	Trt. 1	Trt. 2
1	437, 453	473	462
2	394, 371	386	397
3	513, 499	522	517
4	411, 426	401	438
5	727, 718	744	712
6	569, 568	580	567
7	390, 401	377	418
8	643, 672	674	682
9	295, 281	309	276
10	371, 376	392	400
11	436, 427	430	450
12	581, 599	632	636
13	466, 485	467	501
14	643, 617	636	665
15	350, 357	389	370

Block	Differences (Ranks)						
	Control		−Trt. 1		Control		−Trt. 2	
1	−36	(28)	−20	(18)	−25	(21)	−9	(7)
2	8	(6)	−15	(13)	−3	(3)	−26	(22)
3	−9	(7)	−23	(20)	−4	(4)	−18	(15)
4	10	(8)	25	(22)	−27	(23)	−12	(9)
5	−17	(15)	−26	(23)	15	(12)	6	(6)
6	−11	(9)	−12	(10)	2	(2)	1	(1)
7	13	(11)	24	(21)	−28	(24)	−17	(14)
8	−31	(25)	−2	(2)	−39	(28)	−10	(8)
9	−14	(12)	−28	(24)	−19	(16)	5	(15)
10	−21	(19)	−16	(14)	−29	(25)	−24	(20)
11	6	(4)	−3	(3)	−14	(11)	−23	(19)
12	−51	(30)	−33	(27)	−55	(30)	−37	(27)
13	−1	(1)	18	(16)	−35	(26)	−16	(13)
14	7	(5)	−19	(17)	−22	(18)	−48	(29)
15	−39	(29)	−32	(26)	−20	(17)	−13	(10)

$$
\begin{aligned}
\delta(F) &= \Pr(2\varepsilon_1 > \varepsilon_2 + \varepsilon_3, \varepsilon_1 + \varepsilon_6 > \varepsilon_4 + \varepsilon_5), \\
\Omega(F) &= \Pr(2\varepsilon_1 > \varepsilon_2 + \varepsilon_3, 2\varepsilon_4 > \varepsilon_2 + \varepsilon_3), \\
\chi(F) &= \Pr(2\varepsilon_1 > \varepsilon_2 + \varepsilon_3, 2\varepsilon_4 > \varepsilon_2 + \varepsilon_5), \\
\nu(F) &= \Pr(2\varepsilon_1 > \varepsilon_2 + \varepsilon_3, \varepsilon_4 + \varepsilon_5 > \varepsilon_2 + \varepsilon_6), \\
\Lambda(F) &= \Pr(2\varepsilon_1 > \varepsilon_2 + \varepsilon_3, \varepsilon_4 > \varepsilon_2).
\end{aligned}
$$

Table 2 gives bounds for the functions in (2.4) and their values for uniform, normal and exponential distributions. These bounds were derived using techniques similar to those in Spurrier (1991b).

Table 2: Parameter bounds and parameter values for selected distributions.

Parameter	Lower Bound	Exponential	Normal	Uniform	Upper Bound
$\Delta(F)$	1/3	0.4444	0.5000	0.5000	2/3
$\omega(F)$	4/15	0.3264	0.3549	0.3563	13/30
$\eta(F)$	1/5	0.3129	0.3661	0.3833	8/15
$\delta(F)$	2/9	0.2882	0.3169	0.3233	2/5
$\Omega(F)$	1/6	0.2500	0.3041	0.2917	1/2
$\chi(F)$	2/15	0.2222	0.2767	0.2708	7/15
$\nu(F)$	7/36	0.2533	0.2827	0.2792	13/36
$\lambda(F)$	89/315	0.2894	0.2902	0.2909	7/24
$\Lambda(F)$	5/24	0.2667	0.2966	0.2917	3/8
$\mu(F)$	3/10	0.3056	0.3075	0.3083	$[2^{1/2} + 6]/24$

After a lengthy but straightforward argument, if $\alpha_0 = \alpha_i$ then

$$
\begin{aligned}
E(T_i) &= bc/2 + bc(c-1)\Delta(F)/2 + b(b-1)c^2/4 \text{ and} \quad (2.5)\\
\mathrm{Var}(T_i) &= bc[c^3(b-1)(13-6b) + 4c^2(b-1)(2b-3) \\
&\quad + 2c(2b+3) - 4]/24 \\
&\quad + \binom{c}{2}\{[-9 + 5bc - c^2 b(b-1)]\Delta(F) - b\binom{c}{2}\Delta^2(F) \\
&\quad + 8\binom{b}{2}c\omega(F) + 12\binom{b}{3}c^2\lambda(F)\} \\
&\quad + \binom{c}{3}[12\binom{b}{2}c\delta(F) - 12b\Lambda(F) - 3b\Omega(F)] + \binom{c}{4}6b\eta(F)
\end{aligned}
$$

and if $\alpha_0 = \alpha_i = \alpha_{i'}$, for $i \neq i'$ then

$$
\begin{aligned}
\mathrm{Cov}(T_i, T_{i'}) = \ & b\{\binom{c}{2}[-2(c(b-1)+1)\Delta(F) \\
& -(2c-3)\Delta^2(F) + \Omega(F) \\
& +2(c-2)\chi(F) + 4(b-1)c\nu(F) + 4\Lambda(F)] \\
& +(b-1)c^2[c(b-1)-1]\lambda(F) + 2(b-1)c^2\mu(F) \\
& +[-6(b-1)^2c^3 - 5(b-1)c^2 + 2c]/24\}.
\end{aligned}
\tag{2.6}
$$

While these moments can depend on up to ten parameters, the following theorem shows that the limiting distribution depends only upon $\lambda(F)$ for all c.

Theorem 2.1

Let $q \leq p$ be a positive integer. For extended complete block designs, if $\alpha_0 = \alpha_1 = \ldots = \alpha_q$, the joint distribution of

$$
Z_i = [T_i - \binom{bc+1}{2}/2]/\{b^3c^3[(c-1)\lambda(F) + (4-3c)/12]\}^{1/2},
$$

$i = 1, \ldots, q$, converges to a q-variate normal with zero means, unit variances and common correlation

$$
\rho = \frac{12\lambda(F) - 3}{12(c-1)\lambda(F) + 4 - 3c}
$$

as $b \to \infty$. Moreover, if $\alpha_i < \alpha_0$, the probability that Z_i is less than an arbitrary constant $\to 1$ as $b \to \infty$.

Proof of Theorem 2.1

The proof involves considering the $p + c$ observations within a block as a single multivariate observation and the T_i's as linear combinations of U-statistics in terms of these b multivariate observations. For $i = 1, \ldots, q$,

$$
T_i = bU_{i1} + bU_{i2} + \binom{b}{2}U_{i3},
\tag{2.7}
$$

where $U_{i1} = S_{i1}/b$ is a U-statistic of order 1 with kernel $\sum_{h=1}^{c} \psi(Z_{ijh})$, $U_{i2} = S_{i2}/b$ is a U-statistic of order 1 with kernel $\sum_{h<h'}^{c} \psi(Z_{ijh} + Z_{ijh'})$ and $U_{i3} = S_{i3}/\binom{b}{2}$ is a U-statistic of order 2 with kernel $\sum_{h=1}^{c}\sum_{h'=1}^{c} \psi(Z_{ijh} + Z_{ij'h'})$. Now,

$$
\begin{aligned}
Z_i &= \frac{\{b(U_{i1} - c/2) + b[U_{i2} - \binom{c}{2}/2] + \binom{b}{2}(U_{i3} - c^2/2)\}}{\{b^3c^3[(c-1)\lambda(F) + (4-3c)/12]\}^{1/2}} \\
&= \frac{b^{1/2}(U_{i3} - c^2/2)}{(2\{c^3[(c-1)\lambda(F) + (4-3c)/12]\})^{1/2}} + o(1), i = 1, \ldots, q.
\end{aligned}
\tag{2.8}
$$

The result follows by applying Lehmann's (1963) generalized U-statistic theorem. The argument for the case of $\alpha_i < \alpha_0$ is similar. ∎

Let $d_{\alpha,p,\rho}$ denote the upper α probability point of the maximum of p standard normal random variables with common correlation ρ. Values of $d_{\alpha,p,\rho}$ can be found by interpolating in Bechhofer and Dunnett (1988, Table A) with ∞ degrees of freedom.

Lehmann (1964) gives an unbiased, consistent estimate of $\lambda(F)$ for the case of $c = 1$. The estimate $\hat{\lambda}(F)$ is the proportion of sextuples (i, i', i'', j, j', j'') consisting of three distinct treatments and three distinct blocks such that

$$Y_{ij1} + Y_{i'j'1} < Y_{i'j1} + Y_{ij'1} \text{ and } Y_{ij1} + Y_{i''j''1} < Y_{i''j1} + Y_{ij''1}.$$

As the number of sextuples can be quite large, Lehmann noted one might be satisfied with estimating $\lambda(F)$ based on a subset of the sextuples.

For $c > 1$, the estimator $\hat{\lambda}(F)$ can be generalized. A conveniently programmed subset approach is to treat observation h from the control for each block as an observation from "treatment" $p + h$, $h = 1, \ldots, c$ and then compute Lehmann's $\hat{\lambda}(F)$ estimate on the $p + c$ "treatments". For the example data in Table 1, this approach yields

$$\hat{\lambda}(F) = 18910/65520 = 0.2886142.$$

Let r denote the consistent estimate of ρ found by replacing $\lambda(F)$ in the expression for ρ by $\hat{\lambda}(F)$. A large sample decision rule is to declare $\alpha_i > \alpha_0$ if

$$T_i > \binom{bc + 1}{2}/2 + d_{\alpha,p,r}\{b^3c^3[(c - 1)\hat{\lambda}(F) + (4 - 3c)/12]\}^{1/2}. \qquad (2.9)$$

Theorem 2.2

For extended complete block designs, the decision rule given in (2.9) asymptotically controls the experimentwise Type 1 error rate at α.

Proof of Theorem 2.2

Consider the situation where $\alpha_0 = \alpha_1 = \ldots = \alpha_q$, where $q \leq p$, and the rest, if any, of the α_i's are greater than α_0. A Type 1 error occurs if at least one of $T_1, \ldots, T_q$ are greater than the critical point given in (2.9). The critical point depends on the desired level of significance α, the number of experimental treatments p and the value of $\hat{\lambda}(F)$.

The values of the α and p are fixed for all configurations of the α_i's. The estimator $\hat{\lambda}(F)$ is location invariant. That is adding a constant to all observations from a given treatment has no effect on $\hat{\lambda}(F)$. It is also consistent for $\lambda(F)$ which depends on F but not on the configuration of the α_i's.

Recall that the statistic T_i only involves the observations from the control and experimental treatment i. The events

$$B_q = \{\max(T_1, \ldots, T_q) > \text{critical point}\}$$

are nested for $q = 1, \ldots, p$ with $B_1 \rightarrow B_2 \rightarrow \ldots \rightarrow B_p$. It follows that

$$\Pr(B_q | \alpha_0 = \alpha_1 = \ldots = \alpha_q)$$

is an increasing function of q and hence is maximized with $q = p$. From Theorem 2.1, $\Pr(B_p | \alpha_0 = \alpha_1 = \ldots = \alpha_p)$ goes to α asymptotically. The proof is completed by noting that as $b \rightarrow \infty$, the limit of

$$\Pr(B_q | \alpha_0 \geq \alpha_i, i = 1, \ldots, q \text{ with at least one inequality})$$

is less than the limit of $\Pr(B_q | \alpha_0 = \alpha_1 = \ldots = \alpha_q)$. ∎

For the example data, $r = 0.3166458$. With $\alpha = 0.05$, the probability point $d_{.05,2,r} = 1.93706$ and one declares $\alpha_i > \alpha_0$ if $T_i > 343.7$. Thus with $\alpha = 0.05$, both α_1 and α_2 are declared to be larger than α_0.

A simpler approach which is asymptotically conservative is to replace r by the lower bound for ρ, $1/(c + 64/41)$, and $\lambda(F)$ by its upper bound, $7/24$, in (2.9). These substitutions produce a slight increase in the critical point. For the example data, the critical point using this method is 345.2.

3   Generalized Signed Rank Tests: Balanced Control Incomplete Block Designs

In balanced control incomplete block designs, the control appears $c > 0$ times in each of the b blocks and the remaining $k - c < p$ observations are allocated to the experimental treatments according to a balanced incomplete block design. Thus, each experimental treatment appears once in $n = b(k - c)/p$ blocks and does not appear in the other blocks. Each pair of experimental treatments appears together in $n(k - c - 1)/(p - 1)$ blocks.

The generalized signed rank statistic for treatment i and the control is identical to (2.1) except that the outer sum is over the blocks in which experimental treatment i appears. If $\alpha_0 = \alpha_i$, the mean and variance of T_i are found by replacing b in (2.5) by n.

Theorem 3.1

Let $q \leq p$ be as positive integer. For balanced control incomplete block designs, if $\alpha_0 = \alpha_1 = \ldots = \alpha_q$, the joint distribution of

$$Z_i = \frac{T_i - \binom{nc+1}{2}/2}{\{n^3 c^3 [(c-1)\lambda(F) + (4-3c)/12]\}^{1/2}}$$

$i = 1, \ldots, q$, converges to a q-variate normal with zero means, unit variances and common correlation

$$\rho^* = \frac{k - c - 1}{p - 1} \frac{12\lambda(F) - 3}{12(c-1)\lambda(F) + 4 - 3c}$$

as $b \to \infty$. Moreover, if $\alpha_i < \alpha_0$, the probability that Z_i is less than an arbitrary constant $\to 1$ as $b \to \infty$.

Sketch of Proof of Theorem 3.1

Consider the joint limiting distribution of Z_1 and Z_2. The proof involves considering the k observations within a block as a single multivariate observation. The blocks containing both treatments 1 and 2, the blocks containing treatment 1 but not 2, and the blocks containing treatment 2 but not 1 are thought of as three independent samples of multivariate observations.

For $i = 1, 2$

$$T_i = S_{i1} + S_{i2} + S_{i3},$$

where S_{i1}, S_{i2} and S_{i3} are identical to (2.3) except that the outer sums are over the blocks in which experimental treatment i appears. Let R_{11}, R_{12} and R_{13} denote the contribution to S_{11}, S_{12} and S_{13}, respectively, from blocks containing both treatments 1 and 2. Let R_{14}, R_{15} and R_{16} reflect the contribution to S_{11}, S_{12} and S_{13}, respectively from blocks containing treatment 1 but not treatment 2. Let R_{17} reflect the contribution to S_{13} when block j contains both treatment 1 and 2 and block j$'$ contains treatment 1 but not treatment 2. Define $R_{21}, \ldots, R_{27}$ in an analogous manner with the roles of treatments 1 and 2 reversed.

One can now write $T_i = R_{i1} + \ldots + R_{i7}$, as a linear combination of 7 U-statistics in a manner analogous to (2.7). The U-statistics are functions of the three independent samples of multivariate observations. The limiting distribution of (Z_1, Z_2) follows using arguments similar to those in the proof of Theorem 2.1.

The proof of the limiting distribution of $(Z_1, \ldots, Z_q)$ is similar but involves a finer partitioning of S_{i1}, S_{i2} and S_{i3} to reflect the presence or absence of the various experimental treatments in each block. ∎

Let w denote the consistent estimate of ρ^* found by replacing $\lambda(F)$ by $\hat{\lambda}(F)$. A large sample decision rule is to declare $\alpha_i > \alpha_0$ if T_i exceeds

$$\binom{nc+1}{2}/2 + d_{\alpha,p,w}\{n^3 c^3[(c-1)\hat{\lambda}(F) + (4-3c)/12]\}^{1/2}. \qquad (3.1)$$

Theorem 3.2

For balanced control incomplete block designs, the decision rule given in (3.1) asymptotically controls the experimentwise Type 1 error rate at α.

The proof is analogous to the proof of Theorem 2.2.

A simpler approach which is asymptotically conservative is to replace w by the lower bound for ρ^*,

$$\frac{k-c-1}{p-1}\frac{1}{c+64/41}$$

and $\lambda(F)$ by its upper bound, 7/24, in (3.1).

4 Asymptotic Comparison of Designs and Statistics

In designing comparison with control experiments, we wish to choose a design and a statistic so as to maximize the marginal power

$$\Pr(\text{declare } \alpha_i > \alpha_0 | \alpha_i > \alpha_0)$$

subject to an experimentwise error rate of α. Without loss of generality, let us refer to marginal power in terms of treatment 1.

We are left with several questions regarding generalized signed rank statistics. What is the optimal choice of c for use with generalized signed rank statistics in extended complete block designs and in balanced treatment incomplete block designs? Do these optimal designs coincide with the optimal designs for generalized sign statistics and for normal theory statistics? How does the performance of the generalized signed rank statistic under its optimal design compare with the performance of the generalized sign statistic and normal theory statistic under their optimal designs?

Spurrier (1988) introduced the concept of modified asymptotic efficiency for comparing two comparison with control procedures. This is a natural extension of Pitman efficiency to the comparison with a control problem. In comparing designs and statistics we will consider a sequence of alternatives

$$\alpha_1 = \alpha_0 + \theta/(bk)^{1/2}, \qquad (4.1)$$

where θ is a positive constant selected such that the marginal power converges to a constant $1 - \beta$ as $b \to \infty$ and bk is the total sample size. It is desirable to achieve the marginal power $1 - \beta$ for the smallest possible value of θ. If design/statistic 1 yields the constant θ_1 and design/statistic 2 yields the constant θ_2, the modified asymptotic efficiency, MAE, of the first design/statistic relative to the second design/statistic is

$$MAE(1,2) = (\theta_2/\theta_1)^2. \tag{4.2}$$

The interpretation of $MAE(1,2)$ is identical to the interpretation of Pitman efficiency.

Let Φ denote the distribution function of the standard normal and let z_β be the constant such that $\Phi(z_\beta) = \beta$. Unlike Pitman efficiency, the value of $MAE(1,2)$ can depend upon the choice of α and β through $d_{\alpha,p,\rho} - z_\beta$. In order to compare the generalized signed rank, generalized sign and normal theory procedures it is necessary to assume that F and G have densities, denoted by f and g, respectively, such that Pitman efficiencies among the three tests exists in the paired two-sample location problem.

For extended complete block designs, $k = (p+c)$ and from (2.9) marginal power is

$$\Pr[T_1 > \binom{bc+1}{2}/2 \tag{4.3}$$

$$+d_{\alpha,p,r}\{b^3c^3[(c-1)\hat{\lambda}(F) + (4-3c)/12]\}^{1/2}] =$$

$$\Pr[bU_{11} + bU_{12} + \binom{b}{2}U_{13} > \binom{bc+1}{2}/2$$

$$+d_{\alpha,p,r}\{b^3c^3[(c-1)\hat{\lambda}(F) + (4-3c)/12]\}^{1/2}] =$$

$$\Pr[b\{U_{11} - cE[\psi(Z_{111})]\}$$

$$+b\{U_{12} - \binom{c}{2}E[\psi(Z_{111} + Z_{112})]\}$$

$$+\binom{b}{2}\{U_{13} - c^2E[\psi(Z_{111} + Z_{121})]\}$$

$$> bc\{1/2 - E[\psi(Z_{111})]\}$$

$$+b\binom{c}{2}\{1/2 - E[\psi(Z_{111} + Z_{112})]\}$$

$$+\binom{b}{2}c^2\{1/2 - E[\psi(Z_{111} + Z_{121})]\}$$

$$+d_{\alpha,p,r}\{b^3c^3[(c-1)\hat{\lambda}(F) + (4-3c)/12]\}^{1/2}].$$

Dividing both sides of the last probability statement by $(bc)^{3/2}$ and consolidating $o(1)$ terms yields the following expression for marginal power:

$$\Pr[(b \quad - \quad 1)\{U_{13} - c^2 E[\psi(Z_{111} + Z_{121})]\}/2(bc^3)^{1/2}] \qquad (4.4)$$
$$> \quad d_{\alpha,p,r}[(c-1)\hat{\lambda}(F) + (4-3c)/12]^{1/2} + o(1)$$
$$+ \quad (b-1)(c/b)^{1/2}\{1/2 - E[\psi(Z_{111} + Z_{121})]\}/2].$$

With the sequence of alternatives (4.1), the marginal power converges to

$$1 - \Phi\{d_{\alpha,p,\rho} - \theta[c/(p+c)]^{1/2} \qquad (4.5)$$
$$\times \int_{-\infty}^{\infty} g(x)dG(x)/[(c-1)\lambda(F) + (4-3c)/12]^{1/2}\}$$

as $b \to \infty$. Setting (4.5) equal to $1 - \beta$ and solving for θ yields

$$\theta_{SR} = (d_{\alpha,p,\rho} - z_\beta)(p+c)^{1/2} \qquad (4.6)$$
$$\times[(c-1)\lambda(F) + (4-3c)/12]^{1/2}/[c^{1/2}\int_{-\infty}^{\infty} g(x)dG(x)].$$

For fixed α, β and p, one wishes to select c such that θ_{SR} is minimized. At times this minimization depends on the unknown value of $\lambda(F)$.

Table 3 lists the values of c which minimize θ_{SR} for $\alpha = 0.05$, $\beta = 0.10$, $p = 2(1)10$ and $\lambda(F)$ corresponding to the lower and upper bounds and to the Cauchy, exponential, normal and uniform distributions. It should be noted from Table 3 that $\lambda(F)$ has only a minor effect on the optimal choice of c. The modified asymptotic efficiency of the design with the optimal choice of c relative to the design with $c = 1$ for $\alpha = 0.05$ and $\beta = 0.10$ is also given in Table 3. The gain in efficiency by allowing c to be greater than 1 can be considerable for large p and small $\lambda(F)$.

The factor $(d_{\alpha,p,\rho} - z_\beta)$ in θ_{SR} is relatively insensitive to small changes in ρ. An approximation to the asymptotically optimal choice of c is

$$[p(4 - 12\lambda)/(12\lambda - 3)]^{1/2}$$

which maximizes $\theta_{SR}/(d_{\alpha,p,\rho} - z_\beta)$.

Let us now compare the generalized signed rank procedure with the generalized sign procedure and the normal theory procedure. For the generalized sign procedure with extended complete block designs and the sequence of alternatives (4.1), Spurrier (1988) shows that the value of θ necessary to achieve asymptotic marginal power of $1 - \beta$ is

$$\theta_S = (d_{\alpha,p,1/(c+2)} - z_\beta)[(c+2)(c+p)/(12c)]^{1/2}/\int f(x)dF(x) \qquad (4.7)$$

Table 3: Asymptotically optimal c for extended complete block designs with generalized signed rank statistics and modified asymptotic efficiency of design with optimal c relative to design with $c = 1$, $\alpha = 0.05$, $\beta = 0.10$.

		$\lambda(F)$				
		89/315		0.2879		0.2894
		lower bound		Cauchy		exponential
p	c	MAE	c	MAE	c	MAE
2	2	1.07	2	1.02	2	1.01
3	2	1.14	2	1.08	2	1.07
4	2	1.19	2	1.13	2	1.11
5	3	1.24	2	1.16	2	1.14
6	3	1.28	3	1.19	3	1.16
7	3	1.32	3	1.22	3	1.19
8	3	1.35	3	1.24	3	1.22
9	4	1.38	3	1.27	3	1.24
10	4	1.41	3	1.28	3	1.25

		$\lambda(F)$				
		0.2902		0.2909		7.24
		normal		uniform		upper bound
p	c	MAE	c	MAE	c	MAE
2	2	1.00	1	1.00	1	1.00
3	2	1.06	2	1.05	2	1.05
4	2	1.10	2	1.09	2	1.09
5	2	1.13	2	1.12	2	1.11
6	2	1.15	2	1.14	2	1.14
7	3	1.18	3	1.17	3	1.15
8	3	1.20	3	1.19	3	1.18
9	3	1.22	3	1.21	3	1.20
10	3	1.24	3	1.23	3	1.21

Table 4: $MAE(SR, S)$, $MAE(SR, N)$, $MAE(S, N)$ for extended complete block designs using asymptotically optimal values of c for each statistic, α = 0.05 and β = 0.10.

p		Uniform	exponential	normal	Cauchy
				F	
2	MAE(SR,S)	1.204	0.681	1.354	0.772
	MAE(SR,N)	0.888	1.508	0.954	∞
	MAE(S,N)	0.738	2.214	0.705	∞
3	MAE(SR,S)	1.199	0.683	1.357	0.690
	MAE(SR,N)	0.894	1.526	0.966	∞
	MAE(S,N)	0.745	2.236	0.712	∞
4	MAE(SR,S)	1.172	0.667	1.325	0.674
	MAE(SR,N)	0.894	1.526	0.965	∞
	MAE(S,N)	0.763	2.289	0.729	∞
6	MAE(SR,S)	1.133	0.645	1.281	0.657
	MAE(SR,N)	0.894	1.527	0.965	∞
	MAE(S,N)	0.789	2.367	0.754	∞
8	MAE(SR,S)	1.109	0.635	1.258	0.647
	MAE(SR,N)	0.897	1.540	0.972	∞
	MAE(S,N)	0.809	2.426	0.772	∞
10	MAR(SR,S)	1.096	0.627	1.243	0.639
	MAE(SR,N)	0.897	1.541	0.972	∞
	MAE(S,N)	0.819	2.456	0.782	∞

and gives a table of asymptotically optimal c values. These values of c are 2 for $p = 2, 3$; 3 for $p = 4, 5, 6$; and 4 for $p = 7, \ldots, 10$. These optimal c values are approximately equal to $(2p)^{1/2}$ and are greater than or equal to the asymptotically optimal c values given in Table 3.

For normal theory analysis, it follows through arguments analogous to those in (4.3)-(4.6) that the value of θ necessary to achieve asymptotic marginal power of $1 - \beta$ is

$$\theta_N = (d_{\alpha, p, 1/(c+1)} - z_\beta)\sigma/[(p + c)(c + 1)/c]^{1/2}, \qquad (4.8)$$

provided that σ, the standard deviation of F, exists. The asymptotically optimal values of c are 1 for $p = 2$; 2 for $p = 3, \ldots, 6$; and 3 for $p = 7, \ldots, 10$. These agree with the asymptotically optimal c values for the signed rank statistic under the uniform distribution. These optimal c values are approximately $p^{1/2}$.

The modified asymptotic efficiencies of the generalized signed rank procedure relative to the generalized sign procedure and relative to the normal theory procedure and of the generalized sign procedure relative to the normal theory procedure are:

$$
\begin{aligned}
MAE(SR, S) &= (\theta_S/\theta_{SR})^2 \\
MAE(SR, N) &= (\theta_N/\theta_{SR})^2 \\
MAE(S, N) &= (\theta_N/\theta_S)^2
\end{aligned}
\qquad (4.9)
$$

respectively. These MAE values depend on α, β, F, c and p. Table 4 gives values of $MAE(SR, S)$, $MAE(SR, N)$, and $MAE(S, N)$ for $\alpha = 0.05$ and $\beta = 0.10$ where the values of θ_{SR}, θ_S and θ_N are computed using the (possibly unequal) asymptotically optimal values of c for the particular statistics.

The $MAE(SR, N)$ values for $p = 2$ in Table 4 closely approximate the Pitman efficiency values for the two-sample location problem ($p = 1$, $c = 1$). As p increases the performance of the generalized signed rank procedure improves slightly relative to the normal theory procedure. The performance of the generalized sign procedure relative to the generalized signed rank and to the normal theory procedures is better than the analogous Pitman efficiency values for the two-sample location problem. The performance of the generalized sign procedure relative to the generalized signed rank and to the normal theory procedures improves as p increases.

Let us now turn to balanced control incomplete block designs with fixed block size $k \leq p + 1$. It follows by arguments analogous to (4.3)-(4.6) that the value of θ yielding a marginal power of $1 - \beta$ for the generalized signed

Table 5: Asymptotically optimal c for balanced control incomplete block designs with generalized signed rank statistics, $\alpha = 0.05$, $\beta = 0.10$.

		$\lambda(F)$					
		89/315 lower bound	0.2879 Cauchy	0.2894 exponential	0.2902 normal	0.2909 uniform	7/24 upper bound
k	p	c	c	c	c	c	c
4	3-12	1	1	1	1	1	1
5	4	2	2	1	1	1	1
	5-6	2	2	2	1	1	1
	7-9	2	2	2	2	1	1
	10-12	2	2	2	2	2	1
6	5-12	2	2	2	2	2	2
7	6-12	2	2	2	2	2	2
8	7-12	2	2	2	2	2	2
9	8-12	2	2	2	2	2	2

rank procedure with the sequence of alternatives (4.1) is

$$\theta^*_{SR} = \frac{(d_{\alpha,p,\rho^*} - z_\beta)\{kp[(c-1)\lambda(F) + (4-3c)/12]\}^{1/2}}{[c(k-c)]^{1/2}\int_{-\infty}^{\infty} g(x)dG(x)}. \qquad (4.10)$$

The numerator of θ^*_{SR} is increasing in c and the denominator is symmetric in c about $k/2$. Thus, the asymptotically optimal choice of $c \leq k/2$. Hence, the asymptotically optimal choice of c is 1 for $k = 2,3$. For $k > 3$, it is necessary to make a search over $1 < c \leq k/2$.

Table 5 lists the values of c which minimize θ^*_{SR} for $\alpha = 0.05, \beta = 0.10, k = 4(1)9$ and $p = k - 1(1)12$ for the values of $\lambda(F)$ corresponding to the lower and upper bounds and to the Cauchy, exponential, normal and uniform distributions. For $p = k - 1$ one can compute the MAE of the generalized signed rank procedure based on the asymptotically optimal c relative to the generalized signed rank procedure with $c = 1$. Recall that MAE depends upon α, β and $\lambda(F)$. With $\alpha = 0.05$ and $\beta = 0.10$ the

Table 6: Asymptotically optimal c for balanced control incomplete block designs with generalized sign and normal theory statistics, $\alpha = 0.05$, $\beta = 0.10$.

	generalized sign		normal theory	
	c	p	c	p
4	1	3-12	1	3-12
5	2	4-12	2	4
			1	5-12
6	2	5-12	2	5-7
			1	8-12
7	2	6-12	2	7-11
			1	12
8	2	7-12	2	8-12
9	3	8-12	2	8-12

values of MAE for $k = 5$ range from 1.00 at the upper bound for $\lambda(F)$ to 1.06 at the lower bound for $\lambda(F)$ and for $k = 9$ range from 1.12 at the upper bound for $\lambda(F)$ to 1.23 at the lower bound for $\lambda(F)$. Thus, the amount of possible improvement by allowing c to be greater than 1 increases with the block size k.

It follows from Spurrier (1988) that the value of θ needed to achieve the asymptotic marginal power of $1 - \beta$ using the generalized sign procedure with balanced control incomplete block designs is θ_S^* given by

$$(d_{\alpha,p,(k-c-1)/[(p-1)(c+2)]} - z_\beta)\{kp(c+2)/[12c(k-c)]\}^{1/2} / \int f(x)dF(x).$$

$$(4.11)$$

The corresponding value of θ for normal theory analysis is

$$\theta_N^* = (d_{\alpha,p,(k-c-1)/[c(p-2)+k-1]} - z_\beta)\sigma \frac{[k^2 p(c(p-2) + k - 1)]^{1/2}}{[c(k-c)(kp - p - c)]^{1/2}} \quad (4.12)$$

provided that σ exists. Table 6 gives the asymptotically optimal values of c for the generalized sign procedure and the normal theory procedure with $\alpha = 0.05$ and $\beta = 0.10$. The values of c in Tables 5 and 6 are often in agree-

Table 7: MAE(SR,S), MAE(SR,N) and MAE(S,N) for balanced control incomplete block designs using asymptotically optimal values of c for each statistic, $\alpha = 0.05$ and $\beta = 0.10$.

k	p		uniform	exponential	normal	Cauchy
					F	
4	4	MAE(SR,S)	1.345	0.756	1.513	0.755
		MAE(SR,N)	0.801	1.351	0.861	∞
		MAE(S,N)	0.596	1.787	0.569	∞
4	8	MAE(SR,S)	1.337	0.752	1.504	0.752
		MAE(SR,N)	0.691	1.165	0.742	∞
		MAE(S,N)	0.517	1.550	0.493	∞
4	12	MAE(SR,S)	1.335	0.751	1.502	0.751
		MAE(SR,N)	0.657	1.109	0.706	∞
		MAE(S,N)	0.492	1.476	0.470	∞
6	6	MAE(SR,S)	1.197	0.681	1.354	0.689
		MAE(SR,N)	0.828	1.414	0.895	∞
		MAE(S,N)	0.692	2.076	0.661	∞
6	8	MAE(SR,S)	1.195	0.680	1.352	0.689
		MAE(SR,N)	0.788	1.345	0.851	∞
		MAE(S,N)	0.659	1.977	0.629	∞
6	12	MAE(SR,S)	1.194	0.680	1.351	0.688
		MAE(SR,N)	0.716	1.222	0.773	∞
		MAE(S,N)	0.599	1.798	0.572	∞
8	8	MAE(SR,S)	1.199	0.682	1.356	0.690
		MAE(SR,N)	0.837	1.428	0.904	∞
		MAE(S,N)	0.698	2.094	0.666	∞
8	12	MAE(SR,S)	1.196	0.681	1.353	0.689
		MAE(SR,N)	0.783	1.337	0.846	∞
		MAE(S,N)	0.655	1.964	0.625	∞

ment. When there are differences, the generalized signed rank procedure uses a larger value of c than the normal theory procedure and a smaller value of c than the generalized sign procedure. The MAE values for balanced control incomplete block designs are defined in terms of θ_{SR}^*, θ_S^* and θ_N^* in a manner analogous to (4.9).

Table 7 gives $MAE(SR, S)$, $MAE(SR, N)$ and $MAE(S, N)$ for $\alpha = 0.05$, $\beta = 0.10$ and selected values of k and p. The performance of the generalized signed rank procedure relative to the normal theory procedure is worse than in the two-sample location problem. This deterioration is particularly noticeable when p is much larger than k. The performance of the generalized sign procedure relative to the normal theory procedure is worse than in the two-sample location problem for $k = p = 4$ and slightly better than in the two-sample location problem for $k = p = 6$ and 8. For fixed k, the performance of the generalized sign procedure deteriorates relative to the normal theory procedure as p increases. The performance of the generalized signed rank procedure relative to the generalized sign procedure approximates the results in the two sample location problem for $k = 4$. For $k = 6$ and 8, the performance of the generalized signed rank procedure relative to the generalized sign procedure is not as good as in the two-sample location problem.

Acknowledgment

The author thanks the referees for pointing out an error in the original version of this paper.

References

Bechhofer, R.E. (1969). Optimal allocation of observations when comparing several treatments with a control. In *Multivariate Analysis II* (P.R. Krishnaiah, ed.) 465 - 473, Academic Press, New York.

Bechhofer, R.E. and Dunnett, C.W. (1988). Tables of percentage points of multivariate Student t distributions. In *Selected Tables in Mathematical Statistics*, No. 11. American Mathematics Society, Providence, R.I.

Bechhofer, R.E. and Tamhane, A.C. (1981). Incomplete block designs for comparing treatments with a control: general theory. *Technometrics* **23** 45 - 57.

Bechhofer, R.E. and Tamhane, A.C. (1983). Incomplete block designs for comparing treatments with a control (II): optimal designs for one-sided comparisons when $p = 2(1)6$, $k = 2$ and $p = 3$, $k = 3$. *Sankhya B* **45** 193 - 224.

Chernoff, H. and Savage, I.R. (1958). Asymptotic normality and efficiency of certain nonparametric test statistics. *Ann. Math. Statist.* **29** 972 - 994.

Cox, D.R. (1958). *Planning of Experiments.* John Wiley and Sons, New York.

Dunnett, C.W. (1955). A multiple comparison procedure for comparing several treatments with a control. *Jour. Amer. Statist. Assoc.* **50** 1096 - 1121.

Dunnett, C.W. (1964). New tables for multiple comparisons with a control. *Biometrics* **20** 482 - 491.

Hedayat, A.S., Jacroux, M. and Majumdar, D. (1988). Optimal designs for comparing treatments with controls. *Statistical Science* **3** 462 - 491.

Hollander, M. (1966). An asymptotically distribution-free multiple comparison procedure-treatments vs. control. *Ann. Math. Statist.* **37** 735 - 738.

Hollander, M. (1967). Rank tests for randomized blocks when the alternatives have an a priori ordering. *Ann. Math. Statist.* **38** 867 - 877.

Lehmann, E.L. (1963). Robust estimation in analysis of variance. *Ann. Math. Statist.* **34** 957 - 966.

Lehmann, E.L. (1964). Asymptotically nonparametric inference in some linear models with one observation per cell. *Ann. Math. Statist.* **35** 726 - 734.

Nemenyi, P. (1963). *Distribution-Free Multiple Comparisons.* Ph.D. Dissertation, Princeton University.

Pearce, S.C. (1960). Supplemented balance. *Biometrika* **47** 263 - 271.

Spurrier, J.D. (1988). Generalizations of Steel's treatments-versus-control multivariate sign test. *Jour. Amer. Statist. Assoc.* **83** 471 - 476.

Spurrier, J.D. (1991a). Generalizations of Steel's many-one rank sum test. *Jour. Nonpara. Statist.* To appear.

Spurrier, J.D. (1991b). Improved bounds for moments of some rank statistics. *Communicat. Statist. - Theory and Methods* **20** 2603 - 2608.

Spurrier, J.D. and Nizam, A. (1990) Sample size allocation for simultaneous inference in comparison with control experiments. *Jour. Amer. Statist. Assoc.* **85** 181 - 186.

Steel, R.G.D. (1959a). A multiple comparison rank sum test: treatments versus control. *Biometrics* **15** 560 - 572.

Steel, R.G.D. (1959b). A multiple comparison sign test: Treatments versus control. *Jour. Amer. Statist. Assoc.* **54** 767 - 775.

Wilcoxon, F. (1945). Individual comparisons by ranking methods. *Biometrics* **1** 80 - 83.

Chapter 8

Plackett's Identity, Its Generalizations, and Their Uses

SATISH IYENGAR   Department of Mathematics and Statistics, University of Pittsburgh, Pittsburgh, Pennsylvania

Abstract   Let $\phi_p(x - \mu; \Sigma)$ be the density of a multivariate normal distribution in p dimensions with mean μ and covariance $\Sigma = (\rho_{ij})$. The following partial differential equation,

$$\frac{\partial}{\partial \rho_{ij}} \phi_p(x - \mu; \Sigma) = \frac{\partial^2}{\partial x_i \partial x_j} \phi_p(x - \mu; \Sigma),$$

satisfied by ϕ_p is known as Plackett's identity. It has been used to prove probability inequalities, and to motivate approximations to certain probabilities. In this paper, we review these applications, discuss generalizations to elliptically contoured distributions and to discrete distributions. We also discuss connections between Plackett's identity and classical work on spherical simplices.

1   The Identity

Let $\phi_p(x - \mu; \Sigma)$ be the density of a p-dimensional multivariate normal distribution with mean vector μ and covariance matrix $\Sigma = (\rho_{ij})$.

121

While investigating the problem of evaluating certain multivariate normal probabilities, Plackett (1954) proved that ϕ_p satisfies the following partial differential equation: for $i \neq j$,

$$\frac{\partial}{\partial \rho_{ij}} \phi_p(x - \mu; \Sigma) = \frac{\partial^2}{\partial x_i \partial x_j} \phi_p(x - \mu; \Sigma). \tag{1.1}$$

This equation is now known as Plackett's identity (Tong 1980, 1990). Its proof follows from the straightforward, but lengthy, evaluation of both sides of (1.1). The following proof is simpler. Since the density is the inverse of its characteristic function,

$$\phi_p(x - \mu; \Sigma) = \frac{1}{(2\pi)^p} \int exp[-it'x + it'\mu - \frac{1}{2}t'\Sigma t]\, dt, \tag{1.2}$$

Plackett's identity follows by differentiating (1.2) with respect to ρ_{ij} and the pair x_i, x_j, and interchanging the integral and derivative, which is valid by standard theorems of analysis. This method of proof avoids the problem of handling derivatives of the elements of Σ^{-1}, and it readily generalizes to other cases, as we show below.

Plackett noted that the result for $p = 2$ had long been known. He used this identity to develop a dimensionality reduction formula for evaluating the multivariate normal distribution function. Since that time, (1.1) has been used to derive probability inequalities, suggest other approximation techniques, and to study some geometrical problems, all associated with the normal distribution. In this paper, we review applications and generalizations of Plackett's identity, and discuss connections with certain classical geometrical problems.

There is a companion identity for the case $i = j$:

$$\frac{\partial}{\partial \rho_{ii}} \phi_p(x - \mu; \Sigma) = \frac{1}{2} \frac{\partial^2}{\partial x_i^2} \phi_p(x - \mu; \Sigma). \tag{1.3}$$

The appearance of the "$\frac{1}{2}$" in (1.3) is not surprising if we recall that the normal density is a fundamental solution to the heat equation—see Durrett (1984) for further details. This companion, however, is less interesting than (1.1) for applications because probability calculations often standardize the variables by fixing $\rho_{ii} = 1$. We therefore assume below (unless otherwise stated), that Σ is a correlation matrix, and that $\mu = 0$; furthermore, we only deal with the off-diagonal elements of Σ.

2   Applications

Let $X \sim N_p(\mu, \Sigma)$ denote that X is a p-dimensional normal variate with mean μ and covariance Σ. Consider the problem of evaluating the probability that X is in a given set A:

$$P(X \in A) = \int_A \phi_p(x - \mu; \Sigma)dx. \tag{2.1}$$

The multiple integral in (2.1) is in general very difficult to evaluate. In fact, in his recent book on statistical computing, Thisted (1988) identifies this research problem as one that "has not yet been solved satisfactorily." Only for a few special cases is a closed form expression available. Brute force numerical integration is prohibitive unless p is small. Monte Carlo methods with appropriate variance reduction techniques show much promise. These and other techniques for evaluating (2.1) benefit greatly from an analytic study of the integral. For instance, the determination of an appropriate control variate, or an effective sampling function for importance sampling both rely on properties of A and Σ (see Evans and Stewart 1988 and Thisted 1988, and the references there).

A common approach to the analysis of (2.1) is to study its behavior as Σ varies. In one way or another, all the applications of Plackett's identity address this problem. In fact, Plackett's original application contains many of the key features of all others. We describe Plackett's calculation in some detail, and then turn to other uses of (1.1). For $X \sim N_p(0, \Sigma)$, consider

$$\Phi_p(a; \Sigma) = P(X \geq a) = P(X_1 \geq a_1, \ldots, X_p \geq a_p). \tag{2.2}$$

From (1.1), the partial derivatives of Φ_p with respect to ρ_{ij} are readily obtained, because the partial derivatives with respect to x_i, x_j are undone by integration with respect to those variables: for instance,

$$\frac{\partial}{\partial \rho_{12}}\Phi_p(a; \Sigma) = \int_{a_3}^{\infty} \int_{a_4}^{\infty} \cdots \int_{a_p}^{\infty} \phi_p(a_1, a_2, x_3, \ldots, x_p; \Sigma)\, dx_3\, dx_4 \ldots dx_p. \tag{2.3}$$

Factoring out the marginal density of (X_1, X_2) at (a_1, a_2), we get

$$\frac{\partial}{\partial \rho_{12}}\Phi_p(a; \Sigma) = \frac{1}{2\pi\sqrt{1 - \rho_{12}^2}}exp\{-\frac{1}{2}\frac{(a_1^2 - 2\rho_{12}a_1a_2 + a_2^2)}{1 - \rho_{12}^2}\}\Phi_{p-2}(a'; \Sigma_{p-2}), \tag{2.4}$$

where

$$a_i' = a_i - \frac{(\rho_{1i} - \rho_{2i}\rho_{12})a_1 + (\rho_{2i} - \rho_{1i}\rho_{12})a_2}{1 - \rho_{12}^2}, \quad i = 3, \ldots, p,$$

and Σ_{p-2} is the conditional covariance matrix of $(X_3, \ldots, X_p)$ given (X_1, X_2) (not correlation matrix). Now let $T = (\tau_{ij})$ be some other correlation matrix, and consider the line between Σ and T: $L(t) = tT + (1 - t)\Sigma$ with

$0 \le t \le 1$. Plackett's reduction formula is gotten by integrating the differential element Φ_p along that line between T and Σ:

$$\Phi_p(a; \Sigma) = \Phi_p(a; T) + \sum_{i<j} \int_{\tau_{ij}}^{\rho_{ij}} \frac{\partial}{\partial \lambda_{ij}} \Phi_p(a; L) \, d\lambda_{ij}, \qquad (2.5)$$

where $L = (\lambda_{ij})$ is a point, depending on t, on the line between T and Σ. Plackett gives guidelines on the choice of T; of course, it is important that $\Phi_p(a; T)$ be fairly easily evaluated, for it is still a p-dimensional integral; the integrand in (2.5) is a $(p - 2)$-dimensional integral, hence the term reduction. Of course, for large p this reduction formula is of limited use, but the calculations above are useful in other contexts.

Plackett's identity also sheds light on an early technique for computing multivariate normal probabilities. M. G. Kendall (1941) proposed the use of the tetrachoric series for evaluating $P(X \ge a)$. His method of derivation is to first write the normal density as the inverse of its characteristic function (as in (1.2)), regard the characteristic function as a function of Σ, then expand it in a multivariate Taylor series around identity matrix, and finally integrate the series term by term. The result is the following formal power series for evaluating $P(X \ge a)$:

$$\sum_{n_{12}=0}^{\infty} \cdots \sum_{n_{p-1,p}=0}^{\infty} \prod_{1 \le i < j \le p} \frac{\rho_{ij}^{n_{ij}}}{n_{ij}!} \prod_{i=1}^{p} H_{n_i - 1}(a_i) \phi(a_i), \qquad (2.6)$$

where ϕ is the standard normal density,

$$n_i = \sum_{j:i<j} n_{ij} + \sum_{j:j<i} n_{ji},$$

$H_{-1}(x)$ is Mill's ratio, $\Phi(-x)/\phi(x)$, and for $i \ge 0$, H_i are the Hermite polynomials,

$$H_i(x) = (-1)^i e^{-\frac{x^2}{2}} \frac{d^i}{dx^i} e^{\frac{x^2}{2}}. \qquad (2.7)$$

For many years, it was assumed that the tetrachoric series converged for all Σ. However Harris and Soms (1980) showed that the region of convergence for this series is quite small. They showed that the tetrachoric series converged when $| \rho_{ij} | < (p - 1)^{-1}$, for all $i \ne j$, and that the series diverged whenever p is even and $| \rho_{ij} | > \rho > (p - 1)^{-1}$, for all $i \ne j$. Now, given the derivations of Plackett's identity and the tetrachoric series, it should come as no surprise that the tetrachoric series is just the Taylor series of the function $P(X \ge a)$ about the identity matrix, I. The use of I is motivated by the fact that the calculation of Hermite polynomials and probability calculations are often straightforward for the independent case. On the other hand, the result of Harris and Soms is intuitively clear from our observation, since for equicorrelation matrices, $\rho_{ij} \equiv \rho$, a well-known condition for positive definiteness is that $\rho > -(p-1)^{-1}$: any Taylor series will diverge when it hits that point, or more generally, the boundary of the set of positive definite correlation matrices. One resolution is that for most values of Σ, the expansion about a matrix closer than I would be better. This issue also arose in Plackett's reduction formula, albeit in a different form.

One approach for finding a more appropriate correlation matrix to expand about is given by Iyengar (1982, 1988) who exploited the symmetry of a certain class of problems. Let $d = p(p-1)/2$, and string out the correlations into a vector $\rho = (\rho_{12}, \rho_{13}, \ldots, \rho_{23}, \ldots, \rho_{2p}, \ldots)'$ in d-space; also, let $\bar{\rho}\mathbf{1}_d$ be the vector with d components, each of which is the average of the components of ρ. Let E_α be an p-by-p equicorrelation matrix with parameter α, where $-(p-1)^{-1} < \alpha < 1$. Let $\mathbf{P}_p$ be the set of p-by-p permutation matrices, an element of which is denoted π and let Π be a random matrix which is independent of X and is uniformly distributed over $\mathbf{P}_p$. Call a set A permutation-symmetric if it satisfies $A = \pi A = \{\pi a : a \in A\}$ for all $\pi \in \mathbf{P}_p$.

Since $A = \pi A$ for all $\pi \in \mathbf{P}_p$, $P(X \in A) = P(\pi X \in A)$ for all $\pi \in \mathbf{P}_p$, and $P(X \in A) = P(\Pi X \in A)$ also. Now ΠX is a scale mixture of normals with exchangeable components and density

$$\psi_p(x; \Sigma) = \frac{1}{p!} \sum_\pi \phi_p(x; \pi \Sigma \pi'). \tag{2.8}$$

Its first two moments are $E(\Pi X) = 0$ and $\mathrm{var}(\Pi X) = \frac{1}{p!} \sum_\pi \pi \Sigma \pi' = E_{\bar{\rho}}$. We get our matrix to expand about by fitting to ψ_p the normal density $\phi_p(x; E_{\bar{\rho}})$, which shares the same first two moments. Charles Dunnett (1986) has used the same approximation for multiple comparisons using the multivariate t distribution.

An attractive feature of the use of $E_{\bar{\rho}}$ is that computations for it are also tractable, since they reduce to single integrals, because of the following representation. If $Z = (Z_1, \ldots, Z_p)'$ has density $\phi_p(x; I)$, Z_0 is a standard normal independent of Z, $\mathbf{1}_p$ is a vector of ones, and $\alpha \geq 0$, then $V = \sqrt{1-\alpha}Z + \sqrt{\alpha}Z_0\mathbf{1}_p$ has density $\phi_p(x; E_\alpha)$. Upon conditioning on Z_0, we get the single integral

$$P(V \in A) = \int_{-\infty}^{\infty} P[Z \in \frac{A - t\alpha^{1/2}\mathbf{1}_n}{(1-\alpha)^{1/2}}]\phi(t)\,dt. \tag{2.9}$$

When A is an orthant or a cube, the integrand in (2.9) is just a product of one-dimensional normal marginal probabilities, and if A is a sphere, it is a noncentral χ^2 probability. The one-dimensional integration can then be done by Gaussian quadrature. Thus, for many cases of interest, the right side of (2.9) is easily evaluated.

Equicorrelation matrices are not the only ones for which low-dimensional integrals obtain. For instance, Curnow and Dunnett (1962) considered the following structure: $\rho_{ij} = \lambda_i \lambda_j$, with $-1 \leq \lambda_i \leq 1$, $i = 1, \ldots, p$. In factor analysis, this is known as the single factor model, for this structure can be generated thus: $X_i = \sqrt{1 - \lambda_i^2} Z_i + \lambda_i Z_0$, where Z_0 is the single factor. It also arises in computations for Dunnett's test of several treatments against a single control. Conditioning on Z_0 once again yields single integrals involving $\Phi(x)$. Two, three, and four factor matrices are also good candidates for computational purposes, for they involve at most 4-dimensional integrals.

Higher order terms of the expansion about a matrix other than the identity is problematic, however. They involve multivariate Hermite polynomials that do not separate out into tractable products as in (2.6). The only systematic treatment of this is given by Royen (1987), who investigated convergence properties of such expansions. Henery (1981) has provided another approach based on Gram-Charlier series for certain bivariate normal probabilities. Whether such expansions are computationally feasible, and are competitive with modern Monte Carlo methods remains to be seen.

One of the first qualitative results derived from Plackett's identity is the inequality due to Slepian (1962). Let $X \sim N_p(0, \Sigma = (\rho_{ij}))$ and $Y \sim N_p(0, T = (\tau_{ij}))$ with $\rho_{ij} \geq \tau_{ij}$ and $\rho_{ii} = \tau_{ii}$ then for any vector a, $P(X \geq a) \geq P(Y \geq a)$. The method of proof is borrowed from Plackett's reduction formula. It first draws a line between Σ and T. The directional derivative along that line is a weighted sum of partial derivatives of the type given by (2.3). The weights, $(\rho_{ij} - \tau_{ij})$, are non-negative; since the integrand in (2.3) is positive, the directional derivative is non-negative, and the inequality follows. This intuitively clear fact has been generalized in many directions (see, for example, Das Gupta et al. 1972, Gordon 1987, Joag-Dev et al. 1983, and Tong 1980) using this kind of analytical or other geometrical argument.

Slepian's inequality is useful for the construction of one-sided confidence sets for the mean of a normal population. However, it is commonly assumed that the variance is unknown, so that the t-distribution, rather than the normal, is appropriate. Dunnett and Sobel were among the first to study the multivariate t. For example, using the single factor correlation structure, Dunnett and Sobel (1955) provided lower bounds for multivariate t orthant probabilities that involved only one and two-dimensional marginal probabilities. It is also more common to have two-sided sets, so the study of the probability content of rectangular regions is of interest. However, a direct analog of Slepian's result does not hold for rectangles, so a variety of more limited results have been obtained: for example, if $\rho_{ij} = \lambda_i \lambda_j$ for $i \neq j$ and $\rho_{ii} \geq \lambda_i^2$ for some numbers $\lambda_1, \ldots, \lambda_p$, then increasing any one λ_i will increase the probability content of a rectangle. For an overview of related results, see Tong (1980, 1990).

More generally, if the correlation between one of the variables and the rest increase proportionately, then rectangular probability contents increase: this statement was proved originally by Sidák (1968), but a direct analytical proof using Plackett's identity is due to Jogdeo (1970). Plackett's identity has been useful in deriving one such result, due to Bolviken and Joag-dev (1982). To describe their result, we need some definitions. A covariance matrix is said to be of the positive partial correlation (PPC) type if for every pair $i \neq j$ the partial correlation between X_i and X_j, denoted ρ_{ij}, is non-negative. Bolviken and Joag-dev show that if there is a signature matrix D (a diagonal matrix with ± 1 on the diagonal), such that $D\Sigma D$ is of the PPC type, then $P(|X_i| \leq (\geq)a_i)$ is non-decreasing in $|\rho_{ij}|$. The fact that the PPC condition appears in the statement is not surprising when we recall (2.4), or its analog for rectangular probabilities. There, the covariance matrix is conditional on holding $(X_1, X_2) = (a_1, a_2)$, so that the appearance of partial covariances and correlations is a natural consequence of the use of Plackett's identity.

The inequalities mentioned above yield first-order properties of the probabilities, in the sense that they refer to the first derivative with respect to the elements of Σ. Second-order properties are also of interest, for they indicate convexity and concavity of the functions of interest. This investigation was done by Iyengar and Tong (1990), who used Plackett's identity to study the convexity (in the correlation coefficients) of the cumulative distribution of the multivariate normal and its absolute value. Their results are most explicit for the bivariate case. Let

$$\Phi_2(x; \rho) = P(X_1 \leq x_1, X_2 \leq x_2), \quad \text{and} \tag{2.10}$$

$$G_2(x; \rho) = P(|X_1| \leq x_1, |X_2| \leq x_2), \quad \text{for } x_1 \geq 0, \, x_2 \geq 0. \tag{2.11}$$

Also let $m = \min(\frac{x_1}{x_2}, \frac{x_2}{x_1})$, and $M = \max(\frac{x_1}{x_2}, \frac{x_2}{x_1})$. They showed that (a) if $x_1 x_2 > 0$, then Φ_2 is convex in $\rho \in [0, m]$; if $x_1 x_2 < 0$, then Φ_2 is concave in $\rho \in [M, 0]$; (b) G_2 is convex in $\rho \in [-m, m]$. In particular, when $x_1 = x_2 = c$, Φ_2 is convex for all $\rho \geq 0$, and when $c \geq 0$, G_2 is convex for all ρ in $(-1, 1)$. For Φ_2, even more detailed information is available: when $c \geq \sqrt{2} - 1$, then Φ_2 is convex in $\rho \in (-1, 1)$. Applications of these results (and higher dimensional analogs) are also given in that paper.

3   Elliptically Contoured Distributions

A standardized p-dimensional vector X has an elliptically contoured distribution (ECD) if there is a (correlation) matrix Σ such that the characteristic function of X is $\hat{f}(t) = \psi(t'\Sigma t)$, where ψ is a real-valued function on $R_+ = [0,\infty)$. Then X has the representation

$$X = \tau \Sigma^{\frac{1}{2}} U_p, \qquad (3.1)$$

where the radial part τ is a non-negative random variable, U_p is uniformly distributed on Ω_p, the surface of the unit sphere in p-dimensions, and $\Sigma^{\frac{1}{2}}$ is the non-negative definite symmetric square root of Σ; furthermore, τ and U_p are independent. When X has a density f, it is of the form

$$f(x; \Sigma) = |\Sigma|^{-\frac{1}{2}} g(x'\Sigma^{-1}x), \qquad (3.2)$$

where $g : R_+ \to R_+$,

$$a_p \int_0^\infty v^{p-1} g(v^2)\, dv = 1, \qquad (3.3)$$

and a_p is the area of Ω_p. In this case, τ has the density $h_\tau(v) = a_p v^{p-1} g(v^2)$. For further discussion of ECDs, see Das Gupta et al. (1972) and Cambanis et al. (1981).

The investigation of Das Gupta, et al. has a strong geometric flavor. Among other results, they extended Slepian's inequality to elliptically contoured densities with a sufficiently smooth density. They also showed that within the class of ECDs with density, Plackett's identity characterized the multivariate normal. The first extension of Plackett's identity to ECDs is due to Joag-dev et al. (1983), who showed that if g is sufficiently smooth, then

$$\frac{\partial}{\partial \rho_{ij}} f(x; \Sigma) = -\frac{\partial}{\partial x_j} \left(\sum_{k=1}^p \rho^{ik} x_k \right) f(x; \Sigma), \qquad (3.4)$$

where ρ^{ij} is the ij element of Σ^{-1}. They used (3.4) to provide an analytic proof of the validity of Slepian's inequality for certain ECDs with such densities and showed that if $h(x_1, \ldots, x_n) = h_1(x_1, \ldots, x_k) h_2(x_{k+1}, \ldots, x_n)$ where h_1 and h_2 are concordant (both increasing or both decreasing coordinatewise) then $Eh(X)$ is increasing in ρ_{ij} for $i \neq j$.

Although (3.4) is a Plackett-like identity for ECDs, it did not involve a mixed partial derivative on the right hand side, so it was of interest to find a subfamily of the ECDs which yielded that. Iyengar (1984) and Iyengar and Tong (1990) considered the following family of densities:

$$f_{p,k}(x;\Sigma,\beta) = \frac{\Gamma(\frac{p}{2})}{\Gamma(\frac{p}{2}+k)} \mid \beta\pi\Sigma \mid^{-\frac{p}{2}} (\frac{x'\Sigma^{-1}x}{\beta})^k \exp(-\frac{x'\Sigma^{-1}x}{\beta}), \quad (3.5)$$

where $\beta > 0$ and k is a non-negative integer. When $k = 0$, (3.5) yields a normal distribution. It is easy to see that if $X = \tau_{k,\beta}\Sigma^{\frac{1}{2}}U_p$ has the density $f_{p,k}(x;\Sigma,\beta)$, then $\tau_{k,\beta}^2$ has a gamma density with mean $\beta(\frac{p}{2}+k)$ and variance $\beta^2(\frac{p}{2}+k)$. For the bivariate case, Kotz (1974) studied this family. However, his focus was different: he showed that this family is an instance of a generalization of the Pearson family to higher dimensions. For a discussion of the geometrical uses of this family, see the next section.

Plackett's identity for this family of functions is given by the following:

$$\frac{\partial}{\partial \rho_{ij}} f_{p,k}(x;\Sigma,\beta) = \frac{\beta}{2} \sum_{m=0}^{k} \frac{k!}{m!} \frac{\Gamma(\frac{p}{2}+m)}{\Gamma(\frac{p}{2}+k)} \frac{\partial^2}{\partial x_i \partial x_j} f_{p,m}(x;\Sigma,\beta). \quad (3.6)$$

When $k = 0$ and $\beta = 2$, (3.6) specializes to Plackett's identity for the normal distribution. A proof of this result using the characteristic function requires the following expression, which is of independent interest: the characteristic function of $f_{p,k}(x;\Sigma,\beta)$ is $\psi_{p,k}(t'\Sigma t;\beta)$, where

$$\psi_{p,k}(u;\beta) = e^{-\frac{\beta u}{4}} \sum_{m=0}^{k} \binom{k}{m} \frac{\Gamma(\frac{p}{2})}{\Gamma(\frac{p}{2}+m)} (-\frac{\beta u}{4})^m. \quad (3.7)$$

Finally, Gordon (1987) proved a definitive version of Plackett's identity for ECDs. He showed that the following two statements about functions g and h, each mapping R_+ into itself and vanishing at ∞, are equivalent:

$$h(t) = \frac{1}{2} \int_t^\infty g(r)dr \quad (3.8)$$

and

$$\frac{\partial}{\partial \rho_{ij}} g_\Sigma(x) = \frac{\partial^2}{\partial x_i \partial x_j} h_\Sigma(x), \quad (3.9)$$

where $g_\Sigma(x) = \mid \Sigma \mid^{-\frac{1}{2}} g(x'\Sigma^{-1}x)$, and similarly for h. It is easy to see that if g is either an exponential or an appropriately chosen gamma density the identities of Plackett and Iyengar, (1.1) and (3.6), respectively, are immediate. Another case of interest is $g(r) = (1+r)^{-a}$, where $a > 2$, which deals with the multivariate t-distribution (see Tong 1980 for precise definitions).

4   Geometrical Considerations

The connection between multivariate normal probabilities and the content of sets on the surface, Ω_p of the unit ball has long been recognized. H. Ruben (1954), for instance, has perhaps done the most to explicate this relationship, while adding to earlier work by Schläfli (1858, 1860), Coxeter (1973), and Somerville (1929). See also the paper by Abrahamson (1964) and the references contained therein. In this paper it is not possible to delve into may details, so we will just give a flavor of this work by showing the relationship between Plackett's identity for the multivariate normal, and Schläfli's identity for hyperspherical simplices.

For $X \sim N_p(0, \Sigma)$ with non-singular Σ consider $P(X \geq 0)$, the probability of the positive orthant. Write $X = \tau \Sigma^{-\frac{1}{2}} U_p$, where $\Sigma^{-\frac{1}{2}}$ is a symmetric square root of Σ, to get that

$$P(X \geq 0) = P(b_i' U_p \geq 0, \ i = 1, \ldots, p), \qquad (4.1)$$

where the unit vector b_i is the ith column of Σ, and $b_i' b_j = \rho_{ij}$ is the cosine of the angle between b_i and b_j. Thus, the probability of the positive orthant is the content of the intersection of Ω_p and $\cap_{i=1}^p \{x : b_i' x \geq 0\}$. This region is known as a spherical simplex, which was extnsively studied by Schläfli. By spherical symmetry, this content is a function of the angles between the vectors b_i, rather than the vectors themselves. Denote the angle between b_i and b_j by (ij), and let the content of the spherical simplex be

$$C_p(b) = C_p[(12), (13), \ldots, (\overline{p-1p})]. \qquad (4.2)$$

In the course of his investigations, Schläfli proved the following differential relationship: the infinitesimal change dC_p in the content of the spherical simplex due to an infinitesimal change $d(ij)$ in the angles (ij) is

$$dC_p = \frac{1}{2\pi} \sum_{i<j} C_{p-2}(\overline{ij}) \, d(ij), \quad p = 2, 3, \ldots. \qquad (4.3)$$

In this equation, C_0 is interpreted as 1, $(\overline{ij})$ is the edge (which is also a spherical simplex) of dimension $p-3$ formed by intersecting Ω_p with $b_i' x = 0$ and $b_j' x = 0$; $C_{p-2}(\overline{ij})$ is the content of that edge.

The first three values of C_p are $C_0 = 1$, $C_1 = \frac{1}{2}$, $C_2 = \frac{(12)}{2\pi}$, and $C_3 = \frac{[(12)+(13)+(23)-\pi]}{4\pi}$. Now Plackett's proof of Schläfli's identity (4.3) is a consequence of his identity via (2.5), and by repeated application of the formula for partial correlations,

$$\rho_{ij.kl} = \frac{\rho_{ij.l} - \rho_{ik.l}\rho_{jk.l}}{\sqrt{(1 - \rho_{ik.l}^2)(1 - \rho_{jk.l}^2)}}. \qquad (4.4)$$

Another illustration of the interplay between the geometric and analytical aspects of Plackett's identity is given by Iyengar and Tong (1990). They provide a method of extending certain results from subclasses of ECDs to the entire family of ECDs. The method, which uses the representation $X = \tau \Sigma^{\frac{1}{2}} U_p$, consists of the following steps. First, using analytical methods, prove a statement for a sufficiently rich class of ECDs with the same Σ. Second, use that statement to derive a geometrical fact about Ω_p. And third, use that geometrical fact to show that the statement is true for all ECDs with that Σ. A simple example illustrates these steps. For instance, suppose that $P(X \in A) \geq P(X \in B)$ for all X which have an ECD with density $f(x; I)$. Then, conditioning on τ, we get

$$\int_0^\infty P(U_p \in \frac{A}{v}) h_\tau(v) \, dv \geq \int_0^\infty P(U_p \in \frac{B}{v}) h_\tau(v) \, dv. \qquad (4.5)$$

Letting h_τ tend to a point mass at v, we have the geometrical fact that $P(U_p \in \frac{A}{v}) \geq P(U_p \in \frac{B}{v})$, provided that $P(U_p \in \frac{A}{v})$ is a continuous function of v, as it will be for many applications. Thus, we have established that the area of $\frac{A}{v} \cap \Omega_p$ exceeds the area of $\frac{B}{v} \cap \Omega_p$ for all $v \geq 0$. Now, for any non-negative random variable τ', we have

$$\int_0^\infty P(U_p \in \frac{A}{v}) P(\tau' \in dv) \geq \int_0^\infty P(U_p \in \frac{B}{v}) P(\tau' \in dv) \qquad (4.6)$$

so that $P(X \in A) \geq P(X \in B)$ is true for all ECDs with $\Sigma = I$.

This argument was used by Iyengar and Tong to provide a proof of Slepian's inequality for all ECDs along the lines of Slepian's original proof, avoiding the lengthy geometrical arguments needed otherwise. They also extended the convexity properties of bivariate normal distributions to ECDs (see the discussion following (2.10) and (2.11)), and showed how those convexity results could be used to get improved computational procedures for constructing confidence regions for the normal mean using the multivariate t distribution.

5 Other Generalizations

Not long after Plackett derived (1.1), Price (1958) used Dirac's δ-function to independently derive the following closely related result, which we state in the form given in Patil and Boswell (1970). Let X be a standardized p-variate multivariate normal, and let $g_1(X_1), \ldots, g_p(X_p)$ be differentiable functions of the components of X, each admitting a Laplace transform. Then

$$\frac{\partial}{\partial \rho_{ij}} E[\prod_1^p g_k(X_k)] = E[\frac{\partial^2}{\partial x_i \partial x_j} \prod_1^p g_k(X_k)]. \tag{5.1}$$

Conversely, if this identity holds for arbitrary $g_1, \ldots, g_p$ (with both expectations above defined) then X has a multivariate normal distribution.

Price and other electrical engineers (see, for example, Pawula 1967) used this theorem to facilitate studies in signal processing. In particular, they consider a zero-memory non-linear input-output device with Gaussian input X_i that yields output $g_i(X_i)$. A quantity of interest is the nth-order correlation coefficient of the outputs, which requires the computation of $\prod_1^p g_k(X_k)$. The differential equation of Price's theorem provides a useful computational tool for such calculations: see the work of Pawula, and Papoulis (1965) for a wide range of examples.

Patil and Boswell use Plackett's identity to provide an extension of Price's theorem: subject to certain conditions on the tail of g,

$$\frac{\partial}{\partial \rho_{ij}} E[g(X_1, \ldots, X_n)] = E[\frac{\partial^2}{\partial x_i \partial x_j} g_k(X_1, \ldots, X_n)], \tag{5.2}$$

although the only examples that they provide concern product moments. Patil and Boswell also prove that if Plackett's identity holds for a density, then that density must be multivariate normal; this result is stronger than that of Das Gupta et al., who restricted their attention to elliptically contoured distributions. Then, in a related investigation, (Patil and Boswell 1973), they give characterizations of certain discrete distributions using analogous differential equations. For instance, it is easy to see that for the Poisson distribution with parameter λ, $f(k; \lambda), k = 0, 1, \ldots$,

$$\frac{\partial}{\partial \lambda} f(x; \lambda) = f(k-1; \lambda) - f(k; \lambda). \tag{5.3}$$

When the distribution is concentrated on the positive axis, this equation characterizes the Poisson. The authors give similar results for the negative binomial and multinomial distributions.

Perhaps a more natural extension of Plackett's identity from a geometrical standpoint is to multivariate densities with other contours as their level curves. Cambanis et al. (1983) have defined and extensively studied α-symmetric distributions, whose characteristic functions have the following structure:

$$E(e^i t' X) = h(| t_1 |^\alpha + \ldots + | t_p |^\alpha) \tag{5.4}$$

where $0 < \alpha \leq 2$. Iyengar (1984) considered the more general class of convex contoured distributions, and also showed that the analog of Slepian's inequality does not hold for 1-symmetric distributions. Whether an analog of Plackett's identity applies to these families remains an open problem.

Acknowledgment

This research was supported by the Office of Naval Research Grant N00014-89-J-1496.

References

Abrahamson, I. (1964) Orthant probabilities for the quadrivariate normal distribution. *Ann. Math. Statist.* **35** 1685 - 1703.

Bolviken, E. and Joag-dev, K. (1982). Monotonicity of the probability of a rectangular region under a multivariate normal distribution. *Scand. J. Statist.* **9** 171 - 174.

Cambanis, S., Huang, S. and Simons, G. (1981). On the theory of elliptically contoured distributions. *J. Mult. Analysis* **11** 368 - 385.

Cambanis, S., Keener, R. and Simons, G. (1983). On α-symmetric multivariate distributions. *J. Mult. Analysis* **13** 213 - 233.

Coxeter, H. (1973) *Regular Polytopes.* Dover, New York.

Curnow, R. and Dunnett, C.W. (1962). The numerical evaluation of certain multivariate normal integrals. *Ann. Math. Statist.* **33** 571 - 579.

Das Gupta, S., Eaton, M., Olkin, I., Perlman, M., Savage, J. and Sobel, M. (1972) Inequalities on the probability content of convex regions for elliptically contoured distributions. *Proc. Sixth Berk. Symp. Math. Statist. Prob.* **2** 241 - 264.

Dunnett, C.W. (1986). Multiple comparisons between several treatments and a specified treatment. In *Linear Statistical Inference* (T. Calinski and W. Klonecki, eds.), 432 - 438.

Dunnett, C. and Sobel, M. (1955). Approximations to the probability integral and certain percentage points to a multivariate analog of Student's *t* distribution. *Biometrika* **42** 258 - 260.

Durrett, R. (1984). *Brownian Motion and Martingales in Analysis.* Wadsworth, Belmont, CA.

Evans, M. and Swartz, T. (1988). Sampling from Gauss rules. *SIAM J. Scient. Statist. Comp.* **9** 950 - 961.

Gordon, Y. (1987). Elliptically contoured distributions. *Prob. Theory and Rel. Fields* **76** 429 - 438.

Harris, B. and Soms, A. (1980). The use of the yetrachoric series for evaluating multivariate normal probabilities. *J. Mult. Analysis* **10** 252 - 267.

Henery, R. (1981). An approximation to certain multivariate normal probabilities. *J. Roy. Statist. Soc.* B **43** 81 - 85.

Iyengar, S. (1982). On the Evaluation of Certain Multivariate Normal Probabilities. Dissertation, Stanford U.

Iyengar, S. (1984). A Geometric Approach to Probability Inequalities. Tech. Rep. # 84-04, University of Pittsburgh.

Iyengar, S. (1988). Evaluation of normal probabilities of symmetric regions. *SIAM J. Scient. and Statist. Comp.* **9** 418 - 424.

Iyengar, S. and Tong, Y. (1990). Convexity of elliptically contoured distributions with zpplications. *Sankhya* A **51** 13 - 29.

Joag-Dev, K., Perlman, M. and Pitt, L. (1983). Association of normal random variables and Slepian's inequality. *Ann. Prob.* **11** 451 - 455.

Jogdeo, K. (1970). A simple proof of an Inequality for multivariate normal probabilities of rectangles. *Ann. Statist.* **41** 1357 - 1359.

Kendall, M. (1941). Proof of relations connected with the tetrachoric series and its generalization. *Biometrika* **32** 196 - 198.

Kotz, S. (1974). Multivariate distributions at a cross-road. In *Statistical Distributions in Scientific Work. Vol. 1* (G.P. Patil, S. Kotz and J.K. Ord, eds.), 247 - 270.

Papoulis, A. (1965). *Probability, Random Variables, and Stochastic Processes.* McGraw-Hill, New York.

Patil and G. Boswell, M. (1970). A characteristic property of the multivariate normal density function and some of its applications. *Ann. Math. Statist.* **41** 1970 - 1977.

Patil, G. and Boswell, M. (1973). Characterization of certain discrete distributions by differential equations with respect to their parameters. *Austral. J. Statist.* **15** 128 - 131.

Pawula, R. (1967). A modified version of Price's theorem. *IEEE Trans. on Information Theory* **IT-13** 285 - 288.

Plackett, R. (1954). Reduction formula for multivariate normal integrals. *Biometrika* **41** 351 - 360.

Price, R. (1958). A useful theorem for nonlinear devices having Gaussian inputs. *IRE Trans. on Information Theory* **IT4** 69 - 72.

Royen, T. (1987). An approximation for multivariate normal probabilities of rectangular regions. *Statistics* **18** 389 - 400.

Ruben, H. (1954). On the moments of order statistics in samples from normal populations. *Biometrika* **41** 200 - 227.

Schläfli, L. (1858,1860). On the multiple integral $\int^n dx dy \ldots dz$, whose limits are $p_1 = a_1 x + b_1 y + \ldots + h_1 z > 0, p_2 > 0, \ldots, p_m > 0, x^2 + y^2 + \ldots z^2 < 1$. *Quart. J. Pure Appl. Math.* **2** 261 - 301; **3** 54 - 68; **3** 97 - 107.

Sidák, Z. (1968). On multivariate normal probabilities of rectangles: their dependence on correlations. *Ann. Statist.* **39** 1425 - 1434.

Slepian, D. (1962). The one-sided barrier problem for Gaussian noise. *Bell System Tech. J.* **41** 463 - 501.

Sommerville, D. (1929). *An Introduction to the Geometry of N Dimensions*. Methuen, London.

Thisted, R. (1988). *Elements of Statistical Computing*. Chapman and Hall, New York.

Tong, Y. (1980). *Probability Ineqalities for Multivariate Distributions*. Academic Press, New York.

Tong, Y. (1990). *The Multivariate Normal Distribution*. Springer, New York.

Chapter 9

Some Minimax Test Procedures for Comparing Several Normal Means

ANTHONY J. HAYTER   School of Industrial and Systems Engineering, Georgia Institute of Technology, Atlanta, Georgia

WEI LIU   Department of Mathematics, University of Southampton, Southampton, United Kingdom

Abstract   The problem of testing the equality of k normal means such as the k treatment effects in a one-way fixed effects analysis of variance model is addressed. Power assessment in terms of constraints on certain range functions of the means is considered. Minimax test procedures are developed which allow power level requirements to be met most efficiently. Comparisons are made between the minimax test procedures and various standard test procedures.

1   Introduction

Consider the balanced one-way fixed effects analysis of variance model

$$Y_{ij} = \mu_i + \epsilon_{ij}, \qquad 1 \le i \le k,\ 1 \le j \le n,$$

137

where the μ_i, $1 \leq i \leq k$, are k unknown treatment means and the ϵ_{ij} are i.i.d. $N(0, \sigma^2)$ random variables. Suppose that an experimenter intends initially to perform a size α hypothesis test of the null hypothesis that the treatment means are all equal, $H_0 : \mu_1 = \ldots = \mu_k$, against a general alternative hypothesis, and consider the power assessment of such a test.

Suppose that the variance σ^2 is known. If the sum of squares test statistic $\sum_{i=1}^{k}(X_i - \bar{X})^2$ is used, where $X_i = \bar{Y}_i = \sum_{j=1}^{n} Y_{ij}/n$, $1 \leq i \leq k$, and $\bar{X} = \sum_{i=1}^{k} X_i/k$, then the power of the test procedure depends on the treatment means only through the quantity $\Delta^2 = \sum_{i=1}^{k}(\mu_i - \bar{\mu})^2$ where $\bar{\mu} = \sum_{i=1}^{k} \mu_i/k$. It is generally agreed, however, that power assessment in terms of Δ^2 is not easily interpretable by the experimenter, and that it is more useful to assess power levels in terms of certain range measures of the treatment means such as $r_1 = \max_{1 \leq i \leq k} |\mu_i - \bar{\mu}|$ or $r_2 = \max_{1 \leq i,j \leq k} |\mu_i - \mu_j|$ (these measures have been proposed, for example, by Pearson and Hartley 1951, Scheffé 1959, section 3.3, and Kastenbaum et al. 1970). The approach to power assessment recommended in many text books, is to find the minimum value of Δ^2 for fixed values of r_1 or r_2, and then to use the test procedure based on the sum of squares statistic and to calculate the power for that minimum value of Δ^2. More recently, it has been shown by Hayter and Liu (1990) how the power of a test procedure based on the range statistic $\max_{1 \leq i,j \leq k} |X_i - X_j|$ may be assessed using the measure r_1 or r_2. These test procedures generalize to the F-test and to the Studentized range test procedure, respectively, when the variance σ^2 is unknown.

The purpose of this paper is to answer the following question. Among all suitable tests of size α, which one maximizes the minimum power level within the region of the parameter space defined by either $r_1 \geq \delta$ or $r_2 \geq \delta$ for a fixed value of δ? The derivation of such a minimax test procedure, besides being of theoretical interest, is important for two reasons. Firstly, an experimenter may decide to choose the sample size n so that for specified values of α, β and δ, a size α test procedure will have power no smaller than β whenever $r_1 \geq \delta$ (or $r_2 \geq \delta$). This approach to sample size determination offers a simple, useful and easily interpretable method of power assessment, and the use of the minimax test procedure will allow the probability requirements to be met with the smallest possible sample size. Secondly, comparing the minimax test with the two test procedures mentioned above, and with other common test procedures such as the slippage test procedure which is based on $\max_{1 \leq i \leq k} |X_i - \bar{X}|$, will provide some interesting insights into the power properties of these standard test procedures, which will have implications for the more general situation where σ^2 is not assumed known.

The minimax test procedure will be found among tests $\psi(\mathbf{x})$ (which take the values 0 or 1 as the null hypothesis is respectively accepted or rejected) within the class Ψ of test procedures defined by

$$\Psi = \{\psi(\mathbf{x}) : \psi \text{ is symmetric, exchangeable, shift invariant, convex}\}.$$

Here, symmetric means $\psi(\mathbf{x}) = \psi(-\mathbf{x})$, $\forall \mathbf{x} \in \mathbf{R}^k$, exchangeable means $\psi(\pi \mathbf{x}) = \psi(\mathbf{x})$, $\forall \mathbf{x} \in \mathbf{R}^k$, $\pi \in \Pi$, where Π is the set of all $k!$ permutation transformations, and shift invariant means $\psi(\mathbf{x} + c\mathbf{1}) = \psi(\mathbf{x})$, $\forall \mathbf{x} \in \mathbf{R}^k$, $c \in \mathbf{R}$ where $\mathbf{1} = (1, \ldots, 1) \in \mathbf{R}^k$. The condition of convexity is taken to mean that the test procedure has a convex acceptance region. These properties are all necessary for a sensible test procedure.

Some notation which will be used is $c_1(j) = ((j-k)/k, \ldots, (j-k)/k, j/k, \ldots, j/k)' \in \mathbf{R}^k$, $1 \le j \le k/2$, where the first j and last $k - j$ terms are identical, and $c_2(j) = (-1, \ldots, -1, 0, \ldots, 0, 1, \ldots, 1)' \in \mathbf{R}^k$, $1 \le j \le k/2$, where there are j terms each of -1 and 1. Also, define the "alternative set"

$$\Theta_A(\mathbf{c}, \delta) = \{\mu : \max_{\pi \in \Pi} |(\pi \mathbf{c})' \mu| \ge \delta\} \subseteq \mathbf{R}^k.$$

Thus, $\{\mu : r_1 \ge \delta\} = \Theta_A(c_1(1), \delta)$ and $\{\mu : r_2 \ge \delta\} = \Theta_A(c_2(1), \delta)$. In general, for a given contrast $\mathbf{c}$ and positive real number δ, a minimax size α test procedure is defined to be the size α test procedure which maximizes the minimum power level within the alternative set $\Theta_A(\mathbf{c}, \delta)$.

In Section 3 the minimax test procedures within the class Ψ are found for the distributional assumption that $\mathbf{X} \sim N_k(\mu, \Sigma)$ for covariance matrices Σ which have equal diagonal and equal off-diagonal elements, and for alternative sets $\Theta_A(c_i(j), \delta)$, for $i = 1, 2$ and $1 \le j \le k/2$. The calculations in Section 3 depend upon the results of Section 2, where various least favorable configurations of the treatment means μ_i are found for the class of test procedures Ψ. For a given test procedure $\psi(\mathbf{x})$, a least favorable configuration in $\Theta_A(\mathbf{c}, \delta)$ is defined to be a vector $\mu^* \in \Theta_A(\mathbf{c}, \delta)$ for which the power anywhere within $\Theta_A(\mathbf{c}, \delta)$ is no smaller than at μ^*. The derivations in Section 2 are carried out under the even weaker conditions that the probability density function of $\mathbf{X}$ is $f(\mathbf{x} - \mu)$, where $f(\mathbf{x})$ is a member of the class of densities $\mathbf{F}$ defined by

$$\mathbf{F} = \{f(\mathbf{x}) : f(\mathbf{x}) \text{ is symmetric, exchangeable, unimodal}\}.$$

In this case unimodal is taken to mean that the set $\{\mathbf{x} : f(\mathbf{x}) \ge c\}$ is a convex set for all $c \in \mathbf{R}$. The results of Section 2 generalize previous work on the derivations of least favorable configurations for particular test procedures $\psi(\mathbf{x})$ and densities $f(\mathbf{x})$. Finally, Section 4 contains a summary and discussion of the results.

2   Least Favorable Configurations

In this section the identification of the least favorable configurations of the parameters $\mu \in \mathbf{R}^k$, is discussed. For alternative sets of the form $\Theta_A(c_i(j), \delta)$, a least favorable configuration is found which is common to all densities $f(\mathbf{x}) \in \mathbf{F}$ and to all test procedures $\psi(\mathbf{x}) \in \Psi$. For a given test procedure $\psi(\mathbf{x})$, let $A \subset \mathbf{R}^k$ be the acceptance region, and denote the power of the test procedure for a given set of parameters μ and density $f(\mathbf{x})$ by $\beta(\psi, f, \mu)$, so that

$$\beta(\psi, f, \mu) = 1 - \int I_A(\mathbf{x}) f(\mathbf{x} - \mu) d\mathbf{x}, \qquad (2.1)$$

where $I_A(\mathbf{x})$ is the indicator function of the set A. Notice that $\psi(\mathbf{x}) \in \Psi$ implies that the acceptance region A is symmetric about the origin, exchangeable, convex and shift invariant. The following Lemma, where $\mathbf{C}\{\mathbf{x}_1, \ldots, \mathbf{x}_r\}$ denotes the convex hull of $\{\mathbf{x}_1, \ldots, \mathbf{x}_r\}$, establishes some basic properties of the power function $\beta(\psi, f, \mu)$.

Lemma 2.1

Let $f(\mathbf{x}) \in \mathbf{F}$, $\psi(\mathbf{x}) \in \Psi$ and $\mu \in \mathbf{R}^k$. Then

$$(i) \qquad\qquad \beta(\psi, f, \mu) = \beta(\psi, f, -\mu)$$
$$(ii) \qquad\quad \beta(\psi, f, \mu) = \beta(\psi, f, \pi\mu) \qquad \forall \pi \in \Pi$$
$$(iii) \qquad \beta(\psi, f, \mu) = \beta(\psi, f, \mu + c\mathbf{1}) \qquad \forall c \in \mathbf{R}$$
$$(iv) \qquad \beta(\psi, f, \mu) \geq \beta(\psi, f, \xi) \qquad \text{if } \xi \in \mathbf{C}\{\pm\pi\mu : \pi \in \Pi\}.$$

Proof of Lemma 2.1

Properties (i)-(iii) of the Lemma are immediate from the definitions of the classes Ψ and $\mathbf{F}$ and the representation (2.1). Property (iv) then follows from the convolution theorem of Mudholkar (1966) (see, for example, Eaton 1987, p.90), since the functions $f(\mathbf{x})$ and $I_A(\mathbf{x})$ are both non-negative, unimodal and G-invariant, where G is the group of transformations generated by the permutations $\pi \in \Pi$ and a reflection in the origin. ∎

If, in the definition of the class of densities $\mathbf{F}$, the condition of unimodality is replaced by the stronger requirement of logconcavity, then the power function $\beta(\psi, f, \mu)$ may also be shown to be a log-concave function of μ. However, this extra condition is not necessary in the determination of the least favorable configurations, although the proofs below are essentially similar to those based on logconcavity given in Hayter and Liu (1990) for the Studentized range test procedure.

The following Theorem establishes a least favorable configuration for alternative sets $\Theta_A(c_1(j), \delta)$ which is common to all densities $f(\mathbf{x}) \in \mathbf{F}$ and tests $\psi(\mathbf{x}) \in \Psi$.

Theorem 2.1

Let $f(\mathbf{x}) \in \mathbf{F}$ and $\psi(\mathbf{x}) \in \Psi$. Define $\mu_1(j) = (-\delta/j, \ldots, -\delta/j, \delta/(k - j)), \ldots, \delta/(k - j))' \in \Theta_A(c_1(j), \delta)$ where the first j and last $k - j$ places are identical. Then for any j, $1 \le j \le k/2$,

$$\beta(\psi, f, \mu_1(j)) \le \beta(\psi, f, \mu)$$

for any $\mu \in \Theta_A(c_1(j), \delta)$.

Proof of Theorem 2.1

It follows from Lemma 2.1 that it is sufficient to show that for any $\mu \in \Theta_A(c_1(j), \delta)$, there is a real number c such that $\mu_1(j) + c\mathbf{1} \in \mathbf{C}\{\pm\pi\mu : \pi \in \Pi\}$. This is easily seen, since if $c_1(j)'\mu = \delta^* \ge \delta$, say, then the vector whose first j elements are each equal to $\delta\bar{\mu}_{1j}/\delta^*$ and whose last $k - j$ elements are each equal to $\delta\bar{\mu}_{j+1,k}/\delta^*$, where $\bar{\mu}_{ab} = \sum_{i=a}^{b} \mu_i/(b + 1 - a)$, is a member of $\mathbf{C}\{\pm\pi\mu : \pi \in \Pi\}$, and of the form $\mu_1(j) + c\mathbf{1}$ as required. ∎

Finally, the following Theorem establishes a least favorable configuration for alternative sets $\Theta_A(c_2(j), \delta)$ which is common to all densities $f(\mathbf{x}) \in \mathbf{F}$ and test procedures $\psi(\mathbf{x}) \in \Psi$.

Theorem 2.2

Let $f(\mathbf{x}) \in \mathbf{F}$ and $\psi(\mathbf{x}) \in \Psi$. Define $\mu_2(j) = (-\delta/2j, \ldots, -\delta/2j, 0, \ldots, 0, \delta/2j, \ldots, \delta/2j)' \in \Theta_A(c_2(j), \delta)$ where $-\delta/2j$ and $\delta/2j$ each occupy j places. Then for any j, $1 \le j \le k/2$,

$$\beta(\psi, f, \mu_2(j)) \le \beta(\psi, f, \mu)$$

for any $\mu \in \Theta_A(c_2(j), \delta)$.

Proof of Theorem 2.2

It follows from Lemma 2.1 that it is sufficient to show that $\mu_2(j) \in \mathbf{C}\{\pm\pi\mu : \pi \in \Pi\}$ for any $\mu \in \Theta_A(c_2(j), \delta)$. In this case, if $c_2(j)'\mu = \delta^* \ge \delta$, the vector ξ whose first j elements are each equal to $\delta\bar{\mu}_{1j}/\delta^*$, whose last j elements are each equal to $\delta\bar{\mu}_{k+1-j,k}/\delta^*$, and whose other elements are each equal to $\delta\bar{\mu}_{j+1,k-j}/\delta^*$ (where $\bar{\mu}_{ab}$ is defined in the proof of Theorem 2.1), is a member of $\mathbf{C}\{\pm\pi\mu : \pi \in \Pi\}$. So also is $\zeta = (-\xi_k, \ldots, -\xi_1)$, and hence so is $\mu_2(j) = (\xi + \zeta)/2$, which completes the proof. ∎

The results of this section are useful in themselves, since for a given test procedure $\psi(\mathbf{x})$ and density $f(\mathbf{x})$, an experimenter may assess the power level within a given alternative set by evaluating the power at the least favorable configuration. Furthermore, the fact that the least favorable configurations are the same, regardless of the particular test procedure and density, allows the simple derivation of the minimax test procedures in the next section. Also, if Studentized versions of the test statistics are used, to account for an unknown variance in the analysis of variance model say, then the derivation of the least favorable configurations can be reduced to the form considered above by conditioning on the Studentizing variable, and hence the least favorable configurations will be identical.

3   Derivation of Minimax Test Procedures

In this section minimax test procedures are derived for alternative sets of the form $\Theta_A(\mathbf{c}_i(j), \delta)$ when $X \sim N_k(\boldsymbol{\mu}, \Sigma)$ where Σ is a $k \times k$ matrix with diagonal elements all equal to σ^2, and off-diagonal elements all equal to $\rho\sigma^2$. Thus, $f(\mathbf{x}) = \phi(\mathbf{x}) \propto \exp\{-\mathbf{x}'\Sigma^{-1}\mathbf{x}\,/2\}$ which, of course, is a member of the class of densities **F**. The analysis of variance model discussed in the introduction corresponds to the case with $\rho = 0$ and σ^2 replaced by σ^2/n. Theorem 3.1 below gives the minimax test procedure for an alternative set of the form $\Theta_A(\mathbf{c}_1(j), \delta)$.

Theorem 3.1

Let $f(\mathbf{x}) = \phi(\mathbf{x})$. Then within the class of test procedures $\psi(\mathbf{x}) \in \Psi$ which have size α, the test procedure which maximizes the minimum power within $\Theta_A(\mathbf{c}_1(j), \delta)$ is

$$\psi(\mathbf{x}) = 1 \qquad \Longleftrightarrow \qquad \sum_J \cosh\{w(\bar{x}_J - \bar{x})\} \geq c, \qquad (3.1)$$

where the sum is over all subsets J of $\{1, \ldots, k\}$ of size j, $\bar{x}_J$ is the average of the elements of $\mathbf{x}$ with indices in J, $\bar{x}$ is the average of all the elements of $\mathbf{x}$, $w = \delta k/\sigma^2(k - j)(1 - \rho)$ and the constant c is chosen so that the test procedure has size α.

Proof of Theorem 3.1

It follows from Theorem 2.1 that the minimax test is the test procedure which maximizes

$$L(\psi) = \int_{\mathbf{R}^k} \psi(\mathbf{x})\phi(\mathbf{x} - \boldsymbol{\mu}_1(j))d\mathbf{x}$$

over all test procedures $\psi(\mathbf{x}) \in \Psi$ which satisfy the size condition

$$\alpha = \int_{\mathbf{R}^k} \psi(\mathbf{x})\phi(\mathbf{x})d\mathbf{x}.$$

Now, $\phi(\mathbf{x})$ and $\psi(\mathbf{x}) \in \Psi$ are symmetric and exchangeable, so that

$$L(\psi) = \frac{1}{2k!} \int_{\mathbf{R}^k} \psi(\mathbf{x}) G(\mathbf{x}) d\mathbf{x},$$

where

$$G(\mathbf{x}) = \sum_{\pi \in \Pi} (\phi(\pi \mathbf{x} - \mu_1(j)) + \phi(\pi \mathbf{x} + \mu_1(j))).$$

It then follows from the Neyman-Pearson Lemma that the test procedure given by

$$\psi(\mathbf{x}) = 1 \qquad \Longleftrightarrow \qquad G(\mathbf{x})/\phi(\mathbf{x}) \geq c' \qquad (3.2)$$

where the constant c' is chosen to satisfy the size requirement, will be the minimax test procedure as long as it is in the class Ψ. However,

$$G(\mathbf{x})/\phi(\mathbf{x}) \propto \sum_{\pi \in \Pi} \cosh\{(\pi \mathbf{x})' \Sigma^{-1} \mu_1(j)\}$$

and

$$\Sigma^{-1} \mu_1(j) = \mu_1(j)/\sigma^2(1 - \rho),$$

and so it can be seen that the test procedures given in equations (3.1) and (3.2) are identical. The proof of Theorem 3.1 is then completed by noting that the test procedure given in equation (3.1) is in the class Ψ. ∎

Taking $j = 1$ in Theorem 3.1, the minimax test procedure for the alternative set $\{\mu : r_1 \geq \delta\}$ is seen to be based on the statistic

$$\sum_{i=1}^{k} \cosh\{\delta k(X_i - \bar{X})/\sigma^2(k - 1)(1 - \rho)\}.$$

It is interesting to compare the minimax test procedure with standard test procedures. Let $\epsilon = \delta/\sigma$ and define $\mathbf{Y} = k\mathbf{X}/\sigma(1 - \rho)(k - j)$. Then the minimax test procedure is based on the statistic

$$\sum_{J} \cosh\{\epsilon(\bar{Y}_J - \bar{Y})\}.$$

Notice that under the null hypothesis, the distribution of $\mathbf{Y}$ is independent of both δ and σ and hence of ϵ. For small values of ϵ, the minimax test statistic can be written

$$\sum_{J} \cosh\{\epsilon(\bar{Y}_J - \bar{Y})\} = C_j^k + \frac{\epsilon^2}{2} \sum_{J} (\bar{Y}_J - \bar{Y})^2 + O(\epsilon^4),$$

where $C_j^k = k!/j!(k-j)!$. Then since for $1 \leq j \leq k-1$,

$$\frac{k(k-1)j}{(k-j)C_j^k} \sum_J (\bar{Y}_J - \bar{Y})^2 = \sum_{i=1}^k (Y_i - \bar{Y})^2 \propto \sum_{i=1}^k (X_i - \bar{X})^2,$$

it is apparent that for small values of ϵ, the minimax test procedure is approximately the same as the test procedure based on the sum of squares statistic. On the other hand, for very large values of ϵ, all terms in the expression

$$\sum_J \cosh\{\epsilon(\bar{Y}_J - \bar{Y})\} = \frac{1}{2} \sum_J \{\exp(\epsilon(\bar{Y}_J - \bar{Y})) + \exp(-\epsilon(\bar{Y}_J - \bar{Y}))\}$$

become negligible in comparison with the term

$$\exp\{\epsilon \max_J |\bar{Y}_J - \bar{Y}|\} = \exp\{w \max_J |\bar{X}_J - \bar{X}|\}.$$

Hence, for very large values of ϵ, the minimax test procedure is approximately the same as the range test procedure based on $\max_J |\bar{X}_J - \bar{X}|$. It can also be seen that as ϵ approaches zero, the power level of the minimax test procedure at the least favorable configuration approaches the size α, and as ϵ approaches infinity, the power level approaches one. Therefore, for the alternative set $\{\mu : r_1 \geq \delta\}$, say, the minimax test procedure is approximately the same as the slippage test procedure based on $\max_{1 \leq i \leq k} |X_i - \bar{X}|$ as the guarantied power level within the alternative set approaches one. Similarly, the minimax test procedure is approximately equal to the test procedure based on the sum of squares statistic as the guarantied power level approaches the size α.

Next, in Theorem 3.2, the minimax test procedures for alternative sets of the form $\Theta_A(c_2(j), \delta)$ are derived.

Theorem 3.2

Let $f(\mathbf{x}) = \phi(\mathbf{x})$. Then within the class of test procedures $\psi(\mathbf{x}) \in \Psi$ which have size α, the test which maximizes the minimum power within $\Theta_A(c_2(j), \delta)$ is

$$\psi(\mathbf{x}) = 1 \quad \Longleftrightarrow \quad \sum_{J_1, J_2} \cosh\{z(\bar{x}_{J_1} - \bar{x}_{J_2})\} \geq c \qquad (3.3)$$

where the sum is over all pairs of disjoint subsets J_1, J_2 of $\{1, \ldots, k\}$ both containing j elements, $\bar{x}_{J_i}$ is the average of the element of $\mathbf{x}$ with indices in J_i, $z = \delta/2\sigma^2(1-\rho)$, and the constant c is chosen so that the test procedure has size α.

Proof of Theorem 3.2

The proof is similar to the proof of Theorem 3.1. Here $\Sigma^{-1}\mu_2(j) = \mu_2(j)/\sigma^2(1-\rho)$ so that the test

$$\psi(\mathbf{x}) \qquad \Longleftrightarrow \qquad \sum_{\pi \in \Pi} \cosh\{(\pi\mathbf{x})'\Sigma^{-1}\mu_2(j)\} \geq c',$$

is identical to the test procedure given in equation (3.3) which is in the class Ψ. ∎

Taking $j = 1$ in Theorem 3.2, the minimax test for the alternative set $\{\mu : r_2 \geq \delta\}$, is seen to be based on the statistic

$$\sum_{1 \leq i < j \leq k} \cosh(\delta(X_i - X_j)/2\sigma^2(1-\rho)).$$

Again, it is interesting to consider the minimax test procedure for limiting values of $\epsilon = \delta/\sigma$. In this case define $\mathbf{Y} = \mathbf{X}/2\sigma(1-\rho)$, so that, as before, the null distribution of $\mathbf{Y}$ is independent of ϵ. Then for small values of ϵ, the minimax test statistic can be written

$$\sum_{J_1,J_2} \cosh\{\epsilon(\bar{Y}_{J_1} - \bar{Y}_{J_2})\} = C_j^k C_j^{k-j}/2 + \frac{\epsilon^2}{2} \sum_{J_1,J_2} (\bar{Y}_{J_1} - \bar{Y}_{J_2})^2 + O(\epsilon^4).$$

Then since for $1 \leq j \leq k/2$,

$$\frac{j(k-1)}{2C_j^k C_j^{k-j}} \sum_{J_1,J_2} (\bar{Y}_{J_1} - \bar{Y}_{J_2})^2 = \sum_{i=1}^{k} (Y_i - \bar{Y})^2 \propto \sum_{i=1}^{k} (X_i - \bar{X})^2,$$

it is again apparent that for small values of ϵ, the minimax test procedure is approximately the same as the test procedure based on the sum of squares statistic. For very large values of ϵ, all terms in the expression

$$\sum_{J_1,J_2} \cosh\{\epsilon(\bar{Y}_{J_1} - \bar{Y}_{J_2})\} = \frac{1}{2} \sum_{J_1,J_2} \{\exp(\epsilon(\bar{Y}_{J_1} - \bar{Y}_{J_2})) + \exp(-\epsilon(\bar{Y}_{J_1} - \bar{Y}_{J_2}))\}$$

become negligible in comparison with the term

$$\exp\{\epsilon \max_{J_1,J_2} |\bar{Y}_{J_1} - \bar{Y}_{J_2}|\} = \exp\{z \max_{J_1,J_2} |\bar{X}_{J_1} - \bar{X}_{J_2}|\}.$$

Hence, for very large values of ϵ, the minimax test procedure is approximately the same as the range test procedure based on $\max_{J_1,J_2} |\bar{X}_{J_1} - \bar{X}_{J_2}|$.

These results show that for the alternative set $\{\mu : r_2 \geq \delta\}$, the minimax test procedure is approximately the same as the test procedure based on the range statistic $\max_{1 \leq i,j \leq k} |X_i - X_j|$ as the guarantied power level approaches one, and the minimax test procedure is approximately the same as the test procedure based on the sum of squares statistic as the guarantied power level approaches the size α.

These observations on the approximate minimaxity of the sum of squares test procedure for small values of ϵ are consistent with the finding of Cohen et al. (1985) that the sum of squares test procedure is the locally most powerful unbiased test procedure for this problem among the class of all exchangeable test procedures (that is, the difference in power levels between the sum of squares test procedure and any other exchangeable unbiased test procedure is greater than or equal to zero in neighborhoods about points on the line in the parameter space corresponding to the null hypothesis within planes orthogonal to this line).

Finally, it may be the case that in defining an alternative set $\Theta_A(\mathbf{c}, \delta)$, the experimenter is quite clear as to which contrast $\mathbf{c}$ will provide an alternative set with a useful shape, but is not so clear as to the choice of the width δ. The choice of δ is important since the minimax test procedures given in equations (3.1) and (3.3) depend on the width δ. One way to avoid this problem is for the experimenter to specify a positive function $h(\delta)$, $\delta \geq 0$, and then to use a test procedure which maximizes the worst power within the alternative set $\Theta_A(\mathbf{c}, \delta)$ averaged over all values of δ, $0 \leq \delta < \infty$, weighted with respect to the function $h(\delta)$. In particular, $h(\delta)$ may just be taken to be a constant function. Thus, if $\mu(\delta)$ is a least favorable configuration for the alternative set $\Theta_A(\mathbf{c}, \delta)$ for all test procedures $\psi(\mathbf{x}) \in \Psi$, then the minimax test procedure will be the test procedure $\psi(\mathbf{x}) \in \Psi$ which maximizes

$$\int_{\delta=0}^{\infty} \int_{\mathbf{R}^k} \psi(\mathbf{x}) \phi(\mathbf{x} - \mu(\delta)) h(\delta) d\mathbf{x} d\delta$$

subject to the size constraint. The minimax test procedures may be derived in a manner similar to above. If $h(\delta)$ is a constant function, for example, then the minimax test procedure for the alternative set $\{\mu : r_1 \geq \delta\}$ is

$$\psi(\mathbf{x}) = 1 \qquad \Longleftrightarrow \qquad \sum_{i=1}^{k} \exp\{k(x_i - \bar{x})^2 / 2(k-1)\sigma^2(1-\rho)\} \geq c,$$

and the minimax test procedure for the alternative set $\{\mu : r_2 \geq \delta\}$ is

$$\psi(\mathbf{x}) = 1 \qquad \Longleftrightarrow \qquad \sum_{1 \leq i < j \leq k} \exp\{(x_i - x_j)^2 / 4\sigma^2(1-\rho)\} \geq c.$$

4   Summary and Discussion

The minimax test procedures have been derived for guaranteeing power levels under restrictions on the treatment means using range measures such as r_1 and r_2. As discussed in the introduction, the use of these measures for power assessment and experimental design is common, and the results indicate how this method of power assessment may be tackled more efficiently. Comparisons of the minimax test procedures with standard test procedures show that if the guaranteed power level is close to one, then test procedures based on appropriate range statistics are approximately minimax. Usually, an experiment will be designed to achieve a power level within the specified alternative set close to one, and so this suggests that using the appropriate range statistic will be close to minimax. Minimax invariant tests are often most stringent. However, it it not clear whether the minimax test procedures derived in this paper are most stringent since they depend on the value of δ (see Lehmann 1986, chapter 9 for more details).   When the variance σ^2 is unknown, the derivation of a minimax test procedure is more difficult, but it is expected that the Studentized versions of these range tests will remain close to minimax, at least when the degrees of freedom of the estimate S^2 of σ^2 are not too small. When the variance σ^2 is unknown, the width δ of the alternative set must be specified as a multiple of σ in order to be able to meet the power requirements by using a single-stage experiment as proposed in this paper. In the known variance case, it is shown in this paper that the test procedure based on the sum of squares statistic turns out to be approximately minimax only when the guarantied power level is very small which is not so useful. However, this holds for all of the alternative sets considered, and this is consistent with previous observations that the power of the F-test maintains reasonable power levels at all points of the parameter space.

The minimax test procedures may be used in practice with suitable critical points being calculated by, for example, simulation methods. However, the actual gains in power levels over the corresponding range tests may not be too great. For example, in the known variance case, a simulation study of the power levels of the minimax test procedure for the range measure r_2 and the range test procedure and the sum of squares test has been performed by Hayter and Hurn (1992). This simulation study confirms the theoretical results presented here, and also reveals that for power levels larger than about 0.5, the power functions of the minimax test procedure and the range test procedure are almost identical. Thus, in practice, it is probably sufficient to use the test procedures based on the range statistics.

References

Cohen, A., Sackrowitz, H.B. and Strawderman, W.E. (1985). Multivariate locally most powerful unbiased tests. In *Multivariate Analysis VI* (P.R. Krishnaiah, ed.), 121 - 144. Elsevier Science Publishers.

Eaton, M.L. (1987). *Lectures on topics in probability inequalities.* CWI Tract 35, Amsterdam, The Netherlands.

Hayter, A.J. and Hurn, M. (1992). Power comparisons between the F-test, the Studentized range test, and an optimal test of the equality of several normal means. *Jour. Statist. Comput. and Simul.* **42** 3 - 4, 173 - 185.

Hayter, A.J. and Liu, W. (1990). The power function of the Studentized range test. *Ann. Statist.* **18** 465 - 468.

Kastenbaum, M.A., Hoel, D.G. and Bowman, K.O. (1970). Sample size requirements: One-way analysis of variance. *Biometrika* **57** 421 - 430.

Lehmann, E.L. (1986). *Testing Statistical Hypotheses.* John Wiley and Sons, Inc., New York.

Mudholkar, G. (1966). The integral of an invariant unimodal function over an invariant convex set - an inequality and applications. *Proc. Amer. Math. Soc.* **17** 1327 - 1333.

Pearson, E.S., and Hartley, H.O. (1951). Charts of the power function for analysis of variance tests, derived from the non-central F-distribution. *Biometrika* **38** 112 - 130.

Scheffé, H. (1959). *The Analysis of Variance.* John Wiley and Sons, Inc., New York.

Chapter 10

Approximate Simultaneous Confidence Intervals

JOSEPH GLAZ   Department of Statistics, University of Connecticut, Storrs, Connecticut

Abstract   The difficulty in carrying out simultaneous inference procedures is that we are quite often unable to evaluate the distribution of a multidimensional random vector. In this article the Bonferroni-type and the product-type inequalities that have been used in the statistical literature to approximate multivariate distributions will be reviewed. The dependence structures of the multivariate distributions that guarantee the validity of the product-type inequalities will be presented. The inequalities play an important role in the area of simultaneous estimation and prediction. Special attention will be given to simultaneous prediction of future observations in time series models. Numerical examples for two simple time series will be given. The numerical results show that using higher order probability inequalities improves the accuracy of the simultaneous confidence intervals. The product-type inequalities produce more accurate prediction intervals than the Bonferroni-type inequalities.

1   Introduction and Summary

In this article I will review the probability inequalities that are used in simultaneous inference procedures and in particular in the derivation of simultaneous confidence intervals. The difficulty in deriving simultaneous confidence intervals has its roots in the evaluation of the probability distribution of a random vector. As exact results are usually not available, to facilitate these computations various probability inequalities have been developed. The tighter the probability inequality, the more accurate simultaneous confidence intervals can be obtained.

In Section 2 of this article, *product-type* inequalities will be discussed. These inequalities are valid when the random vector has a specified dependence structure. I survey the development of the product-type inequalities in the statistical literature from the appearance of the article by Kimball (1951). In fact one can trace the origin of product-type inequalities to Chebyshev (see the remarks and references in Hardy et al. 1952, p. 43). The focus in Section 2 is on the product-type inequalities that are used in constructing simultaneous confidence intervals for the mean of a multivariate normal distribution.

In Section 3 of this article a survey is presented for the Bonferroni-type inequalities. The origin of these inequalities can be traced back to Boole (1854). I will present only a certain class of Bonferroni-type inequalities, that have been suggested by Hunter (1976) (later rediscovered by Worsley 1982) and have been extended recently by Hoover (1990). These inequalities are attractive in applications as they are valid without any assumption on the distribution of the random vector that is utilized in the simultaneous inference procedures. A comparison of the product-type and the Bonferroni-type inequalities discussed in this article is presented in Section 3.

In Section 4 the product-type and the Bonferroni-type inequalities are applied to simultaneous prediction intervals for forecasting future observations in time series models. For two simple time series models numerical examples, presented in Tables 1 and 2, respectively, compare the performance of these inequalities in deriving the simultaneous prediction intervals.

2   Product-Type Inequalities

The first use of a product-type inequality to approximate a multivariate cumulative distribution function is in Kimball (1951).

Theorem 2.1 (Kimball 1951)

Let X be a random variable with the density function $f(x)$ and let $g_i(x)$, $i = 1, \cdots, k$ be nonnegative monotone functions of the same type. Then,

$$E\left[\prod_{i=1}^{k} g_i(X)\right] \geq \prod_{i=1}^{k} E[g_i(X)]. \tag{2.1}$$

Proof of Theorem 2.1

The proof is given for $k = 2$ only as it extends easily by induction for any value of k. Let x_0 be a real number such that $x_0 = \sup\{x; g_2(x) \leq E[g_2(X)]\}$. Then it is easy to verify that

$$E[g_1(X)g_2(X)] - E[g_1(X)]E[g_2(X)]$$
$$= E\{[g_1(X) - g_1(x_0)][g_2(X) - E[g_2(X)]]\}$$
$$= \int_{-\infty}^{x_0} [g_1(x) - g_1(x_0)][g_2(x) - E[g_2(X)]]f(x)dx$$
$$+ \int_{x_0}^{\infty} [g_1(x) - g_1(x_0)][g_2(x) - E[g_2(X)]]f(x)dx \geq 0.$$

This completes the proof of Theorem 2.1. ∎

This result has been applied in Kimball (1951) to a two way analysis of variance problem (no interaction) in which two hypotheses (for row effect R and column effect C) are tested simultaneously. The test statistics for testing these two hypotheses are

$$F_1 = MSR/MSE \quad \text{and} \quad F_2 = MSC/MSE,$$

where MSE is the mean square for error. Let α be the probability of type I error for each of the tests. Then, it follows from Theorem 2.1 that the probability of making no type I error is

$$P\{F_1 \leq F_{r-1;\alpha}, F_2 \leq F_{c-1;\alpha}\} \geq P\{F_1 \leq F_{r-1;\alpha}\}P\{F_2 \leq F_{c-1;\alpha}\} = (1-\alpha)^2,$$

where $F_{r-1;\alpha}$ and $F_{c-1;\alpha}$ are the α quantiles of the distributions of F_1 and F_2, respectively (assuming we have r levels of R and c levels of C). The overall probability of type I error for carrying out both testing procedures is bounded above by $1 - (1 - \alpha)^2$. For example, if $\alpha = .05$ then in this situation the overall probability of type I error is bounded above by .0975.

Later, the product-type inequalities have been utilized in constructing rectangular simultaneous confidence intervals for the means of a multivariate normal distribution (Dunn 1958, Khatri 1967, Sidak 1967, 1971, and Scott 1967). The following result of Sidak (1967) is central to this development:

Theorem 2.2

Let $\boldsymbol{X} = (X_1, \cdots, X_n)'$ have a multivariate normal distribution with mean $= 0$ and a covariance matrix $\sum$ Let S be a positive random variable independent of $\boldsymbol{X}$ Then for any positive constants $c_1, \cdots, c_k$

$$P[|X_1|/S \leq c_1, \cdots, |X_k|/S \leq c_k] \geq \prod_{i=1}^{k} P[|X_i|/S \leq c_i]. \qquad (2.2)$$

The above result is useful in the following experimental setup. Let $Y_1, \cdots, Y_n$ be independent, indentically distributed observations (i.i.d.) from a multivariate normal distribution with mean $= \mu = (\mu_1, \cdots, \mu_k)'$ and covariance matrix $\sum$ Assume that all the components of the random vectors Y_i have the same variance, which is estimated by a sample variance S^2 (Sidak 1967, Section 5). Then the simultaneous confidence intervals for μ_i, $i = 1, \cdots, k$, are given by

$$\bar{Y}_i \pm c \, S/n^{-\frac{1}{2}}, \qquad i = 1, \cdots, k.$$

The value c is determined from the inequality (2.2) with $X_i = n^{\frac{1}{2}}(\bar{Y}_i - \mu_i)$ and the marginal distribution of X_i/S, $i = 1, \cdots, k$.

To extend the applicability of the rectangular simultaneous confidence intervals for the means of a multivariate normal distribution, in the case of the unequal variances, Sidak (1971) generalized the inequality in Theorem 2.2. The generalized inequality is valid for a restricted covariance matrix of the multivariate normal distribution (Sidak 1971, Section 2).

Since the validity of these product-type inequalities is not easy to determine a more general approach was needed. Esary et al. (1967) introduced the following concept of positive dependence.

Definition   The random vector $X = (X_1, \cdots, X_n)'$ is *associated* if for all coordinatewise nondecreasing functions f and g such that $E[f(X)]$ and $E[g(X)]$ are finite

$$\text{Cov}[f(X), g(X)] = E[f(X)g(X)] - E[f(X)]E[g(X)] \geq 0.$$

If X is associated then the random variables $X_1, \cdots, X_n$ are said to be associated. The following results are true (Esary et al. 1967) :

a. If $X = (X_1, \cdots, X_n)'$ is associated then for any set of indices $1 \leq i_1 < \cdots < i_k \leq n$, $(X_{i_1}, \cdots, X_{i_k})'$ is associated.

b. The random vector with components being a union of components of associated and independent random vectors, is associated.

c. Any random variable is associated (Kimball 1951).

d. If $X = (X_1, \cdots, X_n)'$ is associated and f_i are coordinatewise nondecreasing functions then $Y = (Y_1, \cdots, Y_m)'$ is associated, where $Y_i = f_i(X)$, $i = 1, \cdots, m$.

e. A random vector of independent components is associated.

Theorem 2.3 (Esary et al. 1967)

If $X = (X_1, \cdots, X_n)'$ is associated then for all $c_1, \cdots, c_n$

$$P[X_1 \leq c_1, \cdots, X_n \leq c_n] \geq \prod_{i=1}^{n} P[X_i \leq c_i] \tag{2.3}$$

and

$$P[X_1 \geq c_1, \cdots, X_n \geq c_n] \geq \prod_{i=1}^{n} P[X_i \geq c_i]. \tag{2.4}$$

Examples

a. If $X = (X_1, \cdots, X_n)'$ has a multivariate normal distribution then X is associated if and only if all the correlations are nonnegative (Pitt 1982 and Joag-Dev et al. 1983).

b. If $X = (X_1, \cdots, X_n)'$ has a multivariate exponential distribution (Marshall and Olkin 1967) then it is associated.

c. Order statistics of a sequence of associated (independent) random variables are associated.

d. Partial sums of associated (independent) random variables are associated.

In general it may be difficult to check if a sequence of random variables are associated. For example, it is still an open problem to present necessary and sufficient conditions for $|X_1|, \cdots, |X_n|$ to be associated, when $X = (X_1, \cdots, X_n)'$ has a multivariate normal distribution. Still, results in this direction are needed to construct rectangular simultaneous confidence intervals for the means. The following stronger concept of positive dependence plays a major role in the development of product-type inequalities.

Definition (Karlin 1968) A nonnegative real valued function of two variables, $f(x_1, x_2)$, is totally positive of order two (TP_2) if for all $x_1 < x_2$ and $x_1^* < x_2^*$,

$$f(x_1, x_2)f(x_1^*, x_2^*) - f(x_1, x_2^*)f(x_1^*, x_2) \geq 0.$$

Definition (Barlow and Proschan 1975) A nonnegative real valued function of n variables, $f(x_1, \cdots, x_n)$, is totally positive of order two in pairs (TP_2 in pairs) if for any pair of arguments x_i and x_j, the function f viewed as a function of x_i and x_j only, with the other arguments kept fixed, is TP_2.

Definition (Karlin and Rinott 1980a)   A nonnegative real valued function of n variables, $f(x_1, \cdots, x_n)$, is multivariate totally positive or order two (MTP_2), if for all $\boldsymbol{x} = (x_1, \cdots, x_n)'$, $\boldsymbol{y} = (y_1, \cdots, y_n)'$

$$f(\boldsymbol{x} \wedge \boldsymbol{y}) f(\boldsymbol{x} \vee \boldsymbol{y}) \geq f(\boldsymbol{x}) f(\boldsymbol{y}),$$

where

$$\boldsymbol{x} \wedge \boldsymbol{y} = (\min(x_1, y_1), \cdots, \min(x_n, y_n))'$$

and

$$\boldsymbol{x} \vee \boldsymbol{y} = (\max(x_1, y_1), \cdots, \max(x_n, y_n))'.$$

If f is MTP_2 then it is also TP_2 in pairs. But, if the support of f is a product space then MTP_2 and TP_2 in pairs are equivalent (Block and Ting 1981).

A random vector is said to be MTP_2 (respectively, absolute value MTP_2 or $AMTP_2$) if its joint density function is MTP_2 (respectively, the joint density function of the absolute value of its components is MTP_2). The following result is of use in showing that a random vector is associated.

Theorem 2.4 (Sarkar 1969)

If the random vector $\boldsymbol{X}$ is MTP_2 then it is associated.

The following interesting examples of MTP_2 random variables are listed in Karlin and Rinott (1980a): multivariate normal with nonnegative partial correlations, absolute value multivariate normal with only nonnegative off-diagonal elements in $-D \sum^{-1} D$, where $\sum$ is the covariance matrix and D is some diagonal matrix with elements ± 1, multivariate logistic, absolute value multivariate Cauchy, certain classes of multivariate t, negative multinomial, certain classes of partial sums of i.i.d. random variables, and order statistics of i.i.d. random variables.

Example   Let $\boldsymbol{Y}_1, \cdots, \boldsymbol{Y}_n$ be i.i.d. observations from a multivariate normal distribution with mean $= \mu$ and covariance matrix $\sum$. Assume that the components of the random vectors $\boldsymbol{Y}_i$ have unequal variances, which are estimated by the sample variances S_i^2, $i = 1, \cdots, k$ (Sidak 1971, Corollary 2). If the density function of the multivariate normal distribution is $AMTP_2$ then it follows from Karlin and Rinott (1981, Section 6) that

$$P[|X_1|/S_1 \leq c_1, \cdots, |X_k|/S_k \leq c_k] \geq \prod_{i=1}^{k} P[|X_i|/S_i \leq c_i], \qquad (2.5)$$

where $X_i = n^{\frac{1}{2}}(\bar{Y}_i - \mu_i)$. Then the simultaneous confidence intervals for μ_i, $i = 1, \cdots, k$, are given by

$$\bar{Y}_i \pm c S_i / n^{-\frac{1}{2}}, \qquad i = 1, \ldots, k.$$

The value c is determined from the inequality (2.5) with $X_i = n^{\frac{1}{2}}(\bar{Y}_i - \mu_i)$ and the marginal t-distribution of X_i/S_i, $i = 1, \cdots, k$. For examples of $AMTP_2$ multivariate normal distributions see Karlin and Rinott (1981, Section 4).

Inequalities (2.2) - (2.5) are referred to as *first order product-type inequalities*, since only the one dimensional marginal distributions have been used. An approximation or an inequality for the probability of an intersection or union of n events is said to be of *order k*, if j dimensional marginal distributions are used in computing it, where $j \leq k$. While the first order product-type approximations and inequalities have the advantage of ease of computation, they are often quite inaccurate (Glaz and Johnson 1984, 1986). The reason for that being that the dependence structure inherent in the random process is exploited only to a minimal degree. Therefore, Glaz and Johnson (1984) proposed to study product-type inequalities of degree $k \geq 2$. The following result is central to the study of these inequalities:

Theorem 2.5 (Glaz and Johnson 1984)

Let $\boldsymbol{X} = (X_1, \cdots, X_n)'$ be an MTP_2 random vector and $I_1, \cdots, I_n$ infinite intervals of the same type. Then,

a.

$$P[X_i \in I_i; \ i = 1, \cdots, n] \geq \gamma_k \geq \gamma_1 \qquad (2.6)$$

b. γ_k is increasing in k, where

$$\gamma_k = P[X_i \in I_i; \ i = 1, \cdots, k] \qquad (2.7)$$
$$\times \prod_{i=k+1}^{n} P[X_i \in I_i | X_j \in I_j; j = i - k + 1, \cdots, i - 1].$$

If the X_i's are stationary and $I_i = I$ then the above equation simplifies to:

$$\gamma_k = P[X_i \in I_i; i = 1, \cdots, k](P[X_k \in I_k | X_j \in I_j; \ j = 1, \cdots, k - 1])^{n-k}.$$

In Section 4 of this article these product-type inequalities will be utilized in constructing simultaneous confidence intervals for the prediction of future observations in time series models.

Remarks

a. The MTP_2 condition in Theorem 2.5 can be relaxed (Glaz 1990) but not as easy to verify.

b. If $\boldsymbol{X} = (X_1, \cdots, X_n)'$ has a multivariate normal distribution and we are interested in product-type inequalities of order 2 or 3, Block et al. (1988a) relax the MTP_2 condition in Theorem 2.5.

c. Improved product-type bounds are discussed in Block et al. (1988b) and Ravishanker et al. (1991).

d. The related product-type inequalities for negative dependence are discussed in Glaz and Johnson (1984) and Glaz (1990).

3   Bonferroni-Type Inequalities

The classical Bonferroni inequalities for the probability of a union of n events have been introduced in Bonferroni (1937). Let $A_1, \cdots, A_n$ be a sequence of events and define the event $A = \bigcup_{i=1}^{n} A_i$. Then for $2 \leq k \leq n$,

$$\sum_{j=1}^{k}(-1)^{j-1}S_j \leq P(A) \leq \sum_{j=1}^{k-1}(-1)^{j-1}S_j$$

where k is an even integer and for $j = 1, \cdots, n$

$$S_j = \sum_{1 \leq i_1 < \cdots < i_j \leq n} P(\bigcap_{m=1}^{j} A_{i_m}). \tag{3.1}$$

The first order Bonferroni upper bound is referred to as Boole's inequality and has been introduced earlier in Boole (1854). Boole's inequality has been extensively used in statistical inference to construct conservative simultaneous confidence intervals and simultaneous testing procedures (Angers and McLaughlin 1979, Bailey 1980, Bhansali 1974, Bohrer et al. 1981, Dunn 1958, Fuchs and Sampson 1980, Paulson 1952, and Ringland 1983). Higher order classical Bonferroni inequalities have not been used in statistical inference because of the computational complexity. Moreover, the classical upper (lower) Bonferroni bounds do not necessarily decrease (increase) with the order of the inequality. For an example see Schwager (1984), who presents an interesting discussion of these inequalities.

As the classical Bonferroni inequalities can be very inaccurate attempts have been made to improve them. Refer to these as Bonferroni-type inequalities. In this article I will concentrate on one special class of Bonferroni-type inequalities, that was proposed by Hunter (1976). The basic idea of this approach is to express

$$A = A_1 \cup (A_2 \cap A_1^c) \cup (A_3 \cap A_2^c \cap A_1^c)$$
$$\cdots \cup \cdots \cup (A_n \cap A_{n-1}^c \cap \cdots \cap A_1^c) \tag{3.2}$$

and obtain the following inequality

$$P(A) \leq P(A_1) + \sum_{i=2}^{n} P(A_i \cap A_{i-1}^c) \tag{3.3}$$

$$= \sum_{i=1}^{n} P(A_i) - \sum_{i=2}^{n} P(A_i \cap A_{i-1}).$$

This inequality is a member of the following class of second order Bonferroni-type inequalities that have been derived in Hunter (1976). Let $v_1, \cdots, v_n$ be the vertices of a tree T, representing the events $A_1, \cdots, A_n$ respectively. The vertices v_i and v_j are joined by an edge e_{ij} if and only if $A_i \cap A_j \neq \emptyset$.

Theorem 3.1 (Hunter 1976)

$$P(A) \leq \sum_{i=1}^{n} P(A_i) - \sum_{\{(i,j); e_{ij} \in T\}} P(A_i \cap A_j). \tag{3.4}$$

The inequality (3.3) is the most stringent within the class of inequalities (3.4) if the events $A_1, \cdots, A_n$ are exchangeable or are ordered in such a way that for $1 \leq i_1 < i_2 \leq n$, $P(A_{i_1} \cap A_{i_2})$ is maximized for $i_2 - i_1 = 1$ (see Worsley 1982, Examples 3.1 and 3.2).

Recently, Hoover (1990) extended this class of Bonferroni-type inequalities to order $k \geq 3$. Consider again the identity (3.2). Then,

$$P(A) \leq S_1 - \sum_{i=1}^{n-1} P_{i,i+1} - \sum_{j=2}^{k-1} \sum_{i=1}^{n-j} p^*_{i,i+1,\cdots,i+j} \tag{3.5}$$

where S_1 is defined in equation (3.1),

$$p_{i,i+1} = P(A_i \cap A_{i+1})$$

and

$$p^*_{i,i+1,\cdots,i+j} = P(A_i \cap A^c_{i+1} \cap \cdots \cap A^c_{i+j-1} \cap A_{i+j}).$$

The inequality (3.5) is a member of the following class of inequalities.

Theorem 3.2 (Hoover 1990)

If $1 \leq k \leq n - 1$, then

$$P(A) \leq P[\bigcup_{i=1}^{k} A_i] + \sum_{j=k+1}^{n} P\{A_j \cap [\bigcap_{\substack{1 \leq i_1 < \cdots < i_{k-1} \leq n \\ i_1, \cdots, i_{k-1} \in T_j}} (\bigcup_{j=1}^{k-1} A_{i_j})^c]\}, \tag{3.6}$$

where T_j is a subset of $\{1, 2, \cdots, j-1\}$ of size $k-1$ and $j \geq k+1$.

Remarks

a. It is easy to verify that the inequality (3.6) is equivalent to the following inequality:

$$P(A) \leq P[\bigcup_{i=1}^{k} A_i] + \sum_{j=k+1}^{n} P\{A_j \cap [\bigcap_{s=1}^{k-1} A_{i_{s,j}}^c]\}, \qquad (3.7)$$

where $1 \leq i_{s+1,j} < i_{s,j} \leq j - 1$.

b. For $k = 1$ and $k = 2$ the inequality (3.6) reduces to the Boole and the Hunter inequalities, respectively.

c. In the case the events $A_1, \cdots, A_n$ are naturally ordered in such a way that $P(\bigcap_{j=1}^{k-1} A_{i_j})$ is maximized for $i_j - i_{j-1} = 1$, $2 \leq j \leq k - 1$ and $3 \leq k \leq n - 1$, the natural ordering with $T_j = \{j - 1, \cdots, j - k + 1\}$ is recommended. In this case (3.6) reduces to (3.5).

d. If $A_1, \cdots, A_n$ are exchangeable events then (3.6) reduces to (3.5) and has the following simple form:

$$P(A) \leq nP(A_1) - (n - 1)p_{1,2} - \sum_{j=2}^{k-1}(n - j)p_{1,2,\cdots,j+1}^*. \qquad (3.8)$$

e. The classical Bonferroni upper (lower) bounds do not necessarily decrease (increase) in k (Schwager 1984). The Bonferroni-type upper bound given in equations (3.5) - (3.8) decrease in k.

f. For $k \geq 3$ there is no simple algorithm to obtain the optimal inequality from the class of inequalities (3.6), or equivalently, (3.7). One can use the approach in Seneta (1988) to obtain a tighter inequality of the following form:

$$P(A) \leq P[\bigcup_{i=1}^{k} A_i] + \sum_{j=k+1}^{n} P\{A_j \cap [\bigcap_{s=1}^{k-1} A_{i_{s,j}^*}^c]\}, \qquad (3.9)$$

where

$$P\{A_j \cap [\bigcap_{s=1}^{k-1} A_{i_{s,j}^*}^c]\} = \min_{1 \leq i_{s+1,j} < i_{s,j} \leq j-1} P\{A_j \cap [\bigcap_{s=1}^{k-1} A_{i_{s,j}}^c]\}$$

and $1 \leq i_{s+1,j}^* < i_{s,j}^* \leq j - 1$.

g. Other interesting approaches to obtain improved Bonferroni-type inequalities are discussed in Hoppe (1985), Hoppe and Seneta (1990), Prekopa (1988), Rescei and Seneta (1987), Seneta (1988), and Tomescu (1986).

The computational effort in evaluating the product-type inequalities given in equations (2.6) - (2.7) and the Bonferroni-type inequalities (3.5) is the same. To see that this is the case, define for $i = 1, \cdots, n$

$$\alpha_{m,n} = P(\bigcap_{i=m}^{n} E_i),$$

where $E_i = (X_i \in I_i)$. Then it is tedious but routine to verify that the kth order Bonferroni-type inequality (3.5) can be expressed as

$$P(\bigcap_{i=1}^{n} E_i) \geq \sum_{i=1}^{n+1-k} \alpha_{i,i+k-1} - \sum_{i=1}^{n-k} \alpha_{i+1,i+k-1} = \beta_k. \qquad (3.10)$$

The kth order product-type inequality given in (2.6) — (2.7) can be written as

$$P(\bigcap_{i=1}^{n} E_i) \geq \alpha_{1,k} \prod_{i=k+1}^{n} (\alpha_{i-k+1,i}/\alpha_{i-k+1,i-1}) = \gamma_k. \qquad (3.11)$$

The advantage in using the Bonferroni-type inequalities is that they are valid without any assumptions on the distribution of the random vector $\boldsymbol{X}$. On the other hand, if the kth order product-type inequalities are valid the following result supports their use.

Theorem 3.3 (Glaz 1990)

Let $\boldsymbol{X} = (X_1, \cdots, X_n)'$ be a random vector and $E_i = (X_i \in I_i)$, $i = 1, \cdots, n$. If γ_k and β_k are the product-type and the Bonferroni-type inequalities, given by (3.10) and (3.11) respectively, then

$$\gamma_k \geq \beta_k.$$

Block et al. (1988b) developed an optimized version of the second order product-type inequality under conditions of positive dependence. As part of their work, they show that the second-order product-type inequality developed in Glaz and Johnson (1984) is superior to the corresponding second-order Bonferroni type inequality. Their proof of the result is analytical in nature. Hoover (1989) independently used a similar approach to the one in Glaz (1990) to derive the proof of Theorem 3.3.

In Section 4 of this article I will illustrate the performance of the Bonferroni-type and the product-type inequalities that were discussed above for the problem of predicting future observations in time series models.

4 Simultaneous Prediction for Time Series Models

To illustrate the use of the probability inequalities, discussed in Section 2 and Section 3, for simultaneous prediction we will present here two simple numerical examples from Glaz and Ravishanker (1991).

Assume that a series of observations X_t, $t = 1, \cdots, n$ have been recorded. We are interested in constructing rectangular simultaneous confidence intervals for the future observations X_{n+i}, $i = 1, \cdots, m$. The following simultaneous confidence intervals will be considered:

$$X_{n+i}(i) \pm c\sqrt{\text{Var}(e_n(i))}, \qquad i = 1, \cdots, m, \tag{4.1}$$

where for $i = 1, \cdots, m$ $X_{n+i}(i)$ are the forecasts for X_{n+i}, and $e_n(i) = X_{n+i} - X_{n+i}(i)$ are the forecast errors. The constant c is obtained from the following equation:

$$P\{|U_n(i)| \le c;\ i = 1, \cdots, m\} = 1 - \alpha, \tag{4.2}$$

where $U_n(i) = e_n(i)/\sqrt{\text{Var}(e_n(i))}$, $i = 1, \cdots, m$, are the standardized forecasts. I am assuming here that $\{\epsilon_t\}$ is a sequence of i.i.d. normal random variables with mean $= 0$ and variance $= \sigma^2$. Thus $U_n = (U_n(1), \cdots, U_n(m))'$ has a multivariate normal distribution with mean $= 0$ and correlation matrix $\sum$. To evaluate the approximate value of c for the simultaneous prediction intervals (4.1), via equation (4.2), one can employ the product-type or the Bonferroni-type inequalities. To employ the Bonferroni-type inequalities (of order $1 \le k \le 5$) one has to evaluate the elements of $\sum$, or equivalently the elements of the covariance matrix of $e_n = (e_n(1), \cdots, e_n(n))'$, to be denoted by Γ, and use the Schervish (1984) algorithm for evaluating the multivariate normal probabilities. The methods for evaluating the elements of the covariance matrices are discussed in Glaz and Ravishanker (1991) and Ravishanker et al. (1991). To apply the product-type inequalities for approximating the value of c, one has to examine the conditions of Karlin and Rinott (1981, Theorem 3.1) for the elements of the covariance matrix Γ. In what follows, two simple numerical examples will be presented.

Example 1 Series D in Box and Jenkins (1976). The data consists of $n = 310$ hourly readings of viscosity of a chemical process. One possibility is to fit an AR(1) model:

$$X_t = \phi X_{t+1} + \epsilon_t, \quad |\phi| < 1,$$

where ϕ is the autoregressive parameter of the model. For more details on the AR(1) model and the more general ARIMA models see Box and Jenkins

(1976). It follows from Glaz and Ravishanker (1991) that U_n is $AMTP_2$ and therefore the product-type inequalities (of order $1 \leq k \leq 5$) can be utilized to approximate the value of c. For this example the estimated parameters are $\hat{\phi} = .87$ and $\hat{\sigma}_\epsilon = .09$. In Table 1 the values of c used in simultaneous prediction intervals (4.1) are given for $m = 10$ using both types of inequalities. These values are compared with the values of c for the marginal prediction intervals. The simulated values of c were obtained from generating 50,000 multivariate normal random vectors using the IMSL subroutine RNMVN.

Example 2 Series D data in Box and Jenkins (1976). As an alternative to the AR(1) model, an ARIMA(0,1,1) model (first order differencing and first order moving average model) is fitted:

$$X_t = X_{t-1} - \theta \ \epsilon_{t-1} + \epsilon_t, \ |\theta| < 1,$$

where θ is the moving average parameter of the model. Denote the elements of $-\Gamma^{-1} = \{\eta_{ij}/\sigma^2\}$. For an ARIMA(0,1,1) model Glaz and Ravishanker (1991) have evaluated

$$\eta_{ij} = \left\{ \begin{array}{ll} (1 - \theta)[1 + \theta^{2(m-i)-1}]/(1 + \theta), & j - i = 1, 1 \leq i \leq m - 1 \\ \theta^{j-i-1}(1 - \theta)[1 + \theta^{2(m-j+1)-1}]/(1 + \theta), & j - i \geq 2, 2 \leq j \leq m, \end{array} \right.$$

and showed that for the situation at hand U_n is MTP_2 if $0 \leq \theta < 1$. For this data set (the estimate of θ) $\hat{\theta} = .06$, and $\hat{\sigma}_{\hat{\theta}} = .06$ so we cannot be sure that U_n is MTP2 and employ the product-type inequalities (of order $1 \leq k \leq 5$) to approximate the value of c. One can still use the product type expressions given in (2.7) as approximations. In Table 2 the values of c used in simultaneous prediction intervals (4.1) are presented for $m = 10$ using Bonferroni-type inequalities and product-type approximations. These values are compared with values of c for the marginal prediction intervals. The simulated values of c were obtained from generating 50,000 multivariate normal random vectors using the IMSL subroutine RNMVN.

From the numerical results presented in Tables 1 and 2 one can conclude that there is merit in using the higher order product-type inequalities (or approximations) and the Bonferroni-type inequalities for simultaneous prediction intervals in time series models. The product-type inequalities (or approximations) have produced more accurate values of c. When the conditions that guarantee the validity of the product-type inequalities do not hold one can safely use the Bonferroni-type inequalities to obtain the values of c, as they produce quite accurate values of c as well. For the performance of the optimized product-type and the Bonferroni-type inequalities in computing the value of c, see Ravishanker et al. (1991). The following remarks suggest a few directions for future research.

Table 1: Comparison of four values of c for series D data, $AR(1)$ model, $m = 10$.

Method	Order	Simultaneous Confidence Levels			
		95	90	80	60
Product-type	1	2.80	2.56	2.29	1.96
	2	2.68	2.41	2.11	1.73
	3	2.66	2.39	2.08	1.70
	4	2.65	2.38	2.07	1.68
	5	2.65	2.38	2.07	1.68
Bonferroni-	1	2.81	2.58	2.33	2.05
Type	2	2.69	2.43	2.14	1.81
	3	2.67	2.40	2.10	1.75
	4	2.66	2.39	2.09	1.73
	5	2.65	2.38	2.08	1.71
Simulation		2.64	2.37	2.06	1.65
Marginal		1.96	1.64	1.28	0.84

Table 2: Comparison of four values of c for series D data, $ARIMA(0, 1, 1)$ model, $m = 10$.

Method	Order	Simultaneous Confidence Levels			
		95	90	80	60
Product-type	1	2.80	2.56	2.29	1.96
	2	2.62	2.35	2.03	1.64
	3	2.59	2.31	1.99	1.59
	4	2.58	2.29	1.97	1.57
	5	2.58	2.29	1.96	1.56
Bonferroni-	1	2.81	2.58	2.33	2.05
Type	2	2.63	2.36	2.06	1.71
	3	2.59	2.32	2.01	1.64
	4	2.58	2.30	1.98	1.61
	5	2.58	2.29	1.97	1.59
Simulation		2.55	2.28	1.95	1.56
Marginal		1.96	1.64	1.28	0.84

Remarks

In general, the process parameters in time series models are unknown and are replaced by their maximum likelihood estimates. For large samples, the resulting simultaneous prediction intervals will converge to the simultaneous prediction intervals based on estimated process parameters, because of the invariance and asymptotic properties of the maximum likelihood estimators. It would be interesting to investigate the effect of the use of the estimated parameters in simultaneous estimation for moderate size time series models. A recent article by Stine (1987) discusses correction terms for Γ to account for the use of estimated parameters, which can be implemented in the procedure that was discussed above.

The variances of the forecast errors are also unknown and for small or moderate size of time series data one might consider using the multivariate t-distribution, instead of the multivariate normal, in calculating the constant c for the simultaneous prediction intervals. There are no numerical algorithms for evaluating the probabilities for the multivariate t-distribution needed for this problem. Simulation studies could be helpful in evaluating the simultaneous prediction intervals in these cases.

Acknowledgment

The author wishes to thank the referee for many useful suggestions that improved the presentation of the results in this article.

References

Angers, C. and McLaughlin, G. (1979). A note on prediction intervals for large future samples. *Technometrics* **21** 383 - 385.

Bailey, B.J.R. (1980). Large sample simultaneous confidence intervals for multinomial probabilities based on transformations of the cell frequencies. *Technometrics* **22** 583 - 586.

Barlow, R.E. and Proschan, F. (1975). *Statistical Theory of Reliability and Life Testing*. Holt, Rinehart and Winston, New York.

Bhansali, R.J. (1974). Asymptotic mean square error of predicting more than one-step ahead using the regression model, *Appl. Statist.* **23** 35 - 42.

Block, H.W., Costigan, T. and Sampson, A. (1988a). Product-type probability bounds of higher order. Department of Mathematics and Statistics, University of Pittsburgh, Technical Report No. 88-08.

Block, H.W., Costigan, T. and Sampson, A. (1988b). Optimal second order product-type probability bounds. Department of Mathematics and Statistics, University of Pittsburgh, Technical Report No. 88-07.

Block, H.W. and Ting, M.L. (1981). Some concepts of multivariate dependence. *Commun. Statist.* A 10 749 - 762.

Bohrer, R., Chow, W., Faith, R., Joshi, V.M., and Wu, C.F. (1981). Multiple three decision rules for factorial simple effects: Bonferroni wins again! *Jour. Amer. Statist. Assoc.* **76** 119 - 124.

Bonferroni, C.E. (1937). Teoria statistica delle classi e calcolo delle probabilita. In *Volume in Onore di Ricarrdo Dalla Volta*, Universita di Firenze, 1 - 62.

Boole, G. (1854). *An Investigation of the Laws of Thought on Which are Founded the Mathematical Theories of Logic and Probabilities.* MacMillan, London. (Reprinted by Dover, New York, 1958).

Box, G.E.P. and Jenkins, G.M. (1976). *Time Series Analysis: Forecasting and Control.* Holden Day, San Francisco.

Dunn, O.J. (1959). Confidence intervals for the means of dependent normally distributed variables. *Jour. Amer. Statist. Assoc.* **54** 613 - 621.

Esary, J.D., Proschan, F., and Walkup, D.W. (1967). Association of random variables with applications. *Ann. Math. Statist.* **38** 1466 - 1474.

Fuchs, C. and Sampson, A.R. (1987). Simultaneous confidence intervals for the general linear model. *Biometrics* **43** 457 - 469.

Glaz, J. (1990). A comparison of Bonferroni-type and product-type inequalities in presence of dependence. In *Topics in Statistical Dependence*, IMS Lecture Notes-Monograph Series, Vol. 16, 223 - 235.

Glaz, J. and Johnson, B.McK. (1984). Probability inequalities for multivariate distribution with dependence structures. *Jour. Amer. Statist. Assoc.* **79** 436 - 441.

Glaz, J. and Johnson, B.McK. (1988). Boundary crossing for moving sums. *Jour. Appl. Probab.* **25** 81 - 88.

Glaz, J. and Ravishanker, N. (1989). Simultaneous prediction intervals for multiple forecasts in time series models. IBM Research Division RC 14385 (#64405).

Glaz, J. and Ravishanker, N. (1991). Simultaneous prediction intervals for multiple forecasts based on Bonferroni and product-type inequalities. *Statist. Probab. Lett.* **12** 57 - 63.

Hardy, G.H., Littlewood, J.E. and Polya, G. (1952). *Inequalities* (Second edition). Cambridge University Press, Cambridge, England.

Hochberg, Y. and Tamhane, A.C. (1987). *Multiple Comparison Procedures.* John Wiley and Sons, Inc., New York.

Hoover, D.R. (1989). Comparison of improved Bonferroni and Sidak/Slepian bounds with applications to normal Markov processes. Department of Statistics, Stanford University, Technical Report No. 412.

Hoover, D.R. (1990). Subset complement addition upper bounds — an improved inclusion-exclusion method. *Jour. Statist. Plan. Infer.* **24** 195 - 202.

Hoppe, F.M. (1985). Iterating Bonferroni bounds. *Statist. Probab. Lett.* **3** 121 - 125.

Hoppe, F.M. and Seneta, E. (1990). A Bonferroni-type identity and permutation bounds. *Internat. Statist. Rev.* **58 (3)** 253 - 261.

Hunter, D. (1976). An upper bound for the probability of a union. *Jour. Appl. Probab.* **13** 597 - 603.

Joag-Dev, K., Perlman, M.D. and Pitt, L.D. (1983). Association of normal random variables and Slepian's inequality. *Ann. Probab.* **11** 451 - 455.

Karlin, S. (1968). *Total Positivity.* Stanford University Press, Stanford.

Karlin, S. and Rinott, Y. (1980a). Classes of orderings of measures and related correlation inequalities I. Multivariate totally positive distributions. *Jour. Multivar. Anal.* **10** 467 - 498.

Karlin, S. and Rinott, Y. (1981). Total positivity properties of absolute value multinormal variables with applications to confidence interval estimates and related probabilistic inequalities. *Ann. Statist.* **9** 1035 - 1049.

Khatri, C.G. (1967). On certain inequalities for normal distributions and their applications to simultaneous confidence bounds. *Ann. Math. Statist.* **38** 1853 - 1867.

Kimball, A.W. (1951). On dependent tests of significance in the analysis of variance. *Ann. Math. Statist.* **22** 600 - 602.

Marshall, A.W. and Olkin, I. (1967). A multivariate exponential distribution. *Jour. Amer. Statist. Assoc.* **62** 30 - 44.

Paulson, E. (1952). On the comparison of several experimental categories with a control. *Ann. Math. Statist.* **23** 239 - 246.

Pitt, L.D. (1982). Positively correlated normal random variables are associated. *Ann. Probab.* **10** 496 - 499.

Prekopa, A. (1988). Boole-Bonferroni inequalities and linear programming. *Oper. Res.* **36** 146 - 162.

Ravishanker, N., Wu, L.S.Y. and Glaz. J. (1991). Multiple prediction intervals for time series: Comparison of simultaneous and marginal intervals. *Jour. Forecast.* **10** 445 - 463.

Rescei, E. and Seneta, E. (1987). Bonferroni-type inequalities. *Adv. App. Probab.* **19** 508 - 511.

Ringland, J.T. (1983). Robust multiple comparisons. *Jour. Amer. Statis. Assoc.* **78** 145 - 151.

Schwager, S.J. (1984). Bonferroni sometimes loses. *The American Statistician* **38** 192 - 197.

Scott, A. (1967). A note on conservative confidence regions for the mean of a multivariate normal distribution. *Ann. Math. Statist.* **38** 278 - 280. Correction: *Ann. Math. Statist.* **39** 2161.

Seneta, E. (1988). Degree, iteration and permutation in improving Bonferroni-type bounds. *Austral. Jour. Statist.* **30A** 27 - 38.

Sidak, Z. (1967). Rectangular confidence regions for means of multivariate normal distributions. *Jour. Amer. Statist. Assoc.* **62** 626 - 633.

Sidak, Z. (1971). On probabilities on rectangles in multivariate student distributions: Their dependence on correlations. *Ann. Math. Statist.* **42** 169 - 175.

Stine, R.A. (1987). Estimating properties of autoregressive forecasts. *Jour. Amer. Statist. Assoc.* **82** 1072 - 1078.

Tomescu, I. (1986). Hypertrees and Bonferroni inequalities. *Jour. Combin. Theory* **B 41** 209 - 217.

Worsley, K.J. (1982). An improved Bonferroni inequality and applications. *Biometrika* **69** 297 - 302.

Chapter 11

Stepwise Multiple Comparisons of Repeated Measures Means Under Violations of Multisample Sphericity

H.J. KESELMAN   Department of Psychology, University of Manitoba, Winnipeg, Manitoba, Canada

Abstract   Stepwise multiple comparison procedures (MCPs) were studied in a groups by trials repeated measures design. The investigation compared 37 stepwise MCPs which included adaptations of: (a) Hayter's (1986) modification of Fisher's (1935) two-stage LSD procedure, (b) Hochberg's (1988) and Shaffer's (1986) modifications of Holm's (1979) sequentially rejective Bonferroni procedure, (c) Peritz's (1970) procedure, (d) multiple step-down procedures, using the Ryan (1960)-Welsch (1977a) method of Type 1 error control, (e) Shaffer's (1979) modification to range procedures, and (f) a step-up procedure due to Welsch (1977a). The stepwise MCPs involved the use of adjusted univariate F-tests, a multivariate test and/or a nonpooled t statistic due to Keselman et al. (1991). To investigate the operating characteristics of the stepwise procedures, the following factors were varied: (a) the sphericity pattern, (b) the equality/inequality of the covariance matrices, (c) the group size equality/inequality, (d) the nature of the pairing of unequal covariance matrices and unequal group sizes, and (e) the nature of the null hypothesis (i.e., complete or partial). Numerous procedures were found to be generally robust to departures from multisample sphericity.

1 Introduction

The analysis of repeated measures means with omnibus and/or multiple comparison statistics requires that certain parametric conditions exist in order to obtain valid tests. In particular, it is assumed that the variances of a set of orthonormalized contrasts among the levels of the repeated measures factor are equal; this requirement is referred to as the sphericity or circularity assumption. If the design contains a between-subjects grouping variable, then it is assumed further that this constant variance is homogeneous across the levels of the between-subjects variable. Jointly, these assumptions are referred to as multisample sphericity (Huynh 1978).

The literature suggests that for omnibus repeated measures main effect hypotheses, either an adjusted degrees of freedom univariate F-test or a multivariate test (Hotelling's 1931 T^2 test) may be used to maintain the number of Type 1 errors close to the nominal value of α, even when the design is unbalanced (See Huynh 1978; Keselman and Keselman 1990). Further, for hypotheses concerning pairwise contrasts of repeated measures means, Keselman et al. (1991) found that robust tests can be obtained in unbalanced nonspherical repeated measures designs by using a nonpooled statistic with error degrees of freedom estimated using Satterthwaite's (1941,1946) solution. Specifically, Keselman et al. (1991) employed a statistic which does not pool across the levels of either the within-subjects repeated measures factor or the between-subjects grouping factor, in arriving at an estimate of the standard error of the contrast.

The purpose of the present study was to determine whether these omnibus and multiple comparison statistics could provide robust testing of pairwise contrasts of repeated measures means for nonspherical unbalanced data when employed in various stepwise multiple comparison procedures (MCPs).

2 Stepwise Multiple Comparison Procedures

To introduce the MCPs, consider the two-way mixed design involving a single between-subjects factor and a single within-subjects factor, in which subjects ($i = 1, \cdots, n_j$, $\sum_j n_j = N$) are randomly selected for each level of the between-subjects factor ($j = 1, \cdots, J$) and observed and measured under all levels of the within-subjects factor ($k = 1, \cdots, K$). In this design, the repeated measures data are modelled by assuming that the observations X_{ijk} are normal, independent, and identically distributed random variables within each level j with common mean vector μ_j and covariance matrix $\sum_j$.

The stepwise strategies discussed below limit the maximum familywise Type 1 error rate (MFWER) (Hayter 1986) to α; proofs can be found in the original articles.

2.1   A Two-Stage Least Significant Difference (LSD) Procedure

A popular approach to pairwise testing is that due to Fisher (1935). In this approach, an omnibus test is conducted at stage one and, if declared nonsignificant, all pairwise contrast hypotheses are regarded as null. On the other hand, if the omnibus test is declared significant at stage one, then all pairwise contrast hypotheses are tested using t statistics, each assessed at an α level of significance. Hayter (1986) showed that this two-stage procedure does not limit the MFWER to α when $K > 3$. Hayter (1986) further showed, however, that by modifying the critical value of the stage two tests of the pairwise differences (that is, by using $q_{\alpha;K-1,\nu}/\sqrt{2}$, where q_α is a value from the Studentized range distribution), Fisher's (1935) two-stage approach provides exact Type 1 error control.

2.2   Sequentially Rejective Bonferroni Procedures

Hochberg's (1988) Step-Up Bonferroni Procedure

In this procedure, the p-values corresponding to the m statistics for testing hypotheses $H_1, \ldots, H_m$ are ordered from smallest to largest, i.e., $p_1 \leq p_2 \leq \cdots \leq p_m$, where $m = K(K-1)/2$, for pairwise contrasts. Then, for any $j = m, m-1, \ldots, 1$, if $p_j \leq \alpha/(m-j+1)$, the Hochberg procedure rejects all $H_{j'}$ ($j' \leq j$). According to this procedure, therefore, one begins by examining the largest p-value, p_m. If $p_m \leq \alpha$, all hypotheses are rejected. If $p_m > \alpha$, then H_m is retained and one proceeds to compare $p_{(m-1)}$ to $\alpha/2$. If $p_{(m-1)} \leq \alpha/2$, then H_j ($j = m-1, \ldots, 1$) are rejected. If not, then $H_{(m-1)}$ is retained and one proceeds to compare $p_{(m-2)}$ with $\alpha/3$, and so on. It is important to note that Hommel (1988), Rom (1990), and Dunnett and Tamhane (1992) have improved Hochberg's (1988) procedure. Dunnett and Tamhane, however, found that the modifications offered by Hommel and Rom resulted in only marginal increases in power. Similarly, Dunnett and Tamhane's (1992) modifications resulted in only modest increases in power: the power differences between their procedure and that of

Hochberg's ranged from 0.0 to 0.035 with an average difference of 0.0134.

Shaffer's (1986) Modification of Holm's (1979) Sequentially Rejective Bonferroni Procedure

As with Hochberg's (1988) procedure, the p-values associated with the test statistics are rank-ordered. In Shaffer's (1986) procedure, however, one begins by comparing the smallest p-value, p_1, to α/m. If $p_1 \geq \alpha/m$, statistical testing stops and all pairwise contrast hypotheses (H_j, $1 \leq j \leq m$) are retained; if $p_1 < \alpha/m$, however, H_1 is rejected and one proceeds to test the remaining hypotheses in a similar step-down fashion by comparing the associated p-values to α/m^*, where m^* equals the maximum number of true null hypotheses, given the number of hypotheses rejected at previous steps. Appropriate denominators for each α-stage test can be found in Shaffer (1986, Table 2) for designs containing up to ten treatment levels.

A Sequentially Rejective Bonferroni Procedure Using an Omnibus Test at Stage One

Shaffer (1986) proposed a modification to her sequentially rejective Bonferroni procedure which involves beginning this procedure with an omnibus test. If the omnibus test is declared nonsignificant, statistical testing stops and all pairwise differences are declared nonsignificant. On the other hand, if one rejects the omnibus null hypothesis one proceeds to test pairwise contrasts using the sequentially rejective Bonferroni procedure previously described with the exception that p_m, the smallest p-value, is compared to a significance level which reflects the information conveyed by the rejection of the omnibus null hypothesis. For example, for $m = 6$, rejection of the omnibus null hypothesis implies at least one inequality of means and therefore p_6 is compared to $\alpha/3$, rather than $\alpha/6$.

2.3   Multiple Step-Down Procedures

One of the most popular stepwise strategies for examining pairwise differences between means is that due to Newman (1939) and Keuls (1952). In this procedure, the means are rank ordered from smallest to largest and the difference between the smallest and largest means is first subjected to an α-level test of significance, typically with a range statistic. If this difference is not significant testing stops and all pairwise differences are regarded as null. If, on the other hand, this first range test is found to be statistically significant, one 'steps-down' to examine two $K - 1$ subsets of ordered means, that is, the smallest mean versus the next-to-largest mean and the largest mean versus the next-to-smallest mean, each tested at an α level of significance. At each stage of statistical testing, only significant subsets of ordered means are subjected to further testing.

Although the Newman-Keuls (NK) procedure is very popular among applied researchers, it is well known that it does not limit the MFWER to α when $K > 3$ (see Hochberg and Tamhane 1987, p. 69). Ryan (1960) and Welsch (1977a), however, showed how to adjust the subset levels of significance in order to limit the MFWER to α. Specifically, in order to maintain strict Type 1 error control (when assumptions are satisfied), a set of p $(p = 2, \cdots, K)$ means should be tested for significance at a level equal to

$$\alpha_p = 1 - (1 - \alpha)^{\frac{p}{K}}, \qquad \text{for} \quad 2 \leq p \leq K - 2, \qquad \alpha_{K-1} = \alpha_K = \alpha. \quad (2.1)$$

This approach to arriving at subset significance levels is used in the following multiple step-down procedures.

Peritz's (1970) Procedure

Peritz (1970) developed a stepwise procedure that can be used to test pairwise differences among means (see Begun and Gabriel 1981). Peritz's procedure employs both NK and Ryan-Welsch (RW) critical values to examine all possible pairwise contrasts in a step-down procedure. That is, one first tests the all possible subset hypotheses $\left\{ \binom{K}{K}, \binom{K}{K-1}, ..., \binom{K}{2} \right\}$ for statistical significance with NK and RW critical values. Subset hypotheses are rejected if they exceed the respective RW critical values and retained if they do not exceed the respective NK critical values. One then proceeds to classify as contentious those hypotheses that are retained using a RW critical value and rejected using a NK critical value. Beginning with the largest contentious set(s), decisions are made about the contentious hypothesis(es) using the following decision rule: Retain a contentious hypotheses if: (a) it is part of another contentious hypothesis previously retained or (b) it is retained according to the RW criterion and if at least one complementary hypothesis is retained according to the RW criterion; otherwise, reject the contentious hypothesis.

Multiple F- or T^2-tests

Here, one computes either F- or T^2-tests on the $\binom{K}{K}, \binom{K}{K-1}, \cdots, \binom{K}{2}$ sets of means. For a pairwise contrast to be declared significant, it: (a) cannot be contained in a larger set of means that previously was deemed homogeneous; and (b) must exceed its critical value. Statistical significance is assessed with the RW α_ps (See Hochberg and Tamhane 1987, pp. 111-112).

Multiple Range Tests

As with the multiple F- and T^2-test procedures, one starts with the largest set of means, testing the largest and smallest means in the set, and stepping-down to lower subsets only when the previous subset(s) was(were) declared significant. With range tests, one can follow the NK short-cut method of testing.

Multiple Range Tests Beginning with an Omnibus F-test

Again, the means are rank ordered from smallest to largest and an omnibus test initially is computed. If the omnibus test is nonsignificant, testing stops and all pairwise contrast hypotheses are retained. If the stage one omnibus test is significant, however, one proceeds to stage two, which involves an examination of the K-range test that is assessed for significance with a $q_{\alpha;K-1,\nu}/\sqrt{2}$ critical value. Thereafter, additional testing ($K - 1$, $K - 2$, $\cdots$, range tests) proceeds in the usual step-down manner, employing RW adjusted significance levels (Shaffer 1979, 1986).

2.4   Welsch's (1977a,b) Step-Up Range Procedure

Welsch proposed a stepwise procedure where ranked and adjacent means (2-range tests) initially are examined and, if significant, larger sets of means (i.e., 3-range, etc.) which contain the significant subset(s) are declared significant by implication. If at least one 2-range test is nonsignificant, one proceeds to test larger range differences (e.g., 3-range, etc.). If this larger range difference is significant all sets of treatments containing this subset are declared significant by implication, and so on. Unlike the step-down procedures previously described, Welsch's (1977a,b) step-up procedure uses slightly more conservative α_p's ($\alpha_p = p\alpha/K$, for $2 \leq p \leq K - 2$, $\alpha_{K-1} = \alpha_K = \alpha$) to assess statistical significance.

3   Stepwise Multiple Comparison Procedures For Repeated Measures Means

In an attempt to obtain robust stepwise MCPs of repeated measures means for nonspherical unbalanced data, the following test statistics were used in the stepwise procedures previously enumerated. As mentioned previously, the repeated measures literature indicates robust omnibus tests $(K > 2)$ of repeated measures main effects in nonspherical unbalanced repeated measures designs generally may be achieved by adopting any one of three adjusted degrees of freedom univariate F-tests ($\hat{\epsilon}$, $\tilde{\epsilon}$, or $\bar{\epsilon}$) or a multivariate test statistic-T^2 (Hotelling 1931) (e.g., See Huynh 1978, Keselman and Keselman 1990). Specifically, these tests are:

1. The $\hat{\epsilon}$-approximate F test (Greenhouse and Geisser 1959), where

$$F_K \overset{\cdot}{\sim} F\{(K-1)\hat{\epsilon},\ (N-J)(K-1)\hat{\epsilon}\} \text{ and} \qquad (3.1)$$

$$\hat{\epsilon} = \frac{[\text{tr}(C^T S\ C)]^2}{(K-1)\text{tr}[C^T S\ C)]^2}. \qquad (3.2)$$

The statistic $\hat{\epsilon}$ is an estimate of the unknown sphericity parameter ϵ, where

$$\epsilon = \frac{[\text{tr}(C^T \sum C)]^2}{(K-1)\text{tr}[C^T \sum C)]^2}, \qquad (3.3)$$

$\sum = \sum_j \frac{(n_j-1)}{(N-J)} \sum_j$ and $S = \sum_j \frac{(n_j-1)}{(N-J)} S_j$, the pooled population and sample covariance matrices, respectively, and $N = \sum_j n_j$.

2. The $\tilde{\epsilon}$-approximate F test (Huynh and Feldt 1976), where

$$F_K \overset{\cdot}{\sim} F\{(K-1)\tilde{\epsilon},\ (N-J)(K-1)\tilde{\epsilon}\} \text{ and} \qquad (3.4)$$

$$\tilde{\epsilon} = \min\left[1,\ \frac{N(K-1)\hat{\epsilon} - 2}{(K-1)[N-J-(K-1)\hat{\epsilon}]}\right]. \qquad (3.5)$$

3. The $\bar{\epsilon}$-approximate F test (Quintana and Maxwell 1985), where

$$F_K \overset{\cdot}{\sim} F\{(K-1)\bar{\epsilon},(N-J)(K-1)\bar{\epsilon}\} \text{ and} \qquad (3.6)$$

$$\bar{\epsilon} = \min[1,\ 1/2(\hat{\epsilon} + \tilde{\epsilon})]. \qquad (3.7)$$

For the three adjusted F tests, F_K is defined as MS_K/MS_E, where $MS_K = \sum_k J\tilde{n}(\bar{X}_k - \tilde{X})^2/(K-1)$, where $\tilde{n} = J/\sum_j(1/n_j)$, $\bar{X}_k = \sum_j \bar{X}_{jk}/J$, and $\tilde{X} = \sum_k \tilde{n}\bar{X}_k/\sum_k \tilde{n}$, and $MS_E = \sum_j[\sum_i \sum_k(X_{ik} - \bar{X}_{.k} - \bar{X}_{i.} - \bar{X}..)^2/((n_j-1)(K-1))]/J$. That is, F_K is a test of the equality of the unweighted means i.e., $H : \sum_j \mu_{j1}/J = \sum_j \mu_{j2}/J = ... = \sum_j \mu_{jK}/J$. (See Maxwell and Delaney 1990, pp. 283-284 or Searle 1987, p. 90).

4. Hotelling's (1931) T^2 statistic (see Morrison 1990 pp. 236-247). The multivariate test of the within-subjects main effect is performed by creating $K - 1$ linearly independent variables and testing the null hypothesis that these variables have a population mean vector equal to the null vector. For unbalanced designs, the means and covariance matrix of the $K-1$ independent variables must be calculated such that the between-subjects groups are weighted equally, in order to arrive at tests of unweighted hypotheses (see O'Brien and Kaiser 1985). A test of this hypothesis is accomplished through the use of Hotelling's (1931) T^2 statistic with upper $100(1 - \alpha)$ percentage points of the T^2 distribution being obtained from the relationship

$$F = \frac{N - J - K + 2}{(N - J)(K - 1)} T^2 \sim F[\alpha; K - 1, N - J - K + 2]. \qquad (3.8)$$

The literature also indicates that robust tests of pairwise contrasts involving nonspherical unbalanced repeated measures data can be obtained using a non pooled statistic (KKS) given by Keselman et al. (1991) and Satterthwaite's (1941, 1946) solution for degrees of freedom. Accordingly, for range tests and non-range pairwise tests, the KKS statistic was employed.

To introduce the KKS statistic, consider the following definitions. Let $\mu_k = \sum_j a_j \mu_{jk}$, where $\sum_j a_j = 1$; i.e., μ_k is a weighted mean of the kth components of the vectors μ_j. A contrast among the levels of the repeated measures means is given by $\psi = \sum_k c_k \mu_k = \sum_j a_j (\sum_k c_k \mu_{jk})$, where $\sum_k c_k = 0$, and is estimated by $\hat{\psi} = \sum_j a_j (\sum_k c_k \bar{X}_{jk})$. Letting $\sum_j$ and S_j denote the respective population and sample covariance matrices for group j, $\sigma^2(\hat{\psi}) = \sum_j a_j^2 (c^T \sum_j c)/n_j$, which is estimated by $\hat{\sigma}^2(\hat{\psi}) = \sum_j a_j^2 (c^T S_j c)/n_j$, where the superscript T denotes the transpose of the indicated vector. For pairwise contrasts, and for $a_j = 1/J$, i.e., for unweighted means, $\hat{\psi} = \sum_j (\bar{X}_{jk} - \bar{X}_{jk'})/J$, and $\hat{\sigma}^2(\hat{\psi}) = \sum_j [(s_{jk}^2 + s_{jk'}^2 - 2s_{jkk'})/n_j]/J^2$.

To test the hypothesis $\psi = 0$, one uses the statistic $\hat{\psi}/\hat{\sigma}(\hat{\psi})$, which, while not distributed as t, can be approximated by Student's t distribution with estimated Satterthwaite (1941, 1946) df given by

$$\nu_s = [\hat{\sigma}(\hat{\psi})]^4 / \sum_j [(a_j^2/n_j)^2 (c^T S_j c)^2 /(n_j - 1)].$$

Applying these robust test statistics to the procedures previously described led to the identification of 37 stepwise MCPs, which were the foci of the present study. The following is an enumeration of these procedures.

(The designated notation for each of the procedures is contained in paren-
thesis after their description).

(1)-(7) Hayter's (1986) modified two-stage LSD procedure, using an omnibus adjusted univariate F-test, multivariate T^2-test or both and followed by KKS t-tests using q_α; $K - 1$, $\nu_s/\sqrt{2}$ as the second stage critical
value. The use of both an adjusted univariate and multivariate test, each
tested at an $\alpha/2$ level of significance, was suggested by Barcikowski and
Robey (1984) and recommended by Looney and Stanley (1989). $[F(\hat{\epsilon}) \rightarrow
H(t_q),\ F(\tilde{\epsilon}) \rightarrow H(t_q),\ F(\bar{\epsilon}) \rightarrow H(t_q),\ F(T^2) \rightarrow H(t_q),\ F(\hat{\epsilon})/T^2 \rightarrow H(t_q),
F(\tilde{\epsilon})/T^2 \rightarrow H(t_q),\ F(\bar{\epsilon})/T^2 \rightarrow H(t_q)]$.

(8)-(9) Hochberg's (1988) and Shaffer's (1986) sequentially rejective Bonferroni procedures, using the KKS statistic. $[HB(t)$ and $S(t)]$.

(10)-(16) Shaffer's (1986) sequentially rejective Bonferroni procedure, us-
ing an adjusted univariate F-test, T^2-test, or both and followed by the
KKS statistic. $[F(\hat{\epsilon}) \rightarrow S(t),\ F(\tilde{\epsilon}) \rightarrow S(t),\ F(\bar{\epsilon}) \rightarrow S(t),\ F(T^2) \rightarrow S(t),
F(\hat{\epsilon})/T^2 \rightarrow S(t),\ F(\tilde{\epsilon})/T^2 \rightarrow S(t),\ F(\bar{\epsilon})/T^2 \rightarrow S(t)]$.

(17) Peritz's (1970) procedure, using the KKS statistic as a range test with
critical values from the Studentized range distribution, $q/\sqrt{2}$, based upon
NK and RW levels of significance. [The algorithm for this procedure can
be obtained from Martin and Toothaker 1989.] $[P(t_q)]$.

(18)-(20) Peritz's (1970) procedure, using adjusted F-tests with NK and
RW levels of significance. For the $\binom{K}{2}$ pairwise tests, the F-tests are computed as t^2_{KKS}. $\{P[F(\hat{\epsilon})],\ P[F(\tilde{\epsilon})],\ \text{and}\ P[F(\bar{\epsilon})]\}$.

(21) Peritz's (1970) procedure, using T^2-tests with NK and RW levels of
significance. For the $\binom{K}{2}$ pairwise tests, the T^2-tests are computed as t^2_{KKS}.
$[P(T^2)]$.

(22)-(24) Multiple F-tests, using the adjusted degrees of freedom univariate F-tests. The error mean square, MS_E, is computed from only the
treatment levels defining the set of means. For the $\binom{K}{2}$ pairwise tests, the
F-tests are computed as t^2_{KKS}. $[F(\hat{\epsilon}),\ F(\tilde{\epsilon}),\ F(\bar{\epsilon})]$.

(25) Multiple omnibus T^2-tests. For the $\binom{K}{2}$ pairwise tests, the T^2-tests
are computed as t^2_{KKS}. $[F(T^2)]$.

(26)-(28) Multiple omnibus adjusted univariate F-tests and T^2-tests, each
tested at an $\alpha/2$ level of significance. For the $\binom{K}{2}$ pairwise tests, the F-
and T^2-tests are computed as t^2_{KKS}. $[F(\hat{\epsilon}/T^2),\ F(\tilde{\epsilon}/T^2),\ \text{and}\ F(\bar{\epsilon}/T^2)]$.

(29) Multiple range tests, using the KKS statistic and critical values from
the Studentized range distribution and RW significance levels. $[RW(t_q)]$.

(30)-(36) Multiple range tests, using an omnibus adjusted univariate F-test, T^2-test, or both as the stage one test and followed by KKS range tests with Studentized range critical values à la Shaffer (1979). $[F(\hat{\epsilon}) \rightarrow S(t_q), F(\tilde{\epsilon}) \rightarrow S(t_q), F(\bar{\epsilon}) \rightarrow S(t_q), F(T^2) \rightarrow S(t_q), F(\hat{\epsilon})/T^2 \rightarrow S(t_q), F(\tilde{\epsilon})/T^2 \rightarrow S(t_q), F(\bar{\epsilon})/T^2 \rightarrow S(t_q)]$.

(37) Welsch's (1977a) step-up procedure, using the KKS statistic and Welsch's (1977b) critical values, $W/\sqrt{2}$. $[W(t_q)]$.

It is important to note that to ensure that MFWER $\leq \alpha$ (when assumptions are satisfied), the range critical values were forced to be monotonically decreasing. (See Hochberg and Tamhane 1987, pp. 114-115).

4   Monte Carlo Study

While some results could be obtained analytically, for example those involving probabilities of linear combinations of independent chi-square distributed variables, the complexity and number of conditions to be compared necessitated a Monte Carlo study. The procedures for testing stepwise repeated measures pairwise multiple comparison hypotheses were compared for a two-way mixed design containing one between-subjects and one within-subjects factor. Across all conditions, the number of levels of the between-subjects and the within-subjects factors were held constant at three and four, respectively.

Five variables were manipulated to investigate the operating characteristics of the various MCPs. These variables were: (1) the sphericity pattern, (2) the equality/inequality of the between-subjects covariance matrices, (3) the group size equality/inequality, (4) the nature of the pairing of unequal covariance matrices and unequal group sizes, and (5) the total sample size.

Box's (1954) correction factor, ϵ, was used to quantify the degree of departure from the overall sphericity assumption. For conditions of overall sphericity, ϵ equals one; with increasing departures from this assumption, ϵ decreases in value to a minimum value of $1/(K-1)$. Matrices with ϵ values of .75 and .40 were employed to investigate the effects of nonsphericity. (See Keselman et al. 1991, Table 1).

Violation of the assumption of homogeneity of between-subjects covariance matrices was investigated by creating between-subjects level matrices whose elements were in different ratios to one another. Specifically, the elements in the between-subjects covariance matrices were in a 1:3:5 ratio. This degree and type of covariance heterogeneity was selected since Keselman et al. (1991) found, of the conditions they investigated, that it resulted in the greatest discrepancies between the empirical and nominal rates of Type 1 error and was a condition under which the effects of covariance heterogeneity could be readily examined.

The MCPs were investigated when the number of observations per between-subjects level was equal or unequal. Group sizes, when equal, were set at either ten or fifteen per group. For a total sample size of $N = 30$, the two cases of group size inequality were: $n_j = 8$, 10, and 12 and $n_j = 6$, 10, and 14. For $N = 45$, the analogous values were: $n_j = 12$, 15, and 18 and $n_j = 9$, 15, and 21. For both $N = 30$ and $N = 45$, the less disparate group size condition had a coefficient of group size variation (C) of .163 while the more disparate condition had $C = .327$.

For those conditions involving both unequal group sizes and unequal covariance matrices, both positive and negative pairings of these group sizes and covariance matrices were investigated. A positive pairing referred to the case in which the largest n_j was associated with the covariance matrix containing the largest element values; a negative pairing referred to the case in which the largest n_j was associated with the covariance matrix with the smallest element values.

To summarize, six pairings of covariance matrices and group sizes were investigated: (a) equal n_j; equal $\sum_j$, (b) equal n_j; unequal $\sum_j$, (c/c') unequal n_j; unequal $\sum_j$ (positively paired), and (d/d') unequal n_j; unequal $\sum_j$ (negatively paired). The c'/d' condition denotes the more disparate unequal group sizes case while the c/d condition designates the less disparate unequal group sizes case.

Pseudorandom observation vectors were generated from a K-variate normal distribution following the procedure enumerated in Keselman et al. (1991).

Empirical familywise Type 1 error rates were collected under conditions where the population mean vectors reflected either a complete null hypothesis (i.e., $\mu_1 = \mu_2 = \cdots = \mu_4 = 0$), or a partial null hypothesis (i.e., $\mu_1 = 0$, $\mu_2 = 0$, $\mu_3 = .8928$, $\mu_4 = .8928$). The familywise Type 1 error rate was defined as the probability that at least one of the pairwise contrasts was statistically significant when the corresponding population contrast was null. Five thousand replications of each condition were performed using a .05 significance level.

5 Results

The empirical percentages of Type 1 errors for $\epsilon = 1.00$, 0.75, and 0.40 are contained in Tables 1, 2 and 3, respectively. In order to conserve space, only a subset of MCPs are presented. Specifically, only those MCPs using $\bar{\epsilon}$, singly or in combination with a multivariate test, are tabled and discussed.

The MCPs that used $\hat{\epsilon}$ and $\tilde{\epsilon}$ gave results which were, respectively, less than or greater than or equal to those obtained when $\bar{\epsilon}$ was employed. In addition, empirical values associated with $N = 45$ are not reported as they were similar, though less deviant from the five percent significance level, than those reported for $N = 30$. Finally, the rates of Type 1 error for the partial null hypothesis condition at each value of ϵ are not tabled since the empirical rates were less than the nominal .05 value for all the procedures except for the Welsch (1977a) procedure. For this procedure, while the empirical values frequently exceeded .05, they rarely exceeded the criterion of robustness used in this study.

In order to help identify robust MCPs, Bradley's (1978) liberal criterion of robustness was employed. According to this criterion, in order to be considered robust, an MCP's empirical rate of Type 1 error $(\hat{\alpha})$ must be contained in the interval $0.5\alpha \leq \hat{\alpha} \leq 1.5\alpha$. For the five percent level of significance used in this study, therefore, an MCP was considered robust in a particular condition if its empirical rate of Type 1 error fell within the interval $.025 \leq \hat{\alpha} \leq .075$. Correspondingly, an MCP was considered to be nonrobust if, for a particular condition, its Type 1 error rate was not contained in this interval. In the tables, asterisks (*) are used to denote these latter values.

5.1 $\epsilon = 1.00$

As seen from Table 1, when the design was balanced and covariance matrices were unequal (condition b), all MCPs effectively controlled the rate of Type 1 error. When unequal covariance matrices were obtained from groups of unequal sizes, however, not all MCPs had rates of Type 1 error which fell within Bradley's (1978) bounds. Specifically, for the conditions where group sizes were moderately unequal and paired with unequal covariance matrices (conditions c and d), all MCPs consistently provided Type 1 error protection within the bounds, except for the multiple omnibus testing strategies using just T^2 {i.e., $P(T^2)$ and $F(T^2)$}. When very unequal group sizes were paired with unequal covariance matrices (conditions c' and d'), however, the subset of procedures which consistently controlled Type 1 errors was reduced to Hochberg's (1988) {$HB(t)$} and Shaffer's (1986) {$S(t)$} sequentially rejective Bonferroni procedures and the multiple omnibus range testing procedures {$P(t_q)$ and $RW(T_q)$}.

5.2 $\epsilon = 0.75$

As seen from Table 2, when $\epsilon = 0.75$, the MCPs' rates of Type 1 error met Brandley's (1978) robustness criterion when the design was bal-

Table 1: Empirical percentages of familywise Type 1 error: Complete null nypothesis ($\epsilon = 1.0$ and $N = 30$).

	a	b	c	c'	d	d'
$F(\bar{\epsilon}) \to H(t_q)$	4.28	3.88	2.92	1.90*	5.24	6.72
$F(T^2) \to H(t_q)$	4.76	5.12	4.06	3.66	7.26	10.68*
$F(\bar{\epsilon})/T^2 \to H(t_q)$	3.18	3.74	2.60	1.86*	5.60	8.70*
$HB(t)$	4.02	4.04	3.88	3.96	4.14	6.10
$S(t)$	4.00	4.04	3.84	3.94	4.14	6.04
$F(\bar{\epsilon}) \to S(t)$	4.08	3.80	2.76	1.88*	4.78	6.16
$F(T^2) \to S(t)$	4.62	5.00	3.96	3.60	6.66	9.54*
$F(\bar{\epsilon})/T^2 \to S(t)$	3.18	3.66	2.60	1.86*	5.24	7.94*
$P(t_q)$	5.02	4.92	4.68	4.66	5.02	6.48
$P[F(\bar{\epsilon})]$	4.66	4.32	3.02	1.90*	6.36	8.80*
$P(T^2)$	5.08	5.76	4.32	3.90	8.68*	15.60*
$F(\bar{\epsilon})$	4.66	4.32	3.02	1.90*	6.36	8.80*
$F(T^2)$	5.08	5.76	4.32	3.90	8.68*	15.60*
$F(\bar{\epsilon})/T^2$	3.26	3.92	2.66	1.88*	6.28	11.06*
$RW(t_q)$	5.02	4.92	4.68	4.66	5.02	6.48
$F(\bar{\epsilon}) \to S(t_q)$	4.02	3.56	2.72	1.86*	4.38	5.26
$F(T^2) \to S(t_q)$	4.48	4.64	3.88	3.40	5.90	7.50
$F(\bar{\epsilon})/T^2 \to S(t_q)$	3.06	3.42	2.56	1.78*	4.80	6.70
$W(t_q)$	5.36	5.56	4.98	5.32	5.96	7.94*

Notes: $F(*) \to H(t_q)$-Hayter's (1986) two-stage procedures; HB-Hochberg's (1988) step-up Bonferroni procedure; $S(t)$-Shaffer's (1986) sequentially rejective Bonferroni procedure; $F(*) \to S(t)$-Shaffer's (1979, 1986) sequentially rejective Bonferroni procedures with an omnibus test statistic; $P(*)$-Peritz's (1970) procedures; $F(*)$-multiple omnibus procedures; $RW(t_q)$-multiple range tests with RW adjusted levels of significance; $F(*) \to S(t_q)$-Shaffer's (1979) multiple range procedures with an initial omnibus test; $W(t_q)$-Welsch's (1977a) step-up procedure.

Table 2: Empirical percentages of familywise Type 1 error: Complete null
hypothesis ($\epsilon = 0.75$ and $N = 30$).

	a	b	c	c'	d	d'
$F(\bar{\epsilon}) \rightarrow H(t_q)$	4.04	4.16	3.48	2.42*	4.90	4.92
$F(T^2) \rightarrow H(t_q)$	4.86	4.86	3.70	3.30	7.22	8.28*
$F(\bar{\epsilon})/T^2 \rightarrow H(t_q)$	4.20	4.48	3.18	2.36*	6.02	7.34
$HB(t)$	4.00	4.08	3.48	3.92	4.46	4.08
$S(t)$	4.00	4.08	3.48	3.92	4.44	4.06
$F(\bar{\epsilon}) \rightarrow S(t)$	3.62	3.76	3.08	2.26*	4.48	4.44
$F(T^2) \rightarrow S(t)$	4.70	4.70	3.52	3.24	6.70	7.44
$F(\bar{\epsilon})/T^2 \rightarrow S(t)$	4.06	4.36	3.06	2.34*	5.60	6.82
$P(t_q)$	3.94	3.86	3.42	3.38	3.94	4.16
$P[F(\bar{\epsilon})]$	5.12	5.68	4.00	2.60	6.84	8.12*
$P(T^2)$	5.52	5.76	4.36	3.52	9.38*	14.22*
$F(\bar{\epsilon})$	5.12	5.68	4.00	2.60	6.84	8.12*
$F(T^2)$	5.52	5.76	4.36	3.52	9.38*	14.22*
$F(\bar{\epsilon})/T^2$	4.48	5.10	3.40	2.44*	7.52*	10.88*
$RW(t_q)$	3.94	3.86	3.42	3.38	3.94	4.24
$F(\bar{\epsilon}) \rightarrow S(t_q)$	3.28	3.26	2.72	2.10*	3.72	3.42
$F(T^2) \rightarrow S(t_q)$	3.90	3.80	2.88	2.70	5.14	5.30
$F(\bar{\epsilon})/T^2 \rightarrow S(t_q)$	3.58	3.70	2.74	2.06*	4.64	4.96
$W(t_q)$	6.34	6.36	5.94	6.32	6.82	6.28

Notes: See Table 1.

Table 3: Empirical percentages of familywise Type 1 error: Complete null hypothesis ($\epsilon = 0.40$ and $N = 30$).

	a	b	c	c'	d	d'
$F(\bar{\epsilon}) \to H(t_q)$	3.34	3.94	3.20	2.60	3.82	4.58
$F(T^2) \to H(t_q)$	2.40*	3.04	2.44*	2.20*	3.62	5.28
$F(\bar{\epsilon})/T^2 \to H(t_q)$	2.82	3.64	2.62	2.18*	3.82	5.38
$HB(t)$	2.18*	2.52	2.30*	2.30*	2.40*	3.22
$S(t)$	1.98*	2.16*	2.12*	2.00*	2.16*	2.98
$F(\bar{\epsilon}) \to S(t)$	2.94	3.38	2.90	2.40*	3.18	3.98
$F(T^2) \to S(t)$	2.20*	2.74	2.28*	2.00*	3.26	4.60
$F(\bar{\epsilon})/T^2 \to S(t)$	2.52	3.28	2.48*	2.04*	3.30	4.60
$P(t_q)$	0.94*	1.28*	1.04*	1.44*	1.34*	1.82*
$P[F(\bar{\epsilon})]$	5.20	6.00	4.48	3.20	7.04	8.70*
$P(T^2)$	4.78	6.06	4.62	4.36	8.80*	15.26*
$F(\bar{\epsilon})$	5.20	6.00	4.48	3.20	7.04	8.70*
$F(T^2)$	4.78	6.06	4.62	4.36	8.80*	15.26*
$F(\bar{\epsilon})/T^2$	4.22	5.50	3.68	3.14	7.36	12.42*
$RW(t_q)$	0.94*	1.28*	1.04*	1.44*	1.34*	1.82*
$F(\bar{\epsilon}) \to S(t_q)$	1.66*	2.00*	1.78*	2.16*	1.80*	2.24
$F(T^2) \to S(t_q)$	1.02*	1.64*	1.12*	1.38*	1.76*	2.40
$F(\bar{\epsilon})/T^2 \to S(t_q)$	1.80*	2.20*	1.78*	1.86*	2.00*	2.44
$W(t_q)$	5.28	5.88	5.46	5.38	5.52	6.08

Notes: See Table 1.

anced and covariance matrices were unequal (condition b). When unequal covariance matrices were combined with moderately unequal group sizes (conditions c and d), however, the multiple omnibus testing strategies that used T^2 either by itself or in conjunction with the $\bar{\epsilon}$ univariate test, i.e., $\{P(T^2),\ F(T^2),\ \text{and}\ F(\bar{\epsilon}/T^2)\}$ did not consistently provide Type 1 error protection. When unequal covariance matrices were paired with the more disparate group sizes (conditions c' and d'), this subset of procedures was further reduced to include only: (1) Hochberg's (1988) procedure $\{HB(t)\}$, (2) Shaffer's (1986) procedures $\{S(t)\ \text{and}\ F(T^2) \to S(t)\}$, (3) the multiple omnibus range testing procedures $\{P(t_q)\ \text{and}\ RW(t_q)\}$, (4) Shaffer's (1979, 1986) multiple omnibus range testing procedure using an omnibus T^2 test $\{F(T^2) \to S(t_q)\}$, and (5) Welsch's (1977a) procedure $\{W(t_q)\}$.

5.3 $\epsilon = 0.40$

Unlike the $\epsilon = 1.00$ and $\epsilon = 0.75$ results, when $\epsilon = 0.40$, not all MCPs were able to limit their rates of Type 1 error within Bradley's (1978) interval, even when group sizes were equal. Specifically, for balanced designs, the procedures due to Shaffer (1979, 1986) $\{S(t),\ F(\bar{\epsilon}) \to S(t_q),\ F(T^2) \to S(t_q)\ \text{and}\ F(\bar{\epsilon})/T^2 \to S(t_q)\}$ and the multiple omnibus range testing strategies $\{P(t_q)\ \text{and}\ RW(t_q)\}$ became quite conservative. This finding is consistent with that of previous studies regarding simultaneous MCPs (see Keselman and Keselman 1988 and Keselman et al. 1991), which reported a decrease in rates of Type 1 error as a function of decreases in the value of ϵ. For the (c) and (d) conditions of moderately unequal group sizes combined with unequal covariance matrices, the robust procedures were: (1) Hayter's (1986) and Shaffer's (1986) procedures using $\bar{\epsilon}$ $\{F(\bar{\epsilon}) \to H(t_q),\ F(\bar{\epsilon})/T^2 \to H(t_q)\ \text{and}\ F(\bar{\epsilon}) \to S(t)$, respectively$\}$, (2) Welsch's (1977a) procedure $\{W(t_q)\}$ and (3) the multiple omnibus testing strategies using $\bar{\epsilon}$ $\{P[F(\bar{\epsilon})]$ Peritz (1970), $F(\bar{\epsilon}),\ F(\bar{\epsilon}/T^2)\}$. For conditions in which unequal covariance matrices were combined with very disparate group sizes (conditions c' and d'), only two procedures consistently provided acceptable Type 1 error control: (1) Hayter's (1986) procedure $\{F(\bar{\epsilon}) \to H(t_q)\}$ and (2) Welsch's (1977a) procedure $\{W(t_q)\}$.

6   Conclusions

Numerous stepwise procedures for testing comparisons of pairwise repeated measures means have been identified which generally limit the familywise rate of Type 1 error to the nominal five percent value when the data do not satisfy the assumption of multisample sphericity. If one considers both positive and negative departures from the nominal significance level, using Bradley's (1978) liberal criterion, then, for the study conditions investigated, the three most promising MCPs are: (1) Welsch's (1977a) step-up procedure, which uses the KKS statistic and Welsch's (1977b) critical values $\{W(t_q)\}$, (2) Hayter's (1986) modified two-stage LSD procedure, using a $\bar{\epsilon}$ adjusted univariate F-test followed by KKS statistics $\{F(\bar{\epsilon}) \rightarrow H(t_q)\}$ and (3) Shaffer's (1986) sequentially rejective Bonferroni procedure, which begins with an adjusted $\bar{\epsilon}$ univariate F-test and is followed by KKS statistics $\{F(\bar{\epsilon}) \rightarrow S(t)\}$. If, on the other hand, one considers only positive deviations from α, then many more MCPs can be considered to be robust.

Regardless of which criterion one prefers, it would be premature to attempt to identify the 'best' MCPs on the basis of the results reported in this study. First, an examination of the robustness of these procedures to departures from multivariate normality is in order. Indeed, previous research has been found the test statistics investigated in this study to be somewhat adversely affected by departures from this assumption (Keselman et al. 1991; Rogan et al. 1979). More work is needed in this area. Second, a comparison of the relative power of these procedures in the presence of multisample nonsphericity is required, in order to arrive at a more complete picture concerning the operating characteristics of these stepwise MCPs. Finally once the 'best' stepwise procedures have been identified on the basis of both robustness and power data, it would be important to compare these procedures to the robust simultaneous MCPs identified by Keselman et al. (1991). While stepwise procedures are generally more powerful than simultaneous MCPs, they are also more complex to compute. Any advantages that these stepwise procedures have, therefore, would have to be weighed against this increased computational complexity.

Acknowledgments

The research reported in this article was supported by the Social Sciences and Humanities Research Council of Canada (Awards No. 410-90-0055 and No. 410-92-0430). I would like to acknowledge the many helpful comments provided by Joanne Keselman and an anonymous reviewer.

References

Barcikowski, R.S., and Robey, R.R. (1984). Decisions in single group repeated measures analysis: Statistical tests and three computer packages. *The American Statistician* **38** 148 - 150.

Begun, J. and Gabriel, K.R. (1981). Closure of the Newman-Keuls multiple comparisons procedure. *Jour. Amer. Statist. Assoc.* **76** 241 - 245.

Box, G.E P. (1954). Some theorems on quadratic forms applied in the study of analysis of variance problems, II. Effects of inequality of variance and correlation between errors in the two-way classification. *Ann. Mathem. Statist.* **25** 484 - 498.

Bradley, J.V. (1978). Robustness? *British Journal of Mathematical and Statistical Psychology* **31** 144 - 152.

Dunnett, C.W. and Tamhane, A.C. (1992) A step-up multiple test procedure. *Jour. Amer. Statist. Assoc.* **87** 162 - 170.

Einot, I. and Gabriel, K.R. (1975). A study of the powers of several methods in multiple comparisons. *Jour. Amer. Statist. Assoc.* **70** 574 - 583.

Fisher, R. A. (1935). *The Design of Experiments.* Oliver and Boyd, London.

Greenhouse, S.W. and Geisser, S. (1959). On methods in the analysis of profile data. *Psychometrika* **24** 95 - 112.

Hayter, A. J. (1986). The maximum familywise error rate of Fisher's least significant difference test. *Jour. Amer. Statist. Assoc.* **81** 1000 - 1004.

Hochberg, Y. (1988). A sharper Bonferroni procedure for multiple tests of significance. *Biometrika* **75** 800 - 802.

Hochberg, Y. and Tamhane, A.C. (1987). *Multiple Comparison Procedures.* John Wiley and Sons, Inc., New York.

Hommel, G. (1988) A stagewise rejective multiple test procedure based on a modified Bonferroni test. *Biometrika* **75** 383-386.

Hotelling, H. (1931). The generalization of Student's ratio. *Ann. of Mathem. Statist.* **2** 360 - 378.

Huynh, H. (1978). Some approximate tests for repeated measurement designs. *Psychometrika* **43** 161 - 175.

Huynh, H. and Feldt, L.S. (1976). Estimation of the Box correction for degrees of freedom from sample data in randomized block and split-plot designs. *Jour. Educat. Statist.* **1** 69 - 82.

Keselman, H.J. and Keselman, J.C. (1988). Repeated measures multiple comparison procedures: Effects of violating multisample sphericity in unbalanced designs. *Jour. Educat. Statist.* **13** 215 - 226.

Keselman, J.C. and Keselman, H.J. (1990). Analyzing unbalanced repeated measures designs. *Brit. Jour. Mathem. Statist. Psych.* **43** 265 - 282.

Keselman, H.J., Keselman, J.C. and Shaffer, J.P. (1991). Multiple pairwise comparisons of repeated measures means under violation of multisample sphericity. *Psych. Bull.* **110** 162 - 170.

Keuls, M. (1952). The use of the 'Studentized range' in conjunction with an analysis of variance. *Euphytica* **1** 112 - 122.

Looney, S.W. and Stanley, W.B. (1989). Exploratory repeated measures analysis for two or more groups: Review and update. *The American Statistician* **43** 220 - 225.

Martin, S.A. and Toothaker, L.E. (1989). Peritz: A fortran program for performing multiple comparisons of means using the Peritz Q method. *Behavior Research Methods, Instruments, and Computers* **21** 465 - 472.

Maxwell, S.E. and Delaney, H.D. (1990). *Designing Experiments and Analyzing Data: A Model Comparison Perspective.* Wadsworth, Belmont, CA.

Morrison, D.F. (1990) *Multivariate Statistical Methods* (Third edition). McGraw-Hill, New York.

Newman, D. (1939). The distribution of the range in samples from a normal population expressed in terms of an independent estimate of standard deviation. *Biometrika* **31** 20 - 30.

O'Brien, R.G. and Kaiser, M.K. (1985). MANOVA method for analyzing repeated measures designs: An extensive primer. *Psych. Bull.* **97** 316 - 333.

Peritz, E. (1970). A note on multiple comparisons, Unpublished manuscript. Hebrew University, Jersusalem, Israel.

Quintana, S.M. and Maxwell, S.E. (1985). *A better-than-average estimate of* ϵ. Paper presented at the annual meeting of the American Educational Research Association.

Rogan, J.C., Keselman, H.J. and Mendoza, J.L. (1979). Analysis of repeated measurements. *Brit. Jour. Mathem. Statist. Psych.* **32** 269 - 286.

Rom, D.M. (1990) A sequentially rejective test procedure based on a modified Bonferroni inequality. *Biometirka* **77** 663-665.

Ryan, T.A. (1960). Significance tests for multiple comparison of proportions, variances and other statistics. *Psych. Bull.* **57** 318 - 328.

Satterthwaite, F.E. (1941). Synthesis of variance. *Psychometrika* **6** 309 - 316.

Satterthwaite, F.E. (1946). An approximate distribution of estimates of variance components. *Biometrics* **2** 110 - 114.

Searle, S.R. (1987). *Linear Models for Unbalanced Data.* John Wiley and Sons, Inc., New York.

Shaffer, J.P. (1979). Comparison of means: An *F* test followed by a modified multiple range procedure. *Jour. Educat. Statist.* **4** 14 - 23.

Shaffer, J.P. (1986). Modified sequentially rejective multiple test procedures. *Jour. Amer. Statist. Assoc.* **81** 826 - 831.

Welsch, R.E. (1977a). Stepwise multiple comparison procedures. *Jour. Amer. Statist. Assoc.* **72** 566 - 575.

Welsch, R.E. (1977b). *Tables for stepwise multiple comparison procedures.* Working paper No. 949-77, Massachusetts Institute of Technology.

Chapter 12

Where Should Multiple Comparisons Go Next?

JOHN W. TUKEY   Department of Mathematics, Princeton University, Princeton, New Jersey

Abstract   The preceding chapters do much to cleanup conventional multiple comparisons: all simple comparisons and the Dunnett case. It is high time to go further. Here I will discuss, more or less briefly, four ways in which we should go.

Ab initio, we will have to recognize that null hypotheses are convenient limiting cases and not something that ever happens. "Non-significant" means that we do not know the direction. And the complicated nature of multiple comparison results urges stress on more and better kinds of graphical presentations.

First, we should recognize that, as measurement in a field evolves, we move from hoping for a few confident directions toward trying to get the sign of every possible simple comparison (every difference) straight. The corresponding sequence of multiple comparison techniques is probably not yet complete.

Second, we need to face the study of two-way tables of results — for instance, interaction tables in the analysis of variance. When we are struggling for a few confident results, two-way differences and the Studentized bi-range offer a natural generalization of one-way differences (simple comparisons) and the Studentized range.

Third, we need useful moderate-measurement-precision techniques for comparing 6 to 50 separate quantities. The Studentized range is likely to lead to long lists of comparisons about whose direction we are confident, often longer lists than we can reasonably apprehend. So we will probably start with one-against-the-midmean, a modernized version of Cornfield et al.'s (Halperin et al. 1955) one-against-the-mean. And we will need to then bring in what we have learned about the first question.

Finally, and crucially, we will need to learn to supplement hard-edged confidence statements with softly guided hints, something that will make us rethink many things.

1 Introduction

The preceding chapters do much to clean up conventional multiple comparisons problems: (a) all simple comparisons and (b) the Dunnett case. It is high time, then, to choose directions in which we can go further. Here I will discuss, more or less briefly, four ways in which I am sure we should go.

fundamentals

Before we start, we need to be clear about certain fundamental points:

- the "null hypothesis" *never* holds.

- omnibus procedures (like F in ANOVA or chi-square in large contingency tables) *cannot* meet our needs; some form of multiple comparisons *can*,

- in almost every application we need to say what we can about the directions in which long-run values lie — *directions* often from some given number, *not* merely from zero,

- purposes for the use of multiple comparisons vary widely, and techniques may need to vary widely as a consequence.

To these we need to add two further points, almost as fundamental:

- multiple comparisons are somewhat complicated — as a result, graphical presentation of multiple comparison results is even more important than graphical presentation of simpler results,

- multiple comparisons is much more sensitive to inhomogeneous variability than are omnibus procedures — so we must routinely seek out, and compensate for, inhomogeneous variability.

* brief discussions *

Our experience with the real world teaches us — if we are willing learners — that, provided we measure to enough decimal places, no two "treatments" ever have identically the same long-run value. Thus "not significant" should *never* mean "accept the null hypothesis". Rather it must mean "we don't know in which direction the long-run difference goes".

Omnibus procedures formally tell us, when "significant", to believe that "something is different", which we knew all along. A significant F fails to tell us what is different — and seduces many innocent users to the entirely unfounded view that "since the effects are (omnibus-wise) statistically significant, we can believe any and all appearances as truth". What could be further from wisdom?

Direction is almost always more important than amount. This is particularly true when, as we shall routinely expect, "direction" encompasses offset direction, (for instance, direction from some break-even value) not just direction from zero.

Multiple comparison questions span a wide range of purposes, from archival storage (perhaps of means and confidence-interval half-length(s)) to immediate description (What should we say that we see in a new set of data?). It would be quite wrong to think that one procedure could meet all needs.

Telling an innocent, however interested and involved, that we can't distinguish B from A, we can't distinguish C from B, we can't distinguish D from C, but we are quite sure that D is less than A, is not an easy idea to get across, but it is far from the hardest that multiple comparisons results may need us to convey. Anything that helps is important — graphical techniques, especially diverse graphical techniques, seem to help more than any other single tool.

* diversity of variability *

The work of Pitman and Welch (see volume 29 of Biometrika, and references therein) taught us (at least those of us who read and pondered) how far from the classical normal distribution we could go and still preserve the *average values* of analysis-of variance mean squares. (We may lose something in variability, but this is usually not a big thing.) Either randomization within blocks (where plots carry an additive effect) or matched distributions, equivalent to two-stage sampling encompassing randomization as a second stage, ensure the usual formulas for average values of mean squares. Averaging over different variabilities of "blocks" is harmless. F-tests, though unhelpful, are robust.

But this does not mean that *treatment*-related differences in variance are harmless. They do not affect average values of mean squares — hence they do not wholly ruin the behavior of observed $F's$. But they can play unholy hob with multiple comparisons once different comparisons have substantially different variances. We dare not use multiple comparisons in practice — for which there is no reasonable substitute — without asking whether standard errors for means (or other summaries) of different treatments are similar or diverse — or without planning, if they are diverse, to do something about this diversity.

We dare not bypass multiple comparisons, because other techniques are insufficiently informative. But we need to be more careful, especially about diversity of variability.

* a far lesser point *

A lesser point, important to how we are about to proceed, is that the distinction between "error rate per batch" and "error rate batchwise" — exemplified by the distinction between the uses of "Bonferroni t" and "the Studentized range", while not negligible, can often be neglected as much smaller than other differences we must take seriously. Thus we shall feel free, as we go on, to neglect this difference whenever neglect simplifies our discussion.

2   Measurement Precision Relative to What is to Be Measured

Those who deal, at least by proxy, with the analysis of measurements — as made in different fields or subfields — cannot fail to recognize great differences in the ratio

$$\frac{\text{standard error of a measure of a difference}}{\text{usual size of differences to be measured}}. \qquad (*)$$

In agriculture or plant breeding this ratio can come close to unity, while in classical physics it is more likely to be between 0.01 and 0.0001. As measurement in a field evolves — perhaps rapidly, perhaps slowly — we can expect this ratio to decrease — though perhaps not in full proportion to the decrease in the standard deviation of measurement. It may be, for instance, that, mainly on economic grounds, fields like agriculture and plant breeding will remain at an early stage, where (*) is large, for the foreseeable future. We must be ready for this, but not at the cost of failing to give adequate support to fields where (*) is smaller, even quite small.

In the use of multiple comparisons, the value of this ratio, (*), is reflected in what sorts of results we can reach — and for which, of those we can reach, we should try to do our best.

For very large ratios, the most useful result we can reach may be an upper bound on the long-run value of any difference — expressible as

(largest observed difference) + (confidence-interval length)

where the confidence-length is probably found from a Studentized range or a Bonferroni t.

As (*) becomes a little smaller, we can strive to supplement such an outer bound with one (or even two or three) statements of confident direction — such as we have 95% simultaneous confidence that the long-run value for treatment F is greater than the long-run value for treatment B. So far as we known, the best thing to do in this region is still the Studentized range — which we shall routinely shorten to *S-range*.

As (*) becomes smaller, we expect to find at least a few simple comparisons about whose sign we may be simultaneously confident. If we are content with directions (of comparisons) from zero, as we often may be, then, in this region, Welsch's (1977) stepwise procedures appear to be the most effective ones we presently have.

As (*) becomes even smaller, we expect to find several simple comparisons about whose sign we can easily be simultaneously confident, and we strive to add still more to our list of confident directions. Psychological experiments seem to offer a fair collection of such data sets, so it may be no accident that psychologists like Ramsay (1981) have played an important role in developing methods apparently appropriate for this range of ratios.

Such methods, which also include methods by Peritz (1970), are often fairly complicated. Fortunately all of this complication may not be necessary, as suggested by the simpler method proposed by Braun and Tukey (1983c) whose performance seemed competitive with — or better than — some of the more complex approaches.

As (*) gets even smaller, we can expect to find significance (which to us means "confident direction from zero") for almost all comparisons. Our effects then have to be devoted to one or both of

- finding confident directions for the few remaining comparisons whose directions are not yet clear and manifest,

- finding shorter confidence intervals (hence confidence in the sign of more offset comparisons) for all comparisons.

So far as we know today, this would seem to be a second appropriate region for the use of the S-range.

We have suggested 4 points along what probably needs to be a longer sequence

S-range $\rightarrow$? $\rightarrow$ Welsch $\rightarrow$? $\rightarrow$ complex techniques $\rightarrow$? $\rightarrow$ S-range.

What other techniques need to be included deserve careful study.

3   Two- and More-Way Tables

Factorial analysis of variance has classically been treated in omnibus terms for each line. Questions like "Is the $A \times B$ interaction significant?" have, regrettably, been thought to have helpful simple answers! With a realistic view of the null hypothesis and the precedent (for one-way tables) of giving up asking "Are the A main effects significant?" and doing multiple comparisons instead, it is clear that we must also do something better for interactions.

If $\{y_{ai}\}$ are values in a two-way table of *either* observations *or* means (over other factors) or interactions (based) on either observations or means), the double comparisons (or cross differences)

$$(ab, ij) = y_{ai} - y_{aj} - y_{bi} + y_{bj}$$

describe — also in an overlapping way — the corresponding interaction. They do this just as the simple comparisons (or simple differences)

$$y_i - y_j$$

describe — also in an overlapping way — the corresponding main effects. Just as the simplest and most natural null model for the latter is

$$y_i = \mu + \epsilon_i$$

with ave $\{\epsilon_i\} = 0$, var $\{\epsilon_i\} = \sigma^2$, cov $(\epsilon_i, \epsilon_j) = 0$ for $i \neq j$, and s^2 an "independent" estimate of σ^2, (and where the value of μ cancels out identically), so too the simplest and most natural null model for 2-factor interactions is

$$y_{ai} = \mu + \alpha_\alpha + \beta_i + \epsilon_{ai}$$

with ave $\{\epsilon_{ai}) = 0$, cov $\{\epsilon_{ai}, \epsilon_{bi}\} = 0$ if "ai" $\neq$ "bj", and s^2 an "independent" estimate of σ^2 (and where the values of μ, the $\{\alpha_a\}$, and the $\{\beta_i\}$ cancel out identically).

Just as

$$\text{range}\{y_i\} = \max\{y_i - y_j\}$$

so we may as well define

$$\text{birange}\{y_{ai}\} = \max\{y_{ai} - y_{aj} - y_{bi} + y_{bj}\}$$

and use the distribution of the S-birange (the Studentized birange) to set limits on

$$\max\{y_{ai} - y_{aj} - y_{bi} + y_{bj}\}$$

leading to simultaneous limits on all the double comparisons

$$(ab, ij) = y_{ai} - y_{aj} - y_{bi} + y_{bj}$$

This is the simplest natural generalization, to 2-way arrays, of the ordinary multiple comparisons for 1-way arrays. The extension to triple comparisons for 3-way arrays, etc. is clear and easy.

Tukey and Hoaglin (1991j) have made calculations that suggest approximating the critical values of the S-birange as $\sqrt{2}$ times an appropriate critical value of the S-range, namely that for

$$\# \text{ of candidates} = 1 + 1.1 \ (\# \text{ of df for interaction})$$

and (of course) the obvious denominator degrees of freedom for S. (We will still have to worry about possible diversity of variance from one cell of the 2-way table to another).

4 Presenting Answers: Simple and Double Comparisons

If we have provided one of three answers (< 0, ??, or > 0) for each simple comparison

$$(ij) = y_i - y_j$$

— simple comparisons that redundantly describe the differences among the $\{y_{ji}\}$ — we can do fairly well, even when there are quite a few values of i, by using either of the techniques shown in Figure 1 to tell us about each and every (ij). In the upper part of Figure 1, we do this by using i for one (skew) coordinate and j for the other. In the lower part we plot each treatment on a number line and show the S-range based guide in several positions. Each presentation tells us, relatively clearly, that, at 5% simultaneous, we are to believe that sample 5 is larger (more positive) then samples 1, 4, and 6, and that sample 3 is more positive than sample 6.

We can combine these two presentations and take advantage of simplification, by grouping treatments as in Figure 2. Samples 4 and 1 are to be treated alike (Figure 1) so far as all questions of significance at 5% simultaneous go, so we combine them into a single group D. We then do not have single numerical differences to place in the tilted squares of the half of a 2-way array, so we show only significance or non-significance at 5% simultaneous. The array teaches that group A (B, resp.) is significantly more positive than groups D and E (group E, resp.). When we decode back to individual treatments, we of course obtain just what we had before.

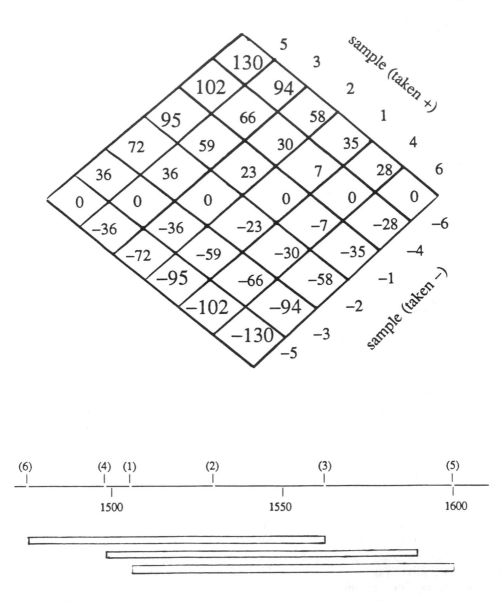

s-range based gauge (length = 92) in 3 crucial positions

Figure 1: Two presentations of the *S*-range based significance: Difference values in a two-way array and number line with multiple positions of an *S*-based significance gauge (both for 5% simultaneous). Large digits mean comparisons significant at 5% simultaneous.

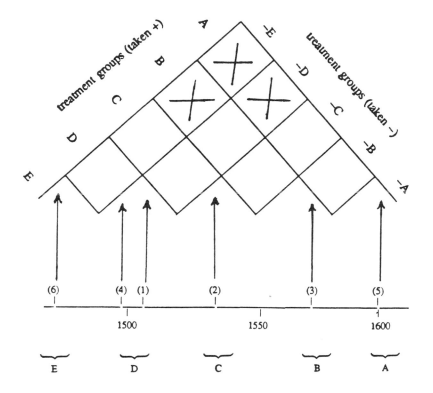

Figure 2: Combined presentation: Grouped significance in half a 2-way array and linked number line.

The importance of grouping as an aid to clarity is often most striking when there are many more treatments to compare.

What do we do with the cross differences (double comparisons) (ab, ij) once we have used the S-birange to choose one of three answers for each one? Plausible, the answer may depend on what kind of interaction we are trying to describe. Thus, until we learn more, it is likely to pay us to make at least two sorts of pictures.

Figure 3 shows (i) an illustrative table of interaction entries, (ii) all row-wise differences (a, ij) of (i), (iii) a table of (ab, ij) for $a < b$, $i < j$ and $(ab) =$ columns, $(ij) =$ rows, and (iv) the result of rearranging rows and columns in (iii) to put large values in the NE corner, and (v) a 45°-rotated version, assuming, for illustration, that "significantly > 0" is "observed ≥ 7". This particular example can be put in a quite understandable form - we do not know how often this is likely to happen with this presentation.

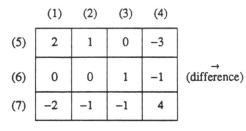

	(1)	(2)	(3)	(4)
(5)	2	1	0	-3
(6)	0	0	1	-1
(7)	-2	-1	-1	4

$\rightarrow$ (difference)

	(12)	(13)	(14)	(23)	(24)	(34)
	1	2	5	1	4	3
	0	-1	1	-1	1	2
	-1	-1	-6	0	-5	-5

$\downarrow$
(second difference)

	(12)	(13)	(14)	(23)	(24)	(34)
(56)	1	3	4	2	3	1
(57)	2	3	11	1	9	8
(67)	1	0	7	-1	6	7

rearrange $\leftarrow$

$\downarrow$

	(23)	(12)	(13)	(24)	(34)	(14)
(57)	1	2	3	9	8	11
(67)	-1	1	0	6	7	7
(56)	2	1	3	3	1	4

(treating 7 or more as significant)

Figure 3: Analysis of a 3×4 intersection table in terms of cross differences = double comparisons.

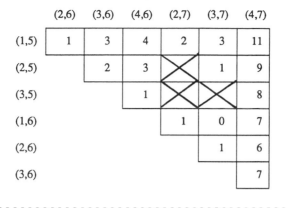

A row-PLUS-column breakdown of the double differences

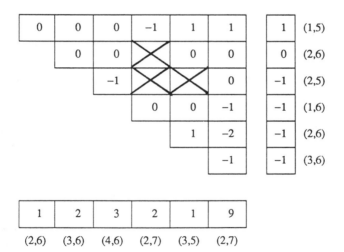

Figure 4: A different presentation of the cross differences of Figure 3; a row-PLUS-column breakdown of this presentation.

Figure 4, instead of taking (ab) and (ij) as coordinates, takes (a,i) and (b,j), keeping only enough of the full table to include all (ab,ij) that are not identically zero. This figure also includes a row-PLUS-column presentation of this breakdown. Notice that only 1 residual, and no row effect, is outside ± 1. The residual of 9 for $(4,7)$ dominates all the rest. For completeness, Figure 5 presents a row-PLUS-column breakdown for the presentation of

-2	-1	-1	1	1	2		2	(57)
-2	0	-2	0	2	0		0	(67)
3	2	2	-1	-2	-1		-1	(56)

1	1	2	5	5	7
(23)	(12)	(13)	(24)	(34)	(14)

Figure 5: A row-PLUS-column breakdown of the presentation of Figure 2.

Figure 2. This time the residuals are not as small, as those in Figure 3, and show more structure, particularly since the (ij) fit shows a substantial difference between "4 present" (6, 5, and 7) and "4 absent" (1, 1, and 2). In this instance the second presentation (Figure 3) seems to be more simply described. How often this will happen seems uncertain.

We need to carry on with both of these approaches until we learn more about how most interactions behave when dissected into double comparisons.

* more graphic versions for larger tables *

If we had, say, a 5 × 10 table, these presentations would call for a 10 × 45 = 450 cell table, as in the style of Figure 3, or for a triangular part of a 4 × 9 by 4 × 9 whose 666 cells consists of the 450 cross differences and 6 triangular blocks, each of 6 x's, as in the style of Figure 4. We need to show what is going on much more simply than either of these.

If there are 5 i's, which means 10 distinct ij's, we can plot the 10 (a, ij) single differences on each of 10 horizontal lines (10 stems), one for each (ij). (The two different 10's are just a numerical coincidence!) Then we can look at the observed range on each horizontal line, and the greatest of these will be the observed birange. It seems worth while to illustrate this kind of plot, while neglecting the birange.

Figure 6 shows such a plot for a somewhat messy data table. We see 8 occurrences (of 10 possible) of K at or near the extreme right. Since all ab have been taken in a positive-going time direction, this shows an upward

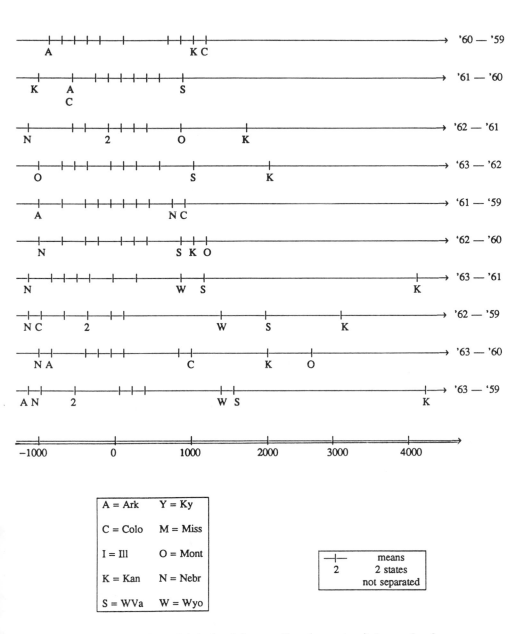

Figure 6: Plot of (ab, i) with one line (one stem) for each ab.

trend in the values for K = Kansas. Since A = Arkansas and N = Nebraska tend to appear on the left suggesting a trend in the opposite direction, it is likely that we would do well to fit (linear) time trends for each state and then repeat this kind of plot, now based upon deviations from trend. (A much larger scale would now be feasible — we could hope to see more things.)

Tendencies for the blocks of values to shift back and forth as wholes in such a plot merely indicate main effects of $\{a\}$ — in this instance main effects of time. Our concern with interaction leaves us looking *only at* the variation among the ticks on each single stem, taking all stems in turn.

5   When the S-range is Too Voluble

When we have 6 to 50 candidates to compare, corresponding to $\binom{6}{2} = 15$ to $\binom{50}{2} = 1225$ simple comparisons, we are   certainly allowed to use the Studentized range to assess all these comparisons. If we are happy with the sort of presentation that derives from Figure 1, we can look at all of these comparisons *en bloc* — this may, indeed, meet many of our needs. To turn our results into words may, however, not be satisfactory.

If, for instance, 10% — or 123 — of the $\binom{50}{2} = 1225$ simple comparisons are significant (i.e. have a well-established sign) it may not be easy to describe these 123 — and distinguish them from the remaining 1102! This is a *description* problem, *not* an *archival one.* Archiving even 50 (or more) treatment values and a significance gauge (based on the S-range) is easy, and is likely to be what we should do to prepare for later reference — especially since our later interest is likely to be confined to a subset of the treatments.

For descriptive presentation, however, we need to focus our attention on fewer differences. If we cannot afford to look at anything proportional to (# of candidates)^2, we need to find something proportional to the # of candidates itself. The most natural choices are of the form

$$(\text{value for treatment}) - (\text{overall reference value})$$

The first such approach seems to have been that of Jerome Cornfield and collaborators (Halperin et al. 1955) who used the mean of the treatment values as the overall reference value.

Some attention can be paid to the identity

$$\text{one} - (\text{mean of all}) \equiv \tfrac{n-1}{n} \, (\text{one} - (\text{mean of all other}))$$

but this seems more a matter of interest than of importance. Using the same reference value for all treatments, thus merely translating treatment values rigidly, *does* seem likely to matter.

Since we do not want to often have one treatment seem deviant because another treatment deviated very strongly in the other direction — so strongly as to move the mean of the treatment values a long way — we favor use of a robust summary for the "overall reference value". A convenient, simple choice is the mid-mean (the mean of the middle-half of the ordered values) of the treatment values.

Use of a more sophisticated, and somewhat more efficient, estimate will buy us little, since the variance of the reference value is roughly

$$1/(n \times \text{ effic})$$

of the variance of the treatment values themselves. (For $n = 10$, moving "effic", the efficiency of the robust summary, from 80% to 90% moves the variance of the deviation from

$$1 + \frac{1}{10(80\%)} = 1.125$$

to

$$1 + \frac{1}{10(90\%)} = 1.111$$

a change in variance of 1.3 parts in 100, hence a change in standard error of 0.6 parts in 100.)

Thus, for description, it seems natural and wise to use, *for description*, when the number of treatments is not small, significance results for

$$\frac{\text{(treatment value)} - \text{(midmean of treatment values)}}{\text{standard error of treatment values}}$$

rather than significance results based on the Studentized range. (We recall that archiving is likely to want to use significance based on the S-range.)

We need to keep the message as intelligible as we can.

If the S-range significance gauge had been, say, 500 in Figure 6, the double comparisons significant (at 5% simultaneous) would have been:

for '60 - '59: C,K > W,O,M,N,Y,I,S,A,
W,O, > Y,I,S,A (24 sig)

for '61 - '60: S > Y,W,M,O,N,I,A,C,K,
Y,W,M,D,N,I > A,C,K (27 sig)

and so on.

Thus we would have perhaps 200 significances to describe if we continue to use the studentized birange. Use of the studentized deviation from the midmean, spending $5\%/10 = 0.5\%$ on each stem (each line in the presentation of Figure 6) would require explaining a much smaller set of significances, and would, in particular, be almost sure to lead to noticing the "trend phenomenon" so strong for Kansas.

All this "in spades" for the 2-way case!

6   Hints, Anyone?

When we look hard at the real world, particularly the world of exploratory data analysis, we find unacceptable a policy of mentioning only those things about which we are confident. The opposite extreme, of mentioning every little appearance we can find, is probably even more unacceptable. We need to live with "de minimus non curat lex" at more than one level of "minimus".

We must do something about shades of uncertainty. The least we can do is to plan to recognize one shade; it seems reasonable to recognize at least two kinds:

- a hint or suggestion

- a leaning.

We could try to make these as sharp as we are used to making "significant at 5%". To me, this year (and also last year), sharpness does not seem the best thing to seek — softness of criteria can open a way for the use of subject matter understanding — but only getting one or more techniques into relatively broad use can provide the experience to make us sure whether a soft approach is better or a crisp approach is better.

Given a measurement whose standard deviation we know (and may as well take to be 1.00) we are used to cutting 1/40 off each end of the distribution and saying that we are confident (at 95%) that what we saw did not come from either of these extreme tails. (We do know of course, that sometimes — roughly one time in twenty — what we see *did* come from one of these extreme tails.)

Those who believe in only speaking about what they are confident about, will say nothing — other, perhaps, than "not significant" — about all results — usually comparisons — whose confidence intervals include zero. But a (95%) confidence interval from -0.03 to $+3.21$ says something quite different from one from -1.33 to $+1.81$! Both of these say something quite different from a confidence interval from -0.05 to $+0.07$! If each author put the confidence intervals in his or her table, and firmly directed the reader to look at them, the situation would not be too badly handled, but, even then, readers who are verbally literate rather than numerically literate will not have been given what they need.

If we are to say something weaker than significance — or confidence — at a standard level, we must take more than 1/20 chance of some analog of a Type 1 error — this means that we shall need to be very careful that our hints are not over-interpreted. (I fear that the Ralph Naders of this world will try to give them more attention — more trumpeting and breast beating — than they deserve.) But if we do not take this chance, we shall fail to tell sensible people things they need to know. Perhaps the right answer is to tool up as if we were going to use new words in an organized way, but then plan to let the new technology guide us whether or not we mention, in old-fashioned sloppy words, particular appearances. We might even say such things are:

- "this data suggests to us that condition A *may* give, on the average, larger (more positive) results than condition B".

The more our thinking slides this way, the less we want a crisp approach, the more we want a soft approach.

7   Choosing Boundaries

We almost surely want to end up with tail areas. But this does not seem to be the place to start. Better we think first about a reasonably conventional single measurement, whose error distribution is vaguely Gaussian (and whose bias is small). Working at 95% means that our standard confidence interval is, roughly,

$$(\text{``observed''} - 2SE, \text{``observed''} + 2SE)$$

where

$$SE = \text{standard error of ``observed''}$$

If confidence begins at $\pm 2SE$, where do hints begin? Plausibly at $\pm 1.0SE$, though also plausibly at somewhat less or at somewhat more. Appreciably less than 1.0 standard error seems weak for a "hint", maybe it should be a "leaning". Anything close to 1.5 standard errors almost must be a "hint". We want something somewhat smaller than 1.5 for an outer guideline. So we take

$$\pm 1SE$$

and

$$\pm 1.33SE$$

as two guidelines. Values smaller than, and *not* close to $\pm 1SE$ almost certainly should *not* be taken as hints. Values larger than, and *not* close to $\pm 1.33SE$ almost certainly should be taken as hints. In between, we use our best judgment.

If we are to have something weaker than a hint, say a *leaning*, we could be motivated by a desire for another soft approach, and two more guidelines. But the world is becoming complicated enough as it is, so we choose to add one more guideline, another guideline for leaning, *near*

$$\pm.67SE$$

and think of the inner hint-guideline, which we already have, as the outer leaning-guideline.

Converting roughly ± 1.33, roughly ± 1.00, and roughly $\pm.67$ into tail areas — most naturally for a situation with many (but not infinite) degrees of freedom — and looking for simple fractions — leads us to $1/11^{th}$, $1/6^{th}$, and $1/4^{th}$, to go with $1/40^{th}$. For a *single measurement*, then, we use

confidence guidelines, at tail area 1/40
outer hint-guidelines, at tail area 1/11
inner hint-guidelines = outer leaning-guidelines, at tail area 1/6
inner leaning-guidelines, at tail area 1/4.

8   Facing Multiplicity

The case of a single measurement is relatively rare. We will not have met our responsibilities if we do not think about the case of many measurements. If a commercial laboratory makes k measurements, one for each of k customers who never interact, that laboratory has no multiplicity problem. Each customer can swallow his or her own 5%.

But if a scientist or technologist makes k measurements, under the usual circumstances where there is a moral obligation to call attention to the most unusual determination (the most unusual single result) or the most unusual comparison (the most unusual difference between two results), that scientist or technologist — physical, behavioral, biological or medical — has a multiplicity problem, often a serious one.

At the level of confidence, we appear to do better by allowing 5% chance that *one or more* confidence intervals will miss the truth. We do rather well to use *simultaneous* confidence or significance procedures. As a result, with m things to look at, and 2 ends for each, we spend roughly $5\%/2m = 1/40m$ chance of error on each end of each of m comparisons or determinations.

Table 1: Type 1 errors, Type 1 mis-suggestions, and Type 1 overemphases as average percentage (entries of 800% or more will usually correspond to situations where not all comparisons larger than a guideline will be mentioned) for analyses with multiplicity $m = \binom{k}{2}$. (Notes: * hint cutoff somewhere between "inner" and "outer"; ** leaning cutoff somewhere between "outer" and "possible overemphases".)

k	m	erroneous confidence	mis-suggestions* inner	outer	possible** overemphases
2	1	5%	18%	37%	55%
4	6	5%	33%	84%	122%
8	28	5%	55%	156%	265%
16	105	5%	86%	342%	512%
30	435	5%	138%	695%	1043%
60	1770	5%	220%	1402%	2104%
120	7140	5%	350%	2817%	4225%
250	31,125	5%	572%	5881%	8821%
500	124,750	5%	908%	11,773%	17,660%
1000	499,500	5%	1443%	23,550%	35,338%

What should we do about hints? The strict analogs — like tail area of $1/11m$ — seem quite unsatisfactory — for a rather qualitative reason. We expect some results that would be individually significant but not simultaneously significant. We cannot treat them as significant, under circumstances where the most extreme will get the attention; we need to *hold* the *significance* line at 5% *simultaneous*. But it seems quite unsatisfactory to have them — at least *some* of them — not be *hints*.

If we try $1/(11m)$ and $1/(6m)$ as the tail areas corresponding to the hint guidelines, we face — for suitable values of m — the uncomfortable combination of individual significance with failure to be a simultaneous hint. So we need to try some other dependence. It turns out to be reasonably satisfactory to use $1/(11m^{2/3})$ and $1/(6m^{1/2})$ for hint-guidelines — and to use $1/(4m^{1/2})$ for inner leaning-guidelines.

As we come to deal with the consequences of such guidelines, we have to face up to more hints being suggested than we wish to mention. Gentle guidelines such as those just suggested will almost surely function best when their use is coupled with a policy of not necessarily mentioning all the hints we could, since too many mentions may be counter-productive. It will often be enough to mention 5 or 10 hints, choosing the apparently largest or most important.

If we are already mentioning as many hints as we feel our audience deserves, we can live with *not* mentioning certain smaller appearances — comparisons, or determinations that would have been individually significant — and, by our standards, would deserve mention if they had not been overshadowed by others. This is important for really large m.

Table 1 gives a general view of how the suggested guidelines would operate over a wide range of numbers of MC candidates and corresponding multiplicities for simple comparisons.

Acknowledgment

This paper was prepared in connection with research at Princeton University sponsored by the Army Research Office.

References

American Petroleum Institute (1965) *Petroleum Facts and Figures*. New York (especially page 43).

Braun, H.I. and Tukey, [1]J. W. (1983c). Multiple comparisons through orderly partitions: The maximum subrange procedure. In *Principals of Modern Psychological Measurement: A Festschrift for Frederic M. Lord* (H. Wainer and S. Messick, eds.), 55 - 65. Erlbaum, Hillsdale, NJ.

Davies, O.L. et. al. (1957). *Statistical Methods in Research and Production*. ICI and Oliver and Boyd, London (especially pages 104 - 106).

Duncan, D.B. (1955). Multiple range and multiple F tests. *Biometrics* **11** 1 - 42.

---

[1]Letters used with years on John Tukey's publications correspond to bibliographies in all volumes of his collected papers

Duncan, D.B. (1965). A Bayesian approach to multiple comparisons. *Technometrics* **7** 171 - 222.

Halperin, M., Greenhouse, S W., Cornfield, J. and Zalokar, J. (1955). Tables of percentage points for the Studentized maximum absolute deviate in normal samples. *Jour. Amer. Statist. Assoc.* **50** 185 - 195.

Peritz, E. (1970). A note on multiple comparisons. Unpublished manuscript, Hebrew University, Jerusalem, Israel. (Cited by Begun and Gabriel 1981, *Jour. Amer. Statist. Assoc.* **76** 241 - 245).

Ramsey, P.H. (1981). Power of univariate pairwise multiple comparison procedures. *Psych. Bull.* **90** 352 - 366.

Tukey, J.W. and Hoaglin, D.C. (1991j). Qualitative and quantitative confidence. In *Fundamentals of Exploratory Analysis of Variance* (D.C. Hoaglin, F. Mosteller, and J.W. Tukey, eds.), To appear, John Wiley and Sons, Inc., New York.

Welsch, R.E. (1977). Stepwise multiple comparison procedures. *Jour. Amer. Statist. Assoc.* **72** 577 - 575.

Chapter 13

Performances of Selection Procedures for 2-Factor Additive Normal Populations with Common Known Variance

ROBERT E. BECHHOFER School of Operations Research and Industrial Engineering, Cornell University, Ithaca, New York

DAVID GOLDSMAN School of Industrial and Systems Engineering, Georgia Institute of Technology, Atlanta, Georgia

MARK HARTMANN Department of Operations Research, University of North Carolina, Chapel Hill, North Carolina

Abstract Paulson (1964), adopting the indifference-zone approach, proposed a closed eliminating sequential procedure for selecting the normal population having the largest mean when the populations have a common known variance. Fabian (1974) improved Paulson's procedure by decreasing the probability overprotection, thereby decreasing the expected number of stages to termination and the corresponding expected total number of observations. Hartmann (1992) generalized the Paulson-Fabian procedure

209

to multi-factor experiments with no factor-level interactions. Using Monte Carlo simulation we study the performance characteristics of the latter procedure relative to other competing procedures when it is employed for 2-factor experiments. We find that although the procedure dominates all competing procedures in terms of minimizing the expected total number of observations in the so-called equal means and contiguous configurations when both factor-levels are moderately large, this dominance is achieved at the expense of an excessively large expected number of stages as compared to other competing procedures. The significance of this finding is discussed.

1   Introduction

It is now well-known that factorial experimentation when employed in ranking and selection problems can result in considerable savings in total sample size relative to independent single-factor experimentation when both guarantee comparable probability requirements. This was pointed out by Bechhofer (1954, Section 4) who proposed a *single-stage* procedure for ranking normal means when there is no interaction between factor-level effects and the common variance of the populations is assumed to be known. Bawa (1972) provided asymptotic calculations of the magnitude of savings in total sample size that could be achieved in such normal mean single-stage settings if one single-stage 2-factor experiment were used in place of the two corresponding independent single-stage single-factor experiments. Bechhofer and Goldsman, henceforth denoted by B-G, (1988a, Section 4.1), showed that the open *sequential* selection procedure of Bechhofer et al. (1968), which was originally devised for *single-factor* experiments involving Koopman-Darmois (K-D) populations, could be generalized to accommodate *multi-factor* experiments involving K-D populations with *additivity* and, in particular, experiments involving normal populations with a common *known* variance and *no interaction* between the factor-levels. Both the original basic open sequential procedure ($\mathcal{P}^*_{\mathrm{B}(2,+)}$) and the truncated version of that procedure ($\mathcal{P}^*_{\mathrm{B_T}(2,+)}$) for *2-factor* experiments involving normal means with *additivity* take observations a matrix-at-a-time. B-G (1988b) conducted Monte Carlo (MC) sampling experiments with $\mathcal{P}^*_{\mathrm{B}(2,+)}$ and $\mathcal{P}^*_{\mathrm{B_T}(2,+)}$ to study the effect of truncation of the procedure on the expected number ($\mathsf{E}\{n\}$) and variance of the number ($\mathsf{Var}\{n\}$) of stages to terminate sampling. Large decreases in $\mathsf{E}\{n\}$ and $\mathsf{Var}\{n\}$ were obtained.

B-G (1989) demonstrated for *single-factor* experiments with common known variance that a considerable reduction in $\mathsf{E}\{n\}$ and expected total number of scalar observations ($\mathsf{E}\{T\}$) (as well as their variances, $\mathsf{Var}\{n\}$ and $\mathsf{Var}\{T\}$) could be achieved by the use of Paulson's (1964) procedure as improved by Fabian (1974) with further improvements by Hartmann (1988). These procedures, as with Paulson's, have the highly desirable property of *eliminating* permanently from further sampling and consideration populations indicated as being non-contending for selection; furthermore, all of these procedures are *closed*, as is $\mathcal{P}^*_{\mathrm{B_T}(2,+)}$.

Hartmann (1992) proposed a *2-factor* selection procedure ($\mathcal{P}_{\mathrm{PFH}(2,+)}$) for normal means with no factor-level interactions and a common known variance; this procedure generalized the Paulson-Fabian (Fabian 1974) *single-factor* procedure. $\mathcal{P}_{\mathrm{PFH}(2,+)}$ guarantees the same probability requirement as do $\mathcal{P}^*_{\mathrm{B}(2,+)}$ and $\mathcal{P}^*_{\mathrm{B_T}(2,+)}$. Because of the highly desirable performance characteristics of $\mathcal{P}^*_{\mathrm{B_T}(2,+)}$ (see B-G 1988b, Section 6) we shall use it as a reference point when evaluating the performance of $\mathcal{P}_{\mathrm{PFH}(2,+)}$. Our choice here was motivated by the fact that for *single-factor* experiments with common known variance, the procedure $\mathcal{P}_{\mathrm{B_T}}$, the Paulson-Fabian procedure, and the latter procedure as improved by Hartmann (1988), dominated all other competing procedures in terms of desirable performance characteristics as reported in B-G (1989). (Note: Hartmann improved the Paulson-Fabian procedure for *single-factor* experiments, and generalized the Paulson-Fabian single-factor procedure to *multi-factor* experiments. The Paulson-Fabian-Hartmann (Hartmann 1988) *single-factor* procedure has not yet been generalized to two or more factors.)

The purpose of the present study is to determine whether $\mathcal{P}_{\mathrm{PFH}(2,+)}$ reduces $\mathsf{E}\{n\}$ and $\mathsf{E}\{T\}$ relative to the corresponding quantities from the truncated sequential procedure $\mathcal{P}^*_{\mathrm{B_T}(2,+)}$. $\mathcal{P}_{\mathrm{PFH}(2,+)}$ takes observations a matrix-at-a-time (as does $\mathcal{P}^*_{\mathrm{B_T}(2,+)}$) but the dimensions of the matrices decrease as populations are eliminated. It was this latter fact that suggested that $\mathsf{E}\{T\}$ might be much lower for $\mathcal{P}_{\mathrm{PFH}(2,+)}$ than for $\mathcal{P}^*_{\mathrm{B_T}(2,+)}$.

The organization of this paper is as follows: In Section 2 we state the problem. The three procedures being studied, namely the 2-factor single-stage procedure $\mathcal{P}_{\mathrm{SS}(2,+)}$ (Bechhofer 1954), the truncated sequential procedure $\mathcal{P}^*_{\mathrm{B_T}(2,+)}$ (B-G 1988b) and the closed sequential procedure with elimination $\mathcal{P}_{\mathrm{PFH}(2,+)}$ (Hartmann 1992) are described in Section 3. We give MC sampling results obtained with these procedures in Section 4, and discuss these results in Section 5. Our conclusions are contained in Section 6.

2   Statement of the Problem

We assume that we have $a \cdot b$ independent normal populations Π_{ij} ($1 \leq i \leq a$, $1 \leq j \leq b$) with unknown population means μ_{ij} and a common known variance σ^2 which without loss of generality we assume to be equal to unity. We denote the sth independent observation from Π_{ij} by X_{ijs}

$(1 \leq i \leq a, 1 \leq j \leq b, s = 1, 2, \ldots)$. Throughout this paper we assume that for $(1 \leq i \leq a, 1 \leq j \leq b, s = 1, 2, \ldots)$ we have

$$\mathsf{E}\{X_{ijs}\} = \mu_{ij} = \mu + \alpha_i + \beta_j \qquad (2.1)$$

where $\sum_{i=1}^{a} \alpha_i = \sum_{j=1}^{b} \beta_j = 0$. This is the standard *additive* (no interaction) fixed-effects ANOVA model. The parameter α_i (β_j) is the "effect" of the i th level of the α-factor (j th level of the β-factor) on μ. The ordered values of the α_i and β_j are denoted by

$$\alpha_{[1]} \leq \alpha_{[2]} \leq \cdots \leq \alpha_{[a]} \quad \text{and} \quad \beta_{[1]} \leq \beta_{[2]} \leq \cdots \leq \beta_{[b]}. \qquad (2.2)$$

We assume that the values of the $\alpha_{[p]}$ ($1 \leq p \leq a$) and $\beta_{[q]}$ ($1 \leq q \leq b$) are unknown, and that the pairings of the $\alpha_{[p]}$ and the $\beta_{[q]}$ with the Π_{ij} ($1 \leq i, p \leq a, 1 \leq j, q \leq b$) are completely unknown. The goal of the experiment and the associated probability requirement are stated below.

Goal

To select that *one* of the $a \cdot b$ populations Π_{ij} ($1 \leq i \leq a, 1 \leq j \leq b$) which is associated with $\alpha_{[a]}$ and $\beta_{[b]}$.

If the population selected by a procedure is indeed associated with $\alpha_{[a]}$ *and* $\beta_{[b]}$, then we say that a *correct selection* (CS) has been made. We limit consideration to procedures that guarantee the following *indifference-zone* requirement on the P{CS}:

Probability Requirement

$$\mathsf{P}\{\mathrm{CS}\} \geq P^* \quad \text{whenever} \quad \alpha_{[a]} - \alpha_{[a-1]} \geq \delta_\alpha^* \quad \text{and} \quad \beta_{[b]} - \beta_{[b-1]} \geq \delta_\beta^*. \quad (2.3)$$

The constants $\{\delta_\alpha^*, \delta_\beta^*, P^*\}$ with $(0 < \delta_\alpha^*, \delta_\beta^* < \infty, 1/ab < P^* < 1)$ are specified prior to the start of experimentation. The three procedures described in Section 3 guarantee (2.3) when σ^2 is known.

3   Three Procedures: 2-Factor Experiments, Additivity and Common Known Variance

We now describe the selection procedures under consideration for 2-factor experiments with additivity when the common variance is known.

3.1   The Bechhofer Single-Stage Procedure ($\mathcal{P}_{\mathrm{SS}(2,+)}$)

Let x_{ijs} ($1 \leq i \leq a, 1 \leq j \leq b, s = 1, 2, \ldots, N$) denote the outcome of the s th observation from Π_{ij}. Let $A_{im} = \sum_{j=1}^{b} \sum_{s=1}^{m} x_{ijs}$ and $B_{jm} =$

$\sum_{i=1}^{a}\sum_{s=1}^{m} x_{ijs}$ $(1 \le i \le a,\ 1 \le j \le b,\ m = 1,2,\ldots)$. For fixed $N = m$ denote the ordered values of the A_{iN} and B_{jN} by $A_{[1]N} < \cdots < A_{[a]N}$ and $B_{[1]N} < \cdots < B_{[b]N}$, respectively. The sampling rule and terminal decision rule of $\mathcal{P}_{\text{SS}(2,+)}$ are stated below.

Sampling rule: Observe random variables $\{X_{ijs}\ (1 \le i \le a, 1 \le j \le b, 1 \le s \le N)\}$ in a *single* stage.

Terminal decision rule: Compute A_{iN} $(1 \le i \le a)$ and B_{jN} $(1 \le j \le b)$. Select the population yielding $A_{[a]N}$ and $B_{[b]N}$ as the one associated with $\alpha_{[a]}$ and $\beta_{[b]}$.

Table 1 in B-G (1988b) gives values of N to guarantee (2.3) for all (a, b) with $2 \le a \le 20$, $1 \le b \le 8$ when $\delta_\alpha^* = \delta_\beta^* = 0.2$, $P^* = 0.9025$ and $\sigma^2 = 1$.

3.2   The Bechhofer-Kiefer-Sobel Truncated Sequential Procedure $(\mathcal{P}_{\text{BT}(2,+)}^*)$

Let x_{ijs} be defined as in Section 3.1. At each stage m $(m = 1,2,\ldots)$ let $A_{im} = \sum_{j=1}^{b}\sum_{s=1}^{m} x_{ijs}$ $(1 \le i \le a)$ denote the row sums and $B_{jm} = \sum_{i=1}^{a}\sum_{s=1}^{m} x_{ijs}$ $(1 \le j \le b)$ denote the column sums of the $a \times b$ matrix $\{\sum_{s=1}^{m} x_{ijs}\ (1 \le i \le a, 1 \le j \le b)\}$. For $m = 1,2,\ldots$ denote the ordered values of the A_{im} and B_{jm} by $A_{[1]m} < \cdots < A_{[a]m}$ and $B_{[1]m} < \cdots < B_{[b]m}$, respectively. Let

$$U_m = \sum_{i=1}^{a-1} \exp\{-\delta_\alpha^*(A_{[a]m} - A_{[i]m})\},$$

$$V_m = \sum_{j=1}^{b-1} \exp\{-\delta_\beta^*(B_{[b]m} - B_{[j]m})\},$$

and

$$Z_m = U_m + V_m + U_m \cdot V_m. \tag{3.1}$$

The sampling rule, stopping rule and terminal decision rule of $\mathcal{P}_{\text{BT}(2,+)}^*$ are stated below.

Sampling rule: At the mth stage of experimentation $(m = 1,2,\ldots)$ observe the random matrix $\{X_{ijm}\ (1 \le i \le a, 1 \le j \le b)\}$.

Stopping rule: After the mth matrix-observation $(m = 1,2,\ldots)$ has been taken, compute Z_m of (3.1). Stop sampling when, for the first time, either

$$Z_n \le (1 - P^*)/P^* \quad \text{or} \quad n = n_0;$$

here n (a random variable) is the value of m at termination, and $n_0 = n_0(a, b; \delta_\alpha^*, \delta_\beta^*, P^*)$ (say) is predetermined as the smallest integer that will guarantee (2.3) when $\mathcal{P}_{\mathrm{B T}(2,+)}^*$ is used.

Terminal decision rule: After stopping select the population that yielded $A_{[a]n}$ and $B_{[b]n}$ as the one associated with $\alpha_{[a]}$ and $\beta_{[b]}$.

There are no formulas for calculating the truncation numbers n_0 exactly. Thus we estimated the n_0 by MC sampling as in B-G (1987, Section 4). Values of the estimated n_0' for our 2-factor experiments are given in B-G (1988b, Table 2) for all (a, b) with $2 \leq a, b \leq 6$ when $\delta_\alpha^* = \delta_\beta^* = 0.2$, $P^* = 0.9025$ and $\sigma^2 = 1$.

3.3 The Paulson-Fabian-Hartmann Closed Sequential Procedure ($\mathcal{P}_{\mathrm{PFH}(2,+)}$)

Before experimentation starts the experimenter specifies P_α^*, $P_\beta^* > 0$ with $P_\alpha^* \cdot P_\beta^* \geq P^*$, λ_α with $0 \leq \lambda_\alpha \leq \delta_\alpha^*/2$ and λ_β with $0 \leq \lambda_\beta \leq \delta_\beta^*/2$. In this article we shall consider only $\lambda_\alpha = \delta_\alpha^*/2$ or $\delta_\alpha^*/4$ and $\lambda_\beta = \delta_\beta^*/2$ or $\delta_\beta^*/4$ because of the desirable properties associated with these choices as demonstrated in B-G (1989). We also only consider here $P_\alpha^* = P_\beta^*$ (although as pointed out in Section 6 this choice is optimal only if $a = b$ and $\delta_\alpha^* = \delta_\beta^*$). For $\lambda_\alpha = \delta_\alpha^*/2$ ($\lambda_\beta = \delta_\beta^*/2$) we define

$$a_{\lambda_\alpha} = \frac{-\sigma^2}{\delta_\alpha^* - \lambda_\alpha} \ln \frac{2(1 - P_\alpha^*)}{a - 1} \quad \left(a_{\lambda_\beta} = \frac{-\sigma^2}{\delta_\beta^* - \lambda_\beta} \ln \frac{2(1 - P_\beta^*)}{b - 1} \right)$$

while for $\lambda_\alpha = \delta_\alpha^*/4$ ($\lambda_\beta = \delta_\beta^*/4$) we define

$$a_{\lambda_\alpha} = \frac{-\sigma^2}{\delta_\alpha^* - \lambda_\alpha} \ln \theta_\alpha \quad \left(a_{\lambda_\beta} = \frac{-\sigma^2}{\delta_\beta^* - \lambda_\beta} \ln \theta_\beta \right)$$

where θ_α (θ_β) solves

$$(1 - P_\alpha^*)/(a - 1) = \theta_\alpha - (1/2)\theta_\alpha^{4/3} \quad \left((1 - P_\beta^*)/(b - 1) = \theta_\beta - (1/2)\theta_\beta^{4/3} \right).$$

This procedure permanently eliminates all populations associated with either α_i or β_j when that α_i (β_j) is no longer indicated as contending for $\alpha_{[a]}$ ($\beta_{[b]}$); the elimination process continues until a single population remains.

Let A_r (B_r) denote the number of remaining α-factor (β-factor) levels before stage r, i.e., indices i (j) for which the populations associated with α_i (β_j) have not been eliminated. Then $A_1 = a$, $B_1 = b$ and sampling is terminated before stage r if we have $A_r = B_r = 1$. If a population has not been eliminated we refer to it as being "active."

Sampling rule: At the rth stage ($r = 1, 2, \ldots$) take a single observation X_{ijr} from each of the $A_r B_r$ active populations; this can be thought of as taking observations a matrix-at-a-time from a matrix $\{X_{i_k j_\ell r},$ $k = 1, \ldots, A_r, \ell = 1, \ldots, B_r\}$ of size $A_r \times B_r$ where $i_1, \ldots, i_{A_r}$ and $j_1, \ldots, j_{B_r}$ are the remaining α- and β-factor levels, respectively. Let $Y_{ir} = \sum_{\ell=1}^{B_r} X_{ij_\ell r}$ denote the row sum associated with α-factor level i, and let $Z_{jr} = \sum_{k=1}^{A_r} X_{i_k j r}$ denote the column sum associated with β-factor level j. Next eliminate the populations associated with α_k if

$$\sum_{s=1}^{r} Y_{ks} < \max_i \left(\sum_{s=1}^{r} Y_{is} \right) - \max \left(a_{\lambda_\alpha} - \lambda_\alpha \sum_{s=1}^{r} B_s, \ 0 \right),$$

where the maximum is taken over the remaining α-factor levels i, and eliminate the populations associated with β_ℓ if

$$\sum_{s=1}^{r} Z_{\ell s} < \max_j \left(\sum_{s=1}^{r} Z_{js} \right) - \max \left(a_{\lambda_\beta} - \lambda_\beta \sum_{s=1}^{r} A_s, \ 0 \right),$$

where the maximum is taken over the remaining β-factor levels j.

Stopping rule: Stop sampling when $A_r = B_r = 1$, i.e., when only one population remains.

Terminal decision rule: Select the remaining population as the one associated with the factor-levels $\alpha_{[a]}$ and $\beta_{[b]}$.

4 Monte Carlo Sampling Results

The performance characteristics of $\mathcal{P}_{\mathrm{PFH}(2,+)}$ were estimated using MC sampling. Experiments with this procedure were carried out for both $\lambda_\alpha = \lambda_\beta = \delta^*/4$ and $\lambda_\alpha = \lambda_\beta = \delta^*/2$ with $\delta^* = 0.2$, $P_\alpha^* = P_\beta^* = 0.95$ ($P_\alpha^* \cdot P_\beta^* = P^* = 0.9025$) and $\sigma^2 = 1$; these latter choices permitted us to make direct comparisons with results reported for $\mathcal{P}_{\mathrm{BT}(2,+)}^*$ in B-G (1988b). $\mathcal{P}_{\mathrm{PFH}(2,+)}$ was studied with three configurations of the population means: (i) equally-spaced δ^* apart in both factors (ES(δ^*)), $\alpha_{[i+1]} - \alpha_{[i]} = \delta^*$ ($1 \leq i \leq a - 1$), $\beta_{[j+1]} - \beta_{[j]} = \delta^*$ ($1 \leq j \leq b - 1$), (ii) least-favorable in both factors (LF), $\alpha_{[1]} = \alpha_{[a-1]} = \alpha_{[a]} - \delta^*$, $\beta_{[1]} = \beta_{[b-1]} = \beta_{[b]} - \delta^*$, and (iii) equal means in both factors (EM), $\alpha_{[1]} = \alpha_{[a]}$, $\beta_{[1]} = \beta_{[b]}$, along with all factor-level combinations (a, b) where $(2 \leq a, b \leq 6)$. All results

Table 1: Estimated performance characteristics of statistical selection procedures. $P^* = 0.9025$, $\delta^* = \delta^*_\alpha = \delta^*_\beta = 0.2$, and $\sigma^2 = 1$.

(a,b)	$\mathcal{P}$	ES(δ^*) in both factors			LF in both factors			EM in both	
		$\bar{p}$	$\bar{n}$	$\bar{t}$	$\bar{p}$ or $\bar{w}$	$\bar{n}$	$\bar{t}$	$\bar{n}$	$\bar{t}$
(2,2)	$\mathcal{P}_{SS(2,+)}$	0.9034	68	272	0.9034	68	272	68	272
	$\mathcal{P}^*_{BT(2,+)}$	0.9040	43.8	175.2	0.9032	43.5	173.9	65.9	263.4
	$n_0 = 86$	(0.0027)	(0.2)	(0.4)	(0.0003)	(<0.1)	(0.2)	(0.2)	(0.4)
	$\mathcal{P}_{PFH(2,+)}(\delta^*/4)$	0.9161	68.0	188.4	0.9148	68.0	188.1	97.4	271.1
		(0.0025)	(0.3)	(0.8)	(0.0018)	(0.2)	(0.5)	(0.4)	(0.9)
	$\mathcal{P}_{PFH(2,+)}(\delta^*/2)$	0.9067	65.3	188.9	0.9105	65.7	189.2	85.2	251.0
		(0.0027)	(0.3)	(0.6)	(0.0018)	(0.2)	(0.4)	(0.3)	(0.6)
(2,3)	$\mathcal{P}_{SS(2,+)}$	0.9338	72	432	0.9034	72	432	72	432
		(0.0023)							
	$\mathcal{P}^*_{BT(2,+)}$	0.9252	39.7	238.1	0.9038	47.3	284.1	74.4	446.4
	$n_0 = 92$	(0.0024)	(0.2)	(0.4)	(0.0003)	(<0.1)	(0.2)	(0.2)	(0.5)
	$\mathcal{P}_{PFH(2,+)}(\delta^*/4)$	0.9308	80.4	259.8	0.9219	93.9	307.9	141.7	458.4
		(0.0023)	(0.4)	(0.9)	(0.0017)	(0.3)	(0.8)	(0.6)	(1.5)
	$\mathcal{P}_{PFH(2,+)}(\delta^*/2)$	0.9330	80.6	271.5	0.9159	93.2	319.7	129.4	440.7
		(0.0023)	(0.4)	(0.8)	(0.0018)	(0.3)	(0.7)	(0.4)	(1.1)
(2,4)	$\mathcal{P}_{SS(2,+)}$	0.9568	79	632	0.9029	79	632	79	632
		(0.0019)							
	$\mathcal{P}^*_{BT(2,+)}$	0.9308	37.4	299.0	0.9037	52.3	418.2	83.7	669.8
	$n_0 = 102$	(0.0023)	(0.2)	(0.5)	(0.0003)	(0.1)	(0.3)	(0.2)	(0.6)
	$\mathcal{P}_{PFH(2,+)}(\delta^*/4)$	0.9442	92.8	324.2	0.9228	120.7	451.5	188.7	678.7
		(0.0021)	(0.5)	(1.1)	(0.0017)	(0.4)	(1.1)	(0.7)	(2.0)
	$\mathcal{P}_{PFH(2,+)}(\delta^*/2)$	0.9410	97.1	351.2	0.9194	123.3	480.3	174.9	668.9
		(0.0022)	(0.4)	(1.0)	(0.0018)	(0.3)	(0.9)	(0.5)	(1.5)
(2,5)	$\mathcal{P}_{SS(2,+)}$	0.9665	87	870	0.9039	87	870	87	870
		(0.0016)							
	$\mathcal{P}^*_{BT(2,+)}$	0.9313	36.2	361.9	0.9027	56.6	565.6	90.8	907.6
	$n_0 = 110$	(0.0023)	(0.2)	(0.5)	(0.0001)	(<0.1)	(0.1)	(0.2)	(0.7)
	$\mathcal{P}_{PFH(2,+)}(\delta^*/4)$	0.9469	101.6	376.1	0.9217	142.9	603.4	227.6	909.4
		(0.0020)	(0.5)	(1.1)	(0.0017)	(0.4)	(1.3)	(0.8)	(2.4)
	$\mathcal{P}_{PFH(2,+)}(\delta^*/2)$	0.9441	109.8	420.1	0.9257	145.7	647.6	208.3	902.3
		(0.0021)	(0.4)	(1.1)	(0.0017)	(0.3)	(1.2)	(0.5)	(1.8)
(2,6)	$\mathcal{P}_{SS(2,+)}$	0.9745	94	1128	0.9042	94	1128	94	1128
		(0.0014)							
	$\mathcal{P}^*_{BT(2,+)}$	0.9362	34.7	416.8	0.9037	61.0	732.0	98.8	1186.0
	$n_0 = 120$	(0.0022)	(0.2)	(0.6)	(0.0003)	(0.1)	(0.4)	(0.3)	(0.9)
	$\mathcal{P}_{PFH(2,+)}(\delta^*/4)$	0.9508	109.7	423.9	0.9257	160.8	758.2	257.1	1137.4
		(0.0020)	(0.5)	(1.2)	(0.0017)	(0.4)	(1.6)	(0.4)	(1.4)
	$\mathcal{P}_{PFH(2,+)}(\delta^*/2)$	0.9466	120.0	481.0	0.9222	163.3	821.3	233.7	1138.2
		(0.0021)	(0.5)	(1.1)	(0.0017)	(0.4)	(1.4)	(0.3)	(1.0)

Table 1: (continued)

(a,b)	$\mathcal{P}$	ES(δ^*) in both factors			LF in both factors			EM in both	
		$\bar{p}$	$\bar{n}$	$\bar{t}$	$\bar{p}$ or $\bar{w}$	$\bar{n}$	$\bar{t}$	$\bar{n}$	$\bar{t}$
(3,3)	$\mathcal{P}_{SS(2,+)}$	0.9463 (0.0021)	62	558	0.9051	62	558	62	558
	$\mathcal{P}^*_{BT(2,+)}$ $n_0=76$	0.9288 (0.0023)	32.8 (0.1)	295.1 (0.4)	0.9053 (0.0003)	41.6 (<0.1)	374.6 (0.2)	66.1 (0.1)	594.9 (0.4)
	$\mathcal{P}_{PFH(2,+)}(\delta^*/4)$	0.9543 (0.0019)	79.3 (0.4)	302.5 (1.0)	0.9241 (0.0017)	90.0 (0.3)	370.8 (0.8)	132.5 (0.5)	548.9 (1.4)
	$\mathcal{P}_{PFH(2,+)}(\delta^*/2)$	0.9576 (0.0018)	76.7 (0.3)	316.9 (0.8)	0.9225 (0.0017)	83.4 (0.2)	377.0 (0.7)	108.3 (0.3)	512.6 (0.9)
(3,4)	$\mathcal{P}_{SS(2,+)}$	0.9582 (0.0018)	60	720	0.9038	60	720	60	720
	$\mathcal{P}^*_{BT(2,+)}$ $n_0=74$	0.9301 (0.0023)	29.4 (0.1)	352.8 (0.4)	0.9048 (0.0003)	41.2 (<0.1)	493.9 (0.3)	65.5 (0.1)	785.7 (0.5)
	$\mathcal{P}_{PFH(2,+)}(\delta^*/4)$	0.9653 (0.0017)	84.4 (0.4)	348.2 (1.1)	0.9229 (0.0017)	103.5 (0.3)	479.4 (1.0)	157.5 (0.6)	717.8 (1.8)
	$\mathcal{P}_{PFH(2,+)}(\delta^*/2)$	0.9619 (0.0017)	83.9 (0.4)	373.1 (0.9)	0.9217 (0.0017)	99.2 (0.3)	498.8 (0.9)	134.2 (0.4)	686.3 (1.3)
(3,5)	$\mathcal{P}_{SS(2,+)}$	0.9667 (0.0016)	62	930	0.9057	62	930	62	930
	$\mathcal{P}^*_{BT(2,+)}$ $n_0=76$	0.9338 (0.0023)	27.2 (0.1)	408.5 (0.5)	0.9047 (0.0002)	41.9 (<0.1)	628.4 (0.2)	67.7 (0.1)	1015.1 (0.5)
	$\mathcal{P}_{PFH(2,+)}(\delta^*/4)$	0.9675 (0.0016)	91.4 (0.5)	394.5 (1.1)	0.9266 (0.0017)	122.8 (0.4)	614.6 (1.3)	193.3 (0.7)	925.1 (2.3)
	$\mathcal{P}_{PFH(2,+)}(\delta^*/2)$	0.9673 (0.0016)	93.7 (0.4)	431.0 (1.0)	0.9249 (0.0017)	121.4 (0.3)	652.8 (1.1)	172.6 (0.5)	909.7 (1.7)
(3,6)	$\mathcal{P}_{SS(2,+)}$	0.9710 (0.0015)	64	1152	0.9036	64	1152	64	1152
	$\mathcal{P}^*_{BT(2,+)}$ $n_0=80$	0.9333 (0.0023)	25.9 (0.1)	465.7 (0.5)	0.9043 (0.0003)	43.5 (<0.1)	783.5 (0.4)	70.9 (0.1)	1275.8 (0.6)
	$\mathcal{P}_{PFH(2,+)}(\delta^*/4)$	0.9708 (0.0015)	99.2 (0.5)	436.6 (1.2)	0.9289 (0.0017)	140.9 (0.4)	760.8 (1.5)	225.9 (0.8)	1144.2 (2.8)
	$\mathcal{P}_{PFH(2,+)}(\delta^*/2)$	0.9719 (0.0015)	104.0 (0.4)	486.7 (1.1)	0.9292 (0.0017)	142.0 (0.4)	820.5 (1.4)	203.4 (0.5)	1140.3 (2.1)
(4,4)	$\mathcal{P}_{SS(2,+)}$	0.9593 (0.0018)	54	864	0.9060	54	864	54	864
	$\mathcal{P}^*_{BT(2,+)}$ $n_0=64$	0.9318 (0.0023)	25.2 (0.1)	403.6 (0.4)	0.9042 (0.0004)	37.3 (<0.1)	597.3 (0.3)	58.5 (<0.1)	936.2 (0.4)
	$\mathcal{P}_{PFH(2,+)}(\delta^*/4)$	0.9717 (0.0015)	84.0 (0.4)	381.6 (1.1)	0.9289 (0.0017)	101.0 (0.3)	547.9 (1.0)	149.8 (0.6)	814.7 (1.7)
	$\mathcal{P}_{PFH(2,+)}(\delta^*/2)$	0.9673 (0.0016)	82.1 (0.4)	410.7 (0.9)	0.9258 (0.0017)	92.2 (0.3)	565.9 (0.9)	118.1 (0.4)	769.5 (1.2)

Table 1: (continued)

(a,b)	$\mathcal{P}$	ES(δ^*) in both factors			LF in both factors			EM in both	
		$\bar{p}$	$\bar{n}$	$\bar{t}$	$\bar{p}$ or $\bar{w}$	$\bar{n}$	$\bar{t}$	$\bar{n}$	$\bar{t}$
	$\mathcal{P}_{\text{SS}(2,+)}$	0.9673 (0.0016)	52	1040	0.9066	52	1040	52	1040
(4,5)	$\mathcal{P}^*_{\text{B}_T(2,+)}$ $n_0 = 61$	0.9299 (0.0023)	23.0 (0.1)	459.8 (0.4)	0.9034 (0.0004)	36.1 (<0.1)	721.3 (0.3)	56.3 (<0.1)	1126.0 (0.4)
	$\mathcal{P}_{\text{PFH}(2,+)}(\delta^*/4)$	0.9752 (0.0014)	88.5 (0.5)	419.9 (1.1)	0.9273 (0.0017)	110.0 (0.4)	652.1 (1.2)	168.4 (0.7)	977.5 (2.1)
	$\mathcal{P}_{\text{PFH}(2,+)}(\delta^*/2)$	0.9718 (0.0015)	86.5 (0.4)	454.6 (1.0)	0.9288 (0.0017)	102.4 (0.3)	681.4 (1.1)	136.5 (0.4)	936.3 (1.6)
	$\mathcal{P}_{\text{SS}(2,+)}$	0.9723 (0.0015)	51	1224	0.9027	51	1224	51	1224
(4,6)	$\mathcal{P}^*_{\text{B}_T(2,+)}$ $n_0 = 61$	0.9386 (0.0022)	21.3 (<0.1)	511.9 (0.5)	0.9033 (0.0004)	35.9 (<0.1)	861.3 (0.3)	56.6 (<0.1)	1359.0 (0.4)
	$\mathcal{P}_{\text{PFH}(2,+)}(\delta^*/4)$	0.9770 (0.0014)	93.7 (0.5)	455.6 (1.2)	0.9292 (0.0017)	124.9 (0.4)	782.9 (1.5)	195.2 (0.7)	1171.0 (2.5)
	$\mathcal{P}_{\text{PFH}(2,+)}(\delta^*/2)$	0.9746 (0.0014)	94.2 (0.4)	502.2 (1.0)	0.9253 (0.0017)	121.0 (0.3)	835.9 (1.3)	167.7 (0.5)	1147.8 (2.0)
	$\mathcal{P}_{\text{SS}(2,+)}$	0.9722 (0.0015)	47	1175	0.9042	47	1175	47	1175
(5,5)	$\mathcal{P}^*_{\text{B}_T(2,+)}$ $n_0 = 55$	0.9328 (0.0023)	20.3 (<0.1)	508.4 (0.4)	0.9037 (0.0004)	33.4 (<0.1)	834.2 (0.3)	51.5 (<0.1)	1288.7 (0.4)
	$\mathcal{P}_{\text{PFH}(2,+)}(\delta^*/4)$	0.9791 (0.0013)	88.8 (0.5)	448.1 (1.2)	0.9316 (0.0016)	107.9 (0.4)	724.1 (1.2)	161.9 (0.6)	1081.1 (2.1)
	$\mathcal{P}_{\text{PFH}(2,+)}(\delta^*/2)$	0.9771 (0.0014)	86.6 (0.4)	488.9 (1.0)	0.9262 (0.0017)	97.6 (0.3)	756.7 (1.0)	122.7 (0.4)	1024.2 (1.4)
	$\mathcal{P}_{\text{SS}(2,+)}$	0.9738 (0.0015)	45	1350	0.9037	45	1350	45	1350
(5,6)	$\mathcal{P}^*_{\text{B}_T(2,+)}$ $n_0 = 53$	0.9374 (0.0022)	19.0 (<0.1)	568.6 (0.5)	0.9048 (0.0004)	32.1 (<0.1)	962.3 (0.3)	49.8 (<0.1)	1495.5 (0.4)
	$\mathcal{P}_{\text{PFH}(2,+)}(\delta^*/4)$	0.9820 (0.0012)	91.2 (0.5)	476.8 (1.2)	0.9320 (0.0016)	114.8 (0.4)	826.7 (1.4)	174.2 (0.7)	1230.5 (2.4)
	$\mathcal{P}_{\text{PFH}(2,+)}(\delta^*/2)$	0.9802 (0.0013)	89.7 (0.4)	526.1 (1.0)	0.9295 (0.0017)	105.6 (0.3)	871.6 (1.2)	137.4 (0.5)	1187.8 (1.7)
	$\mathcal{P}_{\text{SS}(2,+)}$	0.9789 (0.0013)	42	1512	0.9050	42	1512	42	1512
(6,6)	$\mathcal{P}^*_{\text{B}_T(2,+)}$ $n_0 = 48$	0.9398 (0.0022)	17.3 (<0.1)	622.5 (0.5)	0.9029 (0.0002)	30.0 (<0.1)	1080.9 (0.1)	45.7 (<0.1)	1643.5 (0.3)
	$\mathcal{P}_{\text{PFH}(2,+)}(\delta^*/4)$	0.9823 (0.0012)	93.5 (0.5)	506.2 (1.2)	0.9338 (0.0016)	113.5 (0.4)	902.8 (1.5)	169.9 (0.6)	1337.1 (2.3)
	$\mathcal{P}_{\text{PFH}(2,+)}(\delta^*/2)$	0.9832 (0.0012)	89.8 (0.4)	555.7 (1.1)	0.9291 (0.0017)	101.7 (0.3)	949.6 (1.3)	126.5 (0.4)	1286.0 (1.6)

for the ES(δ^*)- and EM-configurations were based on 12,000 independent replications while those for the LF-configuration were based on 24,000 independent replications; every run was independent of all others. The performance characteristics recorded were the estimated achieved probability of a correct selection ($\bar{p}$), the estimated expected number of stages to termination of experimentation ($\bar{n}$), the estimated expected total number of scalar observations to termination of experimentation ($\bar{t}$), and the estimated variance of each of these quantities. (The variances are not reported.) These statistics are recorded in Table 1 in the rows headed $\mathcal{P}_{\mathrm{PFH}(2,+)}(\delta^*/4)$ and $\mathcal{P}_{\mathrm{PFH}(2,+)}(\delta^*/2)$. The number under each entry is the estimated standard error of the number above it.

Also included in Table 1 are statistics for $\mathcal{P}_{\mathrm{SS}(2,+)}$; these results are abstracted from B-G (1988b, Table 2) except that the $\bar{p}$-results for the ES(δ^*)-configuration are newly estimated by MC sampling, each one being based on 12,000 independent replications. Finally, in Table 1 we report statistics for $\mathcal{P}^*_{\mathrm{B}_{\mathrm{T}}(2,+)}$ which are also abstracted from B-G (1988b, Table 2); the entry $\bar{w}$ in the LF-column is an unbiased estimate of the achieved P\{CS|LF\} which always has a smaller variance than $\bar{p}$ for the same number of replications. The $\mathcal{P}_{\mathrm{SS}(2,+)}$ and $\mathcal{P}^*_{\mathrm{B}_{\mathrm{T}}(2,+)}$ performance statistics are given in the appropriate rows of Table 1.

5 Discussion of Procedures and Results

As mentioned earlier, our main reason for conducting these studies was to compare the performance characteristics of the $\mathcal{P}_{\mathrm{PFH}(2,+)}$ procedures with those of $\mathcal{P}^*_{\mathrm{B}_{\mathrm{T}}(2,+)}$. In Table 2 we have listed for each configuration of the population means and factor-level combination (a, b) the procedure that yielded the smallest estimated E$\{n\}$- and E$\{T\}$-values. We have not provided a listing of the procedures according to which achieved the highest estimated P\{CS|LF\}-value although (as we point out below) the more that these values exceed the specified P^*-value of 0.9025, and therefore *overprotect*, the larger the values of E$\{n\}$ and E$\{T\}$.

We note that $\mathcal{P}^*_{\mathrm{B}_{\mathrm{T}}(2,+)}$ yielded the smallest $\bar{n}$-values for the LF- and ES(δ^*)-configurations of the population means, uniformly in (a, b). This is not surprising since in the LF-configuration this procedure is essentially a likelihood ratio test for a so-called "identification" problem (see Bechhofer et al. 1968, Section 4.3.1); because of the way that the truncation numbers were determined, the overprotection afforded by $\mathcal{P}^*_{\mathrm{B}_{\mathrm{T}}(2,+)}$ is all but eliminated. This favorable property of the procedure carries over to the ES(δ^*)-configuration. $\mathcal{P}_{\mathrm{SS}(2,+)}$ yielded the smallest $\bar{n}$-values for the EM-configuration, except for $(a, b) = (2, 2)$.

Table 2: Procedure having minimum $\bar{n}$ or $\bar{t}$ for the ES(δ^*)-, LF-, and EM-configurations. $P^* = 0.9025$, $\delta^* = \delta^*_\alpha = \delta^*_\beta = 0.2$, and $\sigma^2 = 1$.

(a,b)	ES(δ^*)		LF		EM	
	$\bar{n}$	$\bar{t}$	$\bar{n}$	$\bar{t}$	$\bar{n}$	$\bar{t}$
$(2,2)$	$\mathcal{P}^*_{B_T(2,+)}$	$\mathcal{P}^*_{B_T(2,+)}$	$\mathcal{P}^*_{B_T(2,+)}$	$\mathcal{P}^*_{B_T(2,+)}$	$\mathcal{P}^*_{B_T(2,+)}$	$\mathcal{P}_{PFH(2,+)}(\delta^*/2)$
$(2,3)$ to (2.6)	$\mathcal{P}^*_{B_T(2,+)}$	$\mathcal{P}^*_{B_T(2,+)}$	$\mathcal{P}^*_{B_T(2,+)}$	$\mathcal{P}^*_{B_T(2,+)}$	$\mathcal{P}_{SS(2,+)}$	$\mathcal{P}_{SS(2,+)}$
$(3,3)$	$\mathcal{P}^*_{B_T(2,+)}$	$\mathcal{P}^*_{B_T(2,+)}$	$\mathcal{P}^*_{B_T(2,+)}$	$\mathcal{P}_{PFH(2,+)}(\delta^*/4)$	$\mathcal{P}_{SS(2,+)}$	$\mathcal{P}_{PFH(2,+)}(\delta^*/2)$
(3.4) to (6.6)	$\mathcal{P}^*_{B_T(2,+)}$	$\mathcal{P}_{PFH(2,+)}(\delta^*/4)$	$\mathcal{P}^*_{B_T(2,+)}$	$\mathcal{P}_{PFH(2,+)}(\delta^*/4)$	$\mathcal{P}_{SS(2,+)}$	$\mathcal{P}_{PFH(2,+)}(\delta^*/2)$

The $\bar{t}$ results are consistent with $P^* = 0.90$ in B-G (1989, Table 5). Here one should note that the expected number of stages to termination is maximized in the EM-configuration. Also, Corollary 1 of Hartmann (1992) implies that within the class of PFH procedures the choice $\lambda = \delta^*/2$ minimizes the maximum number of stages to termination. $\mathcal{P}_{PFH(2,+)}(\delta^*/4)$ yields smaller $\bar{t}$-values than does $\mathcal{P}_{PFH(2,+)}(\delta^*/2)$ for all (a,b)-values in the ES(δ^*)- and LF-configurations because it permits earlier elimination of non-contending populations. We point out that $\mathcal{P}_{PFH(2,+)}(\delta^*/2)$ and $\mathcal{P}_{PFH(2,+)}(\delta^*/4)$ *overprotect* by a substantial amount in the LF-configuration. This overprotection results in large $\bar{n}$- and $\bar{t}$-values compared to those for $\mathcal{P}^*_{B_T(2,+)}$; it is accentuated in the ES(δ^*)-configuration; and it tends to increase with each of a and b. Of course, if the PFH single-factor procedure could be generalized to a 2-factor procedure the overprotection of the latter in the LF-configuration would be reduced; but based on the findings in B-G (1989) we believe that this reduction would have only a modest effect on the $\bar{n}$- and $\bar{t}$-values.

At first glance it might appear that the experimenter would be well advised to employ $\mathcal{P}_{PFH(2,+)}(\delta^*/2)$ to minimize $E\{T\}$ in the EM-configuration or configurations contiguous to it. However, we note from Table 1 that the small $\bar{t}$-values are purchased at the expense of large $\bar{n}$-values, and the value of $\bar{n}$ increases with increasing a and b in (a,b). In fact, $\bar{n}$ for $\mathcal{P}_{PFH(2,+)}(\delta^*/2)$ is even greater than $\bar{n}$ for $\mathcal{P}_{SS(2,+)}$, uniformly in (a,b) for $1 \le a, b \le 6$ (except for $(2,2)$ in the ES(δ^*)- and LF-configurations). *This phenomenon of*

$\bar{n}$ *increasing rapidly with a, b in the EM-configuration was a surprise to the authors.* The results in Table 5 of B-G (1989) for *single-factor* experiments with $P^* \geq 0.90$ had suggested the possibility of using $\mathcal{P}_{\mathrm{PFH}(2,+)}(\delta^*/2)$ to minimize $\mathrm{E}\{T|\mathrm{EM}\}$ for *2-factor* experiments when a and/or b are moderately large. But $\mathcal{P}_{\mathrm{PFH}(2,+)}(\delta^*/2)$ is unacceptable if one is concerned by excessively large $\mathrm{E}\{n|\mathrm{EM}\}$. The question then arises as to whether $\mathcal{P}^*_{\mathrm{B_T}(2,+)}$ should be used for this purpose.

$\mathcal{P}^*_{\mathrm{B_T}(2,+)}$ has many desirable properties as pointed out in B-G (1988b, Section 8). It is an adaptive procedure which, with high probability, will terminate in only a very small number of stages if the configuration of the $\{\mu_{ij}\}$ is favorable to the experimenter; if, on the other hand, the $\{\mu_{ij}\}$ are in an unfavorable configuration such as the EM- or ones contiguous to it, then the experimenter will be protected against extreme numbers of stages by *truncation*. However, truncation numbers have been determined only for all a, b with $1 \leq a, b \leq 6$ when $\delta^*_\alpha = \delta^*_\beta = \delta^* = 0.2(0.1)0.8$ and $P^* = 0.9025$. Thus if $\delta^*_\alpha \neq \delta^*_\beta$ and/or $P^* > 0.9025$ are specified and/or a and/or $b \geq 7$, then no truncation numbers have thus far been determined; such determinations would be very costly. Since n can with sizeable probability take on excessively large values, particularly in the EM- or contiguous configurations, an experimenter would be well-advised not to employ the *untruncated* $\mathcal{P}^*_{\mathrm{B}(2,+)}$.

6   Concluding Remarks

When we initiated this study we had anticipated that $\mathcal{P}_{\mathrm{PFH}(2,+)}(\delta^*/2)$ would serve as a *conservative* procedure which would minimize $\mathrm{E}\{T|\mathrm{EM}\}$, at least for moderately large a and b. Although this does indeed occur, the negative consequence of very large associated $\mathrm{E}\{n|\mathrm{EM}\}$ would probably preclude its use (except in circumstances where large $\mathrm{E}\{n|\mathrm{EM}\}$ is not an important consideration). The same problem exists if $\mathcal{P}_{\mathrm{PFH}(2,+)}(\delta^*/4)$ is used to minimize $\mathrm{E}\{T|\mathrm{LF}\}$ or $\mathrm{E}\{T|\mathrm{ES}(\delta^*)\}$ for moderately large a and b.

To counter this problem, Hartmann (1992) has proposed that $\mathcal{P}_{\mathrm{PFH}(2,+)}$ be generalized so that matrices are taken in stages (rather than one-at-a-time). This modification may not only reduce the expected number of stages but also make the procedure more competitive with $\mathcal{P}^*_{\mathrm{B_T}(2,+)}$; it has the important virtue of not requiring the determination of truncation numbers. MC simulation of this procedure is underway. Of course, the PFH procedures can also be truncated as was $\mathcal{P}^*_{\mathrm{B}(2,+)}$, but the determination of such truncation numbers would be much more costly than for $\mathcal{P}^*_{\mathrm{B}(2,+)}$.

It should be emphasized that the choice $P^*_\alpha = P^*_\beta$ where $P^*_\alpha \cdot P^*_\beta = P^*$ is *not* optimal when the problem is *not* symmetrical in (a, b) and $(\delta^*_\alpha, \delta^*_\beta)$. We

Table 3: Effect on estimated $E\{n|EM\}$ and $E\{T|EM\}$ for $\mathcal{P}_{\text{PFH}(2,+)}(\delta^*/2)$ by varying P_α^* and P_β^* subject to $P_\alpha^* \cdot P_\beta^* = P^* = 0.9025$. $(a,b) = (2,6)$, $\delta^* = \delta_\alpha^* = \delta_\beta^* = 0.2$, $\lambda_\alpha = \lambda_\beta = \delta^*/2$, and $\sigma^2 = 1$.

P_α^*	P_β^*	$\bar{n}$	$\bar{t}$
0.93000	0.97043	279.3	1345.3
		(0.6)	(2.4)
0.95000	0.95000	233.9	1138.7
		(0.5)	(2.1)
0.97043	0.93000	201.4	1007.2
		(0.5)	(1.9)
0.99176	0.91000	166.1	916.2
		(0.4)	(1.7)
0.99500	0.90703	159.0	913.2
		(0.4)	(1.7)
0.99700	0.90522	153.8	912.4
		(0.4)	(1.6)
0.99800	0.90431	152.6	921.4
		(0.4)	(1.5)
0.99850	0.90386	153.3	932.4
		(0.4)	(1.5)
0.99900	0.90340	157.1	951.1
		(0.4)	(1.7)

show in Table 3 for $\mathcal{P}_{\text{PFH}(2,+)}(\delta^*/2)$ the effect on the estimated $E\{n|EM\}$ and $E\{T|EM\}$ when P_α^* is varied subject to $P_\alpha^* \cdot P_\beta^* = 0.9025$ for $a = 2$, $b = 6$, $\delta_\alpha^* = \delta_\beta^* = 0.2$, $\lambda_\alpha = \lambda_\beta = \delta^*/2$, $\sigma^2 = 1$ and $P^* = 0.9025$. All MC results are based on 12,000 independent replications. It can be seen that the optimal choice to minimize $E\{T|EM\}$ is approximately $P_\alpha^* = 0.997$ while the optimal choice to minimize $E\{n|EM\}$ is approximately $P_\alpha^* = 0.998$. At the present time, given the parameters of the problem, we do not know how to calculate the optimal choice of P_α^*. Throughout this article we have set $P_\alpha^* = P_\beta^* = 0.95$ for convenience for all (a,b) although as noted above this choice is optimal only when $a = b$.

Acknowledgments

The authors gratefully acknowledge the efforts of Ms. Kathy King who typed this manuscript with patience, good humor, and her usual expertise. The research of Robert Bechhofer and David Goldsman was partially supported by NSF grant DDM-90-12020 at the Georgia Institute of Technology, and that of Mark Hartmann was partially supported by NSF grant DMS-89-05645 at the University of North Carolina, Chapel Hill.

References

Bawa, V.S. (1972). Asymptotic efficiency of one R-factor experiment relative to R one-factor experiments for selecting the best normal population. *Jour. Amer. Statist. Assoc.* **67** 660 - 661.

Bechhofer, R.E. (1954). A single-sample multiple decision procedure for ranking means of normal populations with known variances. *Ann. Math. Statist.* **25** 16 - 39.

Bechhofer, R.E. and Goldsman, D.M. (1987). Truncation of the Bechhofer-Kiefer-Sobel sequential procedure for selecting the normal population which has the largest mean. *Commun. Statist. – Simula. Computa.* **16(4)** 1067 - 1092.

Bechhofer, R.E. and Goldsman, D.M. (1988a). Sequential selection procedures for multi-factor experiments involving Koopman-Darmois populations with additivity. In *Statistical Decision Theory and Related Topics, IV* (S.S. Gupta and J.O. Berger, eds.), Vol.2, 3-22, Academic Press, New York.

Bechhofer, R.E. and Goldsman, D.M. (1988b). Truncation of the Bechhofer-Kiefer-Sobel sequential procedure for selecting the normal population which has the largest mean (II): 2-factor experiments with no interaction. *Commun. Statist. – Simula. Computa.* **17(1)** 103 - 128.

Bechhofer, R.E. and Goldsman, D.M. (1989). A comparison of the performances of procedures for selecting the normal population having the largest mean when the variances are known and equal. In *Contributions to Probability and Statistics – Essays in Honor of Ingram Olkin* (L.J. Gleser, M.D. Perlman, S.J. Press and A.R. Sampson, eds.), 303 - 317, Springer-Verlag, New York.

Bechhofer, R.E., Kiefer, J. and Sobel, M. (1968). *Sequential Identification and Ranking Procedures.* The University of Chicago Press, Chicago.

Fabian, V. (1974). Note on Anderson's sequential procedures with triangular boundary. *Ann. Statist.* **2** 170 - 176.

Hartmann, M. (1988). An improvement on Paulson's sequential ranking procedure. *Sequen. Anal.* **7(4)** 362 - 372.

Hartmann, M. (1992). Multi-factor extensions of Paulson's procedures for selecting the best normal population. [This volume, 225 - 245]

Paulson, E. (1964). A sequential procedure for selecting the population with the largest mean from k normal populations. *Ann. Math. Statist.* **35** 174 - 180.

Chapter 14

Multi-Factor Extensions of Paulson's Procedures for Selecting the Best Normal Population

MARK HARTMANN   Department of Operations Research, University of North Carolina, Chapel Hill, North Carolina

Abstract   In this note, we describe new procedures for the problem of selecting the normal population with the largest population mean in an additive model when the populations have a common known variance, and for the problem of selecting the normal population with the largest population mean in an additive model when the populations have a common unknown variance. The procedures are extensions of Paulson's procedures for selecting the normal population which has the largest population mean, with improvements due to Fabian and Hartmann.

1   Introduction

In this note, we describe new procedures for the problem of selecting the normal population with the largest population mean in an additive model when the populations have a common known variance, and for the problem of selecting the normal population with the largest population mean in an additive model when the populations have a common unknown variance.

225

The procedures are extensions of procedures described in Paulson (1964), with improvements due to Fabian (1974), Hartmann (1988) and Hartmann (1991). We will only describe two-factor procedures, but the extensions to multi-factor procedures are straightforward.

Suppose that Π_{ij} for $i = 1, 2, \ldots, a$ and $j = 1, 2, \ldots, b$ are normal populations from which we may observe random variables X_{ijs} with means $\mu + \alpha_i + \beta_j$ and common known or common unknown variance σ^2. We say that a correct selection is made if a procedure selects the population associated with $\alpha_{[a]}$ and $\beta_{[b]}$, where $\alpha_{[1]} \leq \cdots \leq \alpha_{[a]}$ and $\beta_{[1]} \leq \cdots \leq \beta_{[b]}$ are the ordered values of the "effects" of the two factor levels. The procedures all satisfy the indifference-zone probability requirement $P\{CS\} \geq P^*$ whenever $\alpha_{[a]} - \alpha_{[a-1]} \geq \delta_\alpha^*$ and $\beta_{[b]} - \beta_{[b-1]} \geq \delta_\beta^*$. The procedures operate by eliminating all populations associated with either α_i or β_j when that α_i (β_j) is no longer indicated as contending for $\alpha_{[a]}$ $(\beta_{[b]})$; the elimination process continues until a single population remains. So if we let A_r (B_r) denote the number of remaining α-factor $(\beta$-factor$)$ levels, i.e., indices i (j) for which the populations associated with α_i (β_j) have not been eliminated. Then $A_1 = a$, $B_1 = b$, and sampling is terminated before stage r if we have $A_r = B_r = 1$.

In the next section, we describe straightforward extensions of Paulson's procedures for selecting the normal population with the largest population mean when the populations have a common known variance or a common unknown variance. In the final sections, we describe slightly more involved procedures that satisfy a stronger probability requirement and require fewer stages.

2   First Extension for a Known σ^2

Before experimentation starts, the experimenter specifies P_α^* and P_β^* with $P_\alpha^* P_\beta^* \geq P^*$, λ_α with $0 \leq \lambda_\alpha \leq \delta_\alpha^*/2$ and λ_β with $0 \leq \lambda_\beta \leq \delta_\beta^*/2$. Then a_{λ_α} and a_{λ_β} are chosen as in the improved version of Paulson's procedure as described in Fabian (1974) for P_α^*, δ_α^*, λ_α and P_β^*, δ_β^*, λ_β. For example, if $\lambda_\alpha = \delta_\alpha^*/2$, then $a_{\lambda_\alpha} = \frac{-\sigma^2}{\delta_\alpha^* - \lambda_\alpha} \ln \frac{2(1-P_\alpha^*)}{a-1}$ and if $\lambda_\alpha = \delta_\alpha^*/4$, then $a_{\lambda_\alpha} = \frac{-\sigma^2}{\delta_\alpha^* - \lambda_\alpha} \ln \theta_\alpha$, where θ_α solves

$$\frac{1-P_\alpha^*}{a-1} = \theta_\alpha - \frac{1}{2}\theta_\alpha^{4/3}.$$

See Hartmann (1991) for the case $\lambda_\alpha \neq \delta_\alpha^*/2n$ for some $n = 1, 2, \ldots, \infty$. A recommendation is that these values be chosen to satisfy $a_{\lambda_\alpha}/\lambda_\alpha = a_{\lambda_\beta}/\lambda_\beta$ so that the maximum number of observations required to select a best α-factor level and a best β-factor level will be the same.

At the rth stage, we take a single observation X_{ijr} from each of the $A_r B_r$ remaining populations, which can be thought of as taking observations a matrix–at–a–time with a matrix $\{X_{i_k j_l r}, k = 1, \ldots, A_r, l = 1, \ldots, B_r\}$ of size $A_r \times B_r$ where $i_1, \ldots, i_{A_r}$ and $j_1, \ldots, j_{B_r}$ are the remaining α- and β-factor levels, respectively. Let $Y_{ir} = \sum_{k=1}^{B_r} X_{ij_k r}$ denote the row sum associated with α-factor level i, and let $Z_{rj} = \sum_{l=1}^{A_r} X_{i_l j r}$ denote the column sum associated with β-factor level j. Later we will make use of the observation that the differences $Y_{aj} - Y_{ij}$ for $i = 1, 2, \ldots, a-1$ are independent of the differences $Z_{ib} - Z_{ij}$ for $j = 1, 2, \ldots, b-1$, and the means of these random variables do not depend on which of the other factor levels remain, only on the numbers A_r and B_r of remaining factor levels. Next we eliminate the populations associated with α_k if

$$\sum_{j=1}^{r} Y_{kj} < \max_{i} \sum_{j=1}^{r} Y_{ij} - \max\left(a_{\lambda_\alpha} - \lambda_\alpha \sum_{j=1}^{r} B_j, 0\right),$$

where the maximum is taken over the remaining α-factor levels i, and the populations associated with β_k if

$$\sum_{i=1}^{r} Z_{ik} < \max_{j} \sum_{i=1}^{r} Z_{ij} - \max\left(a_{\lambda_\beta} - \lambda_\beta \sum_{i=1}^{r} A_i, 0\right),$$

where the maximum is taken over the remaining β-factor levels j.

To see that the probability requirement is satisfied, suppose that, unknown to the experimenter, $\alpha_{[a]} = \alpha_a \geq \alpha_i + \delta_\alpha^*$ for $i = 1, \ldots, a-1$ and $\beta_{[b]} = \beta_b \geq \beta_j + \delta_\beta^*$ for $j = 1, \ldots, b-1$. We define the event $\{\Pi_{k\cdot} \gg \Pi_{i\cdot}\}$ to be the event that the populations $\Pi_{i\cdot}$ associated with α_i would be eliminated if the procedure were applied to just the populations associated with α_i and α_k, i.e. the event

$$\bigcup_{r=1}^{\infty} \left\{ \sum_{j=1}^{t} Y_{ij} \leq \sum_{j=1}^{t} Y_{kj} + \max\left(a_{\lambda_\alpha} - \lambda_\alpha \sum_{j=1}^{t} B_j, 0\right), \quad 1 \leq t \leq r-1;\right.$$
$$\left. \sum_{j=1}^{r} Y_{kj} < \sum_{j=1}^{r} Y_{ij} - \max\left(a_{\lambda_\alpha} - \lambda_\alpha \sum_{j=1}^{r} B_j, 0\right)\right\}.$$

The event $\{\Pi_{\cdot k} \gg \Pi_{\cdot j}\}$ is analogously defined to be the event that the populations $\Pi_{\cdot j}$ associated with β_j would be eliminated if the procedure were applied to just the populations associated with β_j and β_k. Clearly we have

$$P\{CS\} \geq P\left\{\bigcap_{i=1}^{a-1}\{\Pi_{a\cdot} \gg \Pi_{i\cdot}\} \bigcap_{j=1}^{b-1}\{\Pi_{\cdot b} \gg \Pi_{\cdot j}\}\right\}.$$

Then since the events $\bigcap_{i=1}^{a-1}\{\mathrm{II}_{a\cdot} \gg \mathrm{II}_{i\cdot}\}$ and $\bigcap_{j=1}^{b-1}\{\mathrm{II}_{\cdot b} \gg \mathrm{II}_{\cdot j}\}$ are determined by the numbers $B_1, B_2, \ldots$ and $A_1, A_2, \ldots$ of remaining factor levels at each stage and the differences $Y_{at} - Y_{it}$ for $i = 1, 2, \ldots, a-1$ and $Z_{tb} - Z_{tj}$ for $j = 1, 2, \ldots, b-1$, which are independent, the probability above does not change if we assume that $Y_{i1}, Y_{i2}, \ldots$ for $i = 1, \ldots, a$ and $Z_{1j}, Z_{2j}, \ldots$ for $j = 1, \ldots, b$ are based on *independent* realizations of $\{X_{ijs}\}$ with means $\alpha_1, \ldots, \alpha_a$ and $\beta_1, \ldots, \beta_b$, respectively. Even so, the events $\bigcap_{i=1}^{a-1}\{\mathrm{II}_{a\cdot} \gg \mathrm{II}_{i\cdot}\}$ and $\bigcap_{j=1}^{b-1}\{\mathrm{II}_{\cdot b} \gg \mathrm{II}_{\cdot j}\}$ depend on each other through the numbers $B_1, B_2, \ldots$ and $A_1, A_2, \ldots .$ In order to by-pass these numbers, we consider a hypothetical procedure which takes observations in batches of size $\widehat{B}_1, \widehat{B}_2, \ldots$ and $\widehat{A}_1, \widehat{A}_2, \ldots$ where $\widehat{B}_r$ and $\widehat{A}_r$ are determined at stage $r-1$ as those numbers which minimize the conditional probability of making a correct selection given the batch sizes $\widehat{B}_1, \ldots, \widehat{B}_{r-1}$ and $\widehat{A}_1, \ldots, \widehat{A}_{r-1}$, and the observed partial sums $\widehat{Y}_{i1}, \ldots, \widehat{Y}_{i(r-1)}$ for $i = 1, \ldots, a$ and $\widehat{Z}_{1j}, \ldots, \widehat{Z}_{(r-1)j}$ for $j = 1, \ldots, b$ based on these batch sizes, subject to $1 \le \widehat{B}_r \le b$ and $1 \le \widehat{A}_r \le a$.

To make this precise, suppose the experimenter believes that he/she is observing random variables $X_{\pi_i \nu_j s}$ for some permutations π of $\{1, 2, \ldots, a\}$ and ν of $\{1, 2, \ldots, b\}$. If the experimenter was convinced that $\pi = \hat{\pi}$ and $\nu = \hat{\nu}$, he/she would express this conditional probability as

$$\widehat{P}\left\{\bigcap_{i=1}^{a-1}\{\widehat{\mathrm{II}}_{a\cdot} \gg \widehat{\mathrm{II}}_{i\cdot}\} \bigcap_{j=1}^{b-1}\{\widehat{\mathrm{II}}_{\cdot b} \gg \widehat{\mathrm{II}}_{\cdot j}\} \;\middle|\; \begin{array}{l} \widehat{Y}_{\hat{\pi}_i 1}, \ldots, \widehat{Y}_{\hat{\pi}_i (r-1)}, ; \widehat{B}_1, \ldots, \widehat{B}_r; \\ \widehat{Z}_{1\hat{\nu}_j}, \ldots, \widehat{Z}_{(r-1)\hat{\nu}_j}, ; \widehat{A}_1, \ldots, \widehat{A}_r \end{array}\right\}$$

where the events $\{\widehat{\mathrm{II}}_{a\cdot} \gg \widehat{\mathrm{II}}_{i\cdot}\}$ and $\{\widehat{\mathrm{II}}_{\cdot b} \gg \widehat{\mathrm{II}}_{\cdot j}\}$ are defined with respect to $\widehat{B}_1, \widehat{B}_2, \ldots$ and $\widehat{A}_1, \widehat{A}_2, \ldots$, and the partial sums $\widehat{Y}_{i1}, \widehat{Y}_{2r}, \ldots$ for $i = 1, \ldots, a$ and $\widehat{Z}_{1j}, \widehat{Z}_{2j}, \ldots$ for $j = 1, \ldots, b$ which are based on these batch sizes, and $\widehat{P}$ signifies that this policy for choosing the batch sizes is used to determine the subsequent batch sizes on which the conditional probability is based. Since π and ν are unknown to the experimenter, we must take into account the conditional distribution of π and ν given the observations $\widehat{Y}_{\pi_i 1}, \ldots, \widehat{Y}_{\pi_i (r-1)}$ for $i = 1, \ldots, a$ and $\widehat{Z}_{1\nu_j}, \ldots, \widehat{Z}_{(r-1)\nu_j}$ for $j = 1, \ldots, b$.

This can be incorporated as follows: let $\widehat{B}_r$ and $\widehat{A}_r$ minimize

$$\frac{1}{a!b!}\sum_{\hat{\pi}}\sum_{\hat{\nu}} l_\alpha\left(\sum_{k=1}^{r-1}\widehat{Y}_{\hat{\pi}_i k}, i = 1, \ldots, a; \hat{\pi}\right) l_\beta\left(\sum_{k=1}^{r-1}\widehat{Z}_{k\hat{\nu}_j}, j = 1, \ldots, b; \hat{\nu}\right) \times$$

$$\widehat{P}\left\{\bigcap_{i=1}^{a-1}\{\widehat{\mathrm{II}}_{a\cdot} \gg \widehat{\mathrm{II}}_{i\cdot}\} \bigcap_{j=1}^{b-1}\{\widehat{\mathrm{II}}_{\cdot b} \gg \widehat{\mathrm{II}}_{\cdot j}\} \;\middle|\; \begin{array}{l} \widehat{Y}_{\hat{\pi}_i 1}, \ldots, \widehat{Y}_{\hat{\pi}_i (r-1)}; \widehat{B}_1, \ldots, \widehat{B}_r; \\ \widehat{Z}_{1\hat{\nu}_j}, \ldots, \widehat{Z}_{(r-1)\hat{\nu}_j}; \widehat{A}_1, \ldots, \widehat{A}_r \end{array}\right\}$$

subject to $1 \leq \widehat{B}_r \leq b$ and $1 \leq \widehat{A}_r \leq a$, where

$$l_\alpha \left(\sum_{k=1}^{t} \widehat{Y}_{\hat{\pi}_i k}, i = 1, \ldots, a; \hat{\pi} \right) = \exp \left\{ \frac{1}{\sigma^2} \sum_{i=1}^{a} (\alpha_i - \alpha_{\hat{\pi}_i}) \sum_{k=1}^{t} \widehat{Y}_{\hat{\pi}_i k} \right\}$$

is the likelihood ratio of $\widehat{Y}_{\pi_i 1}, \ldots, \widehat{Y}_{\pi_i t}$ for $i = 1, \ldots, a$ under the probability measure $P_{\hat{\pi}}$ induced by $\hat{\pi}$ relative to P (the probability measure induced by the identity permutation), and

$$l_\beta \left(\sum_{k=1}^{t} \widehat{Z}_{k \hat{\nu}_j}, j = 1, \ldots, b; \hat{\nu} \right) = \exp \left\{ \frac{1}{\sigma^2} \sum_{j=1}^{b} (\beta_j - \beta_{\hat{\nu}_j}) \sum_{k=1}^{t} \widehat{Z}_{k \hat{\nu}_j} \right\}$$

is the likelihood ratio of $\widehat{Z}_{1 \nu_j}, \ldots, \widehat{Z}_{t \nu_j}$ for $i = 1, \ldots, b$ under the probability measure $P_{\hat{\nu}}$ induced by $\hat{\nu}$ relative to P. Since $B_1, B_2, \ldots$ and $A_1, A_2, \ldots$ are always feasible choices, we have

$$P\{CS\} \geq \widehat{P} \left\{ \bigcap_{i=1}^{a-1} \{\widehat{\Pi}_{a\cdot} \gg \widehat{\Pi}_{i\cdot}\} \bigcap_{j=1}^{b-1} \{\widehat{\Pi}_{\cdot b} \gg \widehat{\Pi}_{\cdot j}\} \right\}.$$

Next we will show that the right hand side of the above expression is equal to

$$\widehat{P} \left\{ \bigcap_{i=1}^{a-1} \{\widehat{\Pi}_{a\cdot} \gg \widehat{\Pi}_{i\cdot}\} \right\} \widehat{P} \left\{ \bigcap_{j=1}^{b-1} \{\widehat{\Pi}_{\cdot b} \gg \widehat{\Pi}_{\cdot j}\} \right\}$$

provided that $\lambda_\alpha > 0$ and $\lambda_\beta > 0$. Then, for all $r \geq \max(a_{\lambda_\alpha}/\lambda_\alpha, a_{\lambda_\beta}/\lambda_\beta)$, we have

$$\frac{1}{a!b!} \sum_{\hat{\pi}} \sum_{\hat{\nu}} l_\alpha \left(\sum_{k=1}^{r} \widehat{Y}_{\hat{\pi}_i k}, i = 1, \ldots, a; \hat{\pi} \right) l_\beta \left(\sum_{k=1}^{r} \widehat{Z}_{k \hat{\nu}_j}, j = 1, \ldots, b; \hat{\nu} \right) \times$$

$$\widehat{P} \left\{ \bigcap_{i=1}^{a-1} \{\widehat{\Pi}_{a\cdot} \gg \widehat{\Pi}_{i\cdot}\} \bigcap_{j=1}^{b-1} \{\widehat{\Pi}_{\cdot b} \gg \widehat{\Pi}_{\cdot j}\} \, \middle| \, \begin{array}{c} \widehat{Y}_{\hat{\pi}_i 1}, \ldots, \widehat{Y}_{\hat{\pi}_i r}; \widehat{B}_1, \ldots, \widehat{B}_r; \\ \widehat{Z}_{1 \hat{\nu}_j}, \ldots, \widehat{Z}_{r \hat{\nu}_j}; \widehat{A}_1, \ldots, \widehat{A}_r \end{array} \right\}$$

$$= \left[\frac{1}{a!} \sum_{\hat{\pi}} l_\alpha \left(\sum_{k=1}^{r} \widehat{Y}_{\hat{\pi}_i k}, i = 1, \ldots, a; \hat{\pi} \right) \times \right.$$

$$\left. \widehat{P} \left\{ \bigcap_{i=1}^{a-1} \{\widehat{\Pi}_{a\cdot} \gg \widehat{\Pi}_{i\cdot}\} | \widehat{Y}_{\hat{\pi}_i 1}, \ldots, \widehat{Y}_{\hat{\pi}_i r}, 1 \leq i \leq a; \widehat{B}_1, \ldots, \widehat{B}_r \right\} \right] \times$$

$$\left[\frac{1}{b!}\sum_{\hat{\nu}} l_\beta\left(\sum_{k=1}^{r}\widehat{Z}_{k\hat{\nu}_j}, j=1,\ldots,b; \hat{\nu}\right)\times\right.$$

$$\left.\widehat{P}\left\{\bigcap_{j=1}^{b-1}\{\widehat{\Pi}_{\cdot b}\gg\widehat{\Pi}_{\cdot j}\}|\widehat{Z}_{1\hat{\nu}_j},\ldots,\widehat{Z}_{r\hat{\nu}_j},1\leq j\leq b;\widehat{A}_1,\ldots,\widehat{A}_r\right\}\right]$$

since the events $\bigcap_{i=1}^{a-1}\{\widehat{\Pi}_{a\cdot}\gg\widehat{\Pi}_{i\cdot}\}$ and $\bigcap_{j=1}^{b-1}\{\widehat{\Pi}_{\cdot b}\gg\widehat{\Pi}_{\cdot j}\}$ are completely determined by $\widehat{Y}_{\hat{\pi}_i 1},\ldots,\widehat{Y}_{\hat{\pi}_i r}$ for $i=1,\ldots,a$ and $\widehat{Z}_{1\hat{\nu}_j},\ldots,\widehat{Z}_{r\hat{\nu}_j}$ for $j=1,\ldots,b$.

Now for given $\hat{\pi}$ and $\hat{\nu}$, taking expectations with respect to $\widehat{Y}_{1r},\ldots,\widehat{Y}_{ar}$ and $\widehat{Z}_{r1},\ldots,\widehat{Z}_{rb}$ can be accomplished by replacing the likelihood ratios for $t=r$ by the corresponding ones for $t=r-1$ and taking expectations with respect to $\widehat{Y}_{\hat{\pi}_1 r},\ldots,\widehat{Y}_{\hat{\pi}_a r}$ and $\widehat{Z}_{r\hat{\nu}_1},\ldots,\widehat{Z}_{r\hat{\nu}_b}$. So taking expectations with respect to $\widehat{Y}_{1r},\ldots,\widehat{Y}_{ar}$ and $\widehat{Z}_{r1},\ldots,\widehat{Z}_{rb}$ on both sides of this expression yields

$$\frac{1}{a!b!}\sum_{\hat{\pi}}\sum_{\hat{\nu}} l_\alpha\left(\sum_{k=1}^{r-1}\widehat{Y}_{\hat{\pi}_i k}, i=1,\ldots,a; \hat{\pi}\right) l_\beta\left(\sum_{k=1}^{r-1}\widehat{Z}_{k\hat{\nu}_j}, j=1,\ldots,b; \hat{\nu}\right)\times$$

$$\widehat{P}\left\{\bigcap_{i=1}^{a-1}\{\widehat{\Pi}_{a\cdot}\gg\widehat{\Pi}_{i\cdot}\}\bigcap_{j=1}^{b-1}\{\widehat{\Pi}_{\cdot b}\gg\widehat{\Pi}_{\cdot j}\}\,\middle|\,\begin{matrix}\widehat{Y}_{\hat{\pi}_i 1},\ldots,\widehat{Y}_{\hat{\pi}_i(r-1)};\widehat{B}_1,\ldots,\widehat{B}_r;\\\widehat{Z}_{1\hat{\nu}_j},\ldots,\widehat{Z}_{(r-1)\hat{\nu}_j};\widehat{A}_1,\ldots,\widehat{A}_r\end{matrix}\right\}$$

$$=\left[\frac{1}{a!}\sum_{\hat{\pi}} l_\alpha\left(\sum_{k=1}^{r-1}\widehat{Y}_{\hat{\pi}_i k}, i=1,\ldots,a; \hat{\pi}\right)\times\right.$$

$$\left.\widehat{P}\left\{\bigcap_{i=1}^{a-1}\{\widehat{\Pi}_{a\cdot}\gg\widehat{\Pi}_{i\cdot}\}|\widehat{Y}_{\hat{\pi}_i 1},\ldots,\widehat{Y}_{\hat{\pi}_i r-1},1\leq i\leq a;\widehat{B}_1,\ldots,\widehat{B}_r\right\}\right]\times$$

$$\left[\frac{1}{b!}\sum_{\hat{\nu}} l_\beta\left(\sum_{k=1}^{r-1}\widehat{Z}_{k\hat{\nu}_j},1\leq j\leq b; \hat{\nu}\right)\times\right.$$

$$\left.\widehat{P}\left\{\bigcap_{j=1}^{b-1}\{\widehat{\Pi}_{\cdot b}\gg\widehat{\Pi}_{\cdot j}\}|\widehat{Z}_{1\hat{\nu}_j},\ldots,\widehat{Z}_{(r-1)\hat{\nu}_j},j=1,\ldots,b;\widehat{A}_1,\ldots,\widehat{A}_r\right\}\right]$$

Minimizing this over all $\widehat{B}_r$ and $\widehat{A}_r$ satisfying $1\leq\widehat{B}_r\leq b$ and $1\leq\widehat{A}_r\leq a$ gives the previous equality with r replaced by $r-1$. Thus by induction it must hold for $r=0$, which is the desired equality.

When either $\lambda_\alpha=0$ or $\lambda_\beta=0$, there is no guaranteed maximum number of stages. In this case, we simply truncate the hypothetical procedure at stage R and get upper and lower bounds by the method above, which after letting R tend to infinity yields the desired equality.

Now using Boole's inequality, we have that

$$P\{CS\} \geq (1 - \sum_{i=1}^{a-1} \widehat{P}\{\widehat{\Pi}_i. \gg \widehat{\Pi}_a.\})(1 - \sum_{j=1}^{b-1} \widehat{P}\{\widehat{\Pi}._j \gg \widehat{\Pi}._b\})$$

and so by symmetry it suffices to show that both

$$\widehat{P}\{\widehat{\Pi}_1. \gg \widehat{\Pi}_a.\} \leq \frac{1 - P_\alpha^*}{a - 1} \quad \text{and} \quad \widehat{P}\{\widehat{\Pi}._1 \gg \widehat{\Pi}._b\} \leq \frac{1 - P_\beta^*}{b - 1}.$$

After conditioning on $\widehat{Y}_{i1}, \widehat{Y}_{i2}, \ldots$ for $i = 2, \ldots, a-1$ and $\widehat{Z}_{1j}, \widehat{Z}_{2j}, \ldots$ for $j = 2, \ldots, b-1$, respectively, this can be seen to follow from the lemma in Hartmann (1991) which is given in the appendix, since the stopping times involved are symmetric with respect to $\widehat{Y}_{11}, \widehat{Y}_{12}, \ldots$ and $\widehat{Y}_{a1}, \widehat{Y}_{a2}, \ldots$, and $\widehat{Z}_{11}, \widehat{Z}_{21}, \ldots$ and $\widehat{Z}_{1b}, \widehat{Z}_{2b}, \ldots$, respectively.

3 First Extension for an Unknown σ^2

Before experimentation starts, the experimenter specifies an integer n_0, and numbers λ_α with $0 \leq \lambda_\alpha \leq \delta_\alpha^*/2$ and λ_β with $0 \leq \lambda_\beta \leq \delta_\beta^*/2$. For the sake of simplicity, we will assume that $\lambda_\alpha = \delta_\alpha^*/2n_\alpha$ and $\lambda_\beta = \delta_\beta^*/2n_\beta$ for some integers $n_\alpha, n_\beta = 1, 2, \ldots, \infty$. The procedure begins by taking n_0 observations $X_{ij1}, \ldots, X_{ijn_0}$ from each of the ab populations a matrix at a time. Let

$$S^2 = \sum_{i=1}^{a} \sum_{j=1}^{b} \sum_{l=1}^{n_0} (X_{ijl} - \overline{X}_{ij})^2 / ab(n_0 - 1)$$

be the usual estimate of σ^2 with $f = ab(n_0-1)$ degrees of freedom, in which the quantity $\overline{X}_{ij} = \sum_{l=1}^{n_0} X_{ijl}/n_0$. Then fS^2/σ^2 has the χ_f^2 distribution and is independent of $\overline{X}_{ij}$ for $i = 1, 2, \ldots, a$ and $j = 1, 2, \ldots, b$. Let $\eta_\alpha > 0$ and $\eta_\beta > 0$ satisfy

$$P^* = 1 + (a-1) \sum_{j=1}^{n_\alpha} (-1)^j (1 - \delta_{jn_\alpha}/2)(1 + \frac{(2n_\alpha - j)j\eta_\alpha}{n_\alpha})^{-f/2}$$

$$+ (b-1) \sum_{i=1}^{n_\beta} (-1)^i (1 - \delta_{in_\beta}/2)(1 + \frac{(2n_\beta - i)i\eta_\beta}{n_\beta})^{-f/2}$$

$$+ (a-1)(b-1) \sum_{j=1}^{n_\alpha} \sum_{i=1}^{n_\beta} (-1)^{j+i} (1 - \delta_{jn_\alpha}/2)(1 - \delta_{in_\beta}/2) \times$$

$$(1+\frac{(2n_\alpha - j)j\eta_\alpha}{n_\alpha}+\frac{(2n_\beta - i)i\eta_\beta}{n_\beta})^{-f/2}$$

where δ_{ij} is the Kronecker symbol. The expression on the right of the equality sign is a strictly increasing function of $\eta_\alpha > 0$ and $\eta_\beta > 0$. In case $n_\alpha = \infty$ or $n_\beta = \infty$, the expressions $\frac{(2n_\alpha-j)j\eta_\alpha}{n_\alpha}$ and $\frac{(2n_\beta-i)i\eta_\beta}{n_\beta}$ should be replaced by $2j\eta_\alpha$ and $2i\eta_\beta$, respectively. A recommendation is that these values be chosen to satisfy $\eta_\alpha/\eta_\beta = (\delta_\alpha^*/\delta_\beta^*)^2(n_\alpha/n_\beta)$ so that the maximum number of observations required to select a best α-factor level and a best β-factor level will be the same. Then $a_{\lambda_\alpha}^* = \eta_\alpha f S^2/\delta_\alpha^*$ and $a_{\lambda_\beta}^* = \eta_\beta f S^2/\delta_\beta^*$ as in the improved version of Paulson's procedure as described in Hartmann (1991).

Let $Y_{i0} = \sum_{j=1}^{b}\sum_{l=1}^{n_0} X_{ijl}$ for $1 \le i \le a$ and let $Z_{0j} = \sum_{i=1}^{a}\sum_{l=1}^{n_0} X_{ijl}$ for $j = 1, 2, \ldots, b$. Once again note that the differences $Y_{a0}-Y_{i0}$ are independent of the differences $Z_{0b}-Z_{0j}$ and all of these are independent of S^2. Next we eliminate the populations associated with α_k if

$$Y_{k0} < \max(Y_{10}, \ldots, Y_{a0}) - \max\left(a_{\lambda_\alpha}^* - \lambda_\alpha n_0 b, 0\right),$$

and the populations associated with β_k if

$$Z_{0k} < \max(Z_{01}, \ldots, Z_{0b}) - \max\left(a_{\lambda_\beta}^* - \lambda_\beta n_0 a, 0\right).$$

This concludes the 0th stage of the procedure.

In the subsequent r^{th} stages, we take a single observation $X_{ij(n_0+r)}$ from each of the $A_r B_r$ remaining populations, which can be thought of as taking observations a matrix–at–a–time with a matrix $\{X_{i_k j_l(n_0+r)}, k = 1, \ldots, A_r, l = 1, \ldots, B_r\}$ of size $A_r \times B_r$ where $i_1, \ldots, i_{A_r}$ and $j_1, \ldots, j_{B_r}$ are the remaining α- and β-factor levels, respectively. Denote the row sum associated with α-factor level i by $Y_{ir} = \sum_{k=1}^{B_r} X_{ij_k(n_0+r)}$ and let $Z_{rj} = \sum_{l=1}^{A_r} X_{i_l j(n_0+r)}$ denote the column sum associated with β-factor level j. Later we will make use of the observation that the differences $Y_{aj}-Y_{ij}$ for $i = 1, 2, \ldots, a-1$ are independent of the differences $Z_{ib}-Z_{ij}$ for $j = 1, 2, \ldots, b-1$, and the means of these random variables do not depend on which of the other factor levels remain, only on the numbers A_r and B_r of remaining factor levels. Next we eliminate the populations associated with α_k if

$$\sum_{j=0}^{r} Y_{kj} < \max_i \sum_{j=0}^{r} Y_{ij} - \max\left(a_{\lambda_\alpha}^* - \lambda_\alpha[n_0 b + \sum_{j=1}^{r} B_j], 0\right),$$

where the maximum is taken over the remaining α-factor levels i, and the populations associated with β_k if

$$\sum_{i=0}^{r} Z_{ik} < \max_{j} \sum_{i=0}^{r} Z_{ij} - \max\left(a_{\lambda_\beta}^* - \lambda_\beta[n_0 a + \sum_{i=1}^{r} A_i], 0\right),$$

where the maximum is taken over the remaining β-factor levels j.

To see that the probability requirement is satisfied, suppose that, unknown to the experimenter, $\alpha_{[a]} = \alpha_a \geq \alpha_i + \delta_\alpha^*$ for $i = 1, \ldots, a-1$ and $\beta_{[b]} = \beta_b \geq \beta_j + \delta_\beta^*$ for $j = 1, \ldots, b-1$. We define the event $\{\Pi_{k\cdot} \gg \Pi_{i\cdot}\}$ to be the event that the populations $\Pi_{i\cdot}$ associated with α_i would be eliminated if the procedure were applied to just the populations associated with α_i and α_k, i.e. the event

$$\bigcup_{r=1}^{\infty} \left\{ \sum_{j=1}^{t} Y_{ij} \leq \sum_{j=1}^{t} Y_{kj} + \max\left(a_{\lambda_\alpha} - \lambda_\alpha[n_0 b + \sum_{j=1}^{t} B_j], 0\right), \quad 1 \leq t \leq r-1; \right.$$

$$\left. \sum_{j=1}^{r} Y_{kj} < \sum_{j=1}^{r} Y_{ij} - \max\left(a_{\lambda_\alpha} - \lambda_\alpha[n_0 b + \sum_{j=1}^{r} B_j], 0\right) \right\}.$$

The event $\{\Pi_{\cdot k} \gg \Pi_{\cdot j}\}$ is analogously defined to be the event that the populations $\Pi_{\cdot j}$ associated with β_j would be eliminated if the procedure were applied to just the populations associated with β_j and β_k. After conditioning on the value of S^2, using the same methods as for the case of a common known variance, we get that

$$P\{CS\} \geq E\left[(1 - \sum_{i=1}^{a-1} \hat{P}\{\hat{\Pi}_{i\cdot} \gg \hat{\Pi}_{a\cdot} | S^2\})(1 - \sum_{j=1}^{b-1} \{\hat{\Pi}_{\cdot j} \gg \hat{\Pi}_{\cdot b} | S^2\}) \right]. \quad (3.1)$$

In the appendix we show that the upper bound

$$\sum_{j=1}^{n_\alpha} (-1)^{j+1} (1 - \delta_{jn_\alpha}/2) \exp\{-\frac{(2n_\alpha - j)j}{2n_\alpha} \eta_\alpha \frac{fS^2}{\sigma^2}\}$$

given in Hartmann (1991) for $\hat{P}\{\hat{\Pi}_{i\cdot} \gg \hat{\Pi}_{a\cdot} | S^2\}$ when $i = 1, 2, \ldots, a-1$ is a strictly decreasing function of $\eta_\alpha S^2$. Substituting this and the analogous bound for $\hat{P}\{\hat{\Pi}_{\cdot j} \gg \hat{\Pi}_{\cdot b} | S^2\}$ when $j = 1, 2, \ldots, b-1$ into (3.1) yields an expression of the form

$$P\{CS\} \geq E[f(\eta_\alpha S^2) g(\eta_\beta S^2)]$$

where both f and g are strictly increasing functions. Taking expectations with respect to S^2 therefore gives a strictly increasing function of $\eta_\alpha > 0$ and $\eta_\beta > 0$, and we see that $P\{CS\} \geq P^*$ by our choice of η_α and η_β. Since

$$E[f(\eta_\alpha S^2)g(\eta_\beta S^2)] \geq E[f(\eta_\alpha S^2)]\,E[g(\eta_\beta S^2)]$$

by Chebyshev's inequality (Esary et al. 1967), the resulting lower bound on the $P\{CS\}$ is better than just the product of the single factor bounds as in the known variance case.

4   Extensions which Overcome Stage Fright

Since the procedures of the previous section allow for the early elimination of inferior populations, the matrix of observations taken at each stage gets increasingly smaller, which has the unpleasant side effect of requiring a large number of stages. In this section, we describe procedures which always take a full matrix of observations at each stage. The procedures take observations $\{X_{i_k j_l s_{kl}}, 1 \leq k \leq a, 1 \leq l \leq b\}$ a matrix–at–a–time for some indices $i_1, \ldots, i_a$ and $j_1, \ldots j_b$ which will in general not be distinct, so the indices s_{kl} do not necessarily reflect an absolute order in which the observations are made, since for example the random variables X_{ijs} and $X_{ij(s+1)}$ could be observed as part of the same matrix. Note that we are making the assumption that before taking a matrix of observations, the experimenter can set the $a + b$ factor levels arbitrarily. In what follows, we use the notation $n(m)$ to denote the least common multiple of the numbers $1, 2, \ldots, m$, so that for example $n(5) = n(6) = 60$. The next section discusses the procedures for a common known variance and the final section discusses how they are modified for the situation in which the variance is unknown.

5   Second Extension for a Known σ^2

Before experimentation starts, the experimenter specifies P_α^* and P_β^* with $P_\alpha^* P_\beta^* \geq P^*$, λ_α with $0 \leq \lambda_\alpha \leq \delta_\alpha^*/2$ and λ_β with $0 \leq \lambda_\beta \leq \delta_\beta^*/2$. Then a_{λ_α} and a_{λ_β} are chosen as in the improved version of Paulson's procedure as described in Hartmann (1988) for P_α^*, δ_α^*, λ_α and P_β^*, δ_β^*, λ_β. For example, if $\lambda_\alpha = \delta_\alpha^*/2$, then $a_{\lambda_\alpha} = \frac{-\sigma^2}{\delta_\alpha^* - \lambda_\alpha} \ln 2[1 - (P_\alpha^*)^{1/a-1}]$ and if $\lambda_\alpha = \delta_\alpha^*/4$, then $a_{\lambda_\alpha} = \frac{-\sigma^2}{\delta_\alpha^* - \lambda_\alpha} \ln \theta_\alpha$, where θ_α solves

$$1 - (P_\alpha^*)^{1/a-1} = \theta_\alpha - \frac{1}{2}\theta_\alpha^{4/3}.$$

See Hartmann (1991) for the case $\lambda_\alpha \neq \delta_\alpha^*/2n$ for some $n = 1, 2, \ldots, \infty$. A recommendation is that these values be chosen to satisfy $a_{\lambda_\alpha}/\lambda_\alpha = a_{\lambda_\beta}/\lambda_\beta$ so that the maximum number of observations required to select a best α-factor level and a best β-factor level will be the same.

At the rth stage, we take $n(a)n(b)/k_r$ observations from each of the $A_r B_r$ remaining populations, where the integer k_r is defined by

$$
k_r = \begin{cases}
\gcd\{\frac{n(a)}{A_r}, \frac{n(b)}{B_r}, \frac{n(a)n(b)}{ab}\} & (A_r, B_r \geq 2) \\
\gcd\{\frac{n(a)}{A_r}, \frac{n(a)n(b)}{ab}\} & (B_r = 1) \\
\gcd\{\frac{n(b)}{B_r}, \frac{n(a)n(b)}{ab}\} & (A_r = 1)
\end{cases}
$$

This can be thought of as taking observations a matrix–at–a–time with a matrix of size $n(a)/k_\alpha \times n(b)/k_\beta$, where $k_\alpha = \gcd\{k_r, n(a)/a\}$ and $k_\beta = k_r/k_\alpha$, in which each of the A_r remaining α-factor levels alternate $n(a)/k_\alpha A_r$ times as the α-factor level of the $n(a)/k_\alpha$ rows and each of the B_r remaining β-factor levels alternate $n(b)/k_\beta B_r$ times as the β-factor level of the $n(b)/k_\beta$ columns. On the other hand, since a divides $n(a)/k_\alpha$ and b divides $n(b)/k_\beta$ (by our choice of k_α) this larger matrix can be broken down into $n(a)n(b)/k_r ab$ matrices of size $a \times b$ in which the observations are taken in accordance with the form prescribed in the introduction. Ideally, one would like to have $k_r = n(a)n(b)/ab$ so that the procedure would be able to eliminate inferior populations as soon as possible, but this rarely occurs.

For the purpose of illustration, the table below gives the values of k_α, k_β for $a = 6$, $b = 5$ and all possible values of A_r and B_r. Here $n(a)n(b)/ab = 120$, so the procedure observes between 4 and 60 matrices of size 5×6 at each stage.

	5	4	3	2	1
6	2,1	5,1	10,1	10,1	10,1
5	2,6	1,3	2,2	2,3	2,6
4	1,3	5,3	5,1	5,3	5,3
3	2,2	5,1	10,2	10,1	10,2
2	2,3	5,3	10,1	10,3	10,3
1	2,6	5,3	10,2	10,3	—

(In general such a table could be generated prior to taking any observations in $O(ab[a \log a + b \log b])$ time and stored in arrays, so that k_α and k_β could be accessed in constant time.) Note that at the 0th stage, 60 observations must be taken from each of the populations. Depending on the values of P^*, δ_α^* and δ_β^*, this may represent more observations than would be required by the single-stage procedure of Bechhofer (1954). To continue, suppose that

$A_r = 2$ and $B_r = 4$ so that $k_\alpha = 5$ and $k_\beta = 3$. At stage r, the procdure would then take a matrix of observations of size 12×20, which could be broken down into 8 matrices of size 6×5, as illustrated below:

6 x 5	6 x 5	6 x 5	6 x 5
6 x 5	6 x 5	6 x 5	6 x 5

If both $A_r, B_r \geq 2$, then for the purpose of eliminating populations, we will look at the matrix of size $n(a)/k_\alpha \times n(b)/k_\beta$ in two different ways: first as a matrix of size $n(a)/k_r \times n(b)$ in which each of the A_r remaining α-factor levels alternate $n(a)/k_r A_r$ times as the α-factor level of the $n(a)/k_r$ rows and each of the B_r remaining β-factor levels alternate $n(b)/B_r$ times as the β-factor level of the $n(b)$ columns, and secondly as a matrix of size $n(a) \times n(b)/k_r$ in which each of the A_r remaining α-factor levels alternate $n(a)/A_r$ times as the α-factor level of the $n(a)$ rows and each of the B_r remaining β-factor levels alternate $n(b)/k_r B_r$ times as the β-factor level of the $n(b)/k_r$ columns. For example, the 12×20 matrix pictured above can be broken down into 30 matrices X_{ij} of size 2×4, each of which contains a single observation from each of the remaining populations:

X_{11}	X_{12}	X_{13}	X_{14}	X_{15}
X_{21}	X_{22}	X_{23}	X_{24}	X_{25}
X_{31}	X_{32}	X_{33}	X_{34}	X_{35}
X_{41}	X_{42}	X_{43}	X_{44}	X_{45}
X_{51}	X_{52}	X_{53}	X_{54}	X_{55}
X_{61}	X_{62}	X_{63}	X_{64}	X_{65}

The first way of looking at this is as a matrix of size 4×60 built from the matrices of size 2×4 as follows: first concatenate rows three and four and then rows five and six to the first two rows of the above matrix. The second way of looking at this is as a matrix of size 60×4 with the matrices of size 2×4 arranged in one long column.

Let Y_{ij} for $j = \sum_{t=1}^{r-1} \frac{n(a)}{k_t A_t} + 1, \sum_{t=1}^{r-1} \frac{n(a)}{k_t A_t} + 2, \ldots, \sum_{t=1}^{r} \frac{n(a)}{k_t A_t}$ denote the row sums of the $n(a)/k_r \times n(b)$ matrix associated with each α-factor level i, and let Z_{ij} for $i = \sum_{t=1}^{r-1} \frac{n(b)}{k_t B_t} + 1, \sum_{t=1}^{r-1} \frac{n(b)}{k_t B_t} + 2, \ldots, \sum_{t=1}^{r} \frac{n(b)}{k_t B_t}$ denote the column sums of the $n(a) \times n(b)/k_r$ matrix associated with each β-factor level j. Later we will make use of the observation that because the differences of the row and column sums of the matrix of size $n(a)/k_\alpha \times n(b)/k_\beta$ are independent, the differences $Y_{aj} - Y_{ij}$ are independent of the differences $Z_{ib} - Z_{ij}$. Now for $l = \sum_{t=1}^{r-1} \frac{n(a)}{k_t A_t} + 1, \sum_{t=1}^{r-1} \frac{n(a)}{k_t A_t} + 2, \ldots, \sum_{t=1}^{r} \frac{n(a)}{k_t A_t}$ we eliminate the populations associated with α_k if

$$\sum_{j=1}^{l} Y_{kj} < \max_i \sum_{j=1}^{l} Y_{ij} - \max\left(a_{\lambda_\alpha} - \lambda_\alpha n(b)l, 0\right),$$

where the maximum is taken over the remaining α-factor levels i, and for $l = \sum_{t=1}^{r-1} \frac{n(b)}{k_t B_t} + 1, \sum_{t=1}^{r-1} \frac{n(b)}{k_t B_t} + 2, \ldots, \sum_{t=1}^{r} \frac{n(b)}{k_t B_t}$ we eliminate the populations associated with β_k if

$$\sum_{i=1}^{l} Z_{ik} < \max_j \sum_{i=1}^{l} Z_{ij} - \max\left(a_{\lambda_\beta} - \lambda_\beta n(a)l, 0\right),$$

where the maximum is taken over the remaining β-factor levels j.

If $B_r = 1$, then we will look at the matrix of size $n(a)/k_\alpha \times n(b)/k_\beta$ as a matrix of size $n(a)/k_r \times n(b)$ in which each of the A_r remaining α-factor levels alternate $n(a)/k_r A_r$ times as the α-factor level of the $n(a)/k_r$ rows. Let Y_{ij} for $j = \sum_{t=1}^{r-1} \frac{n(a)}{k_t A_t} + 1, \sum_{t=1}^{r-1} \frac{n(a)}{k_t A_t} + 2, \ldots, \sum_{t=1}^{r} \frac{n(a)}{k_t A_t}$ denote the row sums of this $n(a)/k_r \times n(b)$ matrix associated with each α-factor level i. Now for $l = \sum_{t=1}^{r-1} \frac{n(a)}{k_t A_t} + 1, \sum_{t=1}^{r-1} \frac{n(a)}{k_t A_t} + 2, \ldots, \sum_{t=1}^{r} \frac{n(a)}{k_t A_t}$ we eliminate the populations associated with α_k if

$$\sum_{j=1}^{l} Y_{kj} < \max_i \sum_{j=1}^{l} Y_{ij} - \max\left(a_{\lambda_\alpha} - \lambda_\alpha n(b)l, 0\right),$$

where the maximum is taken over the remaining α-factor levels i.

If $A_r = 1$, then for the purpose of eliminating populations, we will look at the matrix of size $n(a)/k_\alpha \times n(b)/k_\beta$ as a matrix of size $n(a) \times n(b)/k_r$ in which each of the B_r remaining β-factor levels alternate $n(b)/k_r B_r$ times as the β-factor level of the $n(b)/k_r$ columns. Let Z_{ij} for $i = \sum_{t=1}^{r-1} \frac{n(b)}{k_t B_t} + 1, \sum_{t=1}^{r-1} \frac{n(b)}{k_t B_t} + 2, \ldots, \sum_{t=1}^{r} \frac{n(b)}{k_t B_t}$ denote the column sums of this $n(a) \times n(b)/k_r$ matrix associated with each β-factor level j. For $l = \sum_{t=1}^{r-1} \frac{n(b)}{k_t B_t} + 1, \sum_{t=1}^{r-1} \frac{n(b)}{k_t B_t} + 2, \ldots, \sum_{t=1}^{r} \frac{n(b)}{k_t B_t}$ we eliminate the populations associated with β_k if

$$\sum_{i=1}^{l} Z_{ik} < \max_{j} \sum_{i=1}^{l} Z_{ij} - \max\left(a_{\lambda_\beta} - \lambda_\beta n(a)l, 0\right),$$

where the maximum is taken over the remaining β-factor levels j.

To see that the probability requirement is satisfied, suppose that, unknown to the experimenter, $\alpha_{[a]} = \alpha_a \geq \alpha_i + \delta_\alpha^*$ for $i = 1, \ldots, a-1$ and $\beta_{[b]} = \beta_b \geq \beta_j + \delta_\beta^*$ for $j = 1, \ldots, b-1$. We define the event $\{\Pi_{k\cdot} \gg \Pi_{i\cdot}\}$ to be the event that the populations $\Pi_{i\cdot}$ associated with α_i would be eliminated if the procedure were applied to just the populations associated with α_i and α_k, i.e. the event

$$\bigcup_{m=1}^{\infty} \left\{ \sum_{j=1}^{l} Y_{ij} \leq \sum_{j=1}^{l} Y_{kj} + \max\left(a_{\lambda_\alpha} - \lambda_\alpha n(b)l, 0\right), \ l = 1, 2, \ldots, m-1; \right.$$

$$\left. \sum_{j=1}^{m} Y_{kj} < \sum_{j=1}^{m} Y_{ij} - \max\left(a_{\lambda_\alpha} - \lambda_\alpha n(b)m, 0\right) \right\}.$$

The event $\{\Pi_{\cdot k} \gg \Pi_{\cdot j}\}$ is analogously defined to be the event that the populations $\Pi_{\cdot j}$ associated with β_j would be eliminated if the procedure were applied to just the populations associated with β_j and β_k. Then since these events depend only on the differences $Y_{aj} - Y_{ij}$ for $i = 1, 2, \ldots, a-1$ and $Z_{ib} - Z_{ij}$ for $j = 1, 2, \ldots, b-1$, which form independent sequences by our earlier observation, and whose means do not depend on the level of the other factor, we have

$$P\{CS\} \geq P\left\{ \bigcap_{i=1}^{a-1} \{\Pi_{a\cdot} \gg \Pi_{i\cdot}\} \bigcap_{j=1}^{b-1} \{\Pi_{\cdot b} \gg \Pi_{\cdot j}\} \right\}$$

$$= P\left\{ \bigcap_{i=1}^{a-1} \{\Pi_{a\cdot} \gg \Pi_{i\cdot}\} \right\} P\left\{ \bigcap_{j=1}^{b-1} \{\Pi_{\cdot b} \gg \Pi_{\cdot j}\} \right\}.$$

Since the procedure can be viewed as applying single-factor procedures taking observations from populations with means $n(b)\alpha_i$ and $n(a)\beta_j$ and variances $n(b)\sigma^2$ and $n(a)\sigma^2$, respectively, to the two factors separately and both of these procedures satisfy the required inequalities (Hartmann 1988) we have that $P\{CS\} \geq P_\alpha^* P_\beta^* \geq P^*$.

6 Second Extension for an Unknown σ^2

Before experimentation starts, the experimenter specifies an integer n_0, and numbers λ_α with $0 \leq \lambda_\alpha \leq \delta_\alpha^*/2$ and λ_β with $0 \leq \lambda_\beta \leq \delta_\beta^*/2$. For the sake of simplicity, we will assume that $\lambda_\alpha = \delta_\alpha^*/2n$ and $\lambda_\beta = \delta_\beta^*/2n$ for some integer $n = 1, 2, \ldots$ (for reasons which will become clear, we recommend that n is chosen to be a small integer, say $n = 1$ or $n = 2$). The procedure begins by taking n_0 observations $X_{ij1}, \ldots, X_{ijn_0}$ from each of the ab populations a matrix at a time. Let

$$S^2 = \sum_{i=1}^{a} \sum_{j=1}^{b} \sum_{l=1}^{n_0} (X_{ijl} - \overline{X}_{ij})^2 / ab(n_0 - 1)$$

be the usual estimate of σ^2 with $f = ab(n_0 - 1)$ degrees of freedom, in which the quantity $\overline{X}_{ij} = \sum_{l=1}^{n_0} X_{ijl}/n_0$. Then fS^2/σ^2 has the χ_f^2 distribution and is independent of $\overline{X}_{ij}$ for $i = 1, 2, \ldots, a$ and $j = 1, 2, \ldots, b$.

Let $\eta_\alpha > 0$ and $\eta_\beta > 0$ satisfy

$$\sum_{\substack{i_0 + \cdots + i_n = a-1 \\ j_0 + \cdots + j_n = b-1}} \frac{(-1)^{\sum_{k=0}^{n}(n-k)(i_k+j_k)}}{2^{i_0+j_0}} \binom{a-1}{i_0, \ldots, i_n} \binom{b-1}{j_0, \ldots, j_n} \times$$

$$\left(1 + \sum_{k=0}^{n} \frac{n^2 - k^2}{n}(\eta_\alpha i_k + \eta_\beta j_k)\right)^{-f/2} = P^*.$$

The expression on the left of the equality sign, which can be evaluated in $O(a^n b^n)$ time, is a strictly increasing function of $\eta_\alpha > 0$ and $\eta_\beta > 0$. A recommendation is that these values be chosen to satisfy $\eta_\alpha/\eta_\beta = (\delta_\alpha^*/\delta_\beta^*)^2$ so that the maximum number of observations required to select a best α-factor level and a best β-factor level will be the same. Then $a_{\lambda_\alpha}^* = \eta_\alpha f S^2/\delta_\alpha^*$ and $a_{\lambda_\beta}^* = \eta_\beta f S^2/\delta_\beta^*$ as in the improved version of Paulson's procedure as described in Hartmann (1991).

Let $Y_{i0} = \sum_{j=1}^{b} \sum_{l=1}^{n_0} X_{ijl}$ for $1 \leq i \leq a$ and let $Z_{0j} = \sum_{i=1}^{a} \sum_{l=1}^{n_0} X_{ijl}$ for $1 \leq j \leq 2b$. Once again note that the differences $Y_{a0} - Y_{i0}$ are independent of the differences $Z_{0b} - Z_{0j}$ and all of these are independent of S^2. Next we eliminate the populations associated with α_k if

$$Y_{k0} < \max(Y_{10}, \ldots, Y_{a0}) - \max\left(a_{\lambda_\alpha}^* - \lambda_\alpha n_0 b, 0\right),$$

and the populations associated with β_k if

$$Z_{0k} < \max(Z_{01}, \ldots, Z_{0b}) - \max\left(a_{\lambda_\beta}^* - \lambda_\beta n_0 a, 0\right).$$

This concludes the 0th stage of the procedure.

In the subsequent r^{th} stages, we take $n(a)n(b)/k_r$ observations from each of the $A_r B_r$ remaining populations, where the integer k_r is defined as before, define Y_{ij} and Z_{ij} as before, and for $l = \sum_{t=1}^{r-1} \frac{n(a)}{k_t A_t} + 1, \sum_{t=1}^{r-1} \frac{n(a)}{k_t A_t} + 2, \ldots, \sum_{t=1}^{r} \frac{n(a)}{k_t A_t}$ we eliminate the populations associated with α_k if

$$\sum_{j=0}^{l} Y_{kj} < \max_{i} \sum_{j=0}^{l} Y_{ij} - \max\left(a_{\lambda_\alpha}^* - \lambda_\alpha[n_0 b + n(b)l], 0\right),$$

where the maximum is taken over the remaining α-factor levels i, and for $l = \sum_{t=1}^{r-1} \frac{n(b)}{k_t B_t} + 1, \sum_{t=1}^{r-1} \frac{n(b)}{k_t B_t} + 2, \ldots, \sum_{t=1}^{r} \frac{n(b)}{k_t B_t}$ we eliminate the populations associated with β_k if

$$\sum_{i=0}^{l} Z_{ik} < \max_{j} \sum_{i=0}^{l} Z_{ij} - \max\left(a_{\lambda_\beta}^* - \lambda_\beta[n_0 a + n(a)l], 0\right),$$

where the maximum is taken over the remaining β-factor levels j.

To see that the probability requirement is satisfied, suppose that, unknown to the experimenter, $\alpha_{[a]} = \alpha_a \geq \alpha_i + \delta_\alpha^*$ for $i = 1, \ldots, a-1$ and $\beta_{[b]} = \beta_b \geq \beta_j + \delta_\beta^*$ for $j = 1, \ldots, b-1$. We define the event $\{\Pi_k. \gg \Pi_i.\}$ to be the event that the populations $\Pi_i.$ associated with α_i would be eliminated if the procedure were applied to just the populations associated with α_i and α_k, i.e. the event

$$\bigcup_{m=0}^{\infty} \left\{ \sum_{j=0}^{l} Y_{ij} \leq \sum_{j=0}^{l} Y_{kj} + \max\left(a_{\lambda_\alpha}^* - \lambda_\alpha[n_0 b + n(b)l], 0\right), \ 1 \leq l \leq m-1; \right.$$

$$\left. \sum_{j=0}^{m} Y_{kj} < \sum_{j=0}^{m} Y_{ij} - \max\left(a_{\lambda_\alpha}^* - \lambda_\alpha[n_0 b + n(b)m], 0\right) \right\}.$$

The event $\{\Pi_{\cdot k} \gg \Pi_{\cdot j}\}$ is analogously defined to be the event that the populations $\Pi_{\cdot j}$ associated with β_j would be eliminated if the procedure were applied to just the populations associated with β_j and β_k.

Conditioning on the values $Y_{a0}, Y_{a1}, \ldots$ and $Z_{0b}, Z_{1b}, \ldots$ we obtain

$$P\{CS|S^2\} \geq P\left\{ \bigcap_{i=1}^{a-1} \{\Pi_{a.} \gg \Pi_{i.}\} \bigcap_{j=1}^{b-1} \{\Pi_{\cdot b} \gg \Pi_{\cdot j}\} \ \middle| \ S^2 \right\}$$

$$= E\left[P\left\{ \bigcap_{i=1}^{a-1} \{\Pi_{a.} \gg \Pi_{i.}\} \bigcap_{j=1}^{b-1} \{\Pi_{\cdot b} \gg \Pi_{\cdot j}\} \right\} \ \middle| \ Y_{a0}, \ldots, Z_{0b}, \ldots, S^2 \right]$$

$$= E\left[\prod_{i=1}^{a-1} P\{\Pi_{a\cdot} \gg \Pi_{i\cdot}|Y_{a0}, \ldots, S^2\} \prod_{j=1}^{b-1} P\{\Pi_{\cdot b} \gg \Pi_{\cdot j}|Z_{0b}, \ldots, S^2\}\right]$$

The last equality holds because the $\bigcap_{i=1}^{a-1}\{\Pi_{a\cdot} \gg \Pi_{i\cdot}\}$ and $\bigcap_{j=1}^{b-1}\{\Pi_{\cdot b} \gg \Pi_{\cdot j}\}$ are conditionally independent given S^2, as they depend only on the differences $Y_{aj} - Y_{ij}$ for $i = 1, 2, \ldots, a-1$ and $Z_{ib} - Z_{ij}$ for $j = 1, 2, \ldots, b-1$, respectively, which form independent sequences by our earlier observation, and whose means do not depend on the level of the other factor. Also, given $Y_{a0}, Y_{a1}, \ldots$ and S^2 the events $\{\Pi_{a\cdot} \gg \Pi_{i\cdot}\}$ for $i = 1, 2, \ldots, a-1$ are conditionally independent, and given $Z_{0b}, Z_{1b}, \ldots$ and S^2 the events $\{\Pi_{\cdot b} \gg \Pi_{\cdot j}\}$ for $j = 1, 2, \ldots, b-1$ are conditionally independent.

As in Hartmann (1991) we can conclude that the minimum of the above expression subject to $\alpha_a \geq \alpha_i + \delta_\alpha^*$ for $i = 1, \ldots, a-1$ and $\beta_b \geq \beta_j + \delta_\beta^*$ for $j = 1, \ldots, b-1$ occurs when the parameters are in the least favorable configuration (LFC): $\alpha_a = \alpha_i + \delta_\alpha^*$ for $i = 1, \ldots, a-1$ and $\beta_b = \beta_j + \delta_\beta^*$ for $j = 1, \ldots, b-1$. In this case, the final expression above simplifies to

$$E\left[\left(P_{\text{LFC}}\{\Pi_{a\cdot} \gg \Pi_{1\cdot}|Y_{a0}, \ldots, S^2\}\right)^{a-1}\left(P_{\text{LCF}}\{\Pi_{\cdot b} \gg \Pi_{\cdot 1}|Z_{0b}, \ldots, S^2\}\right)^{b-1}\right],$$

and using Jensen's inequality we obtain

$$P\{CS\} \geq E\left[\left(P_{\text{LFC}}\{\Pi_{a\cdot} \gg \Pi_{1\cdot}|S^2\}\right)^{a-1}\left(P_{\text{LCF}}\{\Pi_{\cdot b} \gg \Pi_{\cdot 1}|S^2\}\right)^{b-1}\right]. \tag{6.1}$$

Now we make use of the lower bound

$$1 - \sum_{j=1}^{n}(-1)^{j+1}(1-\delta_{jn}/2)\exp\{-\frac{(2n-j)j}{2n}\eta_\alpha \frac{fS^2}{\sigma^2}\}$$

given in Hartmann (1991) for $P_{\text{LFC}}\{\Pi_{a\cdot} \gg \Pi_{1\cdot}|S^2\}$. After changing the index of summation to $k = n-i$ this lower bound can be rewritten as

$$\sum_{k=0}^{n}(-1)^{n-k}(1-\delta_{k0}/2)\exp\{-\frac{n^2-k^2}{2n}\eta_\alpha \frac{fS^2}{\sigma^2}\}.$$

Thus a lower bound for $\left(P_{\text{LFC}}\{\Pi_{a\cdot} \gg \Pi_{1\cdot}|S^2\}\right)^{a-1}$ can be obtained from the multinomial formula, namely,

$$\sum_{i_0+\cdots+i_n=a-1}\frac{(-1)^{\sum_{k=0}^{n}(n-k)i_k}}{2^{i_0}}\binom{a-1}{i_0, \ldots, i_n}\exp\left\{-\sum_{k=0}^{n}\frac{n^2-k^2}{2n}\eta_\alpha i_k \frac{fS^2}{\sigma^2}\right\}. \tag{6.2}$$

As an aside, taking expectations with respect to S^2 in (6.2) gives

$$\sum_{i_0+\cdots+i_n=a-1} \frac{(-1)^{\sum_{k=0}^{n}(n-k)i_k}}{2^{i_0}} \binom{a-1}{i_0,\ldots,i_n} \left(1+\sum_{k=0}^{n} \frac{n^2-k^2}{n} \eta_\alpha i_k\right)^{-f/2},$$

which shows that the procedures described in Hartmann (1991) can choose η to satisfy

$$\sum_{i_0+\cdots+i_j=k-1} \frac{(-1)^{\sum_{l=0}^{j}(j-l)i_l}}{2^{i_0}} \binom{k-1}{i_0,\ldots,i_j} \left(1+\sum_{l=0}^{j} \frac{j^2-l^2}{j} \eta i_l\right)^{-f/2} = P^*,$$

which for $k > 2$ will result in a smaller value of η. For example if $j = 1$, $k = 5$, $P^* = .75$ and $n_0 = 5$ the value of η is 2.4% smaller.

Here substituting (6.2) and the analogous bound for $(P_{\text{LFC}}\{\text{II}._b \gg \text{II}._1|S^2\})^{b-1}$ into (6.1) and taking expectations with respect to S^2 we see that $P\{CS\} \geq P^*$ by our choice of η_α and η_β. As for the basic extension of Paulson's procedure for a common unknown variance, Chebyshev's inequality shows that the resulting lower bound on the $P\{CS\}$ is better than just the product of the single-factor bounds as in the known variance case.

Appendix

In this appendix, we give a slight strengthening of the lemma in Hartmann (1991) which can be proved using the same likelihood ratio arguments. This lemma will then be used to determine operating characteristics for the improved versions of Paulson's procedures.

Lemma

Let $\{W(t), 0 \leq t < \infty\}$ be a Brownian motion with drift $\mu > 0$, and let $m > 0$ and $\lambda > 0$ be given. If T is a symmetric stopping time such that $|W(T)| \geq \lambda(m-T)$, then

$$P_\mu\{W(T) < 0\} \leq \int_{-\infty}^{\infty} \frac{e^{-2\lambda\xi}}{1+e^{-2\lambda\xi}} \phi\left(\frac{\xi-m\mu}{\sqrt{m}}\right) \frac{d\xi}{\sqrt{m}}, \qquad (A.1)$$

where $\phi(x) = (2\pi)^{-1/2}e^{-x^2/2}$. Further, equality holds if and only if $|W(T)| = \lambda(m-T)$ with probability one.

Lawing and David (1966) and Fabian (1974) show that when $\lambda = \mu/2j$ the integral in (A.1) can be expressed in closed form as

$$\sum_{i=1}^{j}(-1)^{i+1}(1-\delta_{ij}/2)\exp\{-2m(2j-i)i\lambda^2\},$$

where δ_{ij} is the Kronecker symbol.

The improved versions of Paulson's procedure for a common known variance in Fabian (1974) and Hartmann (1988) determine a_λ by solving an equation of the form

$$\int_{-\infty}^{\infty} \frac{e^{-2\lambda\xi}}{1+e^{-2\lambda\xi}} \, \phi\left(\frac{\xi - m\delta^*}{\sqrt{m}}\right) \frac{d\xi}{\sqrt{m}} \;=\; f(P^*, k), \tag{A.2}$$

for m and then setting $a_\lambda = 2m\lambda\sigma^2$. Here $f(P^*, k)$ is a function of the desired probability P^* of making a correct selection and the number k of normal populations.

Corollary 1

For fixed P^*, k, δ^* and σ^2, a_λ/λ is a strictly decreasing function of λ and a_λ is a strictly increasing function of λ.

Proof of Corollary 1

It is implicit in the statement of this corollary that m (and hence a_λ) is uniquely determined by (A.2). To see this, let T' be the stopping time $T' = \inf\{t : |W(t)| \geq \lambda(m'-t)\}$ for some $m' > m$. Since $\lambda(m'-t) > \lambda(m-t)$, applying the lemma to T' with both m and m' we obtain

$$\int_{-\infty}^{\infty} \frac{e^{-2\lambda\xi}}{1+e^{-2\lambda\xi}} \, \phi\left(\frac{\xi - m'\delta^*}{\sqrt{m'}}\right) \frac{d\xi}{\sqrt{m'}} \;<\; \int_{-\infty}^{\infty} \frac{e^{-2\lambda\xi}}{1+e^{-2\lambda\xi}} \, \phi\left(\frac{\xi - m\delta^*}{\sqrt{m}}\right) \frac{d\xi}{\sqrt{m}}.$$

Thus for any $\lambda > 0$ and $\delta^* > 0$ the integral in (A.1) is a strictly decreasing function of m, which gives the desired conclusion.

We will establish the monotonicity of a_λ/λ and a_λ using the fact that the lemma implies that for different values of λ, the continuation regions cannot contain one another. If a_λ/λ was not a strictly decreasing function of λ, then there would exist $\lambda' > \lambda$ such that the corresponding solutions m' and m to (A.2) satisfy $m' \geq m$. If $T' = \inf\{t : |W(t)| \geq \lambda'(m_{\lambda'} - t)\}$ then we have

$$|W(T')| \;=\; \lambda'(m'-T') \;>\; \lambda(m'-T') \;\geq\; \lambda(m-T').$$

Applying the lemma to T' with λ and m as well as with λ' and m' gives

$$\int_{-\infty}^{\infty} \frac{e^{-2\lambda'\xi}}{1+e^{-2\lambda'\xi}} \, \phi\left(\frac{\xi - m'\delta^*}{\sqrt{m'}}\right) \frac{d\xi}{\sqrt{m'}} \;<\; \int_{-\infty}^{\infty} \frac{e^{-2\lambda\xi}}{1+e^{-2\lambda\xi}} \, \phi\left(\frac{\xi - m\delta^*}{\sqrt{m}}\right) \frac{d\xi}{\sqrt{m}}$$

which is a contradiction, since equality holds by (A.2). Now if a_λ was not a strictly increasing function of λ, then there would exist $\lambda' < \lambda$ such that $a_{\lambda'} \geq a_\lambda$. Since a_λ/λ is a strictly decreasing function of λ, the corresponding solutions m' and m to (A.2) must satisfy $m' > m$. If $T' = \inf\{t : |W(t)| \geq \lambda'(m'-t)\}$ then we have

$$|W(T')| = \lambda'(m'-T') = \frac{a_{\lambda'}}{2\sigma^2}\lambda'T' \geq \frac{a_\lambda}{2\sigma^2}\lambda'T' > \frac{a_\lambda}{2\sigma^2}\lambda T' = \lambda(m-T').$$

Applying the lemma to T' with λ and m as well as with λ' and m' gives the same contradiction as above. ∎

This corollary also applies to the extensions of Paulson's procedures for a common known variance described in this paper.

The improved versions of Paulson's procedure for a common unknown variance in Hartmann (1991) with the additional improvement described at the end of the last section determine $a_\lambda^* = \eta f S^2/\delta^*$ from the solution η of the equation

$$E\left[\left\{1 - \int_{-\infty}^\infty \frac{e^{-2\lambda\xi}}{1+e^{-2\lambda\xi}}\,\phi\left(\frac{\xi-m\delta^*}{\sqrt{m}}\right)\frac{d\xi}{\sqrt{m}}\right\}^{k-1}\right] = P^* \qquad (A.3)$$

where $m = \frac{\eta}{2\lambda\delta^*}\left(\frac{fS^2}{\sigma^2}\right)$ is a function of the estimate S^2 of σ^2.

Corollary 2

For fixed P^*, k, δ^* and S^2, a_λ^*/λ is a strictly decreasing function of λ and a_λ^* is a strictly increasing function of λ.

Proof of Corollary 2

Since the integral in (A.1) is a strictly decreasing function of m, the expression in (A.3) is a function increasing function of η. The rest of the proof follows that of Corollary 1, since the continuation regions are proportional to the value of S^2. More precisely, if η and η' are the solutions to (A.3) associated with λ and λ' then

$$\int_{-\infty}^\infty \frac{e^{-2\lambda'\xi}}{1+e^{-2\lambda'\xi}}\,\phi\left(\frac{\xi-m'\delta^*}{\sqrt{m'}}\right)\frac{d\xi}{\sqrt{m'}} < \int_{-\infty}^\infty \frac{e^{-2\lambda\xi}}{1+e^{-2\lambda\xi}}\,\phi\left(\frac{\xi-m\delta^*}{\sqrt{m}}\right)\frac{d\xi}{\sqrt{m}}$$

holds for $m = \frac{\eta}{2\lambda\delta^*}\left(\frac{fS^2}{\sigma^2}\right)$ and $m' = \frac{\eta'}{2\lambda'\delta^*}\left(\frac{fS^2}{\sigma^2}\right)$ regardless of the value of S^2, and taking expectations gives a contradiction to (A.3). ∎

This corollary does not apply to the extensions of Paulson's procedures for a common unknown variance described earlier, since these procedures determine a_{λ_α} and a_{λ_β} by substituting the integral in (A.1) into (3.1) and (6.1), taking expectations with respect to S^2, then setting the result equal to P^*. However, the following corollary can be proved in a similar manner.

Corollary 3

For fixed P_α^*, a, δ_α^*, P_β^*, b, δ_β^*, λ_β and S^2, $a_{\lambda_\alpha}^*/\lambda_\alpha$ is a strictly decreasing function of λ_α and $a_{\lambda_\alpha}^*$ is a strictly increasing function of λ_α. Similarly if

λ_α is fixed, then $a^*_{\lambda_\beta}/\lambda_\beta$ is a strictly decreasing function of λ_β and $a^*_{\lambda_\beta}$ is a strictly increasing function of λ_β.

Acknowledgment

Research partially supported by NSF Grant DMS 89-05645.

References

Bechhofer, R.E. (1954). A single-sample multiple decision procedure for ranking means of normal populations with known variances. *Ann. Math. Statist.* **25** 16 - 39.

Esary, J.D., Proschan, F., and Walkup, D.W. (1967). Association of random variables, with applications. *Ann. Math. Statist.* **38** 1466 - 1474.

Fabian, V. (1974). Note on Anderson's sequential procedures with triangular boundary. *Ann. Statist.* **2** 170 - 175.

Hartmann, M. (1988). An improvement on Paulson's sequential ranking procedure. *Sequen. Anal.* **7** 363 - 372.

Hartmann, M. (1991). An improvement on Paulson's procedure for selecting the population with the largest mean from k normal populations with a common unknown variance. *Sequen. Anal.* **10** 1 - 16.

Lawing, W.D. and David, H.T. (1966). Likelihood ratio computations of operating characteristics. *Ann. Math. Statist.* **37** 1704 - 1716.

Paulson, E. (1964). A sequential procedure for selecting the population with the largest mean from k normal populations. *Ann. Math. Statist.* **35** 174 - 180.

Chapter 15

The Least Favorable Configuration of a Two-Stage Procedure for Selecting the Largest Normal Mean

THOMAS J. SANTNER   Department of Statistics, Ohio State University, Columbus, Ohio

ANTHONY J. HAYTER   School of Industrial and Systems Engineering, Georgia Institute of Technology, Atlanta, Georgia

Abstract   Santner and Behaxeteguy (1992) introduce a two-stage procedure for selecting the normal population with the largest mean when all the populations have a common known variance. In their procedure, a random number of populations are allowed to enter the second stage; this number is controlled by an experimenter-specified upper bound. They consider the determination of sample sizes and constants needed to implement the selection procedure, and they conjecture that the infimum of the probability of correct selection over configurations of population means $\mu = (\mu_1, \ldots, \mu_k)$ for which $\mu_{[k]} - \mu_{[k-1]} \geq \delta^\star$, occurs when $\mu_{[1]} = \mu_{[k-1]} = \mu_{[k]} - \delta^\star$, where $\delta^\star$ is a given positive number and $\mu_{[1]} \leq \ldots \leq \mu_{[k]}$ are the ordered population means. Their conjecture is proved in this paper. Some extensions of this two-stage rule and this result are discussed.

1   Introduction

Consider k normal populations $\Pi_1, \ldots, \Pi_k$ where population Π_i has unknown mean μ_i and (common) known variance σ^2 for $1 \leq i \leq k$. The goal is to select the population with the largest mean, referred to hereafter as correct selection. The indifference zone formulation of Bechhofer (1954) is adopted whereby the probability of correct selection is required to be larger than some prespecified probability level P^* whenever the largest and second largest population means are at least some prespecified distance δ^* apart.

This paper considers the class of two-stage procedures introduced in Santner and Behaxeteguy (1992) for solving this problem. These procedures employ the following two features. Firstly, at the conclusion of the first stage sampling, the procedure chooses a *random* number of populations to enter the second stage, allowing the elimination of all populations whose first-stage sample means indicate that they are inferior. Indeed, if one sample mean is sufficiently far above the others the procedure will stop after one stage of sampling. Secondly, these procedures allow the experimenter to control the *maximum* number of populations which will enter the second stage of the experiment.

Formally, this general class of procedures is defined by a set of values (m, h, n_1, n_2) where m is an integer satisfying $1 \leq m \leq k$, $h > 0$, and $n_1, n_2 \geq 1$ are integers. The value m is the upper bound on the number of populations to be sampled in the second stage. The constant h is a 'yardstick' which determines the random number of populations which are to enter the second stage. Lastly, n_1 and n_2 are the first and second stage sample sizes, respectively, which are common to all populations.

The procedure $R(m, h, n_1, n_2)$ corresponding to the values (m, h, n_1, n_2) operates as follows:

Stage 1: Sample $X_{i1}, \ldots, X_{in_1}$ from Π_i and let $X_i = \sum_{j=1}^{n_1} X_{ij}$ be the corresponding first stage sample total for $1 \leq i \leq k$. Let $X_{[1]} < \ldots < X_{[k]}$ be the ordered first stage sample totals, and let $\boldsymbol{X} = (X_1, \ldots, X_k)$. Construct the subset of populations

$$S(\boldsymbol{X}) = \left\{ i : X_i \geq \max\{X_{[k-m+1]}, X_{[k]} - h\} \right\}.$$

If the set $S(\boldsymbol{X})$ contains exactly one population then stop and declare the associated population to be the population with the largest mean. Otherwise proceed with Stage 2.

Stage 2: For populations Π_i with $i \in S(\boldsymbol{X})$, sample $Y_{i1}, \ldots, Y_{in_2}$, let $Y_i = \sum_{j=1}^{n_2} Y_{ij}$ be the corresponding second stage sample total, and let $Z_i = X_i + Y_i$ be the sample total based on the data from *both* stages. Select the population Π_i associated with $\max\{Z_i : i \in S(\boldsymbol{X})\}$ as the population with the largest mean.

The procedure $R(m, h, n_1, n_2)$ can equivalently be defined in terms of the first stage sample means

$$\frac{1}{n_1} \sum_{j=1}^{n_1} X_{ij}$$

at Stage 1, and the overall sample means

$$\frac{1}{n_1 + n_2} \left\{ \sum_{j=1}^{n_1} X_{ij} + \sum_{j=1}^{n_2} Y_{ij} \right\}$$

at Stage 2. However, in this paper it is defined in terms of the sample sums in order to facilitate the comparison of the present work with that of Sehr (1988), whose analysis of the two-stage rule with $m = k$ is generalized in this paper. In the definition of the procedure it was assumed that the first stage sample totals X_i and the cumulative sample totals Z_i are all unequal as ties occur with probability zero for normally distributed data.

Let $\mu_{[1]} \le \ldots \le \mu_{[k]}$ denote the ordered population means and let $P_{\boldsymbol{\mu}}[\text{CS}]$ be the probability of correctly selecting the population associated with the largest population mean $\mu_{[k]}$. The experimental design requirement is based on the indifference zone approach of Bechhofer (1954).

Design Requirement : For fixed $P^\star$ with $1/k < P^\star < 1$ and fixed $\delta^\star > 0$, the rule $R(m, h, n_1, n_2)$ must satisfy the probability condition $P_{\boldsymbol{\mu}}[\text{CS}] \ge P^\star$ whenever $\boldsymbol{\mu} \in \Omega(\delta^\star) = \{\boldsymbol{\mu} \in \mathbb{R}^k : \mu_{[k]} - \mu_{[k-1]} \ge \delta^\star\}$.

In order to implement this design requirement, it is necessary to find the least favorable configuration of population means, i.e. the values $\boldsymbol{\mu} \in \Omega(\delta^\star)$ at which $P_{\boldsymbol{\mu}}[\text{CS}]$ is minimized.

The class of selection procedures $R(m, h, n_1, n_2)$ contains a number of previously studied selection procedures as special cases. The single-stage procedure of Bechhofer (1954) corresponds to $m = 1$ (h, n_1, n_2 arbitrary), the two-stage procedure of Fairweather (1968) corresponds to $h = \infty$ (m, n_1, n_2 arbitrary), and the two-stage procedures of Cohen (1959), Alam (1970), and Tamhane and Bechhofer (1977, 1979) correspond to $m = k$ (h, n_1, n_2 arbitrary). In all of these special cases, it is known that under the specified indifference zone condition, the probability of correct selection $P_{\boldsymbol{\mu}}[\text{CS}]$ is minimized at the (least favorable) configuration of population means given by $\mu_{[1]} = \mu_{[k-1]} = \mu_{[k]} - \delta^\star$. Santner and Behaxeteguy (1992) conjectured that this same configuration would be least favorable for any selection procedure $R(m, h, n_1, n_2)$. Lastly, Han (1987) and Gupta and Han

(1987) consider selection procedures of the form $R(m, h, n_1, n_2)$ for logistic distributions. They do not determine the infimum of the probability of correct selection but use a Bonferroni inequality to derive a probability bound for this quantity.

The purpose of this paper is to prove that for the general class of selection procedures $R(m, h, n_1, n_2)$, the minimum probability of correct selection does occur at the least favorable configuration of population means satisfying $\mu_{[1]} = \mu_{[k-1]} = \mu_{[k]} - \delta^\star$. This result is proved in Section 2, and the paper concludes with a summary and discussion of applications and extensions in Section 3.

2   The Main Result

In this section the main Theorem concerning the general least favorable configuration of population means is presented and proved.

Theorem

For any selection procedure $R(m, h, n_1, n_2)$, any variance $\sigma^2 > 0$ and any $\delta^\star > 0$,

$$\inf_{\mu \in \Omega(\delta^\star)} P_{\mu}[CS] = P_{\mu^\bullet}[CS]$$

where μ^* is any set of population means satisfying $\mu_{[1]} = \mu_{[k-1]} = \mu_{[k]} - \delta^\star$.

Proof of Theorem

The proof of the theorem is a generalization of the proof given by Sehr (1988), who considered the special case $m = k$. The proof given here mimics Sehr's proof closely, but with modifications to cater for the extra complications arising with a general value of m.

Notice that, with probability one, all of the elements of X, Y, and $Z = X + Y$ must be distinct. Also, with probability one, no linear combination of their elements will be equal, e.g. $x_i \neq x_j + h$ or $x_i + y_i \neq x_j + y_j$ for distinct indices i and j. Therefore in the proofs below, the complications arising in such cases will be ignored. For example, we assume that all the order statistics used to define the procedure are distinct. Also, mathematical identities between sets, statements of disjointness, and representations of the event of correct selection will be defined *up to sets of measure zero*. The first several occurences of this difficulty will be explicitly mentioned and thereafter this phenomenon will be ignored.

From now on suppose, without loss of generality, that $\mu_k = \mu_{[k]}$, so that correct selection is equivalent to selecting population Π_k. The procedure $R(m, h, n_1, n_2)$ depends on only the vectors $X = (X_1, \ldots, X_k)$ and $Y = (Y_1, \ldots, Y_k)$ of the first and second stage population sample totals $X_i =$

$\sum_{j=1}^{n_1} X_{ij}$ and $Y_i = \sum_{j=1}^{n_2} Y_{ij}$ for $1 \le i \le k$. In practice, some of the Y_i will not be observed (those corresponding to populations which are eliminated after the first stage). However, for the purposes of this proof, it is useful to regard all of the Y_i as sampled. Then, correct selection occurs if and only if $(\boldsymbol{X}, \boldsymbol{Y}) \in CS = CS(h, m)$ which is defined by

$$\left\{ (\boldsymbol{x}, \boldsymbol{y}) \in \mathbb{R}^{2k} : k \in S(\boldsymbol{x}); \; x_k + y_k > x_i + y_i \text{ for all } i \in S(\boldsymbol{x}) - \{k\} \right\}.$$

Furthermore, it is easy to see that for any $\rho > 0$, $\rho CS(h, m) = CS(\rho h, m)$, so that for every $\rho > 0$

$$
\begin{aligned}
P_{\mu, \sigma^2}[(\boldsymbol{X}, \boldsymbol{Y}) \in CS(h, m)] &= P_{\mu, \sigma^2}[(\rho \boldsymbol{X}, \rho \boldsymbol{Y}) \in \rho CS(h, m)] \\
&= P_{\rho \mu, \rho^2 \sigma^2}[(\boldsymbol{X}, \boldsymbol{Y}) \in CS(\rho h, m)].
\end{aligned}
$$

Consequently, it suffices to prove the Theorem for $\sigma^2 = 1/n_1$, say, and arbitrary values of $\delta^\star > 0$ and $h > 0$ and $m \in \{1, \ldots, k\}$.

Next, let $\nu_i = n_1 \mu_i$, $1 \le i \le k$, and $\lambda = n_2/n_1$, so that $X_i \sim N(\nu_i, 1)$ and $Y_i \sim N(\lambda \nu_i, \lambda)$, $1 \le i \le k$. Also, for a *fixed* $j \in \{1, \ldots, k-1\}$, it is useful to consider the first and second stage observations from Π_j to have different means, namely μ_j^x and μ_j^y respectively. Then let $\nu_j^x = n_1 \mu_j^x$ and $\nu_j^y = n_1 \mu_j^y$ so that $X_j \sim N(\nu_j^x, 1)$ and $Y_j \sim N(\lambda \nu_j^y, \lambda)$. The corresponding distribution of $(\boldsymbol{X}, \boldsymbol{Y})$ will now be characterized by the $(k+1)$ vector of means $\boldsymbol{\nu}^{xy} = (\nu_1, \ldots, \nu_{j-1}, \nu_j^x, \nu_j^y, \ldots, \nu_k)$. Let $P_{\boldsymbol{\nu}^{xy}}[CS]$ denote the probability of correct selection (selecting Π_k) computed under this distribution.

The Theorem then follows from the following two lemmas:

Lemma 1

$\quad P_{\boldsymbol{\nu}^{xy}}[CS]$ is nonincreasing in ν_j^y.

Lemma 2

$\quad$ If

$$\nu_j^y = \nu_{[k]}^{xy} \le \nu_k = \nu_{[k+1]}^{xy},$$

then $P_{\boldsymbol{\nu}^{xy}}[CS]$ is nonincreasing in ν_j^x in the range $\nu_j^x \le \nu_j^y = \nu_{[k]}^{xy}$.

In Lemma 2, notice that ν_j^y is the second largest element of $\boldsymbol{\nu}^{xy}$, the largest element being ν_k. The Theorem is a direct consequence of the two lemmas since, for any $\mu \in \Omega(\delta^\star)$, as any one of the population means μ_j which is not the best is increased to $\mu_{[k]} - \delta^\star$, it can be seen that $P_{\mu}[CS]$ is nonincreasing. This is because Lemma 1 can first be used to increase μ_j^y to $\mu_k - \delta^\star$, and then Lemma 2 can be used to increase μ_j^x to μ_j^y.

Proof of Lemma 1

$\quad$ Let $\tilde{\boldsymbol{Y}} = (Y_1, \ldots, Y_{j-1}, Y_{j+1}, \ldots, Y_k)$ be the vector of all second stage totals except for that from Π_j and let $\tilde{\boldsymbol{y}} = (y_1, \ldots, y_{j-1}, y_{j+1}, \ldots, y_k)$ be an observed value of $\tilde{Y}$. Also, let $Q_1(\cdot)$ be the joint distribution function of $(X, \tilde{Y})$, which does not depend on ν_j^y. For $(x, \tilde{y}) \in \mathbb{R}^{2k-1}$ let $C_1(x, \tilde{y}) = \{y_j \in \mathbb{R} : (x, y) \in \text{CS}\}$ be the set of second stage y_j leading to correct selection. Thus

$$P_{\nu^{xy}}[\text{CS}] = \int P[Y_j \in C_1(x, \tilde{y})] \, dQ_1.$$

For each $(x, \tilde{y}) \in \mathbb{R}^{2k-1}$ the sets $C_1(x, \tilde{y})$ are now determined. First, consider the set P_1 defined by

$$\{(x, \tilde{y}) \in \mathbb{R}^{2k-1} : k \in S(x); x_k + y_k > x_i + y_i$$
$$\text{whenever } i \in S(x) - \{k\} \text{ and } i \neq j\}.$$

If $(x, \tilde{y}) \notin P_1$, then correct selection will be *impossible* no matter what the value of y_j, since either (a) $k \notin S(x)$ or (b) $k \in S(x)$ but there is a population Π_i $(i \neq j, k)$ which qualifies for the second stage and then goes on to beat Π_k at the end of the second stage. Hence correct selection cannot occur, and $C_1(x, \tilde{y}) = \phi$ for all $(x, \tilde{y}) \notin P_1$.

There are two cases to be considered when $(x, \tilde{y}) \in P_1$ depending on whether $j \notin S(x)$ or $j \in S(x)$. In the case when $j \notin S(x)$, there will be correct selection regardless of the value of y_j, and so $C_1(x, \tilde{y}) = \mathbb{R}$. In the case when $j \in S(x)$, correct selection occurs if and only if Π_k beats Π_j at the end of Stage 2, i.e. if and only if $x_k + y_k > x_j + y_j$. Thus,

$$C_1(x, \tilde{y}) = \begin{cases} \phi & (x, \tilde{y}) \notin P_1 \\ \mathbb{R} & (x, \tilde{y}) \in P_1, j \notin S(x) \\ (-\infty, x_k + y_k - x_j) & (x, \tilde{y}) \in P_1, j \in S(x). \end{cases}$$

Finally, since $C_1(x, \tilde{y})$ is of the form $(-\infty, \alpha)$ for all $(x, \tilde{y})$, it can be seen that $P[Y_j \in C_1(x, \tilde{y})]$ is nonincreasing in ν_j^y and hence $P_{\nu^{xy}}[\text{CS}]$ is nonincreasing in ν_j^y. This completes the proof of Lemma 1. ∎

Proof of Lemma 2

Now let $\tilde{X} = (X_1, \ldots, X_{j-1}, X_{j+1} \ldots, X_k)$ be the vector of all first stage totals except for that from Π_j and let $\tilde{x} = (x_1, \ldots, x_{j-1}, x_{j+1} \ldots, x_k)$ be an observed value of $\tilde{X}$. For each $(\tilde{x}, y) \in \mathbb{R}^{2k-1}$ let $C_2(\tilde{x}, y) = \{x_j \in \mathbb{R} : (x, y) \in \text{CS}\}$ be the set of first stage x_j leading to correct selection. Also, let $z = x_k + y_k - y_j$, and let $\tilde{x}_{[1]} < \ldots < \tilde{x}_{[k-1]}$ be the ordered values of $\tilde{x}$.

The initial objective is to calculate $C_2(\tilde{x}, y)$ for each $(\tilde{x}, y) \in \mathbb{R}^{2k-1}$. For $m < k$ let

$$S^\star(\tilde{x}) = \{i \in \{1, \ldots, k\} - \{j\} : x_i \geq \max\{\tilde{x}_{[k-1]} - h, \tilde{x}_{[k-m]}\}\}$$

be the set of Π_i which tentatively go through to Stage 2 subject to modifications caused by the unveiling of x_j. If $m = k$ take $\tilde{x}_{[k-m]}$ to be $\tilde{x}_{[1]}$ in the definition of $S^\star(\tilde{x})$. Remember that the final set entering Stage 2 based on the entire vector of k first stage sample totals is

$$S(x) = \{i \in \{1,\ldots,k\} : x_i \geq \max\{x_{[k]} - h, x_{[k-m+1]}\}\}.$$

Notice that if $i \in S(x)$ and $i \neq j$, then necessarily $i \in S^\star(\tilde{x})$. Conversely, if $i \notin S^\star(\tilde{x})$ then $i \notin S(x)$ for any x_j.

Let

$$P_2 = \{(\tilde{x}, y) \in \mathbb{R}^{2k-1} : k \in S^\star(\tilde{x}); x_i > x_k \Rightarrow x_k + y_k > x_i + y_i \text{ for } i \neq j, k\}.$$

Then if $(\tilde{x}, y) \notin P_2$, correct selection will be *impossible* no matter what the value of x_j, since either (a) $k \notin S^\star(\tilde{x})$, in which case Π_k cannot enter Stage 2 for any x_j, or (b) if $k \in S^\star(\tilde{x})$ (and Π_k will potentially enter Stage 2 for some values of x_j), then there is some $i \neq j, k$ for which Π_i will enter Stage 2 if Π_k enters Stage 2, and then will beat Π_k at the end of Stage 2. The latter holds since there is $i \neq j, k$ with $x_i > x_k$ (hence $k \in S(x) \Rightarrow i \in S(x)$) and with $x_k + y_k < x_i + y_i$. Consequently, $C_2(\tilde{x}, y) = \phi$ for $(\tilde{x}, y) \notin P_2$.

The strategy now is to partition the set P_2 into two disjoint subsets A and B, and then to determine the form of $C_2(\tilde{x}, y)$ within these two subsets. Specifically, define the sets

$$A = \{(\tilde{x}, y) \in P_2 : i \in S^\star(\tilde{x}) - \{k\} \Rightarrow x_k + y_k > x_i + y_i\}$$

and

$$B = \{(\tilde{x}, y) \in P_2 : \text{there exists } q \in S^\star(\tilde{x}) - \{k\} \text{ such that } x_q + y_q > x_k + y_k\}.$$

Notice that in the definition of the set B, the fact that $x_k > x_q$ is a consequence of the definition of the set P_2. The elements $(\tilde{x}, y) \in B$ correspond to scenarios in which some population Π_q is beaten by Π_k at Stage 1, and yet is still good enough to potentially qualify for Stage 2 ($q \in S^\star$), in which case it will then go on to beat Π_k at the end of Stage 2. A necessary condition for correct selection is that x_j is sufficiently large to prevent Π_q from entering Stage 2 ($q \notin S(x)$).

It is straightforward to prove that $A \cap B = \phi$ and $A \cup B = P_2$, up to a set of measure zero, so that

$$
\begin{aligned}
P_{\nu^{xy}}[\text{CS}] &= \int_{\mathbb{R}^{2k-1}} P[X_j \in C_2(\tilde{x}, y)] \, dQ_2 \\
&= \int_{P_2} P[X_j \in C_2(\tilde{x}, y)] \, dQ_2 \\
&= \int_A P[X_j \in C_2(\tilde{x}, y)] \, dQ_2 + \int_B P[X_j \in C_2(\tilde{x}, y)] \, dQ_2,
\end{aligned}
$$

where $Q_2(\cdot)$ is the joint distribution function of $(\tilde{X}, Y)$. It turns out that every $(\tilde{x}, y) \in A$ has an associated set $C_2(\tilde{x}, y)$ of the form ϕ or $(-\infty, \alpha)$, leading to an integrand in the expression for the probability of correct selection which is nonincreasing in ν_j^x. Unfortunately, some elements $(\tilde{x}, y) \in B$ have associated sets $C_2(\tilde{x}, y)$ of the form (a, b) with $-\infty < a < b < \infty$, in which case the associated integrands are *not* monotone in ν_j^x. However, these awkward subsets of B will be paired with appropriate subsets of A, and the probability of correct selection over their union will be shown to be nonincreasing in ν_j^x.

To analyse the set A, consider two types of first stage outcomes $\tilde{x}$: (a) those indicating tentative second stage sampling from exactly m populations ($|S^*(\tilde{x})| = m$), and (b) those indicating tentative second stage sampling from fewer than m populations ($|S^*(\tilde{x})| < m$). With this in mind, the set A can be partitioned into the following two subsets

$$A^{=m} = \{(\tilde{x}, y) \in A : \tilde{x}_{[k-1]} - h < \tilde{x}_{[k-m]}\}$$

and

$$A^{<m} = \{(\tilde{x}, y) \in A : \tilde{x}_{[k-1]} - h > \tilde{x}_{[k-m]}\}.$$

Clearly $A = A^{=m} \cup A^{<m}$, up to a set of measure zero, with $A^{=m} \cap A^{<m} = \phi$.

In order to construct the set $C_2(\tilde{x}, y)$ for elements $(\tilde{x}, y) \in A^{=m}$, there are two further subcases which need to be considered, depending on whether $x_k = \tilde{x}_{[k-m]}$ or $x_k > \tilde{x}_{[k-m]}$. In the first case, when $(\tilde{x}, y) \in A^{=m}$ and $x_k = \tilde{x}_{[k-m]}$, then

$$C_2(\tilde{x}, y) = (-\infty, x_k)$$

since if $x_j < \tilde{x}_{[k-m]} = x_k$ then $j \notin S(x) = S^*(\tilde{x})$ while $k \in S(x) = S^*(\tilde{x})$ and $x_k + y_k > x_i + y_i$ for all $i \in S(x) = S^*(\tilde{x})$. Conversely, if $x_j > \tilde{x}_{[k-m]} = x_k$, then $k \notin S(x)$ (since $x_k < x_{[k-m+1]}$) so that correct selection cannot occur.

In the second case, when $(\tilde{x}, y) \in A^{=m}$ and $x_k > \tilde{x}_{[k-m]}$, then

$$C_2(\tilde{x}, y) = \begin{cases} (-\infty, \tilde{x}_{[k-m]}) & z \le \tilde{x}_{[k-m]} \\ (-\infty, z) & \tilde{x}_{[k-m]} < z \le x_k + h \\ (-\infty, x_k + h) & x_k + h < z. \end{cases}$$

To see these relationships, notice that correct selection will occur only if either $j \notin S(x)$ (i.e. $x_j < \tilde{x}_{[k-m]}$), or if $j, k \in S(x)$ (i.e. $\tilde{x}_{[k-m]} < x_j < x_k + h$) and $x_j + y_j < x_k + y_k$ (i.e. $x_j < z$).

Now consider the set $A^{<m}$, and partition it into the $k-1$ disjoint subsets

$$A_p^{<m} = \{(\tilde{x}, y) \in A^{<m} : x_p = \tilde{x}_{[k-1]}\}$$

for $p \neq j$. In order to determine the sets $C_2(\tilde{x}, y)$, the cases $\tilde{x}_{[k-1]} = x_k$ and $\tilde{x}_{[k-1]} > x_k$ need to considered separately. In the first case, when $(\tilde{x}, y) \in A_k^{\leq m}$, the set $C_2(\tilde{x}, y)$ is given by

$$C_2(\tilde{x}, y) = \begin{cases} (-\infty, x_k - h) & z \leq x_k - h \\ (-\infty, z) & x_k - h < z \leq x_k + h \\ (-\infty, x_k + h) & x_k + h < z. \end{cases}$$

This formula holds because correct selection will occur only if either $j \notin S(x)$ (i.e. $x_j < x_k - h$), or if $j, k \in S(x)$ (i.e. $x_k - h < x_j < x_k + h$) and $x_j + y_j < x_k + y_k$ (i.e. $x_j < z$).

If $(\tilde{x}, y) \in A_p^{\leq m}$ for some $p < k$, then

$$C_2(\tilde{x}, y) = \begin{cases} (-\infty, x_p - h) & z \leq x_p - h \\ (-\infty, z) & x_p - h < z \leq x_k + h \\ (-\infty, x_k + h) & x_k + h < z. \end{cases}$$

Again, this is because, in this case, correct selection will occur only if either $j \notin S(x)$ (i.e. $x_j < x_p - h$), or if $j, k \in S(x)$ (i.e. $x_p - h < x_j < x_k + h$) and $x_j + y_j < x_k + y_k$ (i.e. $x_j < z$).

Now, consider the set B, and partition it into $k - 2$ disjoint subsets B_p, $p \neq j, k$, defined by

$$B_p = \{(\tilde{x}, y) \in B : p \in S^\star(\tilde{x}); \ x_p + y_p > x_k + y_k; \\ x_i \geq x_p \Rightarrow x_k + y_k > x_i + y_i, i \neq p, j, k\}.$$

The interpretation of the set B_p is that if $(\tilde{x}, y) \in B_p$, then of all the populations which potentially will qualify for the second stage (i.e. they are in $S^\star(\tilde{x})$) and will beat Π_k at the end of the second stage, the hardest one to "knock out" of $S(x)$ is Π_p because it has the largest first stage total. There will be at least one such population by the definition of the set B, and necessarily $x_k > x_p$, from the definition of B. Clearly, $B = \bigcup_{p \neq j,k} B_p$ and the sets B_p, $p \neq j, k$, are disjoint.

Now, for each set B_p consider the two subcases corresponding to whether $x_p = \tilde{x}_{[k-m]}$ or $x_p > \tilde{x}_{[k-m]}$. Define the subsets

$$B_p^= = \{(\tilde{x}, y) \in B_p : x_p = \tilde{x}_{[k-m]}\}$$

and

$$B_p^> = \{(\tilde{x}, y) \in B_p : x_p > \tilde{x}_{[k-m]}\}.$$

Clearly, $B_p^= \cap B_p^> = \phi$ and $B_p = B_p^= \cup B_p^>$.

Next, the sets $C_2(\tilde{x}, y)$ are identified for $(\tilde{x}, y) \in B_p^=$ and $(\tilde{x}, y) \in B_p^>$. First, for $(\tilde{x}, y) \in B_p^=$,

$$
C_2(\tilde{x}, y) = \begin{cases} \phi & z \leq x_p \\ (x_p, z) & x_p < z \leq x_k + h \\ (x_p, x_k + h) & x_k + h < z. \end{cases}
$$

As mentioned above, the key issue here is that correct selection is possible only when x_j is sufficiently large to ensure that Π_p doesn't qualify for the second stage. Hence, the sets $C_2(\tilde{x}, y)$ are obtained by noting that correct selection occurs only if $p \notin S(x)$ but $k \in S(x)$ (i.e. $x_p < x_j < x_k + h$), in which case $j \in S(x)$ necessarily, and also only if $x_j + y_j < x_k + y_k$ (i.e. $x_j < z$).

Second, for $(\tilde{x}, y) \in B_p^>$,

$$
C_2(\tilde{x}, y) = \begin{cases} \phi & z \leq x_p + h \\ (x_p + h, z) & x_p + h < z \leq x_k + h \\ (x_p + h, x_k + h) & x_k + h < z. \end{cases}
$$

The key issue in the determination of these sets $C_2(\tilde{x}, y)$ is that correct selection is possible here only when $x_j > x_p + h$, as this is now the only way to ensure that $p \notin S(x)$. Specifically, correct selection occurs only if $p \notin S(x)$ but $k \in S(x)$ (i.e. $x_p + h < x_j < x_k + h$), in which case $j \in S(x)$ necessarily, and also only if $x_j + y_j < x_k + y_k$ (i.e. $x_j < z$).

It can now be seen that for some values $(\tilde{x}, y) \in B_p^=$ and $(\tilde{x}, y) \in B_p^>$, the sets $C_2(\tilde{x}, y)$ are of the form (a, b) with $-\infty < a < b < \infty$, and so conditioning on $(\tilde{X}, Y)$ will not produce integrands which are monotone in ν_j^x. Therefore, as indicated above, it is necessary to pair off the sets $B_p^=$ and $B_p^>$ with some subsets of the set A.

For $p \neq k, j$, the set $B_p^=$ will be paired with the set $A_p^{=m} \subset A^{=m}$ defined by

$$
\begin{aligned}
A_p^{=m} &= \{(\tilde{x}, y) \in A^{=m} : x_k = \tilde{x}_{[k-m]}; \ p \in S^\star(\tilde{x}); \\
&\quad i \in S^\star(\tilde{x}) \Rightarrow x_i + y_i < x_p + y_p \text{ for } i \neq j, p, k\}.
\end{aligned}
$$

Thus, for $(\tilde{x}, y) \in A_p^{=m}$, Π_p is the population in $S^\star(\tilde{x})$ which apart from Π_k has the largest two stage sample total, i.e. $x_p + y_p = \max_{i \in S^\star - \{k\}} \{x_i + y_i\}$. However, remember that $x_p + y_p < x_k + y_k$ (since this is a subset of A). Clearly, the sets $A_p^{=m}$, $p \neq j, k$ are disjoint, and

$$
\bigcup_{p \neq j, k} A_p^{=m} = \{(\tilde{x}, y) \in A^{=m} : x_k = \tilde{x}_{[k-m]}\}
$$

Let $\Phi(\cdot)$ and $\phi(\cdot)$ represent the standard normal cumulative distribution function and probability density function, respectively. It then follows from the definitions of the sets $C_2(\tilde{x}, y)$ given above that

$$
\int_{A_p^{\neq m} \cup B_p^{\neq}} P[X_j \in C_2(\tilde{x}, y)] dQ_2 = \int_{A_p^{\neq m}} \Phi(x_k - \nu_j^x) dQ_2
$$

$$
+ \int_{B_p^{\neq} : x_p < z \leq x_k + h} \{\Phi(z - \nu_j^x) - \Phi(x_p - \nu_j^x)\} dQ_2
$$

$$
+ \int_{B_p^{\neq} : x_k + h < z} \{\Phi(x_k + h - \nu_j^x) - \Phi(x_p - \nu_j^x)\} dQ_2,
$$

and it will be shown that this is nonincreasing in ν_j^x. Differentiating with respect to ν_j^x gives

$$
- \int_{A_p^{\neq m}} \phi(x_k - \nu_j^x) dQ_2
$$

$$
+ \int_{B_p^{\neq} : x_p < z \leq x_k + h} \{-\phi(z - \nu_j^x) + \phi(x_p - \nu_j^x)\} dQ_2
$$

$$
+ \int_{B_p^{\neq} : x_k + h < z} \{-\phi(x_k + h - \nu_j^x) + \phi(x_p - \nu_j^x)\} dQ_2
$$

$$
\leq - \int_{A_p^{\neq m} : x_k - x_p - y_p < -y_j} \phi(x_k - \nu_j^x) dQ_2 + \int_{B_p^{\neq} : x_p < z} \phi(x_p - \nu_j^x) dQ_2
$$

$$
= - \lambda^{-k/2} \int_{A_p^{\neq m} : x_k - x_p - y_p < -y_j} \phi(x_k - \nu_j^x) \phi\left(\frac{y_j - \lambda \nu_j^y}{\sqrt{\lambda}}\right) \tag{2.1}
$$

$$
\times \prod_{i \neq j} \left\{ \phi(x_i - \nu_i) \phi\left(\frac{y_i - \lambda \nu_i}{\sqrt{\lambda}}\right) \right\} d\tilde{x} dy
$$

$$
+ \lambda^{-k/2} \int_{B_p^{\neq} : x_p < z} \phi(x_p - \nu_j^x) \phi\left(\frac{y_j - \lambda \nu_j^y}{\sqrt{\lambda}}\right)
$$

$$
\times \prod_{i \neq j} \left\{ \phi(x_i - \nu_i) \phi\left(\frac{y_i - \lambda \nu_i}{\sqrt{\lambda}}\right) \right\} d\tilde{x} dy.
$$

Now consider the linear transformation $T_1 : \mathbb{R}^{2k-1} \to \mathbb{R}^{2k-1}$ which takes an element $(\tilde{x}, y)$ and interchanges x_k with x_p and y_k with y_p. Then if $T_1(\tilde{x}, y) = (\tilde{\eta}, \xi)$, say, notice that

$$(\tilde{\eta}, \xi) \in B_p^{=} \cap \{(\tilde{\eta}, \xi) : \eta_p < \eta_k + \xi_k - \xi_j\}$$

$$\Leftrightarrow \quad \eta_k > \eta_p = \tilde{\eta}_{[k-m]} > \tilde{\eta}_{[k-1]} - h, \ \eta_p + \xi_p > \eta_k + \xi_k, \ \eta_p < \eta_k + \xi_k - \xi_j,$$
$$\eta_i \geq \eta_p \Rightarrow \eta_k + \xi_k > \eta_i + \xi_i \ \text{for} \ i \notin \{j, p, k\}$$

$$\Leftrightarrow \quad x_p > x_k = \tilde{x}_{[k-m]} > \tilde{x}_{[k-1]} - h, \ x_k + y_k > x_p + y_p, \ x_k < x_p + y_p - y_j,$$
$$x_i \geq x_k \Rightarrow x_p + y_p > x_i + y_i \ \text{for} \ i \notin \{j, p, k\}$$

$$\Leftrightarrow \quad (\tilde{x}, y) \in A_p^{=m} \cap \{(\tilde{x}, y) : x_k - x_p - y_p < -y_j\},$$

so that T_1 performs a $1-1$ mapping between the two sets

$$B_p^{=} \cap \{(\tilde{x}, y) : x_p < z\}$$

and

$$A_p^{=m} \cap \{(\tilde{x}, y) : x_k - x_p - y_p < -y_j\}.$$

All of the equivalences above are easy to verify since the values of the order statistics are unchanged by the transformation T_1. The absolute value of the Jacobian of the transformation T_1 is unity, and the transformation can be applied to the second integral term in the right hand side of equation (2.1), so that equation (2.1) can be rewritten

$$- \quad \lambda^{-k/2} \int \phi(x_k - \nu_j^x) \phi\left(\frac{y_j - \lambda \nu_j^y}{\sqrt{\lambda}}\right)$$

$$\times \prod_{i \neq j} \left\{\phi(x_i - \nu_i) \phi\left(\frac{y_i - \lambda \nu_i}{\sqrt{\lambda}}\right)\right\} d\tilde{x} dy$$

$$+ \quad \lambda^{-k/2} \int \phi(x_k - \nu_j^x) \phi(x_p - \nu_k) \phi(x_k - \nu_p) \phi\left(\frac{y_k - \lambda \nu_p}{\sqrt{\lambda}}\right) \phi\left(\frac{y_j - \lambda \nu_j^y}{\sqrt{\lambda}}\right)$$

$$\times \phi\left(\frac{y_p - \lambda \nu_k}{\sqrt{\lambda}}\right) \prod_{i \neq k, p, j} \left\{\phi(x_i - \nu_i) \phi\left(\frac{y_i - \lambda \nu_i}{\sqrt{\lambda}}\right)\right\} d\tilde{x} dy$$

$$= \quad \lambda^{-k/2} \int \left[\phi(x_p - \nu_k) \phi(x_k - \nu_p) \phi\left(\frac{y_k - \lambda \nu_p}{\sqrt{\lambda}}\right) \phi\left(\frac{y_p - \lambda \nu_k}{\sqrt{\lambda}}\right)\right.$$

$$\left. - \phi(x_k - \nu_k) \phi(x_p - \nu_p) \phi\left(\frac{y_k - \lambda \nu_k}{\sqrt{\lambda}}\right) \phi\left(\frac{y_p - \lambda \nu_p}{\sqrt{\lambda}}\right)\right]$$

$$\times \phi(x_k - \nu_j^x) \phi\left(\frac{y_j - \lambda \nu_j^y}{\sqrt{\lambda}}\right)$$

$$\times \prod_{i \neq k,p,j} \left\{ \phi(x_i - \nu_i)\phi\left(\frac{y_i - \lambda\nu_i}{\sqrt{\lambda}}\right) \right\} d\tilde{x}dy$$

$$\leq 0.$$

Here all the integrals are over the set

$$A_p^{=m} : x_k - x_p - y_p < -y_j$$

and the last inequality holds since for every $(\tilde{x}, y)$ in the integration region

$$\phi(x_k - \nu_k)\phi(x_p - \nu_p)\phi\left(\frac{y_k - \lambda\nu_k}{\sqrt{\lambda}}\right)\phi\left(\frac{y_p - \lambda\nu_p}{\sqrt{\lambda}}\right)$$

$$\geq \phi(x_p - \nu_k)\phi(x_k - \nu_p)\phi\left(\frac{y_p - \lambda\nu_k}{\sqrt{\lambda}}\right)\phi\left(\frac{y_k - \lambda\nu_p}{\sqrt{\lambda}}\right),$$

which is equivalent to

$$x_k\nu_k + x_p\nu_p + y_k\nu_k + y_p\nu_p \geq x_p\nu_k + x_k\nu_p + y_p\nu_k + y_k\nu_p$$

or

$$(\nu_k - \nu_p)(x_k + y_k - x_p - y_p) \geq 0.$$

This is true since $\nu_k \geq \nu_p$, and $(\tilde{x}, y) \in A_p^{=m} \Rightarrow x_k + y_k > x_p + y_p$.

Finally, the sets $B_p^> = \{(\tilde{x}, y) \in B_p : x_p > \tilde{x}_{[k-m]}\}$, $p \neq j, k$, are paired with the sets $A_p^{\leq m} \cap \{(\tilde{x}, y) : z < x_p - h\}$. It follows from the definitions of the sets $C_2(\tilde{x}, y)$ given above that

$$\int_{(A_p^{\leq m} \cap \{(\tilde{x},y): z<x_p-h\}) \cup B_p^>} P[X_j \in C_2(\tilde{x}, y)]dQ_2$$

$$= \int_{A_p^{\leq m} : z<x_p-h} \Phi(x_p - h - \nu_j^x)dQ_2$$

$$+ \int_{B_p^> : x_p+h<z\leq x_k+h} \{\Phi(z - \nu_j^x) - \Phi(x_p + h - \nu_j^x)\}dQ_2$$

$$+ \int_{B_p^> : x_k+h<z} \{\Phi(x_k + h - \nu_j^x) - \Phi(x_p + h - \nu_j^x)\}dQ_2.$$

Differentiating with respect to ν_j^x gives

$$- \int_{A_p^{\leq m} : z<x_p-h} \phi(x_p - h - \nu_j^x)dQ_2$$

$$+ \int_{B_p^{\geq}: x_p+h<z\leq x_k+h} \{-\phi(z-\nu_j^x)+\phi(x_p+h-\nu_j^x)\}dQ_2$$

$$+ \int_{B_p^{\geq}: x_k+h<z} \{-\phi(x_k+h-\nu_j^x)+\phi(x_p+h-\nu_j^x)\}dQ_2$$

$$\leq - \int_{A_p^{\leq m}: z<x_p-h} \phi(x_p-h-\nu_j^x)dQ_2 + \int_{B_p^{\geq}: x_p+h<z} \phi(x_p+h-\nu_j^x)dQ_2$$

$$= - \lambda^{-k/2} \int_{A_p^{\leq m}: z<x_p-h} \phi(x_p-h-\nu_j^x)\phi\left(\frac{y_j-\lambda\nu_j^y}{\sqrt{\lambda}}\right) \qquad (2.2)$$
$$\times \prod_{i\neq j}\left\{\phi(x_i-\nu_i)\phi\left(\frac{y_i-\lambda\nu_i}{\sqrt{\lambda}}\right)\right\} d\tilde{x}dy$$

$$+ \lambda^{-k/2} \int_{B_p^{\geq}: x_p+h<z} \phi(x_p+h-\nu_j^x)\phi\left(\frac{y_j-\lambda\nu_j^y}{\sqrt{\lambda}}\right)$$
$$\times \prod_{i\neq j}\left\{\phi(x_i-\nu_i)\phi\left(\frac{y_i-\lambda\nu_i}{\sqrt{\lambda}}\right)\right\} d\tilde{x}dy.$$

Now consider the linear transformation $T_2 : \mathbb{R}^{2k-1} \to \mathbb{R}^{2k-1}$ which takes an element $(\tilde{x}, y)$ and maps x_p to $x_p - h$, interchanges y_j and y_p, and leaves all the other elements unchanged. Then if $T_2(\tilde{x}, y) = (\tilde{\eta}, \xi)$, say, notice that

$$(\tilde{\eta}, \xi) \in B_p^{\geq} \cap \{(\tilde{\eta}, \xi) : \eta_p + h < \eta_k + \xi_k - \xi_j\}$$
$$\Leftrightarrow \quad \eta_p > \tilde{\eta}_{[k-m]}, \ \eta_p > \tilde{\eta}_{[k-1]} - h, \ \eta_k > \eta_p, \ \eta_p + \xi_p > \eta_k + \xi_k,$$
$$\eta_p + h < \eta_k + \xi_k - \xi_j, \ \eta_i \geq \eta_p \Rightarrow \eta_i + \xi_i < \eta_k + \xi_k \text{ for } i \notin \{j, p, k\}$$
$$\Leftrightarrow \quad x_p - h > \tilde{x}_{[k-m]}, \ x_p = \tilde{x}_{[k-1]}, \ x_k > x_p - h, \ x_p - h + y_j > x_k + y_k,$$
$$x_p < x_k + y_k - y_p, \ x_i \geq x_p - h \Rightarrow x_i + y_i < x_k + y_k \text{ for } i \notin \{j, p, k\}$$
$$\Leftrightarrow \quad (\tilde{x}, y) \in A_p^{\leq m} \cap \{(\tilde{x}, y) : x_k + y_k - y_j < x_p - h\},$$

so that the transformation T_2 is a $1-1$ mapping from the set $A_p^{\leq m} \cap \{(\tilde{x}, y) : z < x_p - h\}$ to the set $B_p^{\geq} \cap \{(\tilde{x}, y) : x_p + h < z\}$. Note that in verifying the equivalence between the second and third set of statements, the values of the order statistics may change, in contrast to the transformation T_1. The absolute value of the Jacobian of the transformation T_2 is unity, and the inverse transformation T^{-1} can be applied to the second integral term in the right hand side of equation (2.2), so that equation (2.2) can be rewritten

$$- \quad \lambda^{-k/2} \int \phi(x_p - h - \nu_j^x) \phi\left(\frac{y_j - \lambda\nu_j^y}{\sqrt{\lambda}}\right)$$

$$\times \prod_{i \neq j} \left\{ \phi(x_i - \nu_i) \phi\left(\frac{y_i - \lambda\nu_i}{\sqrt{\lambda}}\right) \right\} d\tilde{x} dy$$

$$+ \quad \lambda^{-k/2} \int \phi(x_p - \nu_j^x) \phi(x_p - h - \nu_p) \phi\left(\frac{y_j - \lambda\nu_p}{\sqrt{\lambda}}\right) \phi\left(\frac{y_p - \lambda\nu_j^y}{\sqrt{\lambda}}\right)$$

$$\times \prod_{i \neq j, p} \left\{ \phi(x_i - \nu_i) \phi\left(\frac{y_i - \lambda\nu_i}{\sqrt{\lambda}}\right) \right\} d\tilde{x} dy$$

$$= \quad \lambda^{-k/2} \int \left[\phi(x_p - \nu_j^x) \phi(x_p - h - \nu_p) \phi\left(\frac{y_j - \lambda\nu_p}{\sqrt{\lambda}}\right) \phi\left(\frac{y_p - \lambda\nu_j^y}{\sqrt{\lambda}}\right) \right.$$

$$\left. - \phi(x_p - h - \nu_j^x) \phi(x_p - \nu_p) \phi\left(\frac{y_j - \lambda\nu_j^y}{\sqrt{\lambda}}\right) \phi\left(\frac{y_p - \lambda\nu_p}{\sqrt{\lambda}}\right) \right]$$

$$\times \prod_{i \neq j, p} \left\{ \phi(x_i - \nu_i) \phi\left(\frac{y_i - \lambda\nu_i}{\sqrt{\lambda}}\right) \right\} d\tilde{x} dy$$

$$\leq \quad 0.$$

The integrals are all over

$$A_p^{\leq m} : z < x_p - h$$

and the last inequality holds since for every $(\tilde{x}, y)$ in the integration region

$$\phi(x_p - h - \nu_j^x) \phi(x_p - \nu_p) \phi\left(\frac{y_j - \lambda\nu_j^y}{\sqrt{\lambda}}\right) \phi\left(\frac{y_p - \lambda\nu_p}{\sqrt{\lambda}}\right)$$

$$\geq \phi(x_p - \nu_j^x) \phi(x_p - h - \nu_p) \phi\left(\frac{y_j - \lambda\nu_p}{\sqrt{\lambda}}\right) \phi\left(\frac{y_p - \lambda\nu_j^y}{\sqrt{\lambda}}\right),$$

which is equivalent to

$$(x_p - h)\nu_j^x + x_p\nu_p + y_j\nu_j^y + y_p\nu_p \geq x_p\nu_j^x + (x_p - h)\nu_p + y_j\nu_p + y_p\nu_j^y$$

or

$$h(\nu_j^x - \nu_j^y) \leq (y_j - y_p - h)(\nu_j^y - \nu_p).$$

This is true since $\nu_j^y \geq \nu_p$, $\nu_j^y \geq \nu_j^x$, and within the integration region $x_k + y_k > x_p + y_p$ and $z = x_k + y_k - y_j < x_p - h$, so that $y_j - y_p - h > 0$.

This completes the proof of Lemma 2 and hence also of the Theorem. ∎

3 Summary and Discussion

This final section discusses the establishment of the least favorable configuration of the population means, the use of the procedure $R(m, h, n_1, n_2)$ in the presence of block effects, and various generalizations.

For the special case $m = 2$, the least favorable configuration can be established by a much simpler proof than that given in Section 2 for a general value of m. Furthermore, the result will hold not only for normal densities as discussed in this paper, but also for more general sampling distributions where the Stage 1 observations X_{ij} are assumed to be distributed with density function $f_1(x - \mu_i)$, and the Stage 2 observations Y_{ij} are assumed to be distributed with density function $f_2(x - \mu_i)$. All observations are assumed to be independently distributed, but otherwise there are no restrictions on the density functions $f_1(\cdot)$ and $f_2(\cdot)$.

The alternate proof for the case $m = 2$ works as follows. If μ_k is the largest location parameter then, as in Section 2, μ_j is allowed to be two different values μ_j^x and μ_j^y in Stage 1 and Stage 2, respectively, for $1 \leq j \leq k - 1$. Then Lemma 1 in Section 2 can be used to show that the probability of correct selection is nonincreasing as $\mu_1^y, \ldots, \mu_{k-1}^y$ are each increased to $\mu_k - \delta^\star$. Notice that after this has been done, the random variables $Y_1, \ldots, Y_{k-1}$ are identically distributed. It is then possible to show that the function

$$W(x_1, \ldots, x_{k-1}) = P(\text{CS} \mid X_1 = x_1, \ldots, X_{k-1} = x_{k-1})$$

is nonincreasing in each of its arguments x_i, $1 \leq i \leq k - 1$, and this is sufficient to establish that the probability of correct selection is nonincreasing as $\mu_1^x, \ldots, \mu_{k-1}^x$ are each increased to $\mu_k - \delta^\star$, giving the required result. The fact that $m = 2$ is critical in showing that $W(x_1, \ldots, x_{k-1})$ is nonincreasing in each of its arguments and, in general, this will not be true for $m > 2$. For example, if $m = k = 3$, then $W(h/2 - \epsilon, -h/2) < W(h/2 + \epsilon, -h/2)$ for small $\epsilon > 0$ (since on the left hand side it is possible that all three populations will qualify for Stage 2, yet on the right hand side the second population cannot qualify for Stage 2).

Another issue to be considered is that of blocking effects arising during sampling. The probability properties of the procedure $R(m, h, n_1, n_2)$ are not affected if there are block effects within vector observations of the data. Formally, the distributional assumptions can be generalized to

$$X_{ij} \sim N(\mu_i, \sigma^2) + \beta_j$$

and

$$Y_{ij} \sim N(\mu_i, \sigma^2) + \Omega_j$$

where the β_j and Ω_j are differential block effects. Since the procedure $R(m, h, n_1, n_2)$ involves only comparisons between the X_i and the Z_i, the inclusion of the block effects will yield a joint distribution of the treatment effect differences identical to that in Section 2. There are other experimental design structures for which this property also holds. For example, in a balanced incomplete block design, if the best linear unbiased estimators of the treatment effects are used in place of the sample totals in $R(m, h, n_1, n_2)$, then the infimum of the probability of correct selection is attained at the slippage configuration. However, the calculation of the sample sizes will depend upon the particular design in question.

An important extension is how to deal with a common unknown variance σ^2. In Section 1, the procedure $R(m, h, n_1, n_2)$ assumes a common known variance σ^2, and if the variance is completely unknown, then it is impossible to satisfy the design requirement as stated. This is because, for fixed values of $m, h, n_1, n_2, \delta^\star$ and $\mu = (\mu_1, \ldots, \mu_k)$, the probability of correct selection approaches a limiting value of $1/k$ as the variance increases to ∞. One possible remedy is to modify the procedure $R(m, h, n_1, n_2)$ to use hS in place of h where S^2 is an appropriate independent sum of squares estimate of σ^2. In this case, the probability of correct selection will depend on $\mu = (\mu_1, \ldots, \mu_k)$ and σ^2 only through the quantities $\mu_1/\sigma, \ldots, \mu_k/\sigma$. A valid design requirement would then be to guarantee that the probability of correct selection attains at least some specified value whenever $\mu_{[k]} - \mu_{[k-1]} \geq \delta^\star \sigma$. Notice that here the width of the indifference zone must be defined as a multiple of the standard deviation. The least favorable configuration of population means for this modified procedure will occur at the slippage configuration. The proof follows from the Theorem in Section 2 after conditioning on the value of S.

It is also straightforward to generalize the procedure $R(m, h, n_1, n_2)$ to compare k treatments with a standard or an (unknown) control population. In practice, for a treatment to enter the second stage, a third criterion must now be met, namely, there must be sampling evidence that the treatment is possibly superior to the control. If no treatments demonstrate this quality, the sampling is concluded at the end of the first stage and the standard or control is selected. Otherwise a second stage of sampling is performed at the conclusion of which either the control or one of the treatments is selected as best. If the design requirement is similar to that of Bechhofer and Turnbull (1978), the infimum of the probability of correct selection for such a procedure would have to be analysed by arguments similar to those in Section 2.

Finally, consider the more general design requirement that the mean of the selected population be within a specified amount $\delta^\star$ of the largest mean, as discussed by Fabian (1962). If the means are in the preference zone $\Omega(\delta^\star)$, then this goal is the same as that of Section 1, but otherwise represents a strengthening of it. We conjecture that the infimum of the probability that the procedure selects such a near-optimal treatment (over the entire parameter space) is identical to the infimum derived in Section 2.

References

Alam, K. (1970). A two-sample procedure for selecting the population with the largest mean from k normal populations. *Ann. Inst. Statist. Math.* **22** 127 - 136.

Bechhofer, R.E. (1954). A single-sample multiple decision procedure for ranking means of normal populations with known variances. *Ann. Math. Statist.* **25** 16 - 39.

Bechhofer, R.E. and Turnbull, B. (1978). Two $(k + 1)$-decision selection procedures for comparing k normal means with a specified standard. *Jour. Amer. Statist. Assoc.* **73** 385 - 392.

Cohen, D.S. (1959). A two-sample decision procedure for ranking means of normal populations with a common known variance. Unpublished M.S. Thesis, School of Operations Research and Industrial Engineering, Cornell University, Ithaca, NY.

Fabian, V. (1962). On multiple decision methods for ranking population means. *Ann. Math. Statist.* **33** 248 - 254.

Fairweather, W.R. (1968). Some extensions of Somerville's procedure for ranking means of normal populations. *Biometrika* **55** 411-418.

Gupta, S. and Han, S. (1987). An eliminating type two-stage procedure for selecting the population with the largest mean from k logistic populations. Technical Report 87-39. Department of Statistics, Purdue University, West Lafayette, IN.

Han, S. (1987). Contributions to selection and ranking theory with special reference to logistic populations. Technical Report 87-38. Department of Statistics, Purdue University, West Lafayette, IN.

Santner, T.J. and Behaxeteguy, M. (1992). A two-stage procedure for selecting the largest normal mean whose first stage selects a bounded, random number of populations. *Jour. Statist. Plan. Infer.*, To appear.

Sehr, J. (1988). On a conjecture concerning the least favorable configuration of a two-stage selection procedure. *Commun. Statist.- Theory and Methods* **A17**(10), 3221 - 3233.

Tamhane, A.C. and Bechhofer, R.E. (1977). A two-stage minimax procedure with screening for selecting the largest normal mean. *Commun. Statist. - Theory and Methods* **A6** 1003 - 1033.

Tamhane, A.C. and Bechhofer, R.E. (1979). A two-stage minimax procedure with screening for selecting the largest normal mean (ii): an improved PCS lower bound and associated tables. *Commun. Statist. -Theory and Methods* **A8**(4) 337 - 358.

Chapter 16

A Bayesian Approach to Comparing Treatments with a Control

AJIT C. TAMHANE   Departments of Statistics and Industrial Engineering and Management Sciences, Northwestern University, Evanston, Illinois

G. V. S. GOPAL   Z. S. Associates, Evanston, Illinois

Abstract   We consider the problem of comparing treatments with a control in terms of their unknown means under the normal theory setting when conjugate prior distributions are available on all the means. First, some results concerning the optimal allocation of observations are given. Next, Bayesian subset selection rules are studied for certain selection goals with a particular emphasis on the goal of selecting a subset that includes all treatments having means at least as large as that of the control. Constants required for implementing these rules in order that they guarantee a specified requirement on the probability of a correct selection are tabulated. Finally, a related problem is considered under a decision theoretic formulation. The variances are either assumed to be known and possibly unequal or assumed to be all equal but their common value being unknown. In the latter case, a conjugate prior distribution is assumed for the common variance. Bayes decision rules are derived under an additive overall loss function with either constant or linear loss functions for the component losses. These rules are the analogs of the so-called k-ratio t-tests developed by Duncan (1965) and Waller and Duncan (1969) for the problem of pairwise comparisons. A table of constants for implementing these decision rules is given.

1   Introduction

The problem of comparing several treatments with a control was first studied in a systematic way by Dunnett (1955). Since then many authors have considered various aspects of this problem; for references see Hochberg and Tamhane (1987). However, one aspect that does not seem to have received much attention is that of the use of any prior information that might be available on the treatments and the control. For example, the same control might have been used in similar experiments before or the control might be a standard treatment which has been in use for quite some time, and hence much might be known about it. For another example, the same treatments and the control might have been studied in previous experiments as part of an ongoing investigation, and therefore prior information might be available on all of them.

It seems clear that such prior information ought to be used in designing experiments and drawing conclusions from data. For instance, it is well-known that, to compare $k \geq 2$ treatments with a control in a completely randomized design, one should allocate approximately $\sqrt{k}$ times as many observations on the control as are allocated on each one of the treatments (assuming, of course, the usual homoscedastic linear model). However, if there is prior information available on the control then it may be necessary to take fewer observations on it, and perhaps none at all. In this paper we study a Bayesian approach to some of the design and analysis problems that arise when comparing several treatments with a control.

A brief outline of the paper is as follows: Section 2 gives the model and the notation used. Section 3 considers the design problem. The next two sections consider two different formulations of the problem of selecting treatments better than the control: Section 4 uses the subset selection formulation of Gupta and Sobel (1958) while Section 5 uses the decision theoretic formulation of Duncan (1965). Section 6 gives some concluding remarks.

2   Preliminaries

Suppose we wish to compare $k \geq 2$ treatments, indexed $1, 2, \ldots, k$, with a control, indexed 0. We assume that the observations y_{ij} from treatment i form a random sample from a $N(\mu_i, \sigma_i^2)$ distribution ($0 \leq i \leq k, 1 \leq j \leq n_i$). Thus the sample means $\bar{y}_i$ are independently distributed as $N(\mu_i, \sigma_i^2/n_i)$ random variables (r.v.'s). The parameters of interest are $\mu_i - \mu_0$ $(1 \leq i \leq k)$. Denote $\boldsymbol{\mu} = (\mu_0, \mu_1, \dots, \mu_k)$ and $\boldsymbol{\sigma}^2 = (\sigma_0^2, \sigma_1^2, \dots, \sigma_k^2)$.

The focus of the present paper is on the use of prior information on $\boldsymbol{\mu}$. We assume that this prior information can be specified in terms of independent conjugate prior distributions $N(\theta_i, \sigma_i^2/m_i)$ on the μ_i where the θ_i and m_i are known constants; note that the m_i may be regarded as prior sample sizes $(0 \leq i \leq k)$. Then conditional on the observed sample means $\bar{y}_i$, the posterior distributions of the μ_i are independent normal with posterior means

$$\hat{\mu}_i = p_i \theta_i + q_i \bar{y}_i$$

and posterior variances $\sigma_i^2/(m_i + n_i)$ where $p_i = m_i/(m_i + n_i)$ and $q_i = n_i/(m_i + n_i)$ $(0 \leq i \leq k)$. The joint posterior distribution of the contrasts $\mu_i - \mu_0$ is k-variate normal with mean vector $(\hat{\mu}_1 - \hat{\mu}_0, \dots, \hat{\mu}_k - \hat{\mu}_0)$ and covariance matrix $V = (v_{ij})$, whose elements are

$$v_{ij} = \begin{cases} \sigma_i^2/(m_i + n_i) + \sigma_0^2/(m_0 + n_0) & \text{if } i = j \\ \sigma_0^2/(m_0 + n_0) & \text{if } i \neq j. \end{cases}$$

Different assumptions are made in each section about $\boldsymbol{\sigma}^2$.

3   Some Design Problems

Suppose that we have available a total of N observations. The question is how to allocate them among the treatments and the control, i.e., to determine the sample size vector $\boldsymbol{n} = (n_0, n_1, \dots, n_k)$ subject to $\sum_{i=0}^{k} n_i = N$, in order to optimize a suitable criterion. In the absence of prior information, this problem has been considered by many authors (see, e.g., Hochberg and Tamhane 1987, Ch. 6, Section 1.2). The criterion often used is: maximize the joint coverage probability of simultaneous confidence intervals on $\mu_i - \mu_0$ $(1 \leq i \leq k)$ subject to specified upper bounds on the widths of the intervals. In general, this coverage probability depends both on the $\text{var}(\bar{y}_i - \bar{y}_0) = \sigma_i^2/n_i + \sigma_0^2/n_0$ and on the $\text{cov}(\bar{y}_i - \bar{y}_0, \bar{y}_j - \bar{y}_0) = \sigma_0^2/n_0$. However, it can be shown that asymptotically (as $n_i \to \infty$ $\forall$ i), this coverage probability depends only on the former. Thus in the special case $\sigma_0^2 = \sigma_1^2 = \dots = \sigma_k^2 = \sigma^2$ (say), the asymptotically optimal allocation is obtained by minimizing $\text{var}(\bar{y}_i - \bar{y}_0) = \sigma^2(1/n + 1/n_0)$ subject to $n_0 + kn = N$ (where we have assumed that $n_1 = \dots = n_k = n$ (say)), which yields (ignoring the integer restrictions on n_0 and n) the familiar square root allocation rule: $n_0/n = \sqrt{k}$.

In the case of prior information, we shall consider two criteria for minimization which are functions of the posterior covariance matrix V: (i) A-optimality criterion (Kiefer 1958)); (ii) MV-optimality criterion (Jacroux

1987). These criteria only take into account the posterior variances and not the covariances, but may be justified on asymptotic grounds as discussed above; besides they have the advantage of being mathematically tractable.

To simplify the calculations, we will assume that N is sufficiently large so that

$$\beta_i = \frac{m_i}{N} \quad \text{and} \quad \gamma_i = \frac{n_i}{N} \quad (0 \le i \le k)$$

may be regarded as nonnegative continuous variables with $\sum_{i=0}^{k} \gamma_i = 1$, i.e., the "allocation vector" $\gamma = (\gamma_0, \gamma_1, \ldots, \gamma_k)$ lies in the k-dimensional simplex Γ. Also let $\rho_0 = \sqrt{k}$ and $\rho_i = \sigma_i/\sigma_0$ $(1 \le i \le k)$. We first consider the

   (i) A-optimality Criterion:

$$\text{tr}(V) = \frac{k\sigma_0^2}{m_0 + n_0} + \sum_{i=1}^{k} \frac{\sigma_i^2}{m_i + n_i}.$$

Using the continuous approximation introduced above, the A-optimality problem can be stated as

$$\min_{\gamma \in \Gamma} \left\{ \sum_{i=0}^{k} \frac{\rho_i^2}{\beta_i + \gamma_i} \right\}.$$

The solution to this constrained nonlinear optimization problem is given in the following theorem, whose proof follows from the Kuhn-Tucker conditions (McCormick 1983, pp. 218–219).

Theorem 3.1

   Denote by $\mathcal{S} = \{0, 1, \ldots, k\}$. The A-optimal allocation is given by $\hat{\gamma} = (\hat{\gamma}_0, \hat{\gamma}_1, \ldots, \hat{\gamma}_k)$ where

$$\hat{\gamma}_i = \frac{\rho_i(1 + \sum_{j \in \mathcal{S}} \beta_j)}{\sum_{j \in \mathcal{S}} \rho_j} - \beta_i, \quad i \in \mathcal{S} \tag{3.1}$$

if $\hat{\gamma}_i \ge 0 \ \forall \ i \in \mathcal{S}$. If not, set $\hat{\gamma}_j = 0$ where j is such that $\hat{\gamma}_j = \min_{i \in \mathcal{S}} \hat{\gamma}_i < 0$. Also set $\mathcal{S} = \mathcal{S} - \{j\}$, and recalculate $\hat{\gamma}_i$ for $i \in \mathcal{S}$ using (3.1) until $\hat{\gamma}_i \ge 0 \ \forall \ i \in \mathcal{S}$. ∎

   Note that the allocation (3.1) satisfies

$$\frac{m_i + n_i}{m_j + n_j} = \frac{\rho_i}{\rho_j} \quad \forall \ i, j \in \mathcal{S}. \tag{3.2}$$

Furthermore, if a particular m_i is too large then the corresponding n_i is set equal to zero. Thus less observations (compared to what is prescribed by the square root allocation rule under no prior information) are taken on treatments having more prior information and vice versa.

We next consider the

(ii) MV-optimality Criterion:

$$\max_{1 \leq i \leq k} v_{ii} = \max_{1 \leq i \leq k} \left[\frac{\sigma_0^2}{m_0 + n_0} + \frac{\sigma_i^2}{m_i + n_i} \right].$$

Using the continuous approximation introduced above, the MV-optimality criterion can be stated as

$$\min_{\gamma \in \Gamma} \max_{1 \leq i \leq k} \left[\frac{1}{\beta_0 + \gamma_0} + \frac{\rho_i^2}{\beta_i + \gamma_i} \right] = \min_{\gamma_0} \left\{ \frac{1}{\beta_0 + \gamma_0} + \min_{\gamma \in \Gamma(\gamma_0)} \max_{1 \leq i \leq k} \left[\frac{\rho_i^2}{\beta_i + \gamma_i} \right] \right\}$$

where, for fixed γ_0, $\Gamma(\gamma_0) = \{\gamma \in \Gamma : \sum_{i=1}^k \gamma_i = 1 - \gamma_0\}$. The solution to this constrained nonlinear optimization problem is given in the following theorem, whose proof also follows from the Kuhn-Tucker conditions.

Theorem 3.2

Denote by $S = \{1, 2, \ldots, k\}$. The MV-optimal allocation is given by $\hat{\gamma} = (\hat{\gamma}_0, \hat{\gamma}_1, \ldots, \hat{\gamma}_k)$ where

$$\hat{\gamma}_0 = \frac{1 + \beta_0 + \sum_{j \in S} \beta_j}{1 + \sqrt{\sum_{j \in S} \rho_j^2}} - \beta_0, \tag{3.3}$$

and

$$\hat{\gamma}_i = \frac{\rho_i^2 (1 + \beta_0 + \sum_{j \in S} \beta_j)}{\sqrt{\sum_{j \in S} \rho_j^2} \left(1 + \sqrt{\sum_{j \in S} \rho_j^2} \right)} - \beta_i, \quad i \in S \tag{3.4}$$

if $\hat{\gamma}_i \geq 0$ for $i = 0, i \in S$. If $\hat{\gamma}_0 \geq 0$ but $\hat{\gamma}_i < 0$ for some $i \in S$ then set $\hat{\gamma}_j = 0$ for $j \in S$ such that $\hat{\gamma}_j = \min_{i \in S} \hat{\gamma}_i < 0$. Also set $S = S - \{j\}$, and recalculate $\hat{\gamma}_0, \hat{\gamma}_i$ for $i \in S$ using (3.3) and (3.4) until $\hat{\gamma}_0, \hat{\gamma}_i \geq 0 \ \forall \ i \in S$. If at any step $\hat{\gamma}_0 < 0$ then set $\hat{\gamma}_0 = 0$ and calculate

$$\hat{\gamma}_i = \frac{\rho_i^2 (1 + \sum_{j \in S} \beta_j)}{\sum_{j \in S} \rho_j^2} - \beta_i, \quad i \in S \tag{3.5}$$

if $\hat{\gamma}_i \geq 0 \ \forall \ i \in S$. If not, set $\hat{\gamma}_j = 0$ for $j \in S$ such that $\hat{\gamma}_j = \min_{i \in S} \hat{\gamma}_i < 0$. Also set $S = S - \{j\}$, and recalculate $\hat{\gamma}_i$ for $i \in S$ using (3.5) until $\hat{\gamma}_i \geq 0 \ \forall \ i \in S$. ∎

Note that if all the $\hat{\gamma}_i > 0$ then the resulting allocation satisfies

$$\frac{m_0 + n_0}{m_i + n_i} = \frac{\sqrt{\sum_{i=1}^k \rho_i^2}}{\rho_i^2}.$$

Comparing this with (3.2), we see that both the criteria yield the square root allocation (in terms of the "effective" sample sizes $m_i + n_i$) if all the σ_i^2 are equal. Otherwise they yield different allocations. In fact, it can be shown that

$$(\hat{\gamma}_0)_{\text{MV}} \le (\hat{\gamma}_0)_{\text{A}}.$$

Thus the MV-optimality criterion allocates less observations on the control compared to the A-optimality criterion, and allocates more observations on the treatment having the largest variance.

4 Some Subset Selection Problems

4.1 Selecting a Subset that Includes All "Good" Treatments

(a) The Classical Non-Bayesian Approach

Let $K = \{1, 2, \ldots, k\}$ denote the set of all treatments. This set may be divided into two subsets:

$G =$ Subset of "good" treatments relative to control $= \{i \in K : \mu_i \ge \mu_0\}$,

and

$B =$ Subset of "bad" treatments relative to control $= \{i \in K : \mu_i < \mu_0\}$.

Gupta and Sobel (1958) (abbreviated as GS hereafter) considered the following goal.

Goal I: Select a subset $S \subseteq K$ that includes all "good" treatments, i.e., $G \subseteq S$.

Let Correct Selection (CS) denote the event of selecting S that meets the specified goal. Throughout this section we assume that $\sigma_0^2 = \sigma_1^2 = \cdots = \sigma_k^2 = \sigma^2$ (say), and σ^2 is unknown but an unbiased estimate s^2 of σ^2 is available which is distributed as a $\sigma^2 \chi_\nu^2 / \nu$ r.v. independent of the $\bar{y}_i$. GS postulated the following *probability requirement*:

$$P(CS|\mu, \sigma^2) \ge P^* \ \forall \ \mu \text{ and } \sigma^2 > 0 \tag{4.1}$$

where $P^* > (1/2)^k$ is specified.

GS proposed the following rule to select a subset S which meets the probability requirement (4.1):

$$S = \left\{ i \in K : \bar{y}_i - \bar{y}_0 \ge -gs\sqrt{1/n_i + 1/n_0} \right\} \tag{4.2}$$

where g is the upper $1 - P^*$ equicoordinate point of a k-variate central t-distribution with ν d.f. and associated correlation matrix (ρ_{ij}) where

$$\rho_{ij} = \sqrt{\frac{n_i}{n_i + n_0}} \sqrt{\frac{n_j}{n_j + n_0}}.$$

For this product correlation structure, the critical constant g can be computed without much difficulty. If the n_i on the treatments are equal then we have a common correlation, in which case the values of g have been tabulated for a large number of cases by Bechhofer and Dunnett (1988).

(b) A Bayesian Approach

(i) Probability of a Correct Selection

The critical constant g used in the GS-rule is determined under the least favorable configuration (LFC), i.e., μ that minimizes $P(CS|\mu, \sigma^2)$ which is the configuration $\mu_0 = \mu_1 = \ldots = \mu_k$. This makes the GS-rule very conservative resulting in rather large values for $E(|S|)$. This conservatism can be reduced by using the prior information on μ in two ways: (i) By incorporating it in the probability requirement (and in the evaluation of the $P(CS)$) so that the determination of the critical constant based on the LFC of the μ_i is avoided, and (ii) by incorporating it in the subset selection rule. Chen and Pickett (1984) adopted a Bayesian approach to this subset selection problem in which they assumed a uniform prior distribution on the cardinality of set G (denoted by $|G|$) which they incorporated in the probability requirement (and in the evaluation of the $P(CS)$). However, they used the GS-rule (4.2) (with a different critical constant g), which does not use any prior information in its estimates of the μ_i. Here we adopt an approach that is more akin to Dunnett's (1960) Bayesian approach to the indifference zone selection problem of Bechhofer (1954) in that we assume the knowledge of the prior distribution on μ, and use it in both the ways mentioned above.

In analogy with (4.1), we use the following modified probability requirement:

$$P(CS|\sigma^2) = \int P(CS|\mu, \sigma^2) f(\mu) d\mu \geq P^* \ \forall \ \sigma^2 > 0 \qquad (4.3)$$

where $f(\mu)$ denotes the prior distribution on μ. Note that this requirement is less strict than (4.1) because here only the average $P(CS)$ (averaged over all μ) is required to be at least P^*.

Similarly in analogy with the GS-rule (4.2), we propose the following rule:

$$S = \left\{ i \in K : \hat{\mu}_i - \hat{\mu}_0 \geq -hs\sqrt{1/N_i + 1/N_0} \right\} \qquad (4.4)$$

where $h > 0$ is a critical constant to be determined, and $N_i = m_i + n_i$ $(0 \leq i \leq k)$. In the following theorem we give an equation for h derived from an expression for the $P(CS)$ of rule (4.4); note that this $P(CS)$ is independent of σ^2 and hence σ^2 is suppressed from the notation.

Theorem 4.1

For simplicity assume the following symmetry conditions with respect to the treatments: $m_i = m, n_i = n$ and $\theta_i = \theta$, which implies that

$$\gamma_i = \gamma, \Delta_i = \frac{\theta_i - \theta_0}{\sigma} = \Delta, p_i = p, q_i = q = 1 - p \ (1 \leq i \leq k). \qquad (4.5)$$

Then the critical constant h required in the subset selection rule (4.4) to guarantee the probability requirement (4.3) for Goal I is the unique (positive) solution to the equation

$$\int_0^\infty \int_{-\infty}^\infty \int_{-\infty}^\infty [A(u_0) + B(u_0, z, ht)]^k \, d\Phi(z) d\Phi(u_0) dF_\nu(t) = P^*. \qquad (4.6)$$

Here $\Phi(\cdot)$ and $F_\nu(\cdot)$ denote the c.d.f.'s of $N(0,1)$ and $\sqrt{\chi_\nu^2/\nu}$ r.v.'s, respectively,

$$A(u_0) = \Phi\left(u_0\sqrt{m/m_0} - \Delta\sqrt{m}\right) \qquad (4.7)$$

and

$$B(u_0, z, ht) = \int_{u_0\sqrt{m/m_0}-\Delta\sqrt{m}}^\infty \Phi\left(\frac{au - a_0u_0 + b\Delta + cht + \lambda z}{\sqrt{1-\lambda^2}}\right) d\Phi(u). \qquad (4.8)$$

In the above

$$\left.\begin{array}{l}
a = [(p/q)(1 + q_0^2\gamma/q^2\gamma_0)]^{-1/2} \\
a_0 = [(p_0/q_0)(1 + q^2\gamma_0/q_0^2\gamma)]^{-1/2} \\
b = (q_0^2/n_0 + q^2/n)^{-1/2} \\
c = [(q_0\gamma + q\gamma_0)/(q_0^2\gamma + q^2\gamma_0)]^{1/2} \\
\Delta = (\theta - \theta_0)/\sigma \\
\lambda = q_0\sqrt{\gamma}(q_0^2\gamma + q^2\gamma_0)^{-1/2}.
\end{array}\right\} \qquad (4.9)$$

Proof of Theorem 4.1

First condition on μ and $s/\sigma = t$ (say). We then have

$$\begin{aligned}
P(CS|\mu,t) &= P\left\{\hat{\mu}_i \geq \hat{\mu}_0 - hs\sqrt{1/N_i + 1/N_0} \ \forall \, i \in G\right\} \\
&= P\left\{q_0\bar{y}_0 - q_i\bar{y}_i \leq p_i\theta_i - p_0\theta_0 + hs\sqrt{1/N_i + 1/N_0} \ \forall \, i\right\} \\
&= P\left\{z_i \leq d_i \ \forall \, i \in G\right\}
\end{aligned}$$

where we have put

$$z_i = \frac{q_0 \bar{y}_0 - q_i \bar{y}_i - (q_0 \mu_0 - q_i \mu_i)}{\sigma \sqrt{q_0^2/n_0 + q_i^2/n_i}}$$

and

$$d_i = \frac{p_i \theta_i - p_0 \theta_0 + q_i \mu_i - q_0 \mu_0}{\sigma \sqrt{q_0^2/n_0 + q_i^2/n_i}} + \frac{ht\sqrt{1/N_i + 1/N_0}}{\sqrt{q_0^2/n_0 + q_i^2/n_i}}. \tag{4.10}$$

Conditioned on μ, the z_i are $N(0,1)$ r.v.'s having the product correlation structure:

$$\text{corr}(z_i, z_j) = \rho_{ij} = \lambda_i \lambda_j$$

where

$$\lambda_i = q_0 \sqrt{\gamma_0} (q_0^2 \gamma_i + q_i^2 \gamma_0)^{-1/2}.$$

Hence using (1.1a) of Appendix 3 in Hochberg and Tamhane (1987), we then obtain

$$P(CS|\mu, t) = \int_{-\infty}^{\infty} \prod_{i \in G} \Phi\left(\frac{d_i + \lambda_i z}{\sqrt{1 - \lambda_i^2}}\right) d\Phi(z). \tag{4.11}$$

When $k = 1$, we have $G = \phi$ or $\{1\}$ and hence the correlations ρ_{ij} do not arise. However, by putting $\lambda_1 = 0$ in (4.11) we get the correct expression for $P(CS|\mu, t)$ in this case, namely

$$P(CS|\mu, t) = \begin{cases} 1 & \text{if } G = \phi \\ \Phi(d_1) & \text{if } G = \{1\}. \end{cases}$$

By integrating (4.11) with respect to the prior distribution of μ, viz., $f(\mu) = \prod_{i=0}^{k} f_i(\mu_i)$ where $f_i(\mu_i)$ is the prior on μ_i which is a $N(\theta_i, \sigma^2/m_i)$ distribution, we get

$$
\begin{aligned}
P(CS|t) &= \int P(CS|\mu, t) f(\mu) d\mu \\
&= \int_{-\infty}^{\infty} \sum_{G \subseteq K} \left\{ \int_{-\infty}^{\mu_0} \cdots \int_{-\infty}^{\mu_0} \prod_{i \notin G} f_i(\mu_i) d\mu_i \right\} \\
&\quad \times \left\{ \int_{\mu_0}^{\infty} \cdots \int_{\mu_0}^{\infty} \left[\int_{-\infty}^{\infty} \prod_{i \in G} \Phi\left(\frac{d_i + \lambda_i z}{\sqrt{1 - \lambda_i^2}}\right) d\Phi(z) \right] \right. \\
&\quad \times \left. \prod_{i \in G} f_i(\mu_i) d\mu_i \right\} f_0(\mu_0) d\mu_0. \tag{4.12}
\end{aligned}
$$

Define

$$u_i = \frac{(\mu_i - \theta_i)\sqrt{m_i}}{\sigma} \quad (0 \leq i \leq k),$$

which are i.i.d. $N(0,1)$ r.v.'s. Hence (4.12) becomes

$$P(CS|t) = \int_{-\infty}^{\infty} \sum_{G \subseteq K} \left\{ \prod_{i \notin G} \Phi\left(u_0\sqrt{m_i/m_0} - \Delta_i\sqrt{m_i}\right) \right\}$$

$$\times \left\{ \int_{-\infty}^{\infty} \prod_{i \in G} \left[\int_{u_0\sqrt{m_i/m_0}-\Delta_i\sqrt{m_i}}^{\infty} \Phi\left(\frac{d_i + \lambda_i z}{\sqrt{1-\lambda_i^2}}\right) d\Phi(u_i) \right] d\Phi(z) \right\} d\Phi(u_0)$$

$$(4.13)$$

where from (4.10) we have

$$\begin{aligned}
d_i &= [(p_i/q_i)(1 + q_0^2\gamma_i/q_i^2\gamma_0)]^{-1/2} u_i - [(p_0/q_0)(1 + q_i^2\gamma_0/q_0^2\gamma_i)]^{-1/2} u_0 \\
&\quad + (q_0^2/n_0 + q_i^2/n_i)^{-1/2}\Delta_i + [(q_0\gamma_i + q_i\gamma_0)/(q_0^2\gamma_i + q_i^2\gamma_0)]^{1/2} ht \\
&= a_i u_i - a_0 u_0 + b_i \Delta_i + c_i ht \quad \text{(say)}.
\end{aligned} \tag{4.14}$$

Now using the symmetry condition (4.5) and the resulting equations (4.9), the integral in (4.13) simplifies to

$$\int_{-\infty}^{\infty} \int_{-\infty}^{\infty} d\Phi(z) d\Phi(u_0) \sum_{|G|=0}^{k} \binom{k}{|G|} \left[\Phi\left(u_0\sqrt{m/m_0} - \Delta\sqrt{m}\right) \right]^{k-|G|}$$

$$\times \left[\int_{u_0\sqrt{m/m_0}-\Delta\sqrt{m}}^{\infty} \Phi\left(\frac{au - a_0 u_0 + b\Delta + cht + \lambda z}{\sqrt{1-\lambda^2}}\right) d\Phi(u) \right]^{|G|}$$

$$= \int_{-\infty}^{\infty} \int_{-\infty}^{\infty} [A(u_0) + B(u_0, z, ht)]^k \, d\Phi(z) d\Phi(u_0)$$

Table 1: Critical constants h for Bayesian subset selection rule (Goal I); $P^* = 0.95, \nu = \infty, \Delta = 0$, symmetry condition (4.5) and $m_0 = m, n_0 = n$.

$p_0 = p$	k				
	1	2	3	4	5
0.10	0.018	0.472	0.589	0.596	0.603
0.30	0.385	0.722	0.842	0.852	0.859
0.50	0.478	0.741	0.875	0.972	1.038
0.70	0.465	0.658	0.839	0.910	0.980
0.90	0.329	0.402	0.581	0.701	0.767

where $A(u_0)$ and $B(u_0, z, ht)$ are defined in (4.7) and (4.8), respectively. By integrating out t in the above expression, we obtain the final result for $P(CS)$ given in (4.6).

To see that there is a unique positive solution in h to the equation (4.6) for any $P^* < 1$, note that $B(u_0, z, ht)$ (and hence $P(CS|t)$) is strictly increasing in h for $t > 0$, and as $h \to \infty$, $B(u_0, z, ht) \to 1 - A(u_0)$ and hence $P(CS) \to 1$. ∎

The expression (4.6) for $P(CS)$ simplifies considerably for $k = 1$ (in which case, as noted before, we may take $\lambda_1 = \lambda = 0$) to the following:

$$P(CS) = \Phi\left(\frac{-\Delta\sqrt{m_0 m}}{\sqrt{m_0 + m}}\right)$$
$$+ \int_0^\infty \Phi_2\left(\frac{b\Delta + cht}{\sqrt{1 + a_0^2 + a^2}}, \frac{\Delta\sqrt{m_0 m}}{\sqrt{m_0 + m}} \,\Big|\, \frac{a\sqrt{m_0} + a_0\sqrt{m}}{\sqrt{(1 + a_0^2 + a^2)(m_0 + m)}}\right) dF_\nu(t)$$

where $\Phi_2(\cdot, \cdot | \rho)$ denotes the standard bivariate normal c.d.f. with correlation coefficient ρ. By using the same notation, it may be noted that the integral $B(u_0, z, ht)$ given by (4.8) can be expressed as a standard bivariate normal c.d.f. as follows:

$$B(u_0, z, ht) = \Phi_2\left(\frac{-a_0 u_0 + b\Delta + cht + \lambda z}{\sqrt{1 - \lambda^2 + a^2}}, -u_0\sqrt{m/m_0} + \Delta\sqrt{m} \,\Big|\, \sqrt{q}\right),$$

which was the form we used for computational purposes.

The equation (4.6) was solved for h for selected values of k, ν, p_0, p, P^* and $\Delta = 0$ (i.e., assuming that all the prior means are equal). To save space, here only the values for $k = 1\,(1)\,5, \nu = \infty, p_0 = p = 0.1\,(0.2)\,0.9$ and $P^* = 0.95$ are given in Table 1. Note that these values are for the case of equal allocation, i.e., for $n_0 = n_1 = \cdots = n_k$, and may be compared with the values of g for $P^* = 0.95, \nu = \infty, \rho = 1/2$ which are 1.645, 1.916, 2.062, 2.160 and 2.234, for $k = 1\,(1)\,5$, respectively. From Table 1 we see that h increases with k as expected. Somewhat puzzling is the behavior of h as a function of $p = p_0$. We see that as $p = p_0$ (which are the relative proportions of the prior sample sizes) increases, h first increases and then decreases. This can be explained as follows: When $p = p_0$ is small, the $\mu_i - \mu_0$ are likely to be large in magnitude because they have large variances and zero mean ($\Delta = 0$). Therefore small h is needed to distinguish between "good" and "bad" treatments. When $p = p_0$ is large, small h is needed because $\mu_i - \mu_0$ is accurately known (has small variance).

(ii) Expected Subset Size

A measure of performance of a rule for selecting a subset that includes all "good" treatments is the expected size of the subset — the smaller the

Table 2: Expected proportion of treatments selected $(E(|S|)/k)$ by Bayesian subset selection rule; $P^* = 0.95, \nu = \infty, \Delta = 0$, symmetry condition (4.5) and $m_0 = m, n_0 = n$.

$p_0 = p$	k				
	1	2	3	4	5
0.10	0.502	0.563	0.578	0.579	0.580
0.30	0.600	0.682	0.709	0.712	0.713
0.50	0.684	0.771	0.809	0.835	0.850
0.70	0.761	0.843	0.900	0.918	0.933
0.90	0.838	0.886	0.959	0.982	0.989

better (of course, this would depend on how many treatments are "good") for a specified probability requirement (since $P(CS) = 1$ can be achieved trivially by including all treatments in S). An alternative measure of performance that could be used is the expected number of "bad" populations included in the subset.

For the GS-rule the expected subset size for fixed μ is given by

$$E(|S| \,|\mu) \;=\; \sum_{i=1}^{k} P\{i \in S|\mu\}$$

$$=\; \int_0^{\infty} \left\{ \sum_{i=1}^{k} \Phi\left[gt + \frac{\mu_i - \mu_0}{\sigma\sqrt{1/n_i + 1/n_0}} \right] \right\} dF_\nu(t).$$

For the Bayesian rule the expected subset size averaged over all μ, $E(|S|)$, is given by

$$E(|S|) = \int \sum_{i=1}^{k} P\{i \in S|\mu\} f(\mu) d\mu,$$

which can be written as

$$E(|S|) \;=\; \int_0^{\infty} \left\{ \sum_{i=1}^{k} \int_{-\infty}^{\infty} \int_{-\infty}^{\infty} P\{z_i \le d_i\} f_i(\mu_i) f_0(\mu_0) d\mu_i d\mu_0 \right\} dF_\nu(t)$$

$$=\; \int_0^{\infty} \left\{ \sum_{i=1}^{k} \Phi\left(\frac{b_i \Delta_i + c_i h t}{\sqrt{1 + a_0^2 + a_i^2}} \right) \right\} dF_\nu(t).$$

A direct comparison between the expected subset sizes for the two rules is not meaningful because the two rules satisfy different probability requirements, and their expected subset sizes are computed under different assumptions. Under the configuration $\theta_0 = \theta_1 = \cdots = \theta_k$, the expression for $E(|S|)$ simplifies to

$$E(|S|) = k\Phi_\nu \left(\frac{ch}{\sqrt{1 + a_0^2 + a^2}} \right),$$

where $\Phi_\nu(\cdot)$ denotes the c.d.f. of a Student's t r.v. with ν d.f. Table 2 gives the values of $E(|S|)/k$, the expected proportion of the treatments selected, for the Bayesian rule for the values of $p = p_0$ and k for which Table 1 is prepared. From this table we see that for fixed $p = p_0$, the expected proportion increases with k; similarly for fixed k, the expected proportion increases with $p = p_0$. This increase can be attributed to the increase in the expected number of "good" treatments that occurs under the above conditions.

4.2   Alternative Selection Goals

Thus far we have restricted ourselves to the goal $G \subseteq S$. However, in different applications other goals might be of interest. Two such goals are:

Goal II: Select $S \subseteq K$ that excludes all "bad" treatments, i.e., $S \subseteq G$.

Goal III: Select $S \subseteq K$ that includes all "good" treatments and excludes all "bad" ones, i.e., $S \equiv G$.

In this subsection we briefly discuss rules for these two goals. We first consider GS-type (non-Bayesian) rules. For Goal II, the same GS-rule (4.2) can be used with the critical constant g being just the negative of the critical constant g used for Goal I in order to guarantee the same probability requirement (4.1). In this sense Goals I and II are complementary. Also note that for Goal II, it is desirable to have as large an expected subset size as possible (since $P(CS) = 1$ can be achieved trivially by selecting S to be an empty set). The probability requirement (4.1) for Goal III, however, cannot be guaranteed using a GS-type rule because the infimum of $P(CS)$ for this rule is seen to be $(1/2)^k$, which is attained when all the μ_i are equal. To avoid this problem an indifference-zone type approach (Tong 1969) may be used. A Bayesian approach, as detailed below, provides an alternative way of dealing with this problem.

Now consider Bayesian rules for these goals and suppose that we use (4.4) for Goal II. Following the same method of derivation as in Theorem 4.1, it can be shown that, under the symmetry condition (4.5), the h needed to guarantee the probability requirement (4.3) is given by

$$\int_0^\infty \int_{-\infty}^\infty \int_{-\infty}^\infty [1 - A(u_0) + C(u_0, z, ht)]^k \, d\Phi(z) d\Phi(u_0) dF_\nu(t) = P^*$$

$$(4.15)$$

where

$$C(u_0, z, ht)$$

$$= \int_{-\infty}^{u_0\sqrt{m/m_0} - \Delta\sqrt{m}} \Phi\left(\frac{-au + a_0 u_0 - b\Delta - cht - \lambda z}{\sqrt{1 - \lambda^2}}\right) d\Phi(u)$$

$$= \Phi_2\left(\frac{a_0 u_0 - b\Delta - cht - \lambda z}{\sqrt{1 - \lambda^2 + a^2}}, u_0\sqrt{m/m_0} - \Delta\sqrt{m} \mid \sqrt{q}\right). \quad (4.16)$$

Comparing (4.6) and (4.15), we see that if $\Delta = 0$ then h needed to guarantee (4.3) for Goal II is just the negative of the h needed to guarantee the same probability requirement for Goal I when the Bayesian rule (4.4) is used in both cases.

Now suppose that we use the same rule (4.4) for Goal III. Again following the same method of derivation as in Theorem 4.1, it can be shown that, under the symmetry condition (4.5), an expression for $P(CS)$ is given by

$$P(CS) = \int_0^\infty \int_{-\infty}^\infty \int_{-\infty}^\infty [B(u_0, z, ht) + C(u_0, z, ht)]^k \, d\Phi(z) d\Phi(u_0) dF_\nu(t).$$

$$(4.17)$$

Here B and C are given by (4.8) and (4.16), respectively.

Unfortunately, a solution in h to the equation obtained by setting the above expression to P^* may not always exist and therefore the probability requirement (4.3) may not be guaranteed in all cases. To see this, note that while B is increasing in h, C is decreasing in h for $t > 0$, and hence $P(CS)$ is not monotone in h. Although for $\Delta \to +\infty$ or $-\infty$, $P(CS) \to 1$ for any fixed h, $\sup_h P(CS)$ may be less than P^* for any finite value of Δ and given sample sizes. Therefore (4.3) may not be guaranteed. Of course, it should be possible to solve the design problem of determining the smallest total sample size and its allocation among the treatments and the control in order to guarantee(4.3). We have been able to solve this problem analytically only for $k = 1$ (which corresponds to the goal of deciding whether a single given treatment is "better" or "worse" than the control); the details follow.

Theorem 4.2

For Goal III using the Bayesian subset selection rule (4.4), $\sup_h P(CS)$ is attained at $h = 0$ when $k = 1$ and when the symmetry condition (4.5) is assumed.

Proof of Theorem 4.2

Substituting $k = 1$ and $\lambda = 0$ in the $P(CS)$ expression given in (4.17), we obtain after some simplification

$$P(CS) = \int_0^\infty P(CS|t) dF_\nu(t)$$

$$= \int_0^\infty \{\Phi_2(c_1 + c_2 h, c_3 | \rho) + \Phi_2(-c_1 - c_2 h, -c_3 | \rho)\} \, dF_\nu(t) \quad (4.18)$$

where

$$c_1 = \frac{b\Delta}{\sqrt{1 + a_0^2 + a^2}}, c_2 = \frac{ct}{\sqrt{1 + a_0^2 + a^2}}, c_3 = \frac{\Delta\sqrt{mm_0}}{\sqrt{m + m_0}}$$

and

$$\rho = \frac{a\sqrt{m_0} + a_0\sqrt{m}}{\sqrt{(1 + a_0^2 + a^2)(m_0 + m)}}. \quad (4.19)$$

We can write

$$P(CS|t) = \int_{-\infty}^{c_1 + c_2 h} \Phi\left(\frac{c_3 - \rho w}{\sqrt{1 - \rho^2}}\right) d\Phi + \int_{-\infty}^{-c_1 - c_2 h} \Phi\left(\frac{-c_3 - \rho w}{\sqrt{1 - \rho^2}}\right) d\Phi.$$

Hence

$$\frac{d}{dh} P(CS|t) = c_2\phi(c_1 + c_2 h)\Phi\left[\frac{c_3 - \rho(c_1 + c_2 h)}{\sqrt{1 - \rho^2}}\right]$$

$$- c_2\phi(-c_1 - c_2 h)\Phi\left[\frac{-c_3 - \rho(-c_1 - c_2 h)}{\sqrt{1 - \rho^2}}\right]$$

where $\phi(\cdot)$ denotes the standard normal p.d.f. Equating this expression to zero, we obtain the following equation for h

$$h = \frac{c_3 - c_1\rho}{c_2\rho},$$

which can be shown to equal zero. Furthermore, it can be shown that $d^2 P(CS|t)/dh^2|_{h=0} < 0$. Hence $h = 0$ gives the supremum of $P(CS|t)$ and therefore of $P(CS)$. ∎

Therefore the optimum Bayesian subset selection rule for Goal III under the symmetry condition (4.5) is (conjecturing that Theorem 4.2 holds for all $k \geq 1$)

$$S = \{i \in K : \hat{\mu}_i - \hat{\mu}_0 \geq 0\}. \quad (4.20)$$

Note that this rule makes no use of the estimate s, and thus its properties are independent of the d.f. available for this estimate.

For the special case $\Delta = 0$, by substituting $h = 0$ in (4.18) we get

$$\sup_h P(CS) = 2\Phi_2(0, 0 | \rho)$$

$$= \frac{1}{2} + \frac{1}{\pi} \sin^{-1}(\rho) \quad (4.21)$$

where ρ is given by (4.19). It is clear that once $h = 0$ is fixed as the optimal choice of the critical constant, for given prior sample sizes m_i, the only way to guarantee the specified probability requirement (4.3) is by choosing the actual sample sizes n_i large enough. The following theorem gives the optimal choice of the sample sizes for this purpose.

Theorem 4.3

Assume $k = 1, \Delta = 0$, that the symmetry condition (4.5) is satisfied, and that, for the given prior sample sizes m_0 and m_1 (with $M = m_0 + m_1$), the Bayesian rule (4.20) is to be used to achieve Goal III. Then the minimum total sample size $N = n_0 + n_1$ required to guarantee the probability requirement (4.3) and its optimal allocation (n_0, n_1) (ignoring the integer restrictions on n_0 and n_1) is given by

$$
N = \begin{cases}
\frac{\rho^{*2} m_0 M}{m_1 - \rho^{*2} M}, & n_0 = N, n_1 = 0 & \text{if } \frac{m_0}{m_1} < \frac{1 - \rho^{*2}}{1 + \rho^{*2}} \\
\frac{4 m_0 m_1}{M(1 - \rho^{*2})} - M & m_0 + n_0 = m_1 + n_1 & \text{if } \frac{1 - \rho^{*2}}{1 + \rho^{*2}} \leq \frac{m_0}{m_1} \leq \frac{1 + \rho^{*2}}{1 - \rho^{*2}} \\
\frac{\rho^{*2} m_1 M}{m_0 - \rho^{*2} M} & n_0 = 0, n_1 = N & \text{if } \frac{m_0}{m_1} > \frac{1 + \rho^{*2}}{1 - \rho^{*2}}
\end{cases}
$$

$$(4.22)$$

where $\rho^* = \sin\{\pi(P^* - 0.5)\}$. We refer to these situations as Cases 1, 2, and 3 respectively.

Proof of Theorem 4.3

Suppose first that N is fixed. Then from (4.21) we see that maximizing $\sup_h P(CS)$ (with respect to n_0 and n_1, subject to $n_0 + n_1 = N$) is equivalent to maximizing

$$
\rho^2 = 1 - \frac{m_0 m_1 (M + N)}{M} \times \frac{1}{(m_0 + n_0)(m_1 + n_1)}.
$$

Therefore the optimum values of n_0 and n_1 are found by maximizing $(m_0 + n_0)(m_1 + n_1)$ subject to $n_0 + n_1 = N$. Using the Lagrange multipliers, the solution to this problem can be shown to be

$$
\begin{array}{lll}
n_0 = N, n_1 = 0 & \text{if } m_0 - m_1 < -N & \text{(Case 1)} \\
m_0 + n_0 = m_1 + n_1 & \text{if } -N \leq m_0 - m_1 \leq N & \text{(Case 2)} \\
n_0 = 0, n_1 = N & \text{if } m_0 - m_1 > N & \text{(Case 3)}.
\end{array} \quad (4.23)
$$

We now want to show that the three cases in (4.23) correspond to the three cases in (4.22), respectively. First suppose that Case 1 of (4.23) holds. We then have

$$
\rho = \left\{ 1 - \frac{m_0(M + N)}{M(m_0 + N)} \right\}^{1/2}.
$$

By equating this to ρ^* and solving for N, we get the formula for N for Case 1 of (4.22). Using this value of N, we can check that

$$m_0 - m_1 < -N \iff \frac{m_0}{m_1} < \frac{1 - \rho^{*2}}{1 + \rho^{*2}},$$

which is the condition for Case 1 of (4.22). Next suppose that Case 2 of (4.23) holds. We then have

$$\rho = \left\{ 1 - \frac{4m_0 m_1}{M(M + N)} \right\}^{1/2}.$$

By equating this to ρ^* and solving for N we get the formula for N given for Case 2 of (4.22). Using this value of N, we can check that

$$-N \leq m_0 - m_1 \leq N \iff \frac{1 - \rho^{*2}}{1 + \rho^{*2}} \leq \frac{m_0}{m_1} \leq \frac{1 + \rho^{*2}}{1 - \rho^{*2}}.$$

Case 3 can be checked similarly. ∎

We note that for large values of P^*, the bounds $(1 - \rho^{*2})/(1 + \rho^{*2})$ and $(1 + \rho^{*2})/(1 - \rho^{*2})$ on m_0/m_1 are quite wide; hence Case 2 usually applies except when there is extreme imbalance between m_0 and m_1. For example, for $P^* = 0.95$, the bounds are 0.0124 and 80.7257, respectively. It is also of interest to point out that, for a fixed ratio between m_0 and m_1, N given by (4.22) is directly proportional to M. This may seem paradoxical at first. However, a little thought shows that this is the result of assuming $\Delta = 0$. As the prior sample sizes increase with $\Delta = 0$, μ_0 and μ_1 are likely to be very close and hence larger n_0 and n_1 are needed to distinguish between them.

5   Some Decision Theoretic Problems

It may be noted that Goal I for subset selection corresponds to the multiple hypotheses testing problem:

$$H_{0i} : \delta_i = \mu_i - \mu_0 \geq 0 \quad \text{vs.} \quad H_{1i} : \delta_i = \mu_i - \mu_0 < 0 \quad (1 \leq i \leq k), \quad (5.1)$$

while Goal II corresponds to the testing problem:

$$H_{0i} : \delta_i = \mu_i - \mu_0 < 0 \quad \text{vs.} \quad H_{1i} : \delta_i = \mu_i - \mu_0 \geq 0 \quad (1 \leq i \leq k). \quad (5.2)$$

In both cases the type I familywise error rate requirement is

$$P\{\text{any true } H_{0i} \text{ is rejected}\} \leq 1 - P^*.$$

Instead of this classical error rate controlling approach, in this section we consider a decision theoretic approach. This latter approach assumes that the losses associated with wrong decisions can be specified exactly. The development here is along the lines of Duncan (1965) and Waller and Duncan (1969).

5.1   Preliminaries

We consider the multiple hypotheses testing problem (5.2). (Problem (5.1) can be treated in a similar manner.) Let decision d_{0i} (d_{1i}) correspond to accepting H_{0i} (H_{1i}); for every decision $d_i \in \{d_{0i}, d_{1i}\}$, a loss $L_i(\boldsymbol{\mu}, d_i)$ is incurred ($1 \leq i \leq k$). The total loss $L(\boldsymbol{\mu}, \boldsymbol{d})$ for the decision vector $\boldsymbol{d} = (d_1, d_2, \ldots, d_k)$ is assumed to be additive:

$$L(\boldsymbol{\mu}, \boldsymbol{d}) = \sum_{i=1}^{k} L_i(\boldsymbol{\mu}, d_i).$$

We will consider two special cases for the component loss function L_i ($1 \leq i \leq k$).

Constant Loss:

$$L_i(\boldsymbol{\mu}, d_i) = \begin{cases} c_1 & \text{if } \delta_i < 0 \text{ and } d_i = d_{1i} \\ c_2 & \text{if } \delta_i \geq 0 \text{ and } d_i = d_{0i}. \end{cases} \tag{5.3}$$

Linear Loss:

$$L_i(\boldsymbol{\mu}, d_i) = \begin{cases} c_1(-\delta_i) & \text{if } \delta_i < 0 \text{ and } d_i = d_{1i} \\ c_2 \delta_i & \text{if } \delta_i \geq 0 \text{ and } d_i = d_{0i}. \end{cases} \tag{5.4}$$

In both cases, the loss is assumed to be zero for other combinations of δ_i and d_i. The Bayes rules depend on c_1 and c_2 (where $c_1 > c_2 > 0$) only through their ratio $r = c_1/c_2 > 1$.

For a given data vector $\boldsymbol{y}$, consider a nonrandomized decision rule $\boldsymbol{\varphi}(\boldsymbol{y}) = (\varphi_1(\boldsymbol{y}), \ldots, \varphi_k(\boldsymbol{y}))$ where $\varphi_i(\boldsymbol{y}) = 0$ or 1 depending on whether $d_i = d_{0i}$ or d_{1i}, respectively ($1 \leq i \leq k$). The Bayes risk of $\boldsymbol{\varphi}$ with respect to the priors $f(\boldsymbol{\mu})$ and $f(\sigma^2)$ (note that we are using $f(\cdot)$ as a generic notation for the p.d.f.) is given by:

$$\begin{aligned} R^*(\boldsymbol{\varphi}) &= \int\int R(\boldsymbol{\varphi}, \boldsymbol{\mu}) f(\boldsymbol{\mu}) d\boldsymbol{\mu} f(\sigma^2) d\sigma^2 \\ &= \int\int \left\{ \int \left[\sum_{i=1}^{k} L_i(\boldsymbol{\mu}, \varphi_i(\boldsymbol{y})) \right] f(\boldsymbol{y}|\boldsymbol{\mu}, \sigma^2) d\boldsymbol{y} \right\} \\ &\qquad \times f(\boldsymbol{\mu}) d\boldsymbol{\mu} f(\sigma^2) d\sigma^2 \end{aligned}$$

$$= \sum_{i=1}^{k} R_i^*(\varphi_i) \qquad (5.5)$$

where $f(y|\mu, \sigma^2)$ denotes the p.d.f. of the data vector y conditioned on μ and σ^2, and $R_i^*(\varphi_i)$ denotes the Bayes risk of the component decision rule $\varphi_i(y)$.

Since we have assumed the same loss function (either (5.3) or (5.4)) for all treatment vs. control comparisons, it follows that each component $\varphi_i(y)$ of the Bayes rule will have the same form, which can be obtained by minimizing $R_i^*(\varphi_i)$ for any one i. It is not difficult to show that

$$R_i^*(\varphi_i) = \text{constant} + \int \varphi_i(y) Q_i(y) dy$$

where

$$Q_i(y) = \int \int \{L_i(\mu, d_{1i}) - L_i(\mu, d_{0i})\} f(y|\mu, \sigma^2) f(\mu) f(\sigma^2) d\mu d\sigma^2. \qquad (5.6)$$

Therefore the Bayes rule $\varphi_i(y)$ is given by

$$\varphi_i(y) = \begin{cases} 1 & (d_i = d_{1i}) \quad \text{if } Q_i(y) \le 0 \\ 0 & (d_i = d_{0i}) \quad \text{if } Q_i(y) > 0. \end{cases} \qquad (5.7)$$

Thus the main task involved in deriving the Bayes rule is to evaluate an expression for $Q_i(y)$ and solve the inequality (5.7).

5.2   Bayes Decision Rules

(a) Known Variances

The integration with respect to σ^2 is not required in this case when evaluating $Q_i(y)$ since σ^2 is assumed to be known. Also we have

$$f(y|\mu, \sigma^2) f(\mu) \quad \propto \quad \prod_{i=0}^{k} f_i(\bar{y}_i|\mu_i, \sigma_i^2) f_i(\mu_i|\sigma_i^2)$$

$$= \quad \prod_{i=0}^{k} f_i(\mu_i|\bar{y}_i, \sigma_i^2) f_i(\bar{y}_i|\sigma_i^2) \qquad (5.8)$$

where $f_i(\mu_i|\bar{y}_i, \sigma_i^2)$ is the $N(\hat{\mu}_i, \sigma_i^2/N_i)$ p.d.f. It is clear that $Q_i(y)$ depends on (5.8) only through the product of two terms, $f_i(\mu_i|\bar{y}_i, \sigma_i^2)$ and $f_0(\mu_0|\bar{y}_0, \sigma_0^2)$. Using these facts the Bayes decision rules for the two loss functions (5.3) and (5.4) are derived in the following two theorems.

Theorem 5.1

The Bayes decision rule for the constant loss function (5.3) is given by:
For $i = 1, \ldots, k$, decide $d_i = d_{1i}$ that $\delta_i \geq 0$ iff

$$z_i = \frac{\hat{\delta}_i}{\sigma(\hat{\delta}_i)} \geq z_\alpha \qquad (5.9)$$

where $\hat{\delta}_i = \hat{\mu}_i - \hat{\mu}_0, \sigma(\hat{\delta}_i) = (\bar{\sigma}_i^2 + \bar{\sigma}_0^2)^{1/2}, \bar{\sigma}_i^2 = \sigma_i^2/N_i$ and z_α is the upper $\alpha = 1/(r+1)$ point of the standard normal distribution.

Proof of Theorem 5.1

Substituting for the loss function (5.3) in (5.6) we get

$$
\begin{aligned}
Q_i(y) &\propto \int_{-\infty}^{\infty} \left\{ \int_{-\infty}^{\mu_0} c_1 f_i(\mu_i | \bar{y}_i, \sigma_i^2) d\mu_i - \int_{\mu_0}^{\infty} c_2 f_i(\mu_i | \bar{y}_i, \sigma_i^2) d\mu_i \right\} \\
&\qquad \times f_0(\mu_0 | \bar{y}_0, \sigma_0^2) d\mu_0 \\
&= c_1 - (c_1 + c_2) \int_{-\infty}^{\infty} \int_{\mu_0}^{\infty} f_i(\mu_i | \bar{y}_i, \sigma_i^2) d\mu_i \, f_0(\mu_0 | \bar{y}_0, \sigma_0^2) d\mu_0 \\
&\propto r - (r+1) \int_{-\infty}^{\infty} \Phi[(\hat{\mu}_i - \mu_0)/\bar{\sigma}_i)] f_0(\mu_0 | \bar{y}_0, \sigma_0^2) d\mu_0 \\
&= r - (r+1)\Phi[(\hat{\mu}_i - \hat{\mu}_0)/(\bar{\sigma}_i^2 + \bar{\sigma}_0^2)^{1/2}] \\
&= r - (r+1)\Phi[\hat{\delta}_i/\sigma(\hat{\delta}_i)]. \qquad (5.10)
\end{aligned}
$$

Therefore $Q_i(y) \leq 0$ if and only if (5.9) holds. ∎

Theorem 5.2

The Bayes decision rule for the linear loss function (5.4) is given by: For $i = 1, \ldots, k$, decide $d_i = d_{1i}$ iff

$$z_i = \frac{\hat{\delta}_i}{\sigma(\hat{\delta}_i)} \geq z^*(r) \qquad (5.11)$$

where $z^*(r)$ is the unique solution in z to the equation

$$rz - (r-1)[z\Phi(z) + \phi(z)] = 0. \qquad (5.12)$$

Proof of Theorem 5.2

Substituting for the linear loss function (5.4) in (5.6) we get

$$
Q_i(y) \propto \int_{-\infty}^{\infty} \left\{ \int_{-\infty}^{\mu_0} c_1(-\delta_i) f_i(\mu_i | \bar{y}_i, \sigma_i^2) d\mu_i - \int_{\mu_0}^{\infty} c_2 \delta_i f_i(\mu_i | \bar{y}_i, \sigma_i^2) d\mu_i \right\}
$$

$$\times f_0(\mu_0|\bar{y}_0, \sigma_0^2)d\mu_0$$

$$\propto \quad -r\hat{\delta}_i + (r-1)\int_{-\infty}^{\infty}\int_{\mu_0}^{\infty}\delta_i f_i(\mu_i|\bar{y}_i, \sigma_i^2)$$

$$\times f_0(\mu_0|\bar{y}_0, \sigma_0^2)d\mu_i d\mu_0. \tag{5.13}$$

The double integral in the above can be expressed as

$$\hat{\delta}_i \Phi[\hat{\delta}_i/\sigma(\hat{\delta}_i)] + \sigma(\hat{\delta}_i)\phi[\hat{\delta}_i/\sigma(\hat{\delta}_i)]. \tag{5.14}$$

Substituting this expression in (5.13) and simplifying the inequality $Q_i(y) \leq 0$, we get the inequality

$$rz_i - (r-1)\{z_i\Phi(z_i) + \phi(z_i)\} \geq 0.$$

The left hand side of the above inequality can be seen to be increasing in z_i. Therefore the decision rule can be stated in the form (5.11). ∎

The entries in Table 3 corresponding to d.f. $= \infty$ are the values of $z^*(r)$ for $r = 50, 100, 500$ and 1000.

(b) Common Unknown Variance

In this section we assume that $\sigma_i^2 = \sigma^2$ for $i = 0, 1, \ldots, k$ where the common σ^2 is unknown, but we have an estimate s^2 of σ^2 based on ν d.f. such that $\nu s^2/\sigma^2 \sim \chi_\nu^2$ independent of the $\bar{y}_i$. We further assume that a prior distribution $g_\eta^c(\sigma^2|\tau^2)$ on σ^2 is available which is a conjugate chi-squared p.d.f., i.e., τ^2 is a chi-squared distributed prior estimate of σ^2 with η d.f.

To derive the Bayes rule, note that in this case

$$f(y|\mu, \sigma^2) \propto \prod_{i=0}^{k} f_i(\mu_i|\bar{y}_i, \sigma^2)f_i(\bar{y}_i|\sigma^2)g_\nu(s^2|\sigma^2)g_\eta^c(\sigma^2|\tau^2) \tag{5.15}$$

where $f_i(\mu_i|\bar{y}_i, \sigma^2)$ is the posterior density of μ_i conditioned on $\bar{y}_i$ and σ^2 (which is a $N(\hat{\mu}_i, \sigma^2/N_i)$ p.d.f.), $f_i(\bar{y}_i|\sigma^2)$ is the conditional density of $\bar{y}_i$ conditioned on σ^2 but not on μ_i (which is a $N(\theta_i, \sigma^2(1/m_i + 1/n_i))$ p.d.f.) and $g_\nu(s^2|\sigma^2)$ is the p.d.f. of s^2. Note that if $q_\nu(\cdot)$ denotes the p.d.f. of a χ_ν^2 r.v. then

$$g_\nu(s^2|\sigma^2) = q_\nu(\nu s^2/\sigma^2) \times (\nu/\sigma^2)$$

and

$$g_\eta^c(\sigma^2|\tau^2) = q_\eta(\eta\tau^2/\sigma^2) \times (\eta\tau^2/\sigma^4).$$

Substituting the above expressions in (5.15) and simplifying we obtain

$$f(y|\mu,\sigma^2)f(\mu,\sigma^2) \quad \propto \quad \prod_{i=0}^{k} f_i(\mu_i|\bar{y}_i,\sigma^2) \times \exp(-\hat{\nu}\hat{\sigma}^2/2\sigma^2) \times (1/\sigma^2)^{\hat{\nu}/2+1}$$

$$\propto \quad \prod_{i=0}^{k} f_i(\mu_i|\bar{y}_i,\sigma^2) \times q_{\hat{\nu}}(\hat{\nu}\hat{\sigma}^2/\sigma^2) \times (1/\sigma^2)^2 \qquad (5.16)$$

where

$$\hat{\nu} = \nu + \eta + (k+1), \qquad (5.17)$$

and

$$\hat{\sigma}^2 = \left\{ \nu s^2 + \eta\tau^2 + \sum_{i=0}^{k}[(\bar{y}_i - \theta_i)^2/(1/m_i + 1/n_i)] \right\} /\hat{\nu}. \qquad (5.18)$$

Note that $\hat{\sigma}^2$ can be interpreted as a pooled estimate of σ^2 from all the sources with a total of $\hat{\nu}$ d.f. Substituting (5.16) in (5.6) we get

$$Q_i(y) \quad \propto \quad \int \left[\int \{L_i(\mu,d_{1i}) - L_i(\mu,d_{0i})\} \prod_{i=0}^{k} f_i(\mu_i|\bar{y}_i,\sigma^2)d\mu_i \right]$$

$$\times q_{\hat{\nu}}(\hat{\nu}\hat{\sigma}^2/\sigma^2)(1/\sigma^2)^2 d\sigma^2. \qquad (5.19)$$

Using this result the Bayes decision rules for the two loss functions (5.3) and (5.4) are derived in the following two theorems.

Theorem 5.3

The Bayes decision rule for the constant loss function (5.3) is given by: For $i = 1, \ldots, k$, decide $d_i = d_{1i}$ that $\delta_i \geq 0$ iff

$$t_i = \frac{\hat{\delta}_i}{\hat{\sigma}(\hat{\delta}_i)} \geq t_{\alpha,\hat{\nu}} \qquad (5.20)$$

where $\hat{\sigma}(\hat{\delta}_i) = \hat{\sigma}(1/N_i + N_0)^{1/2}, t_{\alpha,\hat{\nu}}$ is the upper $\alpha = 1/(r+1)$ point of Student's t- distribution with $\hat{\nu}$ d.f. and where $\hat{\nu}$ and $\hat{\sigma}^2$ are given by (5.17) and (5.18), respectively.

Proof of Theorem 5.3

From (5.10) we know that, conditional on σ^2, the quantity in square brackets in (5.19) is proportional to

$$r - (r+1)\Phi[\hat{\delta}_i/\sigma(\hat{\delta}_i)].$$

Substitute this expression in (5.19) and make a change of variables $v = \hat{\nu}\hat{\sigma}^2/\sigma^2$, so that

$$q_{\hat{\nu}}(\hat{\nu}\hat{\sigma}^2/\sigma^2) \times (1/\sigma^2)^2 d\sigma^2 \propto q_{\hat{\nu}}(v)dv.$$

Lastly put $t_i = \hat{\delta}_i/\hat{\sigma}(\hat{\delta}_i)$ and thus $\hat{\delta}_i/\sigma(\hat{\delta}_i) = t_i\sqrt{v/\hat{\nu}}$. Then it is easy to see that

$$
\begin{aligned}
Q_i(y) \quad &\propto \quad \int_0^\infty \left\{ r - (r+1)\Phi[\hat{\delta}_i/\sigma(\hat{\delta}_i)] \right\} \times q_{\hat{\nu}}(\hat{\nu}\hat{\sigma}^2/\sigma^2)(1/\sigma^2)^2 d\sigma^2 \\
&\propto \quad \int_0^\infty \left\{ r - (r+1)\Phi[t_i\sqrt{v/\hat{\nu}}] \right\} q_{\hat{\nu}}(v)dv \\
&= \quad r - (r+1)\Phi_{\hat{\nu}}(t_i) \quad\quad\quad\quad\quad\quad\quad\quad\quad\quad (5.21)
\end{aligned}
$$

where $\Phi_{\hat{\nu}}(\cdot)$ is the c.d.f. of a Student's t r.v. with $\hat{\nu}$ d.f. From (5.21) we see that $Q_i(y) \leq 0$ iff (5.20) holds. ∎

Theorem 5.4

Define a function

$$
\begin{aligned}
\psi_\nu(t) \quad &= \quad \left(\frac{\nu}{4\pi}\right)^{1/2} \left(\frac{\nu}{\nu+t^2}\right)^{(\nu-1)/2} \left\{ \frac{\Gamma[(\nu-1)/2]}{\Gamma[\nu/2]} \right\}. \\
&= \quad \left\{ \left(\frac{\nu}{\nu-1}\right) \left(\frac{\nu+t^2}{\nu}\right) \right\} \phi_\nu(t) \quad\quad\quad\quad (5.22)
\end{aligned}
$$

where $\Gamma(\cdot)$ is the gamma function and $\phi_\nu(\cdot)$ is the p.d.f. of a Student's t r.v. with ν d.f. Then for the linear loss function (5.4), the Bayes decision rule is given by: Make decision $d_i = d_{1i}$ iff

$$t_i = \frac{\hat{\delta}_i}{\hat{\sigma}(\hat{\delta}_i)} \geq t^*(r, \hat{\nu}) \quad\quad\quad\quad (5.23)$$

where $t^*(r, \hat{\nu})$ is the unique solution in t to the equation

$$rt - (r-1)\{t\Phi_{\hat{\nu}}(t) + \psi_{\hat{\nu}}(t)\} = 0. \quad\quad\quad\quad (5.24)$$

Proof of Theorem 5.4

From (5.13) and (5.14) we know that, conditional on σ^2, the quantity in square brackets in (5.19) is proportional to

$$-r\hat{\delta}_i + (r-1)\left\{ \hat{\delta}_i\Phi[\hat{\delta}_i/\sigma(\hat{\delta}_i)] + \sigma(\hat{\delta}_i)\phi[\hat{\delta}_i/\sigma(\hat{\delta}_i)] \right\}.$$

Substituting this expression in (5.19) and making the same changes of variables as in the proof of Theorem 5.3, we get

$$Q_i(y) \;\propto\; -rt_i + (r-1)\left\{ t_i \int_0^\infty \Phi\left(t_i\sqrt{\frac{v}{\hat{\nu}}}\right) q_{\hat{\nu}}(v)dv \right.$$
$$\left. + \int_0^\infty \sqrt{\frac{\hat{\nu}}{v}}\,\phi\left(t_i\sqrt{\frac{v}{\hat{\nu}}}\right) q_{\hat{\nu}}(v)dv \right\}. \tag{5.25}$$

It is easy to show that

$$\int_0^\infty \Phi\left(t_i\sqrt{\frac{v}{\hat{\nu}}}\right) q_{\hat{\nu}}(v)dv = \Phi_{\hat{\nu}}(t_i)$$

and

$$\int_0^\infty \sqrt{\frac{\hat{\nu}}{v}}\,\phi\left(t_i\sqrt{\frac{v}{\hat{\nu}}}\right) q_{\hat{\nu}}(v)dv = \psi_{\hat{\nu}}(t_i).$$

Substituting these expressions in (5.25) it follows that $d_i = d_{1i}$ if and only if

$$rt_i - (r-1)\{t_i\Phi_{\hat{\nu}}(t_i) + \psi_{\hat{\nu}}(t_i)\} \geq 0. \tag{5.26}$$

To show that the L.H.S. of the above inequality is increasing in t_i (and hence (5.26) can be equivalently stated in the form (5.23)), we consider its equivalent expression:

$$rt_i - (r-1)t_i \int_0^\infty \left\{ \Phi\left(t_i\sqrt{\frac{v}{\hat{\nu}}}\right) + \sqrt{\frac{\hat{\nu}}{v}}\,\phi\left(t_i\sqrt{\frac{v}{\hat{\nu}}}\right) \right\} q_{\hat{\nu}}(v)dv.$$

Differentiating this expression with respect to t_i we get

$$r - (r-1)\int_0^\infty \left\{ t_i\sqrt{\frac{v}{\hat{\nu}}}\,\phi\left(t_i\sqrt{\frac{v}{\hat{\nu}}}\right) \right.$$
$$\left. + \Phi\left(t_i\sqrt{\frac{v}{\hat{\nu}}}\right) - t_i\sqrt{\frac{v}{\hat{\nu}}}\,\phi\left(t_i\sqrt{\frac{v}{\hat{\nu}}}\right) \right\} q_{\hat{\nu}}(v)dv$$
$$= r - (r-1)\int_0^\infty \Phi\left(t_i\sqrt{\frac{v}{\hat{\nu}}}\right) q_{\hat{\nu}}(v)dv$$
$$= r - (r-1)\Phi_{\hat{\nu}}(t_i),$$

which is greater than zero. This completes the proof of the theorem. ∎

Solutions $t^*(r,\hat{\nu})$ to (5.24) for $r = 50, 100, 500$ and 1000, and for d.f. $\hat{\nu} = 5(5)30, 40, 60, 120, \infty$ are given in Table 3. Note that $t^*(r,\infty) = z^*(r)$. This follows by letting $\hat{\nu} \to \infty$ in (5.24) for $t^*(r,\hat{\nu})$ and observing that $\Phi_{\hat{\nu}}(t) \to \Phi(t)$ and from (5.22), $\psi_{\hat{\nu}}(t) \to \phi(t)$ (since $\phi_{\hat{\nu}}(t) \to \phi(t)$ and the quantity in braces tends to 1), resulting in (5.12) for $z^*(r)$.

Table 3: $t^*(r, \hat{\nu})$ for Bayesian decision rule for the linear loss function.

$\hat{\nu}$	r			
	50	100	500	1000
5	2.053	2.509	3.773	4.430
10	1.716	2.033	2.802	3.152
15	1.630	1.915	2.580	2.869
20	1.590	1.861	2.482	2.746
25	1.568	1.831	2.426	2.677
30	1.553	1.812	2.391	2.633
40	1.535	1.788	2.349	2.581
60	1.518	1.765	2.308	2.530
120	1.501	1.742	2.269	2.482
∞	1.489	1.720	2.231	2.436

6   Concluding Remarks

In this paper we have presented an approach for exploiting the prior information on the means of the treatments and the control in the problems of optimal allocation of observations and of subset selection using either the conventional error rate control formulation or the decision theoretic formulation. We have shown how excess prior information on a treatment may require that less (or none) actual observations be taken from it. For the subset selection problem using the error rate control formulation, we have shown how the prior information may be incorporated into the selection rule and the probability requirement. Finally we have given a decision theoretic formulation of the same problem and derived Bayesian decision rules. It is worth remarking that, while the critical constants required in the former formulation are highly dependent on k, the number of treatment vs. control comparisons, those required in the latter formulation are virtually independent of k (the only dependence on k being through the degrees of freedom $\hat{\nu}$). This is of course the result of assuming that the overall loss function is additive in component losses, which causes the overall decision problem to decompose into component decision problems.

References

Bechhofer, R.E. (1954). A single-sample multiple decision procedure for ranking means of normal populations with known variances. *Ann. Math. Statist.* **25** 17 - 39.

Bechhofer, R.E. and Dunnett, C.W. (1988). Percentage points of multi-variate Student t distributions. In *Selected Tables in Mathematical Statistics* **11** American Mathematical Society, Providence, R. I.

Chen, H.J. and Pickett, J.R. (1984). Selecting all treatments better than a control under a multivariate normal distribution and a uniform prior distribution. *Commun. Statist., Ser. A* **13** 59 - 80.

Duncan, D.B. (1965), A Bayesian approach to multiple comparisons. *Technometrics* **7** 171 - 222.

Dunnett, C.W. (1955). A multiple comparisons procedure for comparing several treatments with a control. *Jour. Amer. Statist. Assoc.* **50** 1096 - 1121.

Dunnett, C.W. (1960). On selecting the largest of k normal populations (with discussion). *Jour. Roy. Statist. Soc., Ser. B* **22** 1 - 40.

Gupta, S.S. and Sobel, M. (1958). On selecting a subset which contains all populations better than a standard. *Ann. Math. Statist.* **29** 235 - 244.

Hochberg, Y. and Tamhane, A.C. (1987). *Multiple Comparison Procedures*. John Wiley and Sons, Inc., New York.

Jacroux, M. (1987). On the determination and construction of MV-optimal block designs for comparing treatments with a standard treatment. *Jour. Statist. Plan. Inf.* **15** 205 - 225.

Kiefer, J. (1958). On the nonrandomized optimality and randomized nonoptimality of symmetrical designs. *Ann. Math. Statist.* **29** 675 - 699.

McCormick, G.P. (1983). *Nonlinear Programming: Theory, Algorithms and Applications.* John Wiley and Sons, Inc., New York.

Tong, Y.L. (1969). On partitioning a set of normal populations by their locations with respect to a control. *Ann. Math. Statist.* **40** 1300 - 1324.

Waller, R.A. and Duncan, D.B. (1969). A Bayes rule for the symmetric multiple comparisons. *Jour. Amer. Statist. Assoc.* **64** 1484 -1503; Corringendum in *Jour. Amer. Statist. Assoc.* **67** 253–255.

Chapter 17

Nonlinear Renewal Theory and Beyond: Reviewing the Roles in Selection and Ranking

NITIS MUKHOPADHYAY Department of Statistics, University of Connecticut, Storrs, Connecticut

Abstract In the beginning, nonlinear renewal theory was introduced in the area of *sequential analysis* in order to obtain second-order expansions of the associated risk functions. See, for example, Lai and Siegmund (1977, 1979) and Woodroofe (1977, 1982). Recently, one will find several important contributions in the area of *selection and ranking* where the authors have extensively used and improvised such sophisticated tools in order to obtain second-order asymptotic characteristics of the underlying methodologies. This literature and the follow-ups including some of the most recent advances in the three-stage and accelerated sequential methodologies are unified and reviewed, in the context of some interesting selection and ranking, and associated estimation problems.

1 Introduction

In the early stages, the so called *nonlinear renewal theory* was developed and fully exploited in the area of sequential analysis in order to derive asymptotic second-order expansions of the associated risk functions for several types of estimation methodologies. See, for example, Lai and Siegmund (1977, 1979), Woodroofe (1977, 1982), Ghosh and Mukhopadhyay (1980, 1981) and others. The *sequential problems* in the area of *selection and ranking* on the other hand flourished somewhat independently after the appearance of the pioneering works of Bechhofer (1954), Bechhofer et al. (1954) and Gupta (1956). During the past decade, however, important contributions have appeared discussing *asymptotic second-order properties* in full generality for the purely *sequential procedure* of Robbins et al. (1968). Here, the authors used certain improvisations of the nonlinear renewal theoretic tools that were known earlier in the sequential estimation literature. During this period, the area of selection and ranking has also experienced other types of growth. For example, one now has an extensive array of literature on second-order characteristics of *three-stage* and *accelerated sequential* methodologies for various interesting problems in selection and ranking. Typically, the major mathematical tools are quite intricate and those theoretical developments together with others dealing with applications in selection and ranking are scattered in many different journals. In the present *review paper*, we first summarize in Section 2, various important "Theorems" in their generalities under certain types of purely sequential and multistage procedures. In Section 3, we discuss the problem of selecting the normal population associated with the largest mean when the common variance in unknown. Section 4 presents brief discussions on analogous problems for the two-parameter exponential distributions. Section 5 includes estimation problems for ranked parameters which arise in the context of our discussions given in Sections 3 and 4. Section 6 consists of a few concluding remarks and observations.

Before we give details, let us add that we consistently work under Bechhofer's (1954) *indifference-zone approach*. The emphasis in this paper is to put forth and unify the available *second-order asymptotics*; and in order to emphasize the future potential of the existing theoretical machinery, we give examples of how we are able, at this time, to handle certain simple and yet important *non-elimination type* sequential selection and ranking problems. The present review article will perhaps energize future researchers to investigate the possibilities of extending the second-order asymptotic results for the corresponding elimination-type versions of the problems discussed here and elsewhere.

The set of references furnished here is perhaps somewhat incomplete, but adequate. Whenever necessary, one should refer to the original sources for further citations. Here, the aim is to review certain specific aspects in the area of selection and ranking. For a general introduction, one should look at the books authored by Bechhofer et al. (1968), Gibbons et al.

(1977) and Gupta and Panchapakesan (1979). The monograph of Gupta and Huang (1981) and the categorized bibliography of Dudewicz and Koo (1981) would also be beneficial.

Throughout the text, we write $\lceil u \rceil$ for the largest integer $\geq u$ and I(A) for the indicator function of the set or event A.

2   The Theory of Second-Order Asymptotics

In this section, we consider very general types of purely sequential, accelerated sequential and three-stage procedures. We primarily summarize the available tools and results without any sort of derivations. Let W_1, W_2, ... be independent and identically distributed (*i.i.d.*) *positive* and *continuous random variables* having *all moments finite*.

2.1   The Nonlinear Renewal Theory: Purely Sequential Procedure

The set up here is similar to that in Woodroofe (1977). See also Lai and Siegmund (1977, 1979). Define

$$N(h^*) = \inf\{n \geq m : \sum_{i=1}^{n} W_i < h^* n^\delta L(n)\}, \tag{2.1}$$

where $\delta > 1$, $h^* > 0$, $m \geq 1$ and $L(u) = 1 + L_0 u^{-1} + o(u^{-1})$ as $u \to \infty$ with $-\infty < L_0 < \infty$. Write $\theta = E(W_1)$, $\tau^2 = E(W_1^2) - \theta^2$, $\beta^* = 1/(\delta - 1)$, $n_0^* = (\theta/h^*)^{\beta^*}$ and $p = \beta^{*2}\tau^2/\theta^2$. Assume that

$$P(W_1 \leq u) \leq Bu^b \text{ for all } u > 0, \tag{2.2}$$

for some $B > 0$ and $b > 0$. Let

$$\nu = \frac{\beta^*}{2\theta}\{(\delta - 1)^2\theta^2 + \tau^2\} \tag{2.3}$$

$$- \sum_{n=1}^{\infty} n^{-1}E\{\max(0, \sum_{i=1}^{n} W_i - n\delta\theta)\},$$

$$\eta = \beta^*\theta^{-1}\nu - \beta^* L_0 - \tfrac{1}{2}\delta\beta^{*2}\tau^2\theta^{-2}. \tag{2.4}$$

The following theorem summarizes some of the most important results available in the purely sequential framework (2.1).

Theorem 2.1

For the stopping variable $N(h^*)$ defined in (2.1), for every non-zero real number ω and $0 < \epsilon < 1$, we have as $h^* \to 0$:

(i) $P\{N(h^*) \le \epsilon n_0^*\} = O(h^{mb})$ if $m \ge 1$;

(ii) $U^* \xrightarrow{\mathcal{L}} p\chi_1^2$ if $m \ge 1$ and U^* is uniformly integrable if $m > \beta^*/b$ where $U^* = \{N(h^*) - n_0^*\}^2/n_0^*$;

(iii) $E\{(N(h^*)/n_0^*)^\omega\} = 1 + \{\omega\eta + \frac{1}{2}\omega(\omega - 1)p\}n_0^{*-1} + o(n_0^{*-1})$ if (a) $m > (3 - \omega)\beta^*/b$ for $\omega \in (-\infty, 2) - \{-1, 1\}$, (b) $m > \beta^*/b$ for $\omega = 1$ and $\omega \ge 2$, and (c) $m > 2\beta^*/b$ for $\omega = -1$. Here b and η come from (2.2) and (2.4) respectively, while $p = \beta^{*2}\tau^2/\theta^2$.

Parts (i), (ii) and (iii), with $\omega = 1$, were all proved in Woodroofe (1977). Part (iii), in its fullest generality was proved in Mukhopadhyay (1988).

2.2 Accelerated Sequential Procedure

Let us write $\bar{W}_n = n^{-1}\sum_{i=1}^n W_i$. First, we choose and fix $0 < \rho < 1$, $q \ge 0$ and let

$$t(h^*) = \inf\{n \ge m : h^* n^\delta \ge \rho \bar{W}_n\}, \tag{2.5}$$

where $\delta > 0$, $h^* > 0$, $m \ge 1$. Based on $W_1, \cdots, W_{t(h^*)}$, we let

$$N_1(h^*) = \lceil \{h^{*-1}\bar{W}_{t(h^*)}\}^{1/\delta} + q \rceil, \tag{2.6}$$

where recall that $\lceil u \rceil$ stands for the largest integer $\ge u$, throughout the text. Define

$$N(h^*) = \max\{t(h^*), N_1(h^*)\}, \tag{2.7}$$

and if $t(h^*) \ge N_1(h^*)$, then sampling terminates at stage $t(h^*)$, that is with $W_1, \cdots, W_{t(h^*)}$. However, we sample the difference $N_1(h^*) - t(h^*)$ in *one single batch* if $N_1(h^*) > t(h^*)$. This $N(h^*)$ is referred to as the *accelerated stopping time*. Hall (1983) studied various second-order asymptotic properties of $N(h^*)$ in a particular situation involving the construction of fixed-width confidence intervals for the mean of a normal population having unknown variance. Here, we summarize the results from Mukhopadhyay and Solanky (1991) where a general unified theory has been developed.

Let $E(W_1) = \theta$ and $E(W_1^2) - \theta^2 = \tau^2$ as before. Assume (2.2) and define $n_0^* = (h^*/\theta)^{-1/\delta}$, $p = \tau^2(\delta^2\theta^2)^{-1}$ and

$$\nu = \tfrac{1}{2}(\delta^2\theta^2 + \tau^2)(\delta\theta)^{-1} \tag{2.8}$$

$$- \sum_{n=1}^{\infty} n^{-1} E\{\max(0, \sum_{i=1}^{n} W_i - n\theta(1+\delta))\}.$$

For positive numbers r, s, r', s' such that $r^{-1} + s^{-1} = r'^{-1} + s'^{-1} = 1$, we write

$$m_0(\delta) = \begin{cases} \max\{\tfrac{1}{2}sr'(\delta b)^{-1},\ ss'(\delta b)^{-1},\ 2/(\delta^3 b), \delta + 1\} \\ \quad \text{if } \delta = 1 \text{ or } \delta \geq 2; \\ \\ \max\{\tfrac{1}{2}sr'(\delta b)^{-1},\ ss'(\delta b)^{-1},\ (3-\delta)(\delta b)^{-1}, \\ \quad 2/(\delta^3 b), \delta + 1\} \\ \quad \text{if } 0 < \delta < 1 \text{ or } 1 < \delta < 2; \end{cases} \tag{2.9}$$

where "b" comes from (2.2). The following theorems are taken from Mukhopadhyay and Solanky (1991).

Theorem 2.2

For the stopping variable $N(h^*)$ defined in (2.7), we have as $h^* \to 0$:

(i) $E\{[N(h^*) - E(N(h^*))]^2\} = \rho^{-1}pn_0^* + o(h^{*-1/\delta})$ if $m > (b\delta)^{-1}$;

(ii) $E\{|N(h^*) - E(N(h^*))|^\omega\} = O(h^{*-\frac{1}{2}\omega/\delta})$ for $\omega > 0$ if $m > \tfrac{1}{2}\omega(b\delta)^{-1}$;

where $p = \tau^2(\delta^2\theta^2)^{-1}$ and $n_0^* = (h^*/\theta)^{-1/\delta}$.

Theorem 2.3

For the stopping variable $N(h^*)$ defined in (2.7), we have as $h^* \to 0$:

(i) $\eta'' \leq \liminf\{E(N(h^*)) - n_0^*\} \leq \limsup\{E(N(h^*)) - n_0^*\} \leq \eta'' + 1$ if $m > m_0(\delta)$ and $0 < \rho < 1$ is arbitrary;

(ii) $E\{N(h^*)\} = n_0^* + \eta' + o(1)$ if $m > m_0(\delta)$ and ρ^{-1} is an integer;

(iii) $E\{(N(h^*) - n_0^*)^2/N(h^*)\} = p\rho^{-1/\delta} + o(1)$ if $m > \max\{m_0(\delta),\ 2(\delta b)^{-1}\}$ and $0 < \rho < 1$ is arbitrary;

(iv) $1 + \{\omega(\eta''+1) + \tfrac{1}{2}\omega(\omega-1)p\rho^{-1/\delta}\}n_0^{*-1} + o(n_0^{*-1}) \leq E\{[N(h^*)/n_0^*]^\omega\} \leq 1 + \{\omega\eta'' + \tfrac{1}{2}\omega(\omega-1)p\rho^{-1/\delta}\}n_0^{*-1} + o(n_0^{*-1})$ for fixed $\omega < 0$ and $0 < \rho < 1$ if (a) $m > \max\{m_0(\delta),\ 2(\delta b)^{-1}\}$ when $\omega = -1$, (b) $m > \max\{m_0(\delta),\ (3-\omega)(\delta b)^{-1}\}$ when $\omega \in (-\infty, 0) - \{-1\}$;

where $\eta'' = \frac{1}{2}(1-\delta)\tau^2\delta^{-2} + q - p\rho^{-1/\delta}$, $\eta' = \eta'' + q^*$ and $q^* = E\{[q + \rho^{-1/\delta}(1-Z)] - 2 + [q + \rho^{-1/\delta}(1-Z)]\}$, Z being a suitable continuous random variable on $(0,1)$. Here, n_0^* and p are same as in Theorem 2.2, and $m_0(\delta)$ comes from (2.9).

2.3 Three-Stage Procedure

Let $m(\geq 1)$ be the starting sample size. Then, choose and fix $0 < \rho < 1$ and let, for known λ,

$$T(\lambda) = \max\{m, \lceil \rho\lambda\bar{W}_m \rceil\}, \tag{2.10}$$

with $0 < \lambda < \infty$. We sample the difference in the second stage if $T > m$. Next, based on $W_1, \cdots, W_{T(\lambda)}$, define

$$N(\lambda) = \max\{T(\lambda), \lceil \lambda\bar{W}_{T(\lambda)} \rceil\}. \tag{2.11}$$

In the third stage, the sample of size T is further augmented, if necessary, with $(N - T)$ extra observations. Let $n_0^* = \lambda\theta$ where $E(W_1) = \theta$ and we assume that $\lambda = \lambda(m) \to \infty$, $\limsup(m/n_0^*) < \rho$, and $n_0^* = O(m^r)$ for $r > 1$, as $m \to \infty$. Let $Y_i = (W_i - \theta)/\theta$, $i = 1, 2, ...$ and $\tau^2 = E(W_1^2) - \theta^2$.

Theorem 2.4

Let ω be a positive integer. Then, for the stopping variable $N(\lambda)$ defined in (2.11), we have as $\lambda \to \infty$:

$$E\{(N(\lambda))^\omega\} = n_0^{*\omega} + \frac{1}{2}\omega n_0^{*\omega-1}\{(\omega-3)\tau^2\theta^{-2} + \rho\}\rho^{-1} + o(\lambda^{\omega-1}),$$

where $n_0^* = \lambda\theta$.

Theorem 2.5

For the stopping variable $N(\lambda)$ defined in (2.11), we have as $\lambda \to \infty$:

(i) $P\{N(\lambda) \leq \epsilon n_0^*\} = O(\lambda^{-\frac{1}{2}p/r})$ for $0 < \epsilon < 1$ and all fixed $p(> 0)$ where $m = O(n_0^{*1/r})$ for $r > 1$;

(ii) $\{N(\lambda) - n_0^*\}^2/n_0^* \xrightarrow{\mathcal{L}} \rho^{-1}\tau^2\theta^{-2}\chi_1^2$ and $\{N(\lambda) - n_0^*\}^2/n_0^*$ is also uniformly integrable;

(iii) Let ω be a positive real number. Then,

$$E\{(N(\lambda))^{-\omega}\} = n_0^{*-\omega} - \frac{1}{2}\omega n_0^{*-\omega-1}\{\rho - (\omega+3)\tau^2\theta^{-2}\}\rho^{-1} + o(\lambda^{-\omega-1});$$

where $n_0^* = \lambda\theta$.

The basic idea of triple sampling was put forth in Mukhopadhyay (1976). Hall (1981) independently proposed a three-stage sampling procedure in order to construct a fixed-width confidence interval for the mean of a normal population having unknown variance. Hall (1981) was able to derive elegant second-order asymptotics in that particular problem. The general unified theory summarized here is taken from Mukhopadhyay (1990).

3   Selecting the Best Normal Population

Suppose we have π_i, $i = 1, \cdots, k(\geq 2)$, independent, normally distributed populations, each with *unknown mean* μ_i and *common unknown variance* σ^2. Let $\mu_{[1]} \leq \cdots \leq \mu_{[k]}$ be the ordered μ-values. Our aim is to *select the population associated with the largest mean* $\mu_{[k]}$. We do not assume any knowledge about the association of the μ_i's with the $\mu_{[i]}$'s. In what follows, we adopt Bechhofer's (1954) "indifference zone" approach. Given $\delta^*(> 0)$, let

$$\Omega(\delta^*) = \{\boldsymbol{\theta} = (\mu_1, \cdots, \mu_k) : \mu_{[k]} \geq \mu_{[k-1]} + \delta^*\}. \tag{3.1}$$

For a given $P^* \in (1/k, 1)$, we are interested in sequential and other relevant selection procedures such that P (Correct selection (CS) of the population associated with $\mu_{[k]}$) is *asymptotically* (as $\delta^* \to 0$) *at least* P^* whenever the true parameter vector $\boldsymbol{\theta}$ belongs to $\Omega(\delta^*)$.

Bechhofer (1954) has shown that when σ^2 is known, we can achieve $P(CS) \geq P^*$ whenever $\boldsymbol{\theta} \in \Omega(\delta^*)$ by taking a sample of size $\geq h^2\sigma^2/\delta^{*2}$ from each population, and selecting the population associated with the largest sample mean, where "h" is the solution of the integral equation

$$\int_{-\infty}^{\infty} \{\Phi(z + h)\}^{k-1}\phi(z)dz = P^*. \tag{3.2}$$

Here, $\phi(t) = (2\pi)^{-\frac{1}{2}} \exp(-\frac{1}{2}t^2)$ and $\Phi(t) = \int_{-\infty}^{t} \phi(u)du$ for $-\infty < t < \infty$. Let $C = h^2\sigma^2/\delta^{*2}$, which is referred to as the "optimal" fixed sample size required from each population had σ^2 been known. For various values of k and P^*, h-values can be obtained from Bechhofer (1954) and Gibbons et al. (1977, Table A.1).

We should add that $P(CS) = P^*$ when $\boldsymbol{\theta}$ is given by the configuration $\mu_{[1]} = \cdots = \mu_{[k-1]} = \mu_{[k]} - \delta^*$, known as the *least favorable configuration* (LFC). Note that C is unknown, and hence C is to be adaptively estimated by a suitable positive integer valued random variable N which is "close" to C. In a classic paper of Bechhofer et al. (1954), a two-stage procedure was proposed along the lines of Stein (1945) for this problem giving an exact solution. Two-stage procedures, however, are known to have tendencies to significantly overestimate C, even asymptotically, and thus we turn to *non-elimination type* purely sequential methodologies and other relevant followups where one can conclude "second-order" efficiency properties and the like.

Suppose, in general, one devises a statistical methodology giving rise to a positive integer valued random variable N estimating C and selects the population associated with the largest sample mean. Having recorded independent random samples $X_{i1}, \cdots, X_{in}$ from π_i, we write $\bar{X}_{in} = n^{-1} \sum_{j=1}^{n} X_{ij}$, $S_{in}^2 = (n-1)^{-1} \sum_{j=1}^{n} (X_{ij} - \bar{X}_{in})^2$ and $U(n) = k^{-1} \sum_{i=1}^{n} S_{in}^2$ for $n \geq 2$. Note that $U(n)$ is the pooled unbiased estimator of σ^2.

Theorem 3.1

Suppose that N satisfies the following conditions. The distribution of N does not depend on θ; $I(N = n)$ is independent of $(\bar{X}_{1n}, \cdots, \bar{X}_{kn})$ for every fixed $n \geq 2$; and $P(N < \infty) = 1$. Then,

$$\inf_{\theta \in \Omega(\delta^*)} P(CS) = E[\int_{-\infty}^{\infty} \{\Phi(y + N^{\frac{1}{2}} \delta^* \sigma^{-1})\}^{k-1} \phi(y) dy]$$

and this infimum is attained when $\mu_{[1]} = \cdots = \mu_{[k-1]} = \mu_{[k]} - \delta^*$.

This theorem is not available anywhere else in this form. On the other hand, it is a crucial unifying result for our development. Hence, we supply a proof even though it is truly straightforward.

Proof of Theorem 3.1

Let $\delta_i = \mu_{[k]} - \mu_{[i]} (\geq \delta^*)$, $i = 1, \cdots, k-1$. $P(CS) = \sum_{n=m}^{\infty} P(CS|N = n)P(N = n) = \sum_{n=m}^{\infty} P\{\bar{X}_{jn}$ comes from the population associated with $\mu_{[k]}\}P(N = n)$ since $I(N = n)$ is independent of $(\bar{X}_{1n}, \cdots, \bar{X}_{kn})$, where $\bar{X}_{jn} = \max_{1 \leq i \leq k} \bar{X}_{in}$. Now, it is easy to see that

$$P(CS) = \sum_{n=m}^{\infty} [\int_{-\infty}^{\infty} \prod_{i=1}^{k-1} \Phi(y + n^{\frac{1}{2}} \delta_i \sigma^{-1}) \phi(y) dy] P(N = n). \qquad (3.3)$$

Bechhofer (1954) had shown that the integral in (3.3) has $\int_{-\infty}^{\infty} \{\Phi(y + n^{\frac{1}{2}} \delta^* \sigma^{-1})\}^{k-1} \phi(y) dy$ as the lower bound whenever $\theta \in \Omega(\delta^*)$ and the lower bound is attained when $\mu_{[1]} = \cdots = \mu_{[k-1]} = \mu_{[k]} - \delta^*$. Since $P(N = n)$ does not depend on θ for any n, we get, for all $\theta \in \Omega(\delta^*)$,

$$P(CS) \geq \sum_{n=m}^{\infty} [\int_{-\infty}^{\infty} \{\Phi(y + n^{\frac{1}{2}} \delta^* \sigma^{-1})\}^{k-1} \phi(y) dy] P(N = n) \qquad (3.4)$$

and this lower bound is attained (for $\theta \in \Omega(\delta^*)$) when $\mu_{[1]} = \cdots = \mu_{[k-1]} = \mu_{[k]} - \delta^*$. We may just add that $P(N < \infty) = 1$ is needed to make the selection rule well-defined. This completes the proof of the theorem. ∎

3.1 Purely Sequential Procedure

Robbins et al.'s (1968) stopping rule can be stated as

$$N = N(\delta^*) = \inf\{n \geq m_1 : n \geq h^2 U(n)/\delta^{*2}\}, \qquad (3.5)$$

where $m_1(\geq 2)$ is the starting sample size from each population. One keeps checking with the stopping rule (3.3) and takes one additional sample from each π whenever needed. It is easy to see that $P(N < \infty) = 1$ for all θ. After stopping, we select the population associated with the largest $\bar{X}_{iN}$. Theorem 3.1 naturally holds, that is the LFC is given by $\mu_{[1]} = \cdots = \mu_{[k-1]} = \mu_{[k]} - \delta^*$ and under this LFC, we have

$$P(CS) = E\left[\int_{-\infty}^{\infty} \{\Phi(y + N^{\frac{1}{2}}\delta^*\sigma^{-1})\}^{k-1}\phi(y)dy\right]. \qquad (3.6)$$

The stopping variable N given by (3.5) can be written as $1 + N(h^*)$ where $N(h^*)$ comes from (2.1) with $m = m_1 - 1$, $\delta = 2$, $h^* = kC^{-1}$, $n_0^* = C$, $L_0 = 1$ and $W_1 \sim \chi_k^2$. The condition (2.2) holds with $b = k/2$ and ν (given by (2.3)) simplifies to $\frac{1}{2}(k + 2) - \sum_{n=1}^{\infty} n^{-1}E\{\max(0, \chi_{nk}^2 - 2nk)\}$. After repeated use of Theorem 2.1, Mukhopadhyay and Judge (1989) proved the following Theorem. See also Mukhopadhyay (1983) for the case "$k = 2$". Let us write

$$g(y) = \int_{-\infty}^{\infty} \{\Phi(u + y)\}^{k-1}\phi(u)du, \ f(y) = g(y^{\frac{1}{2}}), \ y > 0. \qquad (3.7)$$

Theorem 3.2

For the purely sequential selection procedure (3.5), under the LFC, we have as $\delta^* \to 0$:

(i) $E(N) = C + \beta' + o(1)$ if (a) $m_1 \geq 3$ when $k = 2$, (b) $m_1 \geq 2$ when $k \geq 3$;

(ii) $P(CS) = P^* + h^2 C^{-1}\{\beta' f'(h^2) + h^2 f''(h^2)k^{-1}\} + o(\delta^{*2})$ if (a) $m_1 \geq 4$ when $k = 2$, (b) $m_1 \geq 3$ when $k = 3, 4, 5$, (c) $m_1 \geq 2$ when $k \geq 6$;

where $\beta' = k^{-1}(\nu - 2)$.

Further details can be obtained from Mukhopadhyay and Judge (1989). Comments on moderate sample performances are also given in Robbins et al. (1968) and Mukhopadhyay and Judge (1989).

3.2   Accelerated Sequential Procedure

In order to cut sampling operations, one may estimate only a fraction of C first, by means of purely sequential sampling, and then followup by estimating the whole C by batch sampling. Thus, we first fix $0 < \rho < 1$ and $q \geq 0$. Define, with $U^*(n) = n(n-1)^{-1}U(n)$,

$$t = t(\delta^*) = \inf\{n \geq m_1 : n \geq \rho h^2 U^*(n)/\delta^{*2}\}, \tag{3.8}$$

$$N_1 = N_1(\delta^*) = \lceil h^2 U^*(t)\delta^{*-2} + q \rceil, \tag{3.9}$$

$$N = N(\delta^*) = \max\{t(\delta^*), N_1(\delta^*)\}. \tag{3.10}$$

We start with $m_1(\geq 2)$ samples from each π. Then, we sample one observation at a time from each π sequentially and end up with $X_{i1}, \cdots, X_{it}$ from π_i, $i = 1, \cdots, k$. Note that $t(\delta^*)$ estimates ρC, a fraction of C, the optimal fixed sample size. Then, $N_1(\delta^*)$ estimates C based on t samples observed from each π. If $N = t$, we do not take any more samples. If $N > t$, we sample the difference $(N - t)$ from each π in one single batch at this stage. Now, based on $X_{i1}, \cdots, X_{iN}$ from π_i, we select the population associated with the largest $\bar{X}_{iN}$. Theorem 3.1 again holds in the present situation, hence the LFC continues to be $\mu_{[1]} = \cdots = \mu_{[k-1]} = \mu_{[k]} - \delta^*$ and $P(CS)$ under this LFC is given by (3.6). We should remark that our accelerated sequential procedure (3.8) - (3.10), introduced in Mukhopadhyay and Solanky (1992a), cuts sampling operations by approximately $100(1-\rho)\%$ when compared with the purely sequential procedure (3.5).

The present accelerated sequential methodology agrees with (2.5) - (2.7) with $\delta = 1$, $W_1 \sim \chi_k^2$, $\theta = k$, $\tau^2 = 2k$, $b = k/2$, $p = 2/k$, $n_0^* = C$, $m = m_1 - 1$ and $h^* = kC^{-1}$. We can also obtain $m_0(1)$ from (2.9) with $r' = s' = 2$ and $s = 1.01$ and hence $m_0(1)$ is 2.02 or 2 according as $k = 2$ or $k \geq 3$ respectively. Theorems 2.2 - 2.3 were fully exploited in Mukhopadhyay and Solanky (1992a) to derive the following results.

Theorem 3.3

For the accelerated sequential procedure (3.8) - (3.10), under the LFC, we have as $\delta^* \to 0$:

(i) $q - 2(k\rho)^{-1} \leq \liminf E(N-C) \leq \limsup E(N-C) \leq q - 2(k\rho)^{-1} + 1$ if $m_1 \geq 4$;

(ii) $P(CS) \geq P^* + o(\delta^{*2})$ if $q = \max\{0, 2(k\rho)^{-1} - h^2 f''(h^2)[k^2 \rho f'(h^2)]^{-1}\}$ and (a) $m_1 \geq 5$ when $k = 2$, (b) $m_1 \geq 4$ when $k \geq 3$;

where $f(\cdot)$ has been defined in (3.7).

Various types of moderate sample performances were discussed in Muk-hopadhyay and Solanky (1992). Overall, the moderate sample size proper-ties as well as the asymptotic second-order characteristics of the accelerated sequential methodology compared very favorably with those of the Robbins et al. (1968) procedure.

3.3 Three-Stage Procedure

In order to cut sampling operations even further, one may estimate a fraction of C and then C successively by means of batch sampling. We first fix $0 < \rho < 1$. Let $\beta'' = \frac{1}{2} - 2(k\rho)^{-1}$, $m_1^* = -\beta'' - h^2\rho^{-1}\{f''(h^2)/f'(h^2)\}$ and define

$$T = T(\delta^*) = \max\{m_1, \lceil \rho h^2 U(m_1)/\delta^{*2} \rceil\}, \tag{3.11}$$

$$N = N(\delta^*) = \max\{T, \lceil h^2 U(T)\delta^{*-2} + m_1^* \rceil\}. \tag{3.12}$$

Here $m_1 = m_1(\delta^*) = O(C^{1/r})$ for some $r > 1$. We start with $m_1(\geq 2)$ samples from each π. If $T = m_1$, then we do not sample any more from any population at the second stage. However, if $T > m_1$, then we sample the difference at the second stage to extend $X_{i1}, \cdots, X_{im_1}$ to $X_{i1}, \cdots, X_{iT}$ from π_i, $i = 1, \cdots, k$. Based on T samples from each population, we determine $U(T)$ and thus we find N. Note that T and N estimates ρC and C respec-tively. If $N = T$, we do not take any more samples from any population at the third stage. However, if $N > T$, then we sample the difference at the third stage to extend $X_{i1}, \cdots, X_{iT}$ to $X_{i1}, \cdots, X_{iN}$ from π_i, $i = 1, \cdots, k$. Once we have N samples from each π, we select the population associated with the largest $\bar{X}_{iN}$. Theorem 3.1 still holds in this case and hence the LFC and the expression of $P(CS)$ under the LFC remain quite the same as these were in Sections 3.1 - 3.2. We should add that our three-stage procedure (3.11) - (3.12), introduced in Mukhopadhyay and Judge (1992), cuts sampling operations practically to barebone when compared with the purely sequential or accelerated sequential selection methodologies of previ-ous sections. Yet, this three-stage procedure is shown to enjoy very similar second-order asymptotic characteristics indeed. The three-stage sampling methodology (3.11) - (3.12) agrees with (2.10) - (2.11) where $\lambda = h^2/\delta^{*2}$, $n_0^* = C$, $\theta = \sigma^2$, $W_1 \sim \sigma^2 k^{-1}\chi_k^2$ and $\tau^2 = 2\sigma^4/k$. Results similar to those in Theorem 2.5 were exploited in Mukhopadhyay and Judge (1992) to derive the following theorem.

Theorem 3.4

For the three-stage procedure (3.11) - (3.12), under the LFC, we have as $\delta^* \to 0$:

(i) $E(N) = C - h^2 \rho^{-1} \{ f''(h^2)/f'(h^2) \} + o(1);$

(ii) $P(CS) = P^* + o(\delta^{*2});$

where $f(\cdot)$ has been defined in (3.7).

Remark 3.1 There are extensive sets of comments on moderate sample comparisons of purely sequential, accelerated sequential and three-stage selection procedures in Mukhopadhyay and Solanky (1992a), and Mukhopadhyay and Judge (1989, 1992). In Mukhopadhyay and Judge (1992), one will also find comparisons between the three-stage and two-stage selection procedures. The performances of the three-stage procedures discussed in Section 3.3 are found to be remarkably similar to those of the purely sequential and accelerated sequential procedures asymptotically as well as for moderate values of C. The two-stage procedure of Bechhofer et al. (1954) fares significantly "worse" in this comparison. We must add, however, that the two-stage procedure is designed to have $P(CS) \geq P^*$ for all $\theta \in \Omega(\delta^*)$ whereas the procedures discussed here guarantees this P^*-requirement only asymptotically.

Remark 3.2 Among the procedures discussed in Sections 3.1 - 3.3, when we compared the estimated values (and their estimated standard errors) of $E(N)$ and $P(CS)$ for a wide range of moderate values of C, we found that the purely sequential procedure came out first with the accelerated sequential procedure being a close second, while the three-stage procedure turned out to be third in standing. But, overall, the "performances" of all three procedures were very similar and fairly close. In the original papers, one will also find certain guidelines for the choices of m_1 and ρ.

Remark 3.3 When the variances are unknown and unequal, the problem of selecting the normal population associated with the largest mean via double sampling was very cleverly handled in Dudewicz and Dalal (1975). Following this fundamental piece of work, Rinott (1978) discussed certain interesting and key points in this context. Mukhopadhyay (1979) also gave a number of additional results.

4   Selecting the Best Exponential Population

Let π_i, $i = 1, \cdots, k (\geq 2)$, be independent, two-parameter exponential populations having the probability density function $q(x; \mu_i, \sigma)$ where

$$q(t; \mu, \sigma) = \sigma^{-1} \exp\{-(t - \mu)/\sigma\} I(t > \mu), \qquad (4.1)$$

$0 < \sigma < \infty$, $-\infty < \mu_1, \cdots, \mu_k < \infty$. We assume that the *location parameters* $\mu_1, \cdots, \mu_k$ and the *scale parameter* σ *are all unknown*. Let $\mu_{[1]} \le \cdots \le \mu_{[k]}$ be the ordered μ-values. Our aim is to select the population associated with the largest location $\mu_{[k]}$. We do not assume any knowledge about the association of μ_i's with the $\mu_{[i]}$'s. In what follows, we again adopt Bechhofer's (1954) *indifference zone approach*. Given $\delta^*(> 0)$, let $\Omega(\delta^*)$ be the same as in (3.1). For a given $P^* \in (1/k, 1)$, we are interested in *non-elimination type* sequential and other relevant selection procedures such that $P(CS)$ is *asymptotically* (as $\delta^* \to 0$) *at least* P^* whenever the true parameter vector θ belongs to $\Omega(\delta^*)$.

For known σ, results are in Barr and Rizvi (1966), Desu and Sobel (1968), and Raghavachari and Starr (1970). Having recorded $X_{i1}, \cdots, X_{in}$ from π_i, we write $T_{in} = \min\{X_{i1}, \cdots, X_{in}\}$, $V_{in} = (n-1)^{-1} \sum_{j=1}^{n} (X_{ij} - T_{in})$, $V(n) = k^{-1} \sum_{i=1}^{k} V_{in}$ for $n \ge 2$. Note that $V(n)$ is the pooled unbiased estimator of σ. For known σ, it was shown that we could achieve $P(CS) \ge P^*$ whenever $\theta \in \Omega(\delta^*)$ by taking a sample of size $\ge a\sigma/\delta^*$ from each population, and selecting the population associated with the largest T_{in}, where "a" is the solution of the integral equation

$$\int_0^\infty \{1 - \exp(-z - a)\}^{k-1} \exp(-z) dz = P^*. \tag{4.2}$$

For certain given values of k and P^*, a-values can be obtained from Desu and Sobel (1968) and Raghavachari and Starr (1970). We should add that $P(CS) = P^*$ when θ is given by the configuration $\mu_{[1]} = \cdots = \mu_{[k-1]} = \mu_{[k]} - \delta^*$, known as the LFC. Let $C^* = a\sigma/\delta^*$, which is referred to as the "optimal" fixed sample size required from each population had σ been known.

However, note that C^* is unknown, since σ is unknown. Hence, one needs sequential and other adaptive procedures for the present situation along the lines of Section 3. Desu et al. (1977) proposed and studied a remarkable two-stage procedure for this problem along the lines of Bechhofer et al. (1954). Mukhopadhyay (1984) considered the case $k = 2$ having unequal scale parameters and discussed both two-stage as well as purely sequential selection procedures. See also Mukhopadhyay and Hamdy (1984).

Suppose that one constructs a general statistical methodology giving rise to a positive integer valued random variable N estimating C^* and selects the population associated with the largest T_{iN} value.

Theorem 4.1

Suppose that N satisfies the following conditions. The distribution of N does not depend on θ; $I(N = n)$ is independent of $(T_{1n}, \cdots, T_{kn})$ for every fixed $n \geq 2$; and $P(N < \infty) = 1$. Then,

$$\inf_{\theta \in \Omega(\delta^*)} P(CS) = E[\int_0^\infty \{1 - \exp(-z - N\delta^*\sigma^{-1})\}^{k-1} \exp(-z)dz]$$

and this infimum is attained when $\mu_{[1]} = \cdots = \mu_{[k-1]} = \mu_{[k]} - \delta^*$.

Its proof is very similar to that of Theorem 3.1 and hence it is omitted.

4.1 Purely Sequential Procedure

Mukhopadhyay's (1986) stopping rule can be stated as

$$N = N(\delta^*) = \inf\{n \geq m_1 : n \geq aV(n)/\delta^*\}, \tag{4.3}$$

where $m_1(\geq 2)$ is the starting sample size from each population. One takes an additional observation from each π whenever needed according to the stopping rule (4.3). After stopping, we select the population associated with the largest T_{iN}. Theorem 4.1 holds in this case and hence the LFC is given by $\mu_{[1]} = \cdots = \mu_{[k-1]} = \mu_{[k]} - \delta^*$ and under this LFC, we have

$$P(CS) = E[\int_0^\infty \{1 - \exp(-z - N\delta^*\sigma^{-1})\}^{k-1} \exp(-z)dz]. \tag{4.4}$$

Theorem 2.1 can be repeatedly used after suitably expanding the integral inside the expectation in (4.4). Asymptotic second-order expansions of $E(N) - C^*$ and $P(CS) - P^*$ up to the respective orders $o(1)$ and $o(\delta^*)$ can be obtained along the lines of Theorem 3.2 in Mukhopadhyay (1986) for the case of general $k(\geq 2)$. Details are omitted for brevity.

4.2 Accelerated Sequential Procedure

Here, the idea is to cut sampling operations compared with the purely sequential selection methodology (4.3) and yet maintain second-order characteristics of both the average sample size as well as $P(CS)$. We first fix $0 < \rho < 1$ and $q \geq 0$. Define, with $V^*(n) = n(n-1)^{-1}V(n)$,

$$t = t(\delta^*) = \inf\{n \geq m_1 : n \geq \rho aV^*(n)/\delta^*\}, \tag{4.5}$$

$$N_1 = N_1(\delta^*) = \lceil aV^*(t)\delta^{*-1} + q \rceil, \tag{4.6}$$

$$N = N(\delta^*) = \max\{t(\delta^*), N_1(\delta^*)\}. \tag{4.7}$$

We start with $m_1(\geq 2)$ samples from each π. Then, we implement the accelerated sequential procedure (4.5) - (4.7) along the same lines as in the case of (3.8) - (3.10). Finally, based on $X_{i1}, \cdots, X_{iN}$ from π_i, we select the population associated with the largest T_{iN}.

This procedure was introduced and studied in Mukhopadhyay and Solanky (1992b). In view of Theorem 4.1, the LFC is given by $\mu_{[1]} = \cdots = \mu_{[k-1]} = \mu_{[k]} - \delta^*$ and under this LFC, $P(CS)$ has the same expression given in (4.4). Again, the unified theory discussed in Section 2 turns out to be most pertinent, and the results like those summarized in Theorem 3.3 were proved in Mukhopadhyay and Solanky (1992b). Further details are omitted for brevity. We add that moderate sample size performances of this accelerated sequential selection methodology compared very favorably with those of the purely sequential one given by (4.3).

Remark 4.1   For brevity, we merely mention that a three-stage selection procedure for this problem of ours was developed in Mukhopadhyay (1987). This procedure looked very much like the one proposed in (3.11) - (3.12), but adapted to the new situation by mimicking the expression of C^*. Moderate sample comparisons of the three methodologies summarized in this section are available in Mukhopadhyay and Solanky (1992b).

5   Estimating Ordered Parameters

The area of estimating ordered parameters is certainly quite vast in its own right. See, for example, Chapter 21 of Gupta and Panchapakesan (1979) and the books of Barlow et al. (1972) and Dykstra et al. (1986). In this section, however, we explore and summarize certain interesting ordered parameter estimation problems where the unified theories discussed in Section 2 have fruitfully led to various types of intricate *asymptotic second-order results* for the associated *non-elimination type* purely sequential and multistage estimators. We may add that our present specific endeavors may seem somewhat narrow at first, and yet the future scope of research in this otherwise broad area seems almost endless.

5.1   Largest Location of Exponential Populations

Here $\pi_1, \cdots, \pi_k$ stand for the same populations as in Section 4. We assume that $\mu_1, \cdots, \mu_k$ and σ are all unknown. For arbitrary but fixed $d(> 0)$ and $\alpha \in (0, 1)$, let I be a random confidence interval for $\mu_{[k]}$, *having length d*. We address various sampling techniques for constructing intervals

I for $\mu_{[k]}$ such that $P(\mu_{[k]} \in I)$ is *asymptotically* (*as* $d \to 0$) *at least* α *for all* $\mu_1, \cdots, \mu_k$ and σ.

Having recorded $X_{i1}, \cdots, X_{in}$ from π_i, we use previously defined T_{in} and $V(n)$ to estimate μ_i and σ respectively. We propose to estimate $\mu_{[k]}$ by $T_n^* = \max\{T_{1n}, \cdots, T_{kn}\}$ and consider the confidence interval

$$I_n = [T_n^* - d, \, T_n^*] \tag{5.1}$$

for $\mu_{[k]}$. The literature in the case $k = 1$ started to develop with the classic paper of Ghurye (1958), and this area has grown steadily since then. One may look at the paper of Mukhopadhyay (1988) for a recent review.

Our interest lies in the situation when there are *at least two populations* available; we refer to Mukhopadhyay and Kuo (1988). It is easily seen that

$$
\begin{aligned}
P\{\mu_{[k]} \in I_n\} &= \prod_{i=1}^{k}[1 - \exp\{-n(d + (\mu_{[k]} - \mu_i))/\sigma\}] \\
&\geq \{1 - \exp(-nd/\sigma)\}^k,
\end{aligned}
$$

for all θ, while equality holds when $\mu_1 = \cdots = \mu_k$. Thus,

$$\inf_{\theta} P\{\mu_{[k]} \in I_n\} = \{1 - \exp(-nd/\sigma)\}^k. \tag{5.2}$$

We require that $\{1 - \exp(-nd/\sigma)\}^k \geq \alpha$ and thus n has to be chosen as the smallest integer $\geq a^*\sigma/d$ with $a^* = -\ln(1 - \alpha^{1/k})$. Let us write $n^* = a^*\sigma/d$, for the "optimal" fixed sample size required from each population, had σ been known. Of course, n^* is unknown since σ is unknown. Mukhopadhyay and Kuo (1988) proposed two-stage procedures as well as purely sequential and three-stage estimation procedures. In each case, the fixed-width confidence interval I_N was proposed for $\mu_{[k]}$ where "N" is the appropriate random sample size from each π. In each scenario, one has

$$\inf_{\theta} P\{\mu_{[k]} \in I_N\} = E[\{1 - \exp(-Nd/\sigma)\}^k],$$

the LFC being the same as $\mu_1 = \cdots = \mu_k$. Utilizing the unified theories developed in Sections 2.1 and 2.2, Mukhopadhyay and Kuo (1988) obtained asymptotic (as $d \to 0$) second-order expansions of $E(N) - n^*$ and $\inf_{\theta} P\{\mu_{[k]} \in I_N\} - \alpha$ up to respective orders $o(1)$ and $o(d)$ in the case of the purely sequential and three-stage estimation methodologies.

Point Estimation of Largest Location of Exponential Populations

In the paper of Kuo and Mukhopadhyay (1990a), the goal was to construct *minimax* and *bounded maximal risk* point estimators of $\mu_{[k]}$ under a wide variety of *loss functions*. A purely sequential procedure was proposed determining "N", the random sample size from each population, thereby estimating $\mu_{[k]}$ pointwise by means of T_N^*. Again, in view of the available unified theory discussed in Section 2.1, the authors successfully obtained "asymptotic" second-order expansions of certain interesting characteristics (e.g. average sample size, risk functions etc.) associated with the sequential point estimator T_N^* of $\mu_{[k]}$ in a straightforward fashion. A three-stage estimation procedure was also addressed in that same paper.

Certain new two-stage and accelerated sequential point estimation procedures in this context have recently been studied in Mukhopadhyay et al. (1992). The "asymptotic" second-order expansions of the average sample size and the risk functions associated with T_N^* in the case of the accelerated sequential methodology turned out to be quite as expected.

5.2 Largest Mean of Normal Populations

Here $\pi_1, \cdots, \pi_k$ stand for the same populations as in Section 3. We assume that $\mu_1, \cdots, \mu_k$ and σ are all unknown. *Fixed-width confidence interval* problems for estimating $\mu_{[k]}$ were introduced in Saxena and Tong (1969) and Tong (1970). In Kuo and Mukhopadhyay (1990b), "asymptotic" second-order expansions of various characteristics of Tong's (1970) purely sequential fixed-width confidence interval procedure were developed. In the same problem, a suitable three-stage procedure was also introduced in the same paper.

The corresponding point estimation problems for $\mu_{[k]}$ turned out to be more challenging than the exponential counterpart. In the paper of Kuo and Mukhopadhyay (1990b), the authors also constructed *bounded maximal risk* point estimators of $\mu_{[k]}$ under a wide variety of *loss functions* via purely sequential and three-stage methodologies. Again, the thrust in that paper was the second-order "asymptotics". In the same vein, for *fixed sample size analyses*, Cohen and Sackrowitz (1982) did provide several interesting results.

6 Concluding Remarks

We have reviewed certain interesting recent results in the area of selection and ranking where the emphasis lies in obtaining suitable "asymptotic" *second-order properties*. The sampling strategies considered included *non-elimination type* purely sequential, accelerated sequential, and three-stage methodologies. The general theories for such sampling methodologies are indeed very intricate and yet, one can see that these tools can be fruitfully utilized in several selection and ranking problems culminating in many types of sophisticated results. Such general theories can also certainly be used in other complicated problems in this area, which would be explored, we hope, more fully in the future. Having such futuristic ideas in mind, we included in Section 2, various available general second-order expansions under specific "assumptions" for immediate use by the researchers in the area of selection of ranking. Section 3 *details* certain interesting results in *explicit forms* and we have also indicated how the machinery given in Section 2 are applied in selecting the "best" normal population. We refrained from giving such details in Sections 4 and 5 for brevity.

Acknowledgment

The referee's comments have been helpful in revising this manuscript and I am grateful to the referee for the same.

References

Barlow, R.E., Bartholomew, D.J., Bremner, J.M. and Brunk, H.D. (1972). *Statistical Inference Under Order Restrictions.* John Wiley and Sons, Inc., New York.

Barr, D.R. and Rizvi, M.H. (1966). An introduction to ranking and selection procedures. *Jour. Amer. Statist. Assoc.* **61** 640 - 645.

Bechhofer, R.E. (1954). A single-sample multiple decision procedure for ranking means of normal populations with known variances. *Ann. Math. Statist.* **25** 16 - 39.

Bechhofer, R.E., Dunnett, C.W. and Sobel, M. (1954). A two-sample multiple decision procedure for ranking means of normal populations with a common unknown variance. *Biometrika* **41** 170 - 176.

Bechhofer, R.E., Kiefer, J. and Sobel, M. (1968). *Sequential Identification and Ranking Procedures.* University of Chicago Press, Chicago.

Cohen, A. and Sackrowitz, H. (1982). Estimating the mean of the selected population. In *Statistical Decision Theory and Related Topics III* (S.S. Gupta, and J. Berger, eds), 243 - 270. Academic Press, New York.

Desu, M.M., Narula, S.C. and Villarreal (1977). A two-stage procedure for selecting the best of k exponential distributions. *Commun. Statist., Ser. A* **6** 1231 - 1243.

Desu, M.M. and Sobel, M. (1968). A fixed subset-size approach to the selection problem. *Biometrika* **55** 401 - 410.

Dudewicz, E.J. and Dalal, S.R. (1975). Allocation of observations in ranking and selection with unequal variances. *Sankhya, Ser. B* **37** 28 - 78.

Dudewicz, E.J. and Koo, J.O. (1981). *The Complete Categorized Guide to Statistical Selection and Ranking Procedures.* Amer. Sciences Press, Columbus.

Dykstra, R., Robertson, T. and Wright, F.T. (1986). *Advances in Order Restricted Statistical Inference.* Symposium Proceedings (Iowa City, September, 1985). Springer Verlag, Inc., New York.

Ghosh, M. and Mukhopadhyay, N. (1980). Sequential point estimation of the difference of two normal means. *Ann. Statist.* **8** 221 - 225.

Ghosh, M. and Mukhopadhyay, N. (1981). Consistency and asymptotic efficiency of two-stage and sequential procedures. *Sankhya, Ser. A* **43** 220 - 227.

Ghurye, S.G. (1958). Note on sufficient statistics and two-stage procedures. *Ann. Math. Statist.* **29** 155 - 166.

Gibbons, J.D., Olkin, I. and Sobel, M. (1977). *Selecting and Ordering Populations: A New Statistical Methodology.* John Wiley and Sons, Inc., New York.

Gupta, S.S. (1956). On a Decision Rule for a Problem in Ranking Means. Ph.D. dissertation, Univ. of North Carolina, Chapel Hill.

Gupta, S.S. and Huang, D.Y. (1981). *Multiple Statistical Decision Theory: Recent Developments.* Springer Verlag, Inc., New York.

Gupta, S.S. and Panchapakesan, S. (1979). *Multiple Decision Procedures.* John Wiley and Sons, Inc., New York.

Hall, P. (1981). Asymptotic theory of triple sampling for sequential estimation of a mean. *Ann. Statist.* **9** 1229 - 1238.

Hall, P. (1983). Sequential estimation saving sampling operations. *Jour. Roy. Statist. Soc., Ser. B* **45** 219 - 223.

Kuo, L. and Mukhopadhyay, N. (1990a). Point estimation of the largest location of k negative exponential populations. *Sequen. Anal.* **9** 297 - 304.

Kuo, L. and Mukhopadhyay, N. (1990b). Multi-stage point and interval estimation of the largest mean of k normal populations and the associated second-order properties. *Metrika* **37** 291 - 300.

Lai, T.L. and Siegmund, D. (1977). A non-linear renewal theory with applications to sequential analysis I. *Ann. Statist.* **5** 946 - 954.

Lai, T.L. and Siegmund, D. (1979). A non-linear renewal theory with applications to sequential analysis II. *Ann. Statist.* **7** 60 - 76.

Mukhopadhyay, N. (1976). Fixed-width confidence intervals for the mean using a three-stage procedure. Unpublished manuscript.

Mukhopadhyay, N. (1979). Some comments on two-stage selection procedures. *Commun. Statist., Ser. A* **8** 671 - 683.

Mukhopadhyay, N. (1983). Theoretical investigations of some sequential and two-stage procedures to select the larger mean. *Sankhya, Ser. A* **45** 346 - 356.

Mukhopadhyay, N. (1984). Sequential and two-stage procedures for selecting the better exponential population covering the case of unknown and unequal scale parameters. *Jour. Statist. Plan. Inf.* **9** 33 - 43.

Mukhopadhyay, N. (1986). On selecting the best exponential population. *Jour. Ind. Statist. Assoc.* **24** 31 - 41.

Mukhopadhyay, N. (1987). Three-stage procedures for selecting the best exponential population. *Jour. Statist. Plan. Inf.* **16** 345 - 352.

Mukhopadhyay, N. (1988). Sequential estimation problems for negative exponential populations. *Commun. Statist., Theory and Methods (Reviews Sections)* **17** 2471 - 2506.

Mukhopadhyay, N. (1990). Some properties of a three-stage procedure with applications in sequential analysis. *Sankhya, Ser. A* **52** 218 - 231.

Mukhopadhyay, N., Chattopadhyay, S. and Sahu, S.K. (1992). Further developments in estimation of the largest mean of k normal populations. *Metrika* **39**, In press.

Mukhopadhyay, N. and Hamdy, H.I. (1984). Two-stage procedures for selecting the best exponential population when the scale parameters are unknown and unequal. *Sequen. Anal.* **3** 51 - 74.

Mukhopadhyay, N. and Judge, J. (1989). Second-order expansions for a sequential procedure. *Sankhya, Ser. A* **51** 318 - 327.

Mukhopadhyay, N. and Judge, J. (1992). Three-stage procedures for selecting the largest normal mean. In *H. Robbins Volume* (Z. Govindarajulu ed.), In press.

Mukhopadhyay, N. and Kuo, L. (1988). Fixed-width interval estimations of the largest location of k negative exponential populations. *Sequen. Anal.* **7** 321 - 332.

Mukhopadhyay, N. and Solanky, T.K.S. (1991). Second order properties of accelerated stopping times with applications in sequential estimation. *Sequen. Anal.* **10** 99 - 123.

Mukhopadhyay, N. and Solanky, T.K.S. (1992a). Accelerated sequential procedure for selecting the largest mean. *Sequen. Anal.* **11** 137 - 148.

Mukhopadhyay, N. and Solanky, T.K.S. (1992b). Accelerated sequential procedure for selecting the best exponential population. *Jour. Statist. Plan. Inf.*, In press.

Raghavachari, M. and Starr, N. (1970). Selection problems for some terminal distributions. *Metron* **28** 185 - 197.

Rinott, Y. (1978). On two-stage procedures and related probability — inequalities. *Commun. Statist., Ser. A* **8** 799 - 811.

Robbins, H., Sobel, M. and Starr, N. (1968). A sequential procedure for selecting the best of k populations. *Ann. Math. Statist.* **39** 88 - 92.

Saxena, K.M.L. and Tong, Y.L. (1969). Interval estimation of the largest mean of k normal populations with known variances. *Jour. Amer. Statist. Assoc.* **64** 296 - 299.

Stein, C. (1945). A two-sample test for a linear hypothesis whose power is independent of the variance. *Ann. Math. Statist.* **16** 243 - 258.

Tong, Y.L. (1970). Multi-stage interval estimation of the largest mean of k normal populations. *Jour. Roy. Statist. Soc., Ser. B* **32** 272 - 277.

Woodroofe, M. (1977). Second order approximations for sequential point and interval estimation. *Ann. Statist.* **5** 985 - 995.

Woodroofe, M. (1982). *Nonlinear Renewal Theory in Sequential Analysis.* CBMS #39, SIAM Publications, Philadelphia.

Chapter 18

One-Sided Sequential Tests to Establish Equivalence Between Treatments with Special Reference to Normal and Binary Responses

CHRISTOPHER JENNISON   School of Mathematical Sciences, University of Bath, Bath, United Kingdom

BRUCE W. TURNBULL   School of Operations Research and Industrial Engineering, Cornell University, Ithaca, New York

Abstract   Dunnett and Gent (1977) have discussed the problem of establishing the equivalence of a new treatment with a standard one. In the one-sided equivalence problem they considered, it is desired to demonstrate that the new treatment is no worse than the standard. Dunnett and Gent note that it is not sufficient to fail to reject a null hypothesis of equality: this may be due simply to lack of power. In this paper we propose group sequential tests for equivalence based on ideas related to repeated confidence intervals. These tests adapt readily to unpredictable group sizes, to situations without a rigid stopping rule, and to non-normal observations.

For the problem of comparing two binomial distributions, the sample size needed to satisfy given size and power requirements depends strongly on the average success probability. Since this is typically unknown, an adaptive choice of group size is needed to produce efficient tests meeting the specified error probability requirements. Comparisons are made with a non-adaptive sequential procedure recently proposed by Durrleman and Simon (1990). A special case is the experiment where interim analyses are performed, not for the purpose of early termination, but simply to adjust the sample size so that nominal error rates will be guaranteed, despite the presence of a nuisance parameter.

1   Introduction

We consider the problem of equivalence testing where it is desired to demonstrate that the efficacy of a new treatment is no worse than that of a standard treatment. This is the goal of many positive control clinical trials when a proposed new therapy has advantages in toxicity, cost or ease of administration over an effective standard treatment. More details on the background of this problem are given in Dunnett and Gent (1977) and in Durrleman and Simon (1990). (The latter article will be subsequently referred to as "DS".)

As these authors note, in this problem it is not sufficient to fail to reject a null hypothesis of equality: this may be due simply to lack of power in the study. Instead, Dunnett and Gent (1977) proposed testing a specific hypothesis of non-equality and concluding equivalence if this hypothesis is rejected in the direction of equality. Suppose θ represents a measure of the difference between the experimental and standard treatments, with $\theta > 0$ if the standard is superior. The hypothesis H: $\theta = \Delta$ ($\Delta > 0$) is tested and if it is rejected in favor of $\theta < \Delta$, it is concluded that the experimental treatment is equivalent to the standard, in that its efficacy is at most Δ below that of the standard. The choice of "medically significant difference," Δ, is discussed in Jennison and Turnbull (1989, Section 2.3.2). This procedure could be termed a "one-sided" equivalence test. It is the setup considered by Dunnett and Gent (1977) and by DS and is the one we consider here. (The "two-sided" problem where the objective is to show that θ lies in an interval $(-\Delta, \Delta)$, say, has been considered by Jennison and Turnbull 1992.)

There is the usual relationship between the equivalence test and an associated confidence interval. A size α test rejects H: $\theta = \Delta$ in favor of $\theta < \Delta$ if and only if an upper $(1-\alpha)$ confidence interval for θ lies completely below Δ. We shall exploit this relationship to motivate and define a sequential test of equivalence. First we must introduce the sequential analog of a confidence interval. Suppose a group sequential study has a maximum of K analyses. Following Robbins (1970), Jennison and Turnbull (1984,1989), we say $\{\bar{\theta}_k; k = 1, \cdots, K\}$ is a $(1 - \alpha)$ upper confidence sequence if

$$P_\theta[\bar{\theta}_k > \theta \text{ for all } k = 1, \cdots, K] = 1 - \alpha \tag{1.1}$$

for all θ. Similarly $\{\underline{\theta}_k; k = 1, \cdots, K\}$ is a $(1 - \beta)$ lower confidence sequence if

$$P_\theta[\underline{\theta}_k < \theta \text{ for all } k = 1, \cdots, K] = 1 - \beta \tag{1.2}$$

for all θ.

Such upper and lower confidence sequences can be used to test H: $\theta = \Delta$ with Type I error at most α and power at least $1 - \beta$ at $\theta = 0$. The "derived test" proposed by Jennison and Turnbull (1989, Section 2.3.2) and by DS continues to sample as long as *both* 0 and Δ are included in the interval $(\underline{\theta}_k, \bar{\theta}_k)$. The experiment stops at the first analysis k that $(0, \Delta)$ is not wholly contained in $(\underline{\theta}_k, \bar{\theta}_k)$ with equivalence accepted if $\bar{\theta}_k < \Delta$, or rejected if $\underline{\theta}_k > 0$. Termination at analysis K is ensured by choosing the sample size such that $\bar{\theta}_K - \underline{\theta}_K = \Delta$. By construction, we have

$$P[\text{Accept equivalence}|\theta = \Delta] < P[\bar{\theta}_k < \Delta \text{ for some } k|\theta = \Delta] = \alpha \tag{1.3}$$

$$P[\text{Reject equivalence}|\theta = 0] < P[\underline{\theta}_k > 0 \text{ for some } k|\theta = 0] = \beta \tag{1.4}$$

Thus, error probabilities at $\theta = \Delta$ and 0 are bounded by α and β, respectively. The inequalities are strict; for example, if $\underline{\theta}_k > 0$ the test stops to reject equivalence, and acceptance is not possible at a later stage when $\bar{\theta}_{k'} < \Delta$ $(k' > k)$. However, as we shall see (Table 1) this conservatism is slight (see also Jennison and Turnbull 1989, Table 3, and DS, Table 2).

In this paper we consider the construction and performance of the group sequential procedure in two settings. In Section 2 we consider the problem of normal observations with known variance. In Section 3 we extend the methodology to the problem of comparing binomial responses which was the one addressed by Dunnett and Gent (1977) and by DS. Here, because the variances depend on the success probabilities, which are unknown, an adaptive choice of sample size is necessary to control the error probabilities of the test. A special case is the experiment where interim analyses are performed, not for the purpose of early termination, but simply to adjust the sample size so that nominal error rates will be guaranteed, despite the presence of a nuisance parameter (Gould 1992).

2   Normal Responses

Independent observations X_{1i} and X_{2i} ($i \geq 1$) are available from the standard and experimental treatments, respectively. Suppose $X_{1i} \sim N(\mu_S, \sigma^2)$ and $X_{2i} \sim N(\mu_E, \sigma^2)$, $i = 1, 2, \cdots$, where σ^2 is known. Let $\theta = \mu_S - \mu_E$. Analyses are performed after observing every additional $2n$ subjects, n on each treatment, up to a maximum of K analyses. At the kth analysis ($1 \leq k \leq K$), after a total of $n_k = nk$ subject responses have been observed on each treatment, the sufficient statistic for θ is $W_k = \sum_{i=1}^{nk}(X_{1i} - X_{2i})$ which is, marginally, normally distributed with mean $nk\theta$ and variance $2nk\sigma^2$.

Consider first the fixed sample size test of H: $\theta = \Delta$ with Type I error rate α and power $1-\beta$ at $\theta = 0$. If there are n_f observations per treatment, the $1 - \alpha$ confidence interval for θ has endpoints:

$$\bar{\theta} = \frac{W_k}{n_f} + \sigma z_\alpha \sqrt{\frac{2}{n_f}} \quad \text{and} \quad \underline{\theta} = \frac{W_k}{n_f} - \sigma z_\beta \sqrt{\frac{2}{n_f}} \qquad (2.1)$$

where z_α is the upper α point of the standard normal distribution. The sample size needed to satisfy error probability requirements (1.3) and (1.4) is

$$n_f = \frac{2\sigma^2}{\Delta^2}(z_\alpha + z_\beta)^2 \qquad (2.2)$$

and this will serve as a benchmark for the maximum and expected sample sizes of the sequential tests.

In the group sequential case, $K > 1$, the sequential confidence limits have the form

$$\bar{\theta}_k = \frac{W_k}{kn} + \sigma c_k(\alpha) \sqrt{\frac{2}{kn}} \qquad (2.3)$$

$$\underline{\theta}_k = \frac{W_k}{kn} - \sigma c_k(\beta) \sqrt{\frac{2}{kn}}, \qquad (2.4)$$

the constants $\{c_k(\alpha), 1 \leq k \leq K\}$ and $\{c_k(\beta), 1 \leq k \leq K\}$ being chosen such that (1.1) and (1.2) hold. The maximum sample size

$$n_K = n_{\max} = \frac{2\sigma^2}{\Delta^2}(c_K(\alpha) + c_K(\beta))^2 \qquad (2.5)$$

on each treatment group ensures that $\bar{\theta}_K - \underline{\theta}_K = \Delta$ and, thus, the test which stops to accept equivalence if $\bar{\theta}_k < \Delta$ and to reject equivalence if $\underline{\theta}_k > 0$ will terminate at or before stage K.

There are many choices for the sequences of constants $\{c_k(\alpha)\}$ and $\{c_k(\beta)\}$. For example, we can take $c_k(\alpha) = Z_P(K, \alpha)$, $c_k(\beta) = Z_P(K, \beta)$, i.e., constants over $k = 1, \cdots, K$. The subscript "P" here refers to "Pocock" since the critical values are related to the group sequential hypothesis test proposed by Pocock (1977). Alternatively, the critical values, $c_k(\alpha) = Z_B(K, \alpha)\sqrt{K/k}$, $c_k(\beta) = Z_B(K, \beta)\sqrt{K/k}$, can be used, based on the group sequential procedure of O'Brien and Fleming (1979). Values of the constants Z_P and Z_B for various choices of K, α, β are tabulated in, for example, Jennison and Turnbull (1989, Table 1). If the group sizes are unequal or unpredictable, the Lan and DeMets (1983) approach can be used. This method is based on a specified "error spending function," $f_\alpha(t)$, which is non-decreasing with $f_\alpha(0) = 0$ and $f_\alpha(t) = \alpha$ for $t \geq 1$. It also assumes a maximum number of subjects per treatment, $n_{\max}$, which will eventually be reached if early stopping does not occur. Suppose the total number of subjects per treatment observed in the first k groups is n_k, so marginally $W_k \sim N(n_k\theta, 2n_k\sigma^2)$, then the constants $c_k = c_k(\alpha)$ $(k \geq 1)$ are defined successively as the solutions of

$$P[W_i < \sigma c_i\sqrt{2n_i} \ (1 \leq i \leq k - 1), \ W_k > \sigma c_k\sqrt{2n_k} \mid \theta = 0]$$

$$= f_\alpha(\frac{n_k}{n_{\max}}) - f_\alpha(\frac{n_{k-1}}{n_{\max}}) \qquad (2.6)$$

for $k = 1, 2, \cdots, K$. The constants $c_k(\beta)$ are defined in precisely the same way in terms of error spending function $f_\beta(t)$ with $f_\beta(t) = \beta$ for $t \geq 1$. We shall consider error spending functions of the form $f_\gamma(t) = \min(\gamma, \gamma t^\rho)$ $(\gamma = \alpha$ or $\beta)$ with $\rho = 1$ and $\rho = 2$, which are common choices; see Kim and DeMets (1987). In designing a study based on an error spending function, it usually suffices to derive $n_{\max}$ and a target group size $n_{\max}/K$ under the assumption of equal group sizes. Implementation for unequal group sizes is straightforward: variations in the actual group sizes may cause an inconclusive result at stage K, i.e. neither $\bar{\theta}_K < \Delta$ nor $\underline{\theta}_K > 0$. In this case it would seem reasonable to reject equivalence to maintain a Type I error rate $\leq \alpha$ (condition (1.3)). This would result in a slight increase in the Type II error probability of wrongly rejecting equivalence, possibly above the nominal level β in (1.4); however, as long as the attained n_K exceeds or is close to $n_{\max}$, this effect will be negligible.

Table 1 shows error probabilities and expected sample sizes (ASN) for $K = 5$, $\sigma^2 = 1$, $\Delta = 0.2$, $\alpha = 0.05$ and $\beta = 0.05, 0.025, 0.005$. (Since the cost of incorrectly rejecting equivalence can be high, one might often specify $\beta \leq \alpha$.) All tests assume equal group sizes. Results are for four different methods of choosing the critical values c_k, namely those based on constant critical values Z_P (Pocock 1977), those based on decreasing values $Z_B\sqrt{K/k}$ (O'Brien and Fleming 1979), and those based on the error spending functions $f_\gamma(t) = \min(\gamma, \gamma t^\rho)$ $(\gamma = \alpha, \beta)$ for $\rho = 1$ and $\rho = 2$.

Table 1: Sample sizes and error probabilities of the procedure for normal data defined in Section 2 for $K = 5$, $\alpha = 0.05$, $\beta = 0.05$, 0.025 and 0.005, $\Delta = 0.2$ and $\sigma^2 = 1$. (n_{max} and ASN refer to the number of observations on each of the two treatment arms; group sizes for the sequential tests are $n = n_{max}/5$.; and O'B-F refers to O'Brien and Fleming while L-DeM refers to Lan and DeMets.)

	n_{max}	ASN $\theta = \Delta$	ASN $\theta = \Delta/2$	ASN $\theta = 0$	P(accept) $\theta = \Delta$	P(reject) $\theta = 0$
			$\beta = 0.05$			
Fixed	541	541	541	541	0.05	0.05
Pocock	901	341	466	341	0.044	0.044
O'B-F	613	379	466	379	0.046	0.046
L-DeM, $\rho=1$	763	341	458	341	0.044	0.044
L-DeM, $\rho=2$	641	357	457	357	0.046	0.046
			$\beta = 0.025$			
Fixed	650	650	650	650	0.05	0.025
Pocock	1028	414	553	367	0.045	0.022
O'B-F	719	468	557	421	0.047	0.023
L-DeM, $\rho=1$	895	414	545	371	0.045	0.022
L-DeM, $\rho=2$	762	430	545	394	0.046	0.023
			$\beta = 0.005$			
Fixed	891	891	891	891	0.05	0.005
Pocock	1305	585	728	412	0.047	0.0045
O'B-F	956	675	744	502	0.048	0.0045
L-DeM, $\rho=1$	1174	584	723	421	0.046	0.0045
L-DeM, $\rho=2$	1026	600	728	461	0.046	0.0046

The results of the fixed sample test ($K = 1$) are also shown for comparison. Note that, since σ is known, we can take $\sigma = 1$ without loss of generality. For general Δ and σ the maximum and expected sample sizes should be multiplied by $\sigma^2(0.2/\Delta)^2$. We note from the table that the error probabilities are slightly conservative as remarked earlier. The Pocock-based procedure has low expected sample sizes but a high maximum. As noted by DS (p. 333), the method employing the error spending function $\rho = 1$ has about the same expected sample sizes, but a more reasonable maximum sample size. At the cost of a slight increase in expected sample sizes, the error spending function method with $\rho = 2$ reduces the maximum sample size still further — closer to that of the fixed sample procedure.

3 Binomial Responses

3.1 The Durrleman and Simon Approach

Suppose now that responses X_{1i} and X_{2i} are binary (0 or 1) with success ($X = 1$) probabilities π_S on the standard and π_E on the experimental treatment. The problem is to test H: $\pi_S - \pi_E = \Delta$ with Type I error rate $\leq \alpha$ and power $\geq 1-\beta$ when $\pi_S = \pi_E$. It is convenient to reparametrize the problem by $\theta = \pi_S - \pi_E$, the treatment difference, and $\pi = \frac{1}{2}(\pi_S + \pi_E)$, a nuisance parameter which influences the variance of estimates of θ. At the kth analysis ($1 \leq k \leq K$) of a group sequential experiment, as before, we define $W_k = \sum_{i=1}^{n_k}(X_{1i} - X_{2i})$, which is marginally approximately normally distributed with mean $n_k\theta$ and variance $2n_k\sigma^2$, where

$$\sigma^2 = \{\pi_S(1 - \pi_S) + \pi_E(1 - \pi_E)\}/2. \tag{3.1}$$

The variance of W_k depends on π_S and π_E which are unknown. Operationally, DS proposed replacing the unknown σ in the formulae for confidence limits, (2.3) and (2.4), by an estimate $\hat{\sigma}_k$ where

$$2\hat{\sigma}_k^2 = p_{Sk}(1 - p_{Sk}) + p_{Ek}(1 - p_{Ek}) \tag{3.2}$$

and $p_{Sk} = \sum_{i=1}^{n_k} X_{1i}/n_k$, $p_{Ek} = \sum_{i=1}^{n_k} X_{2i}/n_k$, the observed proportions of successes on the two treatments in the first k groups of observations. A group sequential test is then obtained exactly as in Section 2.

However, the group size n and the maximum sample size $n_{\max} = Kn$ also depend on the unknown σ (see (2.5)) and these must typically be set in advance before any estimates of σ are available. DS took $\pi_E = \pi_S = \pi_S^*$, a specified value, e.g. 0.7, when determining $n_{\max}$ and $n = n_{\max}/K$ via (2.5) and (3.1). In DS (Table 2), they presented simulation results showing

Table 2: Error probabilities, probabilities of an inconclusive result, and expected sample sizes for the Durrleman and Simon (1990) procedure for binomial data with O'Brien and Fleming critical values, $K = 5$, $\alpha = \beta = 0.05$ and $\Delta = 0.1$. ($\theta = \pi_S - \pi_E$, ASN = expected sample size. Results are based on 50,000 replications. Standard errors for expected sample sizes are less than 1, and for estimates of probabilities around 0.05 are 0.001.)

		Design			
		$\pi_S^* = 0.6$		$\pi_S^* = 0.8$	
		($n_{max} = 595$)		($n_{max} = 400$)	
True value		$\theta = 0$	$\theta = \Delta$	$\theta = 0$	$\theta = \Delta$
$\pi_S = 0.6$	P(error)	0.047	0.046	0.047	0.050
	P(inconclusive)	0.000	0.003	0.073	0.077
	ASN	366	368	281	283
$\pi_S = 0.8$	P(error)	0.030	0.031	0.045	0.048
	P(inconclusive)	0.000	0.000	0.002	0.017
	ASN	311	330	243	257

acceptance probabilities and expected sample sizes for $K = 5$, $\alpha = \beta = 0.05$, $\Delta = 0.10$, $\pi_E = 0.60$, 0.65, 0.70, 0.75 and $\pi_S = 0.70$, the prespecified "design" value. They used critical values based on the Pocock, O'Brien and Fleming, and Lan and DeMets ($\rho = 1$) procedures as described in the previous section and also two further sets of critical values based on the procedure of Fleming, Harrington and O'Brien (1984).

A problem with this approach is that its operating characteristics are not very robust to poor choices of the design parameter π_S^*, i.e. when $\pi_S^* \neq \pi_S$. Also, when $\pi_S(1 - \pi_S) > \pi_S^*(1 - \pi_S^*)$, i.e. $|\pi_S - \frac{1}{2}| < |\pi_S^* - \frac{1}{2}|$, there can be a substantial probability of an inconclusive result, i.e. neither boundary crossed by the last stage. If this occurs, one can choose to reject equivalence, thereby preserving the Type I error (1.3), but possibly seriously inflating the Type II error (1.4). Alternatively, one could take additional observations beyond n_{max}, but this would be on an ad-hoc basis, and the error probabilities would not necessarily be controlled. To illustrate what can happen when the design π_S^* is different from the true π_S, we refer to Table 2. Using the same parameters as DS ($K = 5$, $\alpha = \beta = 0.05$,

$\Delta = 0.1$) and using boundaries based on O'Brien and Fleming critical values (DS, Table 1), Table 2 gives error rates, expected sample sizes and probabilities of an inconclusive result for $\pi_S = 0.6$, 0.8 and $\pi_S^* = 0.6$, 0.8. Remember that the test is of $\pi_S - \pi_E = \Delta$ versus $\pi_S = \pi_E$, so for $\pi_S = 0.6$, for example, Type I error rates are calculated at $\pi_S = 0.6$, $\pi_E = 0.4$ ($\theta = \Delta$) and Type II error rates at $\pi_S = \pi_E = 0.6$ ($\theta = 0$). The entries are based on a simulation study with 50,000 replications. From Table 2, we can see that if we design for $\pi_S^* = 0.8$ and the true $\pi_S = 0.6$, there is a 7–8% chance of an inconclusive result. If such results are treated as rejections of equivalence, the Type II error rate increases to 12% — much greater than the nominal β of 5%. On the other hand, if the design value is $\pi_S^* = 0.6$ and the true value is $\pi_S = 0.8$, the test is conservative with expected sample sizes 25–30% larger than necessary. This is not just a problem for the O'Brien and Fleming based procedure: further simulation results, not displayed, demonstrate that similar phenomena occur for the procedures based on the other sets of critical values considered by DS.

These problems are caused by the fact that, even though the cutoff values for W_k are allowed to depend on successive estimated values of π_S and π_E, the group sizes are not. We shall propose an adaptive procedure whereby the group sizes can depend on updated estimates of π_S and π_E. This procedure will have the disadvantage that a maximum sample size cannot be prespecified, but it will be able to guarantee prespecified error probabilities α and β (up to the adequacy of the normal approximation) with an efficient use of sample size, irrespective of the true values π_S and π_E. As a preliminary, in Section 3.2 we first consider the artificial problem when $\pi = \frac{1}{2}(\pi_S + \pi_E)$ is known and only $\theta = \pi_S - \pi_E$ is unknown. This leads on to the construction, in Section 3.3, of an adaptive procedure for when both π_S and π_E (and thus π and θ) are unknown.

3.2   The Case When the Average Success Probability is Assumed Known

In this artificial problem we assume $\pi = \frac{1}{2}(\pi_S + \pi_E)$ is known. We shall use an error spending function method to construct the critical values as described in Section 2. Again we denote cumulative sample sizes per treatment as $n_1, \cdots, n_K$ and assume $n_K = n_{max}$, the target maximum sample size. Under $\theta = \Delta$, we use the approximation

$$W_k \sim N(n_k \Delta, 2n_k \sigma_\Delta^2), \quad k = 1, 2, \cdots, K$$

with independent increments, where

$$\sigma_\Delta^2 = \sigma_\Delta^2(\pi) = \frac{1}{2}[(\pi - \frac{\Delta}{2})(1 - \pi + \frac{\Delta}{2}) + (\pi + \frac{\Delta}{2})(1 - \pi - \frac{\Delta}{2})].$$

Thus (2.3) becomes $\bar{\theta}_k = (W_k/n_k) + \sigma_\Delta c_k(\alpha)\sqrt{2/n_k}$ (cf. (2.3)), and the test derived from repeated confidence intervals stops to accept equivalence if $\bar{\theta}_k < \Delta$, i.e., if

$$W_k < n_k\Delta - \sigma_\Delta c_k(\alpha)\sqrt{2n_k}. \qquad (3.3)$$

For specified error spending function f_α, the critical values $\{c_k(\alpha); 1 \leq k \leq K\}$ are defined as in (2.6) using the normal approximation with σ_Δ replacing σ. Similarly under $\theta = 0$, W_k is approximately $N(0, 2n_k\sigma_0^2)$ distributed with $\sigma_0^2 = \sigma_0^2(\pi) = \pi(1-\pi)$. Thus (2.4) becomes $\underline{\theta}_k = (W_k/n_k) - \sigma_0 c_k(\beta)\sqrt{2/n_k}$ (cf. (2.4)) and the derived test stops to reject equivalence at analysis k if

$$W_k > \sigma_0 c_k(\beta)\sqrt{2n_k}. \qquad (3.4)$$

Again (2.6) is employed recursively with specified error spending function f_β to construct the constants $c_k(\beta)$ $(1 \leq k \leq K)$, this time with σ_0 replacing σ.

For equal group sizes $n_k = kn$ $(k = 1, 2, \cdots, K)$, the standardized critical values $\{c_k(\alpha), c_k(\beta); k = 1, \cdots, K\}$ do not depend on n. In this special but important case, we denote the critical values as $\{\tilde{c}_k(\alpha), \tilde{c}_k(\beta); k = 1, \cdots, K\}$ to emphasize that they depend only on K and the error spending functions, and not on the group size, n, nor $n_{\max} = Kn$. Thus, since π is assumed known and Δ has been specified, termination is ensured by analysis K when $n_K = Kn = n_{\max}$, by setting

$$n_{\max} = n_{\max}(\pi) = 2\{\tilde{c}_K(\alpha)\sigma_\Delta(\pi) + \tilde{c}_K(\beta)\sigma_0(\pi)\}^2/\Delta^2. \qquad (3.5)$$

This is a natural extension of (2.5), when the differences $\{X_{1i} - X_{2i}; i \geq 1\}$ have variance $2\sigma_\Delta^2$ under $\theta = \Delta$ but variance $2\sigma_0^2$ under $\theta = 0$. The necessary group size is then $n = n_{\max}/K$.

Table 3 shows expected sample sizes and error probabilities of this procedure for $K = 5$, $\alpha = \beta = 0.05$, $\Delta = 0.1$ and $f_\alpha(t) = f_\beta(t) = 0.05 \min(1, t^2)$ and $\pi = 0.5, 0.6, 0.7, 0.8, 0.9$. We prefer the choice of quadratic error spending function $(\rho = 2)$, rather than the linear one $(\rho = 1)$ of DS, because this leads to a lower maximum sample size, $n_{\max}$, closer to n_f, without much increase in expected sample size. All results are based on 50,000 simulations. The table shows not only the performance when the assumed or "design" value of π is equal to the true value, but also what happens when they are not equal. Note that if the true π does not equal the design value, no attempt is made to estimate the variances of the statistics W_k, $k = 1, \cdots, K$, from the data, rather, variances under $\theta = \Delta$ and $\theta = 0$ are taken to be σ_Δ^2 and σ_0^2 as defined above. Since the boundaries for accepting or rejecting equivalence converge, by design, at analysis K, the problem of inconclusive

Sequential Tests to Establish Equivalence Between Treatments   325

Table 3: Sample sizes and error probabilities of procedures of Sections 3.2 and 3.3 for binomial data with $K = 5$, $\alpha = \beta = 0.05$ and $\Delta = 0.1$ based on the Lan and DeMets error spending function $f(t) = 0.05\min(1, t^2)$, $t \geq 0$. (Results are based on 50,000 replications. Standard errors for expected sample sizes are all less than 1. Standard errors for estimates of probabilities around 0.05 are 0.001.)

Design Value of π	True Value of π	n_{max}	ASN $\theta=\Delta$	ASN $\theta=\Delta/2$	ASN $\theta=0$	P(accept) $\theta=\Delta$	P(reject) $\theta=0$
0.9	0.9	228	130	162	126	0.046	0.045
	0.8	228	122	140	119	0.123	0.112
	0.7	228	116	129	114	0.163	0.153
	0.6	228	113	123	111	0.183	0.172
	0.5	228	111	122	110	0.189	0.180
0.8	0.9	407	229	325	229	0.009	0.010
	0.8	407	226	289	226	0.047	0.045
	0.7	407	221	270	221	0.077	0.074
	0.6	407	217	261	218	0.091	0.093
	0.5	407	216	258	218	0.099	0.099
0.7	0.9	535	302	446	299	0.002	0.004
	0.8	535	303	404	301	0.023	0.025
	0.7	535	299	381	297	0.043	0.048
	0.6	535	297	370	295	0.059	0.062
	0.5	535	295	366	294	0.062	0.064
0.6	0.9	612	344	523	347	0.002	0.002
	0.8	612	346	477	347	0.015	0.017
	0.7	612	345	453	345	0.033	0.036
	0.6	612	343	438	342	0.045	0.044
	0.5	612	343	437	343	0.048	0.051
0.5	0.9	638	357	545	362	0.001	0.002
	0.8	638	359	500	363	0.013	0.015
	0.7	638	358	474	361	0.029	0.032
	0.6	638	358	461	360	0.042	0.043
	0.5	638	356	457	360	0.043	0.046
Adaptive test	0.9		135	171	136	0.044	0.039
	0.8		227	290	226	0.048	0.044
	0.7		301	382	298	0.046	0.046
	0.6		345	436	343	0.047	0.046
	0.5		360	454	359	0.045	0.046

tests does not arise. However, when the true variances of the W_k differ from the assumed values, the error probabilities can be quite different from their nominal values; this is seen in Table 3, where empirical error probabilities considerably larger than the nominal value of 0.05 appear when the true π is nearer to 0.5 than the design value. Our solution, described in detail in Section 3.3, is to modify the method, using values of σ_Δ^2 and σ_0^2 estimated from the accumulating data. These values can be used in defining boundaries for the W_k and also, in order to avoid the problem of inconclusive tests faced by the DS procedure of Section 3.1, to choose group sizes in such a way that the acceptance and rejection boundaries converge at the Kth analysis.

3.3   An Adaptive Procedure

For the case of unknown π, we aim to approximate the test for known π described in Section 3.2. At each analysis, an estimate of π is used to calculate a target maximum sample size and, hence, the next group size. This target maximum sample size appears in the argument of the error spending function. The current estimate of π is also used to estimate σ_0^2 and σ_Δ^2 when computing the next pair of critical values $c_k(\alpha)$, $c_k(\beta)$.

A natural estimate of π at analysis k is

$$\hat{\pi}_k = \frac{1}{2n_k} \sum_{i=1}^{n_k} (X_{1i} + X_{2i}), \quad k = 1, \cdots, K.$$

However, we have found from our simulation results that

$$\tilde{\pi}_k = \max\{0.1, \min(0.9, \hat{\pi}_k)\}, \quad k = 1, \cdots, K \tag{3.6}$$

is preferable. This is because the most serious problems occur when $\hat{\pi}_k(1 - \hat{\pi}_k)$ underestimates $\pi(1 - \pi)$, and in consequence σ_0^2, σ_Δ^2 and $n_{\max}$ are underestimated. The effect is most serious when $\pi > 0.9$ or $\pi < 0.1$. Our test, which we describe below, satisfies the error probability requirements (1.3), (1.4) for $0.1 \leq \pi \leq 0.9$, and is conservative for more extreme values. In general, cutoff values other than 0.1 and 0.9 could be used in (3.6) for different α, β, Δ, f_α, f_β. However, we have found our method, with estimates of π constrained to lie in the interval $[0.1, 0.9]$, to work well in a variety of situations with $\Delta = 0.1$. The limitation, when $\Delta = 0.1$, to conservative tests if $\pi < 0.1$ or $\pi > 0.9$ is not a major problem: if success probabilities are so close to 0 or 1, one would probably choose a much smaller value of Δ.

The adaptive procedure is as follows. An initial group size n_1 is chosen. Since corrections can be made later, it suffices to set $n_1 = n_{\max}(\pi_0)/K$

for any plausible value of π_0; in our simulations we used $n_1 = 100$, corresponding to $\pi_0 = 0.79$ when $\alpha = \beta = 0.05$, $K = 5$, $\Delta = 0.1$ and $f_\gamma(t) = \min(\gamma t^2, \gamma)$, $\gamma = \alpha, \beta$.

At stage k ($2 \leq k \leq K$), the kth group has $(k/K)n_{\max}(\tilde{\pi}_{k-1}) - n_{k-1}$ observations on each treatment, giving $n_k = (k/K)n_{\max}(\tilde{\pi}_{k-1})$. Here, the maximum sample size function, $n_{max}(\pi)$, is given by (3.5), the formula for a test with equally sized groups. The critical values $c_k(\alpha)$ are determined by solving successively

$$P\{W_1 > \sigma_\Delta(\tilde{\pi}_1)c_1(\alpha)\sqrt{2n_1} \quad \text{or} \quad \cdots \quad \text{or} \quad W_k > \sigma_\Delta(\tilde{\pi}_k)c_k(\alpha)\sqrt{2n_k}$$

$$|W_j \sim N(0, 2n_j\sigma_\Delta^2(\tilde{\pi}_k)), \quad j = 1, \cdots, k\}$$

$$= \begin{cases} f_\alpha(n_k/n_{\max}(\tilde{\pi}_k)) & k = 1, \cdots, K-1 \\ \alpha & k = K \end{cases} \tag{3.7}$$

Here $\sigma_\Delta^2(\pi)$ is as defined in Section 3.2. The critical values $c_k(\beta)$ are computed in the same way but with β replacing α, f_β replacing f_α and $\sigma_0(\tilde{\pi}_k)$ replacing $\sigma_\Delta(\tilde{\pi}_k)$. (Note that, since the distributions of the sequences $\{W_1, W_2, \cdots\}$ and $\{\tilde{\pi}_1, \tilde{\pi}_2, \cdots\}$ are approximately independent, we may allow future group sizes and critical values to depend on the current value $\tilde{\pi}_k$, without invalidating error probabilities associated with the sequential boundaries — see Lan and DeMets (1989), Jennison and Turnbull (1991a).) We now have all that is necessary to carry out the decision rule as described in (3.3) and (3.4).

The special treatment of $k = K$ in (3.7) ensures that in the ideal case when $\sigma_0^2(\tilde{\pi}_K) = \sigma_0^2(\pi)$ and $\sigma_\Delta^2(\tilde{\pi}_K) = \sigma_\Delta^2(\pi)$, the total errors spent are indeed precisely α and β. Although constants named $\tilde{c}_K(\alpha)$ and $\tilde{c}_K(\beta)$ appear in the definition of the maximum sample size function, (3.5), these are not the critical values used for the boundaries of the adaptive procedure. While it suffices to calculate the maximum sample size under the simplifying assumption of equal group sizes, the actual boundaries must be recalculated to guarantee the required error probabilities. If $\tilde{\pi}_k$ varies greatly with k, it is possible that $n_{k-1} > (k/K) n_{max}(\tilde{\pi}_{k-1})$ and our prescription gives a negative value for the kth group size. In this case, one could take the kth group size to be zero and omit the kth analysis. Other variations are possible; for example, in our simulations, we retained a minimum group size of 20 throughout. Also, if $\tilde{\pi}_k$ varies between analyses it is possible that $c_k(\alpha) = \infty$ fails to reduce the left hand side of (3.7) to its required value. The problem here is that, under the current estimate of σ_Δ^2, the error probability allocated up to analysis k has already been spent in analyses 1 to $k-1$; the solution is to set $c_k(\alpha) = \infty$ and move on to the next analysis. The same approach is adopted if a similar problem occurs with the solution for $c_k(\beta)$.

Simulation results for this adaptive procedure are shown at the bottom of Table 3. The close agreement of expected sample sizes and error probabilities with those of tests constructed using the true value of π demonstrates the success of this approach. It should be noted that, because π is being estimated, there is a very slight chance that the test will be inconclusive by stage K. In this case we would reject equivalence in order to preserve the Type I error (1.3). However, the simulations demonstrate that the effect is inconsequential and the Type II error is still controlled at level β.

A special case of this design is one where no early termination is permitted, but interim analyses are performed solely to make adjustments to the sample size so that the nominal error rates are guaranteed (Gould 1992). In this case we would use error spending functions $f_\gamma(t) = 0$ for $t < 1$, $f_\gamma(t) = \gamma$ for $t \geq 1$, for $\gamma = \alpha, \beta$. Thus we do not need to solve (3.7); we have $c_k(\alpha) = c_k(\beta) = \infty$ for $1 \leq k \leq K - 1$ and $c_K(\alpha) = \tilde{c}_K(\alpha) = z_\alpha$, $c_K(\beta) = \tilde{c}_K(\beta) = z_\beta$, the upper percentage points of the standard normal distribution (see (2.1)). At each stage only $\tilde{\pi}$ needs to be computed. Since $\tilde{\pi}$ depends only on the cumulative overall success proportion at each stage, the treatment assignment of each subject need not be revealed. In some trials, the blinding of patient assignments at interim looks may be an important consideration.

4   Discussion

We have presented one-sided group sequential tests of equivalence based on ideas related to repeated confidence intervals. This construction leads to simply defined tests which adapt readily to unpredictable group sizes, to non-normal observations, and to the possibility of continuing even though a boundary has been crossed. For further general discussion see Jennison and Turnbull (1989, 1991b).

The binomial example of Section 3 illustrates how a group sequential test can be designed in the presence of unknown nuisance parameters (here π) which affect the required sample size. This problem can arise more generally in other contexts. Examples include the case of response variables with unknown variances or whose variances depend on their means, the case of survival data with unknown baseline failure rate or competing risk censoring rate, and the case of bivariate responses where the correlation coefficient is unknown. The basic ideas of the adaptive approach are quite generally applicable and they offer a way to satisfy size and power constraints simultaneously in such situations.

We have been discussing the so-called one-sided equivalence problem, which was also the one addressed by Dunnett and Gent (1977) and by Durrleman and Simon (1990). As mentioned in Section 1, in a 2-sided problem we wish to reject equivalence if θ lies outside an interval, $(-\Delta, \Delta)$ say. The same basic ideas apply, but the details are now quite different, since acceptance of equivalence occurs if the sequence (k, W_k) crosses an "inner wedge" boundary. This problem is discussed in Jennison and Turnbull (1992).

Acknowledgment

This research was supported by Grant R01 GM28364 from the U.S. National Institutes of Health, by a U.K. SERC Visiting Fellowship GR/F 72864 and by the U.S. Army Research Office through the Mathematical Sciences Institute at Cornell University.

References

Dunnett, C.W. and Gent, M. (1977). Significance testing to establish equivalence between treatments, with special reference to data in the form of 2×2 tables. *Biometrics* **33** 593 - 602.

Durrleman, S. and Simon, R. (1990). Planning and monitoring of equivalence studies. *Biometrics* **46** 329 - 336.

Fleming, T.R., Harrington, D.R. and O'Brien, P.C. (1984). Designs for group sequential tests. *Control. Clin. Trials* **5** 348 - 361.

Gould, A.L (1992). Interim analyses for monitoring clinical trials that do not materially affect the Type I error rate. *Statist. Med.* **11** 55 - 66.

Jennison, C. and Turnbull, B.W. (1984). Repeated confidence intervals for group sequential clinical trials. *Control. Clin. Trials* **5** 33 - 45.

Jennison, C. and Turnbull, B.W. (1989). Interim analyses: the repeated confidence interval approach (with discussion). *Jour. Royal Statist.Soc., Series B* **51** 305 - 361.

Jennison, C. and Turnbull, B.W. (1991a). Group sequential tests and repeated confidence intervals. In *Handbook of Sequential Analysis* (B.K. Ghosh and P.K. Sen, eds.), Chapter 12, 283 - 311. Marcel Dekker, New York.

Jennison, C. and Turnbull, B.W. (1991b). Exact calculations for sequential t, χ^2 and F tests. *Biometrika* **78** 133 - 141.

Jennison, C. and Turnbull, B.W. (1992). Sequential equivalence testing and repeated confidence intervals, with applications to normal and binary responses. *Biometrics* **48**, To appear.

Kim, K. and DeMets, D.L. (1987). Design and analysis of group sequential tests based on the type I error spending rate function. *Biometrika* **74** 149 - 154.

Lan, K.K.G. and DeMets, D.L. (1983). Discrete sequential boundaries for clinical trials. *Biometrika* **70** 659 - 663.

Lan, K.K.G. and DeMets, D.L. (1989). Changing frequency of interim analysis in sequential monitoring. *Biometrics* **45** 1017 - 1020.

O'Brien, P.C. and Fleming, T.R. (1979). A multiple testing procedure for clinical trials. *Biometrics* **35** 549 - 556.

Pocock, S.J. (1977). Group sequential methods in the design and analysis of clinical trials. *Biometrika* **64** 191 - 199.

Robbins, H. (1970). Statistical methods related to the law of the iterated logarithm. *Ann. Math. Statist.* **41** 1397 - 1409.

Chapter 19

Subset Selection Procedures for Binomial Models Based on a Class of Priors and Some Applications

SHANTI S. GUPTA   Department of Statistics, Purdue University, West Lafayette, Indiana

YUNING LIAO   Department of Statistics, Purdue University, West Lafayette, Indiana

Abstract   The problem of selecting a subset containing the best of $k(\geq 2)$ binomial populations is studied. The approach is more general than the classical subset selection procedures studied by Gupta and Sobel (1960) and Gupta and McDonald (1986). In these preceding papers, the infimum of the probability of a correct selection occurs when all the parameters are equal (in the limit) to the largest unknown parameter. Thus it is natural to formulate the problem on the assumption that the largest unknown parameter follows a prior. Several priors have been considered and the associated procedures have been evaluated. Performance comparisons are made between the classical and the new procedures. Applications of this approach to the control problem and Poisson models are provided.

1   Introduction

In general, in order to solve a ranking and selection problem, we usually need to find the LFC, least favorable configuration (more details can be found in Bechhofer 1954, Gupta 1956 and Gupta and Panchapakesan 1979, etc). Fortunately, the LFC is not very difficult to find for many distributions under some general assumptions. Then a lower bound for the probability of a correct selection can be calculated under the LFC and the preassigned probability $P^*(1/k < P^* < 1)$ of a correct selection(CS) is guaranteed by choosing the lower bound under the LFC to be at least equal to the required P^* - value. In some situations the computation of the lower bound on the probability of a correct selection under the LFC does not depend on the parameter space. The location and scale-type parameter models are two well formulated examples of these.

However, if the LFC does not provide a usable lower bound for the probability of a correct selection (like the decision problem where we want to pick the population associated with the largest parameter from several Poisson populations), or if it is very difficult to determine (like the decision problem of the binomial models), then the computation of the lower bound of the probability of a correct selection turns out to be very difficult, and the classical selection approach cannot even be applied. The main reason for this difficulty, in the computation of any usable lower bound for the probability of a correct selection, is its dependence on the unknown parameter vector $\underline{\theta}$ itself.

In the following, we study the subset selection approach for the problem of selecting the best population from among k binomial populations, where we introduce the prior information into the problem. We formulate the problem based on the prior distribution of the largest unknown parameter $\theta_{[k]}$, where the prior distribution can be very loosely defined. This prior is very useful for determining a lower bound on the probability of a correct selection.

2   Binomial Populations

The classical subset selection procedure for selecting the population associated with the largest success probability from several binomial populations was first studied by Gupta and Sobel (1960). Let $\pi_1, \pi_2, \cdots, \pi_k$ denote k independent binomial populations. For each $i = 1, 2, \cdots, k$, let X_i denote the observed number of successes based on n independent observations from population π_i. Then X_i follows a binomial distribution with probability function $f(x|p_i)$, where

$$f(x|p_i) = \binom{n}{x} p_i^x (1 - p_i)^{n-x}, \quad x = 0, 1, \cdots, n; \quad 0 < p_i < 1.$$

Let $p_{[1]} \leq p_{[2]} \leq \cdots \leq p_{[k]}$ be the ranked values of the unknown p_i values, $i = 1, 2, \cdots, k$. The best population is the one associated with the largest parameter $p_{[k]}$. Here, the goal is to select a nonempty subset of $\pi_1, \pi_2, \cdots, \pi_k$, which contains the best population with a minimum guaranteed probability $P^*(1/k < P^* < 1)$, which is usually called the P^* - condition.

In Gupta and Sobel (1960), the following selection rule was proposed:

$$R^{max} : \text{ Select } \pi_i \text{ if and only if : } X_i \geq \max_{1 \leq j \leq k} X_j - d, \quad 0 \leq d \leq n.$$

where d is an integer and is determined to satisfy the P^* - condition. Note d depends on the values of n, k, and P^*.

Let $X_{(i)}$ be the observation of the successes associated with the unknown parameter $p_{[i]}$, $i = 1, 2, \cdots, k$. Then

$$P(CS|R^{max})$$
$$= P\left(X_{(k)} \geq X_{(j)} - d, \ j = 1, 2, \cdots, k - 1\right)$$
$$= \sum_{x=0}^{n} \binom{n}{x} p_{[k]}^x (1 - p_{[k]})^{n-x} \prod_{j=1}^{k-1} \left\{ \sum_{t=0}^{x+d} \binom{n}{t} p_{[j]}^t (1 - p_{[j]})^{n-t} \right\}$$
$$\geq \sum_{x=0}^{n} \binom{n}{x} p_{[k]}^x (1 - p_{[k]})^{n-x} \left\{ \sum_{t=0}^{x+d} \binom{n}{t} p_{[k]}^t (1 - p_{[k]})^{n-t} \right\}^{k-1} \quad (2.1)$$

where the inequality follows from the fact that the sum in the braces is a non-increasing function of $p_{[j]}$, since this summation can be represented by an incomplete beta function: this was also observed by Gupta and Sobel (1960).

In order to meet the P^* - condition, it suffices to find the smallest integer $d(0 \leq d \leq n)$ such that

$$\inf_{0 \leq p \leq 1} P(CS|p, d) \geq P^*,$$

where

$$P(CS|p, d) = \sum_{x=0}^{n} \binom{n}{x} p^x (1-p)^{n-x} \left\{ \sum_{j=0}^{x+d} \binom{n}{j} p^j (1-p)^{n-j} \right\}^{k-1}.$$

Recently, the Gupta-Sobel rule R^{max} has been extensively studied by Gupta and McDonald (1986) and by Gupta et al. (1976). It was shown by Gupta and Sobel (1960) that the infimum value of $P(CS|p,d)$ occurs at $p = 0.5$ for $k = 2$. For $k > 2$, the minimizing value of p never exceeds 0.5 for all n. For larger n, the infimum takes place for the values of p between 0.4 and 0.5. In the limit (as $n \to \infty$), the infimum is, of course, at $p = 0.5$ (Gupta and McDonald 1986). In Gupta et al. (1976), they used the conditional tests to compute the (conservative) values of d. Their non-randomized rule is defined as follows:

$$R_2 : \quad \text{Select } \pi_i \text{ if and only if}: \ X_i \geq \max_{1 \leq j \leq k} X_j - D(t),$$

given $T = \sum_{i=1}^{k} X_i = t$, where $D(t) > 0$ depends on the observed t and is chosen to satisfy the P^* - condition.

Gupta et al. (1976) showed that when $k = 2$, the infimum of $P(CS|R_2)$ is attained when $p_1 = p_2 = p$ and that this infimum is independent of the common value p. For $k > 2$, they proved that $\inf_\Omega P(CS|R_2)$ is always $\geq P^*$, if $D(t)$ is chosen as follows:

$$D(t) = \begin{cases} d(t) & \text{for } k = 2, \\ \max\{d(r) : r = 0, 1, \cdots, \min(t, 2n)\} & \text{for } k > 2, \end{cases}$$

where $d(r)$ is defined as the smallest value such that

$$N(k; d(r), r, n) = \begin{cases} P^* \binom{2n}{r} & \text{for } k = 2, \\ (1 - \frac{1-P^*}{k-1})\binom{2n}{r} & \text{for } k > 2, \end{cases} \qquad (2.2)$$

and

$$N(k; d(r), r, n) = \sum \binom{n}{s_1} \cdots \binom{n}{s_k},$$

with the summation taken over the set of all nonnegative integer s_i such that $\sum_{i=1}^{k} s_i = t$, and $s_k \geq \max_{1 \leq j \leq k-1} s_j - d(r)$.

A conservative value of the constant d for the classical procedure R^{max} is then given by taking $d = \max_{0 \leq t \leq kn} d(t)$ of above. Gupta et al. (1976) have tabulated the smallest value $d(t)$ satisfying (2.2) for $k = 2, 4(1)10$, $n = 1(1)10$, $t = 1(1)10$, and $P^* = 0.75, 0.90, 0.95, 0.99$. These (conservative) d - values for the procedure R^{max} for the same P^* values and $n = 1(1)4$

when $k = 3(1)15$, and $n = 5(1)10$ with $k = 3(1)5$ were also tabulated by these authors.

On examining the inequality (2.1), we see that the infimum of the probability of a correct selection occurs when all the parameters are equal (in the limit) to the largest unknown parameter. Therefore, it is natural to formulate the selection problem on the assumption that the largest unknown parameter follows a prior distribution.

Let $G(\cdot)$ be the prior distribution on the space of the largest parameter $p_{[k]}$. The $P(CS)$, using the natural subset selection rule R^{max}, has the following lower bound:

$$P(CS|G, R^{max})$$

$$= \int_0^1 P\left(X_{(k)} \geq X_{(j)} - d, \ j = 1, 2, \cdots, k - 1 | p_{[k]}\right)$$

$$= \int_0^1 \left(\sum_{x=0}^{n} \binom{n}{x} p_{[k]}^x (1 - p_{[k]})^{n-x} \prod_{j=1}^{k-1} \left\{ \sum_{t=0}^{x+d} \binom{n}{t} p_{[j]}^t (1 - p_{[j]})^{n-t} \right\} |p_{[k]} \right)$$

$$\geq \int_0^1 \sum_{x=0}^{n} \binom{n}{x} p_{[k]}^x (1 - p_{[k]})^{n-x} \left\{ \sum_{t=0}^{x+d} \binom{n}{t} p_{[k]}^t (1 - p_{[k]})^{n-t} \right\}^{k-1} \quad (2.3)$$

(The integrals are with respect to the measure $dG(p_{[k]})$.) Note that this inequality is similar to (2.1).

Now, we proceed to classify the prior information about the largest (unknown) parameter $p_{[k]}$ into three categories,

(A) the prior distribution of $p_{[k]}$ is completely known and is proper,

(B) the prior distribution of $p_{[k]}$ is totally unknown,

(C) the prior distribution of $p_{[k]}$ is partly known.

In the above, the assumption (A) though unrealistic, can still be used to obtain results for the case (B) and case (C).

3   Results Associated with the Prior $G(\cdot)$

From (2.6), we see that the prior distribution of $p_{[k]}$ determines a lower bound on the $P(CS)$. We will consider the problem, when $G(\cdot)$ falls into one of above three categories.

(A) The distribution function $G(\cdot)$ is completely known

For any given $P^*(1/k < P^* < 1)$, if $G(\cdot)$ is a proper distribution, then we can find the smallest integer $d(0 \leq d \leq n)$ satisfying

$$P(CS|G, R^{max}) \geq P^*.$$

Actually, we need only to find the smallest integer $d(0 \le d \le n)$ satisfying

$$P(n, k, d, G) \ge P^*,$$

where

$$P(n, k, d, G) = \int_0^1 \sum_{x=0}^n \binom{n}{x} p^x (1-p)^{n-x} \left\{ \sum_{t=0}^{x+d} \binom{n}{t} p^t (1-p)^{n-t} \right\}^{k-1} dG(p).$$

$$(3.1)$$

For a specific choice of $G(\cdot)$, we give the following proposition.

Proposition 3.1

Let

$$\mathcal{B}_e = \left\{ b_{\alpha,\beta}(p) = \frac{p^{(\alpha-1)}(1-p)^{(\beta-1)}}{B(\alpha,\beta)} \; : \; \alpha, \; \beta > 0 \right\},$$

where $B(\alpha, \beta) = [\Gamma(\alpha)\Gamma(\beta)]/\Gamma(\alpha + \beta)]$ is the beta function. Then, for any beta distribution $b_{\alpha,\beta} \in \mathcal{B}_e$, we have

$$P(n, k, d, b_{\alpha,\beta}) = \frac{1}{B(\alpha,\beta)} \sum_{x=0}^n \binom{n}{x} \sum_{h=0}^{(k-1)(x+d)} \mathcal{H}(h) \mathcal{R}_{\alpha,\beta}(h, x),$$

where

$$\mathcal{H}(h) = \prod_{\substack{j_1, j_2, \cdots, j_{k-1} < h}}^{j_1 + j_2 + \cdots + j_{k-1} = h} \binom{n}{j_l},$$

and

$$\mathcal{R}_{\alpha,\beta}(h, x) = B(h + x + \alpha, nk - h - x + \beta).$$

We note that $P(n, k, d, b_{\alpha,\beta})$ can be written as a summation of a series of weighted beta functions, because the conjugate prior for a binomial distribution is a beta distribution.

(B) The distribution function $G(\cdot)$ is completely unknown

When G is completely unknown, since

$$\inf_{G \in \mathcal{G}} P(n, k, d, G) = \inf_{0 < p < 1} P(CS|p, d),$$

where $\mathcal{G}$ is the class of all distributions defined on the interval $(0, 1)$. Note that the right hand side above is the same expression as that of Gupta and Sobel (1960). So, by using the above lower bound for the $P(CS)$, we have the usual classical subset selection procedure. Or one may use the "*noninformative*" priors, one of them is the uniform distribution,

$$\mathcal{U}(p) = I_{\{0 < p < 1\}},$$

which is a member of $\mathcal{B}_e$, for $\alpha = \beta = 1$. Then, we have

$$P(n, k, d, \mathcal{U}) = \frac{\sum_{x=0}^{n} \sum_{j_1=0}^{x+d} \cdots \sum_{j_{k-1}}^{x+d} \binom{n}{x} \binom{n}{j_1} \cdots \binom{n}{j_{k-1}}}{(nk + 1)\binom{nk}{x+j_1+\cdots+j_{k-1}}}.$$

Another "noninformative" prior is the so-called Jeffreys prior with the following density function

$$
\begin{aligned}
G_J(p) &= \frac{[p(1 - p)]^{-\frac{1}{2}}}{B(\frac{1}{2}, \frac{1}{2})} \\
&= \frac{1}{\pi} [p(1 - p)]^{-\frac{1}{2}}.
\end{aligned}
$$

which, again, is a member of $\mathcal{B}_e$ with $\alpha = \beta = \frac{1}{2}$.

Using the uniform prior and Jeffreys prior, we have computed the values of $P(n, k, d, G)$, for $n = 2(1)10$, $k = 2(1)5$, $d = 0, 1, \cdots, n - 1$ in Tables 1 and 2. Thus for specified values of $P(n, k, d, G)$, one can find d values from these tables. For example, if $n = 6$, $k = 4$, for $P^* \leq 0.9069$ for the uniform prior (and $\leq (0.9346)$ for Jeffreys prior), we have $d_{\mathcal{U}}(d_{G_J}) \leq 2$, where $d_{\mathcal{U}}(d_{G_J})$ is the constant used in the subset selection rule R^{max}, when the prior of $p_{[k]}$ is assumed to be uniform (Jeffreys). Note that for $P^* = 0.90$, from Gupta and Sobel (1960) and Gupta et al. (1976), we have $d = 3$.

From Table 3, note also that the lower bounds are much larger than those of the classical procedure provided, for instance, when $n = 5$, $k = 3$, $d = 1$, the lower bound of the guarantee $P(CS)$ increases from 0.7265 for the classical procedure to 0.8188 for $G = U$, and to 0.8675 for $G = G_J$ (see column of $d = 1$ in Table 3). For other choices of n, k, d, the values of $P(n, k, d, G)$, when $G = U(G_J)$, can also be easily calculated by (3.1).

Remark 3.1   The classical subset selection rules choose the constant assuming a prior parameter distribution which places all its mass on the parameter configuration which algebraically minimizes the $P(CS)$. Thus any other prior distribution on the parameter $p_{[k]}$ will result in a larger $P(CS)$ value for a given selection rule constant. So, if one has knowledge of the prior distribution of $p_{[k]}$, one can use it and get a better result.

(C) The distribution function $G(\cdot)$ is partly known

(i) The support (say $A \subset [0, 1]$) of the distribution $G(\cdot)$ is known, but the exact form of the distribution function $G(\cdot)$ is unknown.

Table 1: $P(n, k, d, G)$ values for $G = U$ and $n = 2(1)7$; $k = 2(1)5$; $0 \le d < n$.

n	d	k			
		2	3	4	5
2	0	0.7667	0.6571	0.5929	0.5504
	1	0.9667	0.9429	0.9246	0.9100
3	0	0.7286	0.6060	0.5350	0.4884
	1	0.9357	0.8929	0.8613	0.8367
	2	0.9929	0.9869	0.9818	0.9773
4	0	0.7032	0.5726	0.4979	0.4491
	1	0.9095	0.8522	0.8111	0.7798
	2	0.9825	0.9687	0.9573	0.9476
	3	0.9984	0.9970	0.9957	0.9945
5	0	0.6847	0.5488	0.4716	0.4214
	1	0.8875	0.8188	0.7708	0.7346
	2	0.9711	0.9493	0.9317	0.9170
	3	0.9953	0.9912	0.9876	0.9844
	4	0.9996	0.9993	0.9990	0.9987
6	0	0.6705	0.5307	0.4518	0.4007
	1	0.8686	0.7910	0.7376	0.6979
	2	0.9596	0.9301	0.9069	0.8878
	3	0.9911	0.9836	0.9771	0.9714
	4	0.9988	0.9976	0.9965	0.9956
	5	0.9999	0.9998	0.9998	0.9997
7	0	0.6591	0.5163	0.4362	0.3845
	1	0.8524	0.7674	0.7098	0.6674
	2	0.9484	0.9119	0.8836	0.8607
	3	0.9861	0.9748	0.9652	0.9568
	4	0.9973	0.9949	0.9927	0.9907
	5	0.9997	0.9994	0.9991	0.9988
	6	1.000	1.000	0.9999	0.9999

Table 1: $P(n,k,d,G)$ values for $G=U$ and $n=8(1)10$; $k=2(1)5$; $0 \leq d < n$ (continued).

n	d	k 2	3	4	5
	0	0.6498	0.5046	0.4236	0.3715
	1	0.8382	0.7470	0.6861	0.6416
	2	0.9377	0.8947	0.8620	0.8358
	3	0.9807	0.9654	0.9525	0.9415
	4	0.9954	0.9919	0.9878	0.9845
8	5	0.9992	0.9985	0.9978	0.9971
	6	0.9999	0.9998	0.9998	0.9997
	7	1.0000	1.0000	1.0000	1.0000
	0	0.6419	0.4948	0.4131	0.3606
	1	0.8257	0.7293	0.6656	0.6194
	2	0.9276	0.8787	0.8421	0.8131
	3	0.9750	0.9556	0.9396	0.9260
	4	0.9931	0.9871	0.9819	0.9771
9	5	0.9985	0.9972	0.9959	0.9947
	6	0.9998	0.9996	0.9994	0.9991
	7	1.0000	1.0000	0.9999	0.9999
	8	1.0000	1.0000	1.0000	1.0000
	0	0.6351	0.4865	0.4042	0.3515
	1	0.8145	0.7137	0.6476	0.6000
	2	0.9181	0.8639	0.8238	0.7923
	3	0.9692	0.9458	0.9268	0.9108
	4	0.9904	0.9823	0.9753	0.9690
	5	0.9976	0.9954	0.9934	0.9916
10	6	0.9995	0.9991	0.9987	0.9983
	7	0.9999	0.9999	0.9998	0.9998
	8	1.0000	1.0000	1.0000	1.0000

Table 2: $P(n, k, d, G)$ values for $G = G_J$ and $n = 2(1)7$; $k = 2(1)5$; $0 \leq d < n$.

n	d	k			
		2	3	4	5
2	0	0.8203	0.7355	0.6839	0.6486
	1	0.9766	0.9600	0.9471	0.9368
3	0	0.7871	0.6898	0.6320	0.5929
	1	0.9541	0.9236	0.9010	0.8832
	2	0.9951	0.9911	0.9876	0.9846
4	0	0.7640	0.6586	0.5968	0.5554
	1	0.9346	0.8931	0.8632	0.8401
	2	0.9879	0.9785	0.9706	0.9639
	3	0.9989	0.9980	0.9971	0.9963
5	0	0.7465	0.6357	0.5708	0.5278
	1	0.9178	0.8675	0.8320	0.8049
	2	0.9799	0.9647	0.9525	0.9422
	3	0.9968	0.9941	0.9916	0.9894
	4	0.9998	0.9995	0.9993	0.9991
6	0	0.7327	0.6172	0.5505	0.5064
	1	0.9032	0.8456	0.8057	0.7757
	2	0.9717	0.9510	0.9346	0.9211
	3	0.9939	0.9888	0.9844	0.9805
	4	0.9991	0.9984	0.9977	0.9970
	5	0.9999	0.9999	0.9998	0.9998
7	0	0.7213	0.6023	0.5341	0.4892
	1	0.8904	0.8268	0.7832	0.7508
	2	0.9636	0.9377	0.9176	0.9013
	3	0.9905	0.9827	0.9761	0.9704
	4	0.9982	0.9966	0.9951	0.9937
	5	0.9998	0.9996	0.9994	0.9992
	6	1.0000	1.0000	1.0000	1.0000

Table 2: $P(n, k, d, G)$ values for $G = G_J$ and $n = 8(1)10$; $k = 2(1)5$; $0 \leq d < n$ (continued).

n	d	k			
		2	3	4	5
	0	0.7118	0.5899	0.5204	0.4748
	1	0.8791	0.8102	0.7637	0.7293
	2	0.9558	0.9251	0.9017	0.8828
	3	0.9867	0.9761	0.9673	0.9596
	4	0.9969	0.9942	0.9917	0.9895
8	5	0.9995	0.9990	0.9985	0.9981
	6	0.9999	0.9999	0.9998	0.9998
	7	1.000	1.000	1.000	1.000
	0	0.7035	0.5793	0.5087	0.4626
	1	0.8689	0.7956	0.7465	0.7105
	2	0.9483	0.9132	0.8868	0.8657
	3	0.9827	0.9692	0.9582	0.9487
	4	0.9953	0.9913	0.9877	0.9845
9	5	0.9990	0.9981	0.9973	0.9965
	6	0.9999	0.9997	0.9996	0.9994
	7	10000	1.000	0.9999	0.9999
	8	1.000	1.000	1.000	1.000
	0	0.6964	0.5701	0.4989	0.4522
	1	0.8597	0.7825	0.7313	0.6939
	2	0.9412	0.9020	0.8729	0.8498
	3	0.9786	0.9623	0.9490	0.9378
	4	0.9935	0.9880	0.9832	0.9789
	5	0.9984	0.9969	0.9956	0.9943
10	6	0.9997	0.9994	0.9991	0.9988
	7	0.9999	0.9999	0.9999	0.9998
	8	1.000	1.000	1.000	1.000

Table 3: $P(n,k,d,G)$ values for various procedures and $n = 5$, $k = 3$.

Procedures	\multicolumn{6}{c}{d}
	0	1	2	3	4	5
$\displaystyle\inf_{0<p<1} P(CS\mid p,d)$	.4673	.7265	.9040	.9797	.9981	1.000
$\displaystyle\inf_{.75<p<1} P(CS\mid p,d)$	.5039	.8045	.9528	.9937	.9997	1.000
$\displaystyle\inf_{0<\varepsilon<1} P(n,k,d,Q^{\varepsilon}_{\frac{1}{2}})$	.4676	.7269	.9041	.9798	.9981	1.000
$\displaystyle\inf_{0<\varepsilon<1} P(n,k,d,Q^{\varepsilon}_{\frac{1}{4}})$	.4844	.7718	.9378	.9888	.9991	1.000
$\displaystyle\inf_{0<\varepsilon<1} P(n,k,d,Q^{\varepsilon}_{\frac{3}{4}})$	.5039	.7975	.9396	.9890	.9991	1.000
$\displaystyle\inf_{0<\varepsilon<1} P(n,k,d,Q^{\varepsilon}_{.0})$	.5143	.7901	.9378	.9888	.9991	1.000
$\displaystyle\inf_{0<\varepsilon<1} P(n,k,d,Q^{\varepsilon}_{1.0})$	.5300	.7975	.9396	.9890	.9991	1.000
$P(n,k,d,U)$	.5488	.8188	.9493	.9912	.9993	1.000
$P(n,k,d,G_J)$	.6357	.8675	.9647	.9941	.9995	1.000

Then, we may use the infimum of $P(CS\mid d,p)$ on A, or a uniform prior on A, etc. Actually, it is reasonable for the experimenter to know that the support of G is not the whole parameter space. For instance, we may know that all the populations have their success rate very high, say at least higher than $\frac{3}{4}$, so the support of the largest (unknown) $p_{[k]}$ is not the whole interval $[0,1]$, but a subset of it (i.e. $[\frac{3}{4},1]$). Note the argument above is similar to that of Gupta and McDonald (1986), where they defined the subset selection rule R'_B. But now we go further and apply any reasonable prior to compute a lower bound instead of computing only the infimum of $P(CS\mid d,p)$ on A as was done by Gupta and McDonald (1986).

(ii) $G(\cdot)$ is in a ε-contamination class of a known distribution G_0, say, G belongs to the class Γ, where

$$\Gamma = \{G : \ G = (1-\varepsilon)G_0 + \varepsilon Q, \ Q \in \mathcal{Q}\}.$$

The distribution G_0 can be any known proper function. For the choice of $\mathcal{Q}$, one may consider the case (B) and case (i) of (C) above.

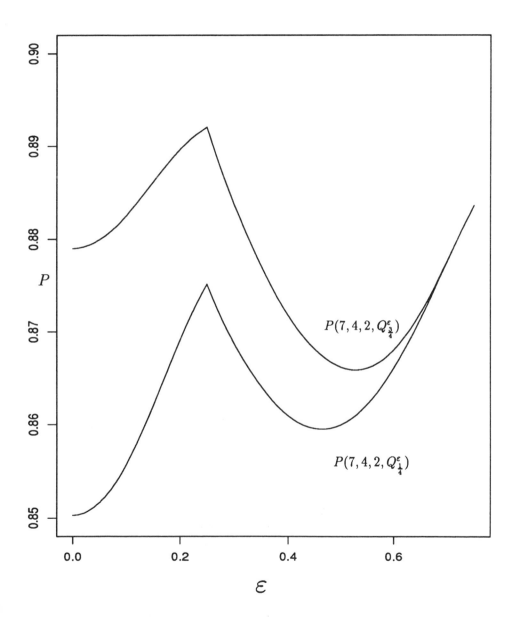

Figure 1: $P(7, 4, 2, Q_{a_0}^\varepsilon)$, for $a_0 = \frac{1}{4}(\frac{3}{4})$.

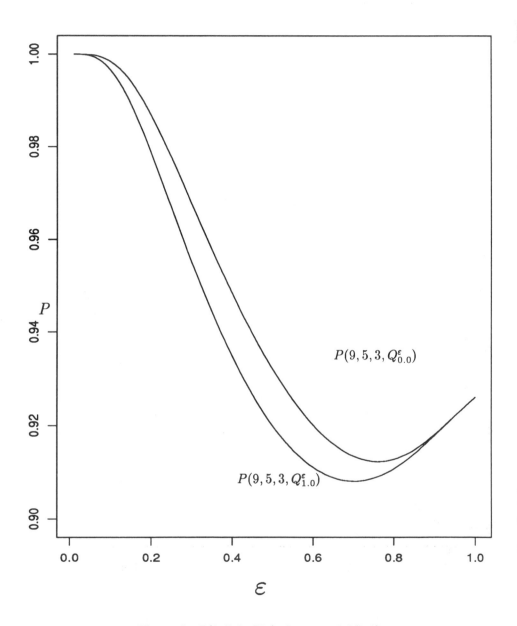

Figure 2: $P(9, 5, 3, Q_{a_0}^{\varepsilon})$, for $a_0 = 0.0(1.0)$.

(iii) Another interesting choice of $G(\cdot)$ is: $G \in \mathcal{Q}_{a_0,b_0}$, where $\mathcal{Q}_{a_0,b_0}$ is

$$Q : Q \text{ is uniform on } (max\{0, a_0 - \varepsilon\}, min\{1, b_0 + \varepsilon\})$$

where $0 \leq a_0 \leq b_0 \leq 1, \varepsilon > 0$. The above choice may be appropriate when the practitioner can specify an interval (or a point), say $[a_0, b_0]$, within which one is reasonably confident that the parameter $p_{[k]}$ most likely falls in.

Figure 1 and Figure 2 show the behavior of $P(n, k, d, Q_{a_0}^\varepsilon)$ as a function of ε, when $a_0 = \frac{1}{4}(\frac{3}{4})$ and $a_0 = 0(1)$ for some selected integer values of n, k, and d, here $Q_{a_0}^\varepsilon$ is the uniform distribution function on the interval $(max\{0, a_0 - \varepsilon\}, min\{1, a_0 + \varepsilon\})$. In Figure 1, as ε increases from 0.0 to 0.25, $P(n, k, d, Q_{a_0}^\varepsilon)$ increases and then it decreases and increases again. This happens for both $a_0 = \frac{1}{4}$ and $a_0 = \frac{3}{4}$. In Figure 2, at $\varepsilon = 0.0$, $P(n, k, d, Q_{a_0}^\varepsilon)$ equals one, as expected, and then decreases and later it increases. As stated earlier, such figures can be used to study the behavior of the lower bound in a neighborhood of $[a_0, b_0]$.

Table 3 shows how different procedures (for different priors) perform. Here we can see that if the largest (unknown) parameter $p_{[k]}$ is believed to be larger than .75, the lower bounds of the $P(CS)$ provided by the classical procedure should be replaced by $\inf\limits_{.75 < p < 1} P(CS|p, d)$, which is much better than the one provided by the classical procedure.

The above approach can be easily applied to the following problems.

4   Populations Better than Control

Let $p_0 \in (0, 1)$ denote the unknown probability of a unit being defective in the control population π_0. Population π_i is said to be *good* if $p_i \leq p_0$, and *bad* if $p_i > p_0$, $i = 1, 2, \cdots, k$ (see Gupta and Sobel 1958). Here, we assume that $\pi_1, \pi_2, \cdots, \pi_k$ are k independent binomial populations, and for each i, $i = 1, 2, \cdots, k$, we have a random observation X_i, which is the number of defectives observed in a sample of n independent observations arising from the population π_i.

For the parameter p_0 associated with the control population π_0, we assume that there is a proper prior distribution $G(\cdot)$ for p_0. In general, if $G(\cdot)$ is not degenerated, then we have to assume that there is a sample X_0 arising from population π_0 which follows a distribution $F(\cdot)$, where $F(\cdot)$ may not be a binomial distribution, but must be characterized only by the control parameter p_0 (i.e. p_0 is the only parameter of F). Further we assume that p_0 can be estimated by X_0, for example, X_0 is the unbiased estimate of p_0. Here, we only consider the case where F, again, is binomial with parameter n and p_0.

Then, using the natural subset selection rule R^{min} defined as follows:

R^{min} : Retain π_i, if and only if : $X_i < X_0 + d$, for some d.

We define a correct selection(CS) to be such that all $k_1(\leq k)$ good populations are selected in the selected subset. So

$$P(CS|R^{min}, G, F)$$

$$= \int_{\mathcal{P}_0} \int_{\mathcal{X}_0} P\{X_i < X_0 + d, \ i = 1, 2, \cdots, k_1 | \ p_0, x_0\} \, dF(x_0) dG(p_0)$$

$$= \int_{\mathcal{P}_0} \int_{\mathcal{X}_0} \prod_{i=1}^{k_1} \left[\sum_{t=0}^{x_0+d} \binom{n}{t} p_i^t (1 - p_i)^{n-t} \mid p_0, x_0 \right] dF(x_0) dG(p_0)$$

$$\geq \int_{\mathcal{P}_0} \int_{\mathcal{X}_0} \left[\sum_{t=0}^{x_0+d} \binom{n}{t} p_0^t (1 - p_0)^{n-t} \right]^{k_1} dF(x_0) dG(p_0)$$

$$\geq \int_{\mathcal{P}_0} \int_{\mathcal{X}_0} \left[\sum_{t=0}^{x_0+d} \binom{n}{t} p_0^t (1 - p_0)^{n-t} \right]^{k} dF(x_0) dG(p_0), \qquad (4.1)$$

where $k = k_1 + k_2$, with k_1 being the number of the populations that is better than the standard. $\mathcal{P}_0$ and $\mathcal{X}_0$ are supports of p_0 and x_0, respectively.

To meet the so-called P^*-requirement, one need only to find the smallest integer $d(0 \leq d \leq n)$ satisfying

$$\int_{\mathcal{P}_0} \int_{\mathcal{X}_0} \left[\sum_{t=0}^{x_0+d} \binom{n}{t} p_0^t (1 - p_0)^{n-t} \right]^{k} dF(x_0) dG(p_0) \geq P^*.$$

Clearly, such a d exists, when both G and F are proper distributions.

In our case, since F is also a binomial distribution $(B(n, p_0))$, so the extreme right hand side of (4.1) becomes

$$\int_{\mathcal{P}_0} \sum_{x=0}^{n} \binom{n}{x} p_0^x (1 - p_0)^{n-x} \left[\sum_{t=0}^{x+d} \binom{n}{t} p_0^t (1 - p_0)^{n-t} \right]^{k} dG(p_0). \qquad (4.2)$$

Furthermore, if the prior G is chosen to minimize the probability function on the right hand side of (4.2), then the whole selection process coincides with that of the classical subset selection approach (Gupta and Sobel 1958), as is expected. That is, the desired value of d is the smallest integer for which

$$\min_{0 \leq p_0 \leq 1} \left\{ \sum_{x=0}^{n} \binom{n}{x} p_0^x (1 - p_0)^{n-x} \left[\sum_{t=0}^{x+d} \binom{n}{t} p_0^t (1 - p_0)^{n-t} \right]^{k} \right\} \geq P^*.$$

For other situations, the following remark applies.

Remark 4.1   For the choice of the prior distribution G, one can proceed as in Section 3.

5   Poisson Populations

It has been shown that the usual R^{max} type of selection procedures for selecting the largest parameter from among k Poisson populations does not exist for some values of the lower bound $P^*(1/k < P^* < 1)$ of a correct selection (Goel 1972). Moreover, Leong and Wong (1977) showed that the infimum of the probability of a correct selection, using the classical type of selection procedure, is $1/k$ (see also Gupta et al. 1979).

Since then, some different approaches and techniques have been proposed by Gupta and Huang (1975), Goel (1975) and Gupta and Wong (1977) etc. In particular, in the paper of Gupta and Huang (1975), they successfully avoided the difficulty of finding

$$P(CS|R^{max}) \geq \inf_{0<\lambda<\infty} P(CS|\lambda,d),$$

where

$$P(CS|\lambda,d) = \sum_{x=0}^{\infty} \frac{e^{-\lambda}\lambda^x}{x!} \left\{ \sum_{t=0}^{x+d} \frac{e^{-\lambda}\lambda^t}{t!} \right\}^{k-1},$$

by using the scale-type rule and the following inequality

$$\sum_{x=0}^{\infty} \frac{e^{-\lambda}\lambda^x}{x!} \left\{ \sum_{t=0}^{\left[\frac{x+1}{c}\right]} \frac{e^{-\lambda}\lambda^t}{t!} \right\}^{k-1} \geq \left\{ \sum_{x=0}^{\infty} \frac{e^{-\lambda}\lambda^x}{x!} \sum_{t=0}^{\left[\frac{x+1}{c}\right]} \frac{e^{-\lambda}\lambda^t}{t!} \right\}^{k-1}.$$

In the following, we apply the same technique which had been used in previous sections to handle the Poisson models, that is, by considering the prior information about the largest (unknown) parameter.

5.1   Formulation of the Problem and Some Results

Let $\pi_1, \pi_2, \cdots, \pi_k$ be k Poisson populations with parameters $\lambda_1, \cdots, \lambda_k$ respectively. For each π_i, let X_i denote a random observation arising from the ith population. It is assumed that X_i follows a Poisson distribution with probability mass function $f(x|\lambda_i)$, where

$$f(x|\lambda_i) = \frac{e^{-\lambda_i}\lambda_i^x}{x!}, \qquad x = 0,1,2,\cdots; \quad \lambda_i > 0.$$

Let $\lambda_{[1]} \leq \lambda_{[2]} \leq \cdots \leq \lambda_{[k]}$ denote the ordered values of the unknown parameters λ_i's, $i = 1, 2, \cdots, k$. We assume that the largest (unknown) parameter $\lambda_{[k]}$ has a proper prior distribution $G(\cdot)$. Our goal is to select a nonempty subset which contains the population associated with the largest parameter $\lambda_{[k]}$ with at least a guaranteed probability $P^*(1/k < P^* < 1)$ of a correct selection.

For the natural subset selection rule $R^{\max}$:

$R^{\max}$: Select π_i if and only if : $X_i \geq \max_{1 \leq j \leq k} X_j - d$, for some $d \geq 0$.

Let $X_{(i)}$ be the random variable associated with the unknown parameter $\lambda_{[i]}$, where $\lambda_{[1]} \leq \lambda_{[2]} \leq \cdots \leq \lambda_{[k]}$ is the ordered values of λ_i's, $i = 1, 2, \cdots, k$, then

$$P(CS|G, R^{\max})$$

$$= \int_0^\infty P\left(X_{(k)} \geq X_{(j)} - d, \ \forall \ j \neq k | \lambda_{[k]}\right) \ dG(\lambda_{[k]})$$

$$= \int_0^\infty \left[\sum_{x=0}^\infty \frac{e^{-\lambda_{[k]}} \lambda_{[k]}^x}{x!} \prod_{i=1}^{k-1} \left\{ \sum_{t=0}^{x+d} \frac{e^{-\lambda_{[i]}} \lambda_{[i]}^t}{t!} \right\} | \lambda_{[k]} \right] dG(\lambda_{[k]})$$

$$\geq \int_0^\infty \sum_{x=0}^\infty \frac{e^{-\lambda} \lambda^x}{x!} \left\{ \sum_{t=0}^{x+d} \frac{e^{-\lambda} \lambda^t}{t!} \right\}^{k-1} dG(\lambda),$$

where the last inequality follows from property (i) of the following lemma.

Lemma 5.1

Let

$$P(CS|\underline{\lambda}, d) = \sum_{x=0}^\infty \frac{e^{-\lambda_{[k]}} \lambda_{[k]}^x}{x!} \prod_{i=1}^{k-1} \left\{ \sum_{t=0}^{x+d} \frac{e^{-\lambda_{[i]}} \lambda_{[i]}^t}{t!} \right\},$$

then the function $P(CS|\underline{\lambda}, d)$ is

(i) non-increasing in $\lambda_{[i]}$, keeping other components of $\underline{\lambda}$ fixed, for $i = 1, 2, \cdots, k-1$; and

(ii) non-decreasing in $\lambda_{[k]}$, keeping other components of $\underline{\lambda}$ fixed.

Proof of Lemma 5.1

We have

$$\frac{\partial}{\partial \lambda} \left(\sum_{i=0}^{x+d} \frac{e^{-\lambda} \lambda^i}{i!} \right) = -\frac{e^{-\lambda} \lambda^{x+d}}{(x+d)!} < 0,$$

therefore

$$P(CS|\underline{\lambda}, d) \geq P(CS|\lambda_{[k]}\underline{1}, d), \quad \text{for all } \underline{\lambda} \in (0, \infty)^k.$$

Hence property (i) follows.

For property (ii), consider

$$a(x + d) = \prod_{i=1}^{k-1} \left\{ \sum_{t=0}^{x+d} \frac{e^{-\lambda_{[i]} \lambda_{[i]}^t}}{t!} \right\},$$

which is increasing in $x + d$. Now

$$P(CS|\underline{\lambda}, d) = \sum_{x=0}^{\infty} \frac{e^{-\lambda_{[k]} \lambda_{[k]}^x}}{x!} a(x + d), \quad \text{for fixed } \lambda_{[i]}, \ i = 1, 2, \cdots, k - 1.$$

So

$$\frac{\partial P(CS|\underline{\lambda}, d)}{\partial \lambda_{[k]}} = \sum_{x=0}^{\infty} \frac{e^{-\lambda_{[k]} \lambda_{[k]}^x}}{x!} [a(x + 1 + d) - a(x + d)] \geq 0. \quad \blacksquare$$

Now, let

$$Q(n, k, d, G) = \int_0^\infty \sum_{x=0}^\infty \frac{e^{-\lambda} \lambda^x}{x!} \left\{ \sum_{t=0}^{x+d} \frac{e^{-\lambda} \lambda^t}{t!} \right\}^{k-1} dG(\lambda).$$

Then, for any given $P^*(1/k < P^* < 1)$ and a proper $G(\cdot)$, we can find the smallest integer $d(d \geq 0)$ satisfying

$$
\begin{aligned}
P(CS|G, R^{max}) &\geq Q(n, k, d, G) \\
&\geq P^*.
\end{aligned}
$$

Proposition 5.1

The conjugate priors for the problem is the gamma distribution function. For any gamma distribution $g_{\alpha,\beta}$, $\alpha, \beta > 0$, i.e.

$$g_{\alpha,\beta}(\lambda) = \frac{\beta^\alpha}{\Gamma(\alpha)} \lambda^{\alpha-1} e^{-\beta\lambda} \quad (\lambda > 0),$$

we have

$$Q(n, k, d, g_{\alpha,\beta}) = \sum_{x=0}^\infty \sum_{j_1=0}^{x+d} \cdots \sum_{j_{k-1}=0}^{x+d} \frac{\mathcal{H}_{\alpha,\beta}(x, k, h)}{x! j_1! \cdots j_{k-1}!},$$

where $h = j_1 + j_2 + \cdots + j_{k-1}$, and

$$\mathcal{H}_{\alpha,\beta}(x,h,k) = \frac{\Gamma(x+h+\alpha)\beta^\alpha}{\Gamma(\alpha)(\beta+k)^{x+h+\alpha}}.$$

Remark 5.1 One may note that (so far) we do not know any proper noninformative prior for the Poisson distribution. However, we can start by using the following:

$$g_{1/2,\beta}(\lambda) = \frac{\beta^{1/2}}{\Gamma(1/2)}\lambda^{-1/2}e^{-\beta\lambda} \quad (\lambda > 0),$$

where $\Pi(\lambda) \propto \lambda^{-1/2}$ is Jeffreys (improper) noninformative prior. The value β has to be determined in advance.

6   Conclusions

The previous discussion shows that the classical type procedure R^{max} can still be used and a lower bound for the $P(CS)$ computed by using a proper prior G defined on the space of the largest (unknown) parameter $\lambda_{[k]}$.

Acknowledgment

The authors are thankful to the referee for several helpful comments and suggestions. This research was supported in part by the NSF Grants DMS-8702620 and DMS-8717799 at Purdue University.

References

Bechhofer, R.E. (1954). A single-sample multiple decision procedure for ranking means of normal populations with known variances. *Ann. Math. Statist.* **25** 16 - 39.

Goel, P.K. (1972). A Note on the Nonexistence of Subset Selection Procedures for Poisson Populations. Mimeograph series No. 303, Department of Statistics, Purdue University, West Lafayette, Indiana.

Goel, P.K. (1975). A Note on Subset Selection Procedures with Poisson Processes. Unpublished Report.

Gupta, S.S. (1956). On a decision rule for a problem in ranking means. Ph.D. Thesis (Mimeograph Series No.150), Institute of Statistics, University of North Carolina, Chapel Hill, NC.

Gupta, S.S. and Huang, D.-Y. (1975). On subset selection procedures for Poisson populations and some applications to the multinomial selection problems. In *Applied Probability* (R.P. Gupta, ed.), North-Holland, Amsterdam, 97 - 109.

Gupta, S.S., Huang, D.-Y. and Huang, W.-T. (1976). On ranking and selection procedures and tests of homogeneity for binomial populations. In *Essays in Probability and Statistics* (S. Ikeda et al., eds.), Shinko Tsusho Co. Ltd., Tokyo, Japan, Chapter 33, 501 - 533.

Gupta, S.S., Leong, Y.K. and Wong, W.-Y. (1979). On subset selection procedures for Poisson populations. *Bull. of the Malay. Math. Soc.* (2) **2** No. 2, 89 - 110.

Gupta, S.S. and Mcdonald, G.C. (1986). A statistical selection approach to binomial models. *Jour. Qual. Tech.* **18** (No. 2) 103 - 115.

Gupta, S.S. and Panchapakesan, S. (1979). *Multiple Decision Procedures: Theory and Methodology of Selecting and Ranking Populations.* John Wiley and Sons, Inc., New York.

Gupta, S.S. and Sobel, M. (1958). On selecting a subset which contains all populations better than a standard. *Ann. Math. Statist.* **29** 235 - 244.

Gupta, S.S. and Sobel, M. (1960). Selecting a subset containing the best of Several Binomial Populations. In *Contributions to Probability and Statistics* (I. Olkin, S.G. Ghurye, W. Hoeffding, W.G. Madow, and H.B. Mann, eds.), Stanford University Press, Stanford CA, Chapter 20, 224 - 248.

Gupta, S.S. and Wong, W.-Y. (1977). On subset selection procedures for Poisson processes and some applications to the binomial and multinomial problems. In *Operations Research Verfahren* (R. Henn et al., eds.), Verlag Anton Hain, Meisenheim am Glan, West Germany, 49 - 70.

Leong, Y.K. and Wong, W.-Y. (1977). On Poisson selection problems. Research Report 23/77, Department of Mathematics, University of Malaya, Kuala Lumpur, Malaysia.

Chapter 20

A Two-Stage Procedure for Selecting the δ^*-Optimal Guaranteed Lifetimes in the Two-Parameter Exponential Model

TACHEN LIANG   Department of Mathematics, Wayne State University, Detroit, Michigan

S. PANCHAPAKESAN   Department of Mathematics, Southern Illinois University, Carbondale, Illinois

Abstract   Let $\pi_1, \ldots, \pi_k$ be k two-parameter exponential populations with π_i having unknown location parameter (guaranteed life time) μ_i and unknown (common) scale parameter σ. The population π_i is called δ^*-optimal (or good) if $\mu_i \geq \mu_{[k]} - \delta^*$, where $\delta^* > 0$ is a specified constant and $\mu_{[k]}$ is the largest μ_i. A two-stage procedure is proposed which selects a nonempty subset of the k populations and guarantees with probability at least P^* that the selected subset includes only δ^*-optimal populations. Besides screening out non-good populations, the selection rule assures that a high proportion of sufficiently good populations is selected.

1   Introduction

Let π_i $(1 \leq i \leq k)$ denote a two-parameter exponential population with unknown location parameter μ_i and unknown common scale parameter σ. The density associated with π_i is

$$f(x; \mu_i, \sigma) = \frac{1}{\sigma} \exp\left\{ -\frac{(x - \mu_i)}{\sigma} \right\},$$

where $\mu_i > 0$, $\sigma > 0$, $x \geq \mu_i$, and $1 \leq i \leq k$. Actually, μ_i can be any real number. However, in the context of $f(x; \mu_i, \sigma)$ representing a life length density, we take $\mu_i > 0$; it is called the *guaranteed lifetime*.

Let $\mu_{[1]} \leq \ldots \leq \mu_{[k]}$ denote the ordered values of the μ_i; the correct pairings of the ordered and the unordered μ_i are assumed to be unknown. Any population π_i is called δ^*-*optimal* (or *good*) population if $\mu_i \geq \mu_{[k]} - \delta^*$. A *correct selection* (CS) is said to have occurred if any selected subset of the k populations contains *only* good populations. We want to define a rule R for which

$$P\{CS \mid \mu, \sigma\} \equiv \Pr\{a \text{ CS occurs}\} \geq P^* \text{ for all } (\mu, \sigma) \in \Omega \qquad (1.1)$$

where $\mu = (\mu_1, \ldots, \mu_k)$ and $\Omega = \{(\mu_1, \ldots, \mu_k, \sigma) \mid \sigma > 0, \ \mu_i > 0, \ i = 1, \ldots, k\}$ is the parameter space.

We propose and study a two-stage procedure for this problem. Our treatment of the problem parallels that of Santner (1976) for the case of normal means; however, our results do not follow from Santner's results. Selection from two-parameter exponential populations has been studied earlier by different authors. Under the indifference-zone (IZ) approach of Bechhofer (1954), Raghavachari and Starr (1970) considered the goal of selecting the t best populations (i.e. the populations associated with the t largest μ_i's), assuming that σ is *known*. In this case of known σ, the procedure of Desu and Sobel (1968) for location parameters applies; in fact, their procedure is devised to select a subset of *fixed* size s so that the t best populations $(1 \leq t \leq s < k)$ are included in the selected subset. When σ is *unknown*, a single-stage procedure for selecting the best under the IZ formulation does not exist. For this case, Desu et al. (1977) studied a non-elimination type two-stage procedure similar to that of Bechhofer et al. (1954) for the normal means problem, whereas Kim and Lee (1984) studied an elimination type two-stage procedure.

Under the subset selection (SS) approach of Gupta (1956), for selecting the best population, procedures can be defined analogous to the normal means problem in the case of known as well as unknown σ. Leu and Liang (1990) have discussed selection of the best with a preliminary test in order to decide whether a selection should be made at all. For the goal of selecting the populations better than a control, Gupta and Leu (1986) studied isotonic subset selection rules, while Bristol et al. (1992) have investigated procedures under an IZ formulation.

In Section 2, we propose a two-stage procedure for our problem of selecting good populations and establish its validity. Evaluation of a constant associated with the procedure is discussed in Section 3. The stopping rule used in the selection procedure involves another constant to be determined so that the selection procedure can guarantee that a (high) specified proportion of "sufficiently good" (to be defined) populations is selected. Sections 4 and 5 deal with this efficiency requirement and associated computational formulas.

2   A Class of Two-Stage Procedures

Given $\delta^* > 0$ and $k^{-1} < P^* < 1$, we define a class of procedures $\{R(N)\}$, where $R(N)$ consists of the following steps:

(1) Choose n (arbitrary) observations $X_{i1}, \ldots, X_{in}$ from π_i, $i = 1, \ldots, k$, and let

$$\hat{\sigma} = \frac{1}{k(n-1)} \sum_{i=1}^{k} \sum_{j=1}^{n} (X_{ij} - Y_i), \qquad (2.1)$$

where $Y_i = \min_{1 \le j \le n} X_{ij}$.

(2) Find $d = d(P^*, \nu, k)$ satisfying

$$\Pr\left[U_l \le U_k + \frac{d}{\nu} W, \quad l = 1, \ldots, k-1\right] = P^* \qquad (2.2)$$

where the U_l are independent and identically distributed (i.i.d.), each having an exponential distribution with unit mean (denoted by E(1)), and W, independent of the U_l, has a chi-square distribution with $\nu = 2k(n-1)$ degrees of freedom (denoted by χ^2_ν).

(3) Fix N by *any* stopping rule of the form

$$N = \max\{n, [\hat{\sigma} \max\{h, d/\delta^*\}] + 1\} \qquad (2.3)$$

where $[y]$ denotes the greatest integer $\le y$. Then take an additional $N - n$ observations from each π_i. A particular choice of h which achieves a second requirement will be given later.

(4) Select all π_i's for which

$$Y_i \ge \max_{1 \le r \le k} Y_r - \delta^* + \frac{d\hat{\sigma}}{N}$$

where Y_i is the smallest of the N observations from π_i, $i = 1, \ldots, k$, but $\hat{\sigma}$ is still given by (2.1).

Let $\{CS(N)\}$ denote the event of a correct selection using $R(N)$ and $\Omega_i(\delta^*)$ the subset of Ω corresponding to "exactly i good populations." Then

$$\Omega_i(\delta^*) = \begin{cases} \{(\mu, \sigma) \mid \mu_{[k-i+1]} \geq \mu_{[k]} - \delta^* > \mu_{[k-i]}\}, & i = 1, \ldots, k-1 \\ \{(\mu, \sigma) \mid \mu_{[1]} \geq \mu_{[k]} - \delta^*\}, & i = k. \end{cases} \quad (2.4)$$

We note that $P\{CS(N) \mid (\mu, \sigma)\} = 1$ for all $(\mu, \sigma) \in \Omega_k(\delta^*)$.

Lemma 2.1

Let $\mu^0 = (0, \ldots, 0, \delta^*)$. Then, for fixed $\sigma > 0$ and any $1 \leq i \leq k$,

$$\inf_{\Omega_i(\delta^*)} P\{CS(N) \mid (\mu, \sigma)\} = P\{CS(N) \mid (\mu^0, \sigma)\}.$$

Proof of Lemma 2.1

We need only consider $i = 1, \ldots, k-1$, since the lemma is obviously true for $i = k$. For any fixed i, $1 \leq i \leq k-1$, let $I \equiv I(i) = \{1, \ldots, k-i\}$. Further, let $Y_{(i)}$ denote the Y_i associated with $\mu_{[i]}$, $i = 1, \ldots, k$, and $Y_{[1]} \leq \ldots \leq Y_{[k]}$ denote the ordered Y_i. Then, for $(\mu, \sigma) \in \Omega_i(\delta^*)$, the probability of a correct selection (PCS) is given by

$$\begin{aligned} \text{PCS} &= \Pr\{Y_{(r)} \leq Y_{[k]} - \delta^* + \frac{d\hat{\sigma}}{N}, \text{ for all } r \in I\} \\ &= \sum_{m \geq n} \Pr\{Y_{(r)} \leq Y_{[k]} - \delta^* + \frac{d\hat{\sigma}}{m}, \text{ for all } r \in I \text{ and } \hat{\sigma} \in B_m\}, \end{aligned}$$

where B_m is defined by $\{\hat{\sigma} \in B_m\} = \{N = m\}$, which implies that $\bigcup_m B_m = (0, \infty)$. Now, let $U_r = m(Y_{(r)} - \mu_{[r]})/\sigma$, $r = 1, \ldots, k$. Then the U_r are i.i.d., each $E(1)$ distributed. We can now write

$$\begin{aligned} \text{PCS} &= \sum_{m \geq n} \Pr\{U_r + \frac{m}{\sigma}\mu_{[r]} \leq \max_{1 \leq j \leq k}(U_j + \frac{m}{\sigma}\mu_{[j]}) - \frac{m\delta^*}{\sigma} + \frac{d\hat{\sigma}}{\sigma} \\ &\qquad\qquad \text{for all } r \in I \text{ and } \hat{\sigma} \in B_m\} \\ &= \sum_{m \geq n} \Pr\{A(\mu, \sigma, m)\}, \quad \text{say,} \end{aligned}$$

where $A \equiv A(\mu, \sigma, m)$ denotes the appropriate event.

The lemma is established if we show that $\Pr\{A\}$ is nondecreasing in $\mu_{[s]}$, $s = k - i + 1, \ldots, k-1$ and nonincreasing in $\mu_{[s]}$, $s = 1, \ldots, k-i$. Towards this end, fix m and $s \in \{k - i + 1, \ldots, k-1\}$. Consider μ and μ' having $\mu'_{[s]} > \mu_{[s]}$ and $\mu'_{[j]} = \mu_{[j]}$ for $j \neq s$. If $A(\mu, \sigma, m)$, then for $r \in I$,

$$U_r + \frac{m}{\sigma}\mu_{[r]} \leq \max_j(U_j + \frac{m}{\sigma}\mu_{[j]}) - \frac{m\delta^*}{\sigma} + \frac{d\hat{\sigma}}{\sigma}$$

$$\leq \max_j(U_j + \frac{m}{\sigma}\mu'_{[j]}) - \frac{m\delta^*}{\sigma} + \frac{d\hat\sigma}{\sigma}$$

$\Rightarrow A(\mu',\sigma,m)$ occurs, since $\mu'_{[r]} = \mu_{[r]}$, $r \in I$, and $\hat\sigma \in B_m$. Thus PCS is nondecreasing in $\mu_{[s]}$ for $s = k - i + 1, \ldots, k - 1$. A similar argument shows that PCS is nonincreasing in $\mu_{[s]}$, $s \in I$. This completes the proof of the lemma.

Theorem 2.1

Every procedure in the class $\{R(N)\}$ satisfies the probability requirement (1.1).

Proof of Theorem 2.1

By Lemma 2.1, for each fixed $\sigma > 0$,

$$\inf_{\mu} P\{CS(N) \mid (\mu,\sigma)\} = \min_{1\leq i\leq k} \inf_{\Omega_i(\delta^*)} P\{CS(N) \mid (\mu,\sigma)\}$$

$$= \min_{1\leq i\leq k} \Pr\{Y_{(r)} \leq Y_{[k]} - \delta^* + \frac{d\hat\sigma}{N} \text{ for all } r \in I(i) \mid (\mu^0,\sigma)\}$$

$$= \Pr\{Y_{(r)} \leq Y_{[k]} - \delta^* + \frac{d\hat\sigma}{N} \text{ for all } r < k \mid (\mu^0,\sigma)\}$$

$$\geq \Pr\{Y_{(r)} \leq Y_{(k)} - \delta^* + \frac{d\hat\sigma}{N} \text{ for all } r < k \mid (\mu^0,\sigma)\}$$

$$= \sum_{m\geq n} \Pr\{Y_{(r)} \leq Y_{(k)} - \delta^* + \frac{d\hat\sigma}{m} \text{ for all } r < k \text{ and } \hat\sigma \in B_m\}$$

$$= \sum_{m\geq n} \Pr\{U_r \leq U_k + \frac{d\hat\sigma}{\sigma}, \ r = 1,\ldots,k-1, \text{ and } \hat\sigma \in B_m\}$$

where the U_i are i.i.d., each having an $E(1)$ distribution. It is known that $Z_\nu = \nu\hat\sigma/\sigma$ has a χ^2_ν distribution and that it is independent of the U_i. Letting $D_m(\sigma) = \{\nu w/\sigma \mid w \in B_m\}$, we have

$$\inf_{\mu} P\{CS(N) \mid (\mu,\sigma)\}$$

$$\geq \sum_{m\geq n} \Pr\{U_r \leq U_k + \frac{d}{\nu}Z_\nu, 1 \leq r \leq k, Z_\nu \in D_m(\sigma)\}$$

$$= \sum_{m\geq n} \int_{D_m(\sigma)} \Pr\{U_r \leq U_k + \frac{dw}{\nu}, \ r = 1,\ldots,k\}dP\{Z_\nu \leq w\}$$

$$= \int_0^\infty \Pr\{U_r \leq U_k + \frac{dw}{\nu}\}dP\{Z_\nu \leq w\}$$

$$= \Pr\{U_r \leq U_k + \frac{dW}{\nu}, \ r = 1,\ldots,k-1\}, \text{ where } W \sim \chi^2_\nu$$

$$= P^*, \text{ by (2.2).}$$

3   Evaluation of Constant d

For given k, n, and P^*, the constant d is given by (2.2), which can easily be shown to be

$$\int_0^\infty \int_0^\infty \left[1 - e^{-(u+dw/\nu)}\right]^{k-1} e^{-u} g_\nu(w) du\, dw = P^* \qquad (3.1)$$

where g_ν is the χ_ν^2 density. The binomial expansion of $[1 - e^{-(u+dw/\nu)}]^{k-1}$ followed by straightforward integration converts (3.1) into

$$\sum_{\alpha=0}^{k-1} (-1)^\alpha \binom{k-1}{\alpha} (1+\alpha)^{-1} \left(1 + \frac{2\alpha d}{\nu}\right)^{-\nu/2} = P^*. \qquad (3.2)$$

For $k = 2$, (3.2) simplifies to

$$1 - \frac{1}{2}\left(1 + \frac{2d}{\nu}\right)^{-\nu/2} = P^*,$$

which yields

$$d = \frac{\nu}{2}\left[\{2(1 - P^*)\}^{-\frac{2}{\nu}} - 1\right].$$

The values of d/ν have been tabulated by Desu et al. (1977) for $k = 2$ (1) 6; $n = 2$ (1) 30; $P^* = 0.95$, and by Bristol et al. (1992, Table 3) for $k = 3$ (1) 16, 21; $n = 2$ (1) 10, 15, 20, 30, 60; $P^* = 0.90, 0.95, 0.99$.

4   An Efficiency Requirement

Let $0 < \Delta < \delta^*$ and $0 < p < 1$, and let $S(N)$ denote the number of populations selected by $R(N)$ having $\mu_i \geq \mu_{[k]} - \Delta$ (*sufficiently good* populations). Let $P(N)$ denote the proportion of sufficiently good populations that are selected, i.e.,

$$P(N) = \frac{S(N)}{i} \text{ for } (\mu, \sigma) \in \Omega_i(\Delta), \ i = 1, \ldots, k.$$

We use $P(N)$ as a measure of efficiency of $R(N)$ and require that

$$E[P(N) \mid (\mu, \sigma)] \geq p \text{ for all } (\mu, \sigma) \in \Omega.$$

For convenience, for $i = 1, \ldots, k$, let $\theta_i = \mu_{[i]}$ and $Q^{(i)} = E[S(N) \mid (\mu, \sigma) \in \Omega_i(\Delta)]$, where $\Omega_i(\Delta)$ is defined by (2.4) with Δ instead of δ^*.

Then

$$
\begin{aligned}
Q^{(i)} &= \sum_{j=k-i+1}^{k} \Pr\{Y_{(j)} \geq Y_{(m)} - \delta^* + \frac{d\hat\sigma}{N}, 1 \leq m \leq k; m \neq j\} \\
&= \sum_{j=k-i+1}^{k} \Pr\{U_m \leq U_j + \frac{N}{\sigma}(\theta_j - \theta_m) + \frac{N\delta^* - d\hat\sigma}{\sigma} \text{ for all } m \neq j\} \\
&= \sum_{j=k-i+1}^{k} Q_j^{(i)}, \text{ say,}
\end{aligned}
$$

where the U_i are i.i.d. $E(1)$ and are independent of $\hat\sigma$.

Lemma 4.1

For $i = 1, \ldots, k-1$,

$$
\inf_{\Omega_i(\Delta)} E[S(N)] = \inf_{\Omega_i^0(\Delta)} E[S(N)],
$$

where $\Omega_i^0(\Delta) = \{(\mu, \sigma) \mid \theta_1 = \ldots = \theta_{k-i} = \theta_k - \Delta\}$.

Proof of Lemma 4.1

The proof is immediate by noting that, for $j = k-i+1, \ldots, k$, $Q_j^{(i)}$ is nonincreasing in θ_s, $s = 1, \ldots, k-i$, when all other θ's are fixed.

Lemma 4.2

For $i = 1, \ldots, k-1$, and $(\mu, \sigma) \in \Omega_i^0(\Delta)$, $E[S(N)]$ is nondecreasing in θ_{k-i+1} when all other θ's are fixed.

Proof of Lemma 4.2

For $(\mu, \sigma) \in \Omega_i^0(\Delta)$ and $j = k-i+1, \ldots, k$,

$$
Q_j^{(i)} = \Pr \left[\begin{array}{l} U_s \leq U_j + \frac{N}{\sigma}(\theta_j - \theta_k + \Delta) + \frac{N\delta^* - d\hat\sigma}{\sigma}, 1 \leq s \leq k-i; \\ U_m \leq U_j + \frac{N}{\sigma}(\theta_j - \theta_m) + \frac{N\delta^* - d\hat\sigma}{\sigma}, k-i+1 \leq m \neq j \leq k; \end{array} \right].
$$

Let $Q_j^{(i)}(\hat\sigma)$ denote the probability of the event associated with $Q_j^{(i)}$ conditioned on $\hat\sigma$. The lemma is proved if we show that $E[S(N) \mid \hat\sigma] = \sum_{j=k-i+1}^{k} Q_j^{(i)}(\hat\sigma)$ is nondecreasing in $\theta_{[k-i+1]}$ when all other θ's are fixed. For convenience, let

$$
\begin{aligned}
C(\theta_j, \theta_k) &= \{N(\theta_j - \theta_k + \Delta) + N\delta^* - d\hat\sigma\}/\sigma, \\
D(\theta_j, \theta_m) &= \{N(\theta_j - \theta_m) + N\delta^* - d\hat\sigma\}/\sigma.
\end{aligned}
$$

First consider $Q_{k-i+1}^{(i)}(\hat\sigma)$. Noting that $N\delta^* - d\hat\sigma > 0$ and $\theta_j - \theta_k + \Delta \geq 0$ for $j = k-i+1, \ldots, k$, we get $Q_{k-i+1}^{(i)}(\hat\sigma) =$

$$\int_0^\infty e^{-x} \left[1 - e^{-x - C(\theta_{k-i+1}, \theta_k)}\right]^{k-i} \prod_{m=k-i+2}^k \left[1 - e^{-x - D(\theta_{k-i+1}, \theta_m)}\right] dx.$$

Differentiating $Q_{k-i+1}^{(i)}$ w.r.t. θ_{k-i+1}, we obtain

$$\frac{\partial Q_{k-i+1}^{(i)}(\hat\sigma)}{\partial \theta_{k-i+1}} = (k-i)I + \frac{N}{\sigma} \sum_{j=k-i+2}^k \int_0^\infty e^{-x} h(x) dx \qquad (4.1)$$

where I denotes the obvious integral and $h(x) =$

$$\left[1 - e^{-x - C(\theta_{k-i+1}, \theta_k)}\right]^{k-i} \prod_{\substack{m=k-i+2 \\ m \neq j}}^k \left[1 - e^{-x - D(\theta_{k-i+1}, \theta_m)}\right] e^{-x - D(\theta_{k-i+1}, \theta_j)}.$$

We need only note that $I > 0$. For $j = k - i + 2, \ldots, k$, we get

$$Q_j^{(i)}(\hat\sigma) = \int_{B_j}^\infty e^{-x} \left[1 - e^{-x - C(\theta_j, \theta_k)}\right]^{k-i} \prod_{\substack{m=k-i+1 \\ m \neq j}}^k \left[1 - e^{-x - D(\theta_j, \theta_m)}\right] dx,$$

where $B_j = \max(0, -D(\theta_j, \theta_k))$, and

$$\frac{\partial Q_j^{(i)}(\hat\sigma)}{\partial \theta_{k-i+1}} = -\frac{N}{\sigma} \int_{B_j}^\infty e^{-x} \left[1 - e^{-x - C(\theta_j, \theta_k)}\right]^{k-i} g(x) dx \qquad (4.2)$$

where

$$g(x) = \prod_{\substack{m=k-i+1 \\ m \neq j}}^k \left[1 - e^{-x - D(\theta_j, \theta_m)}\right] e^{-x - D(\theta_k, \theta_{k-i+1})}.$$

In order to show that $E[S(N) \mid \hat\sigma]$ is nondecreasing in θ_{k-i+1}, it is sufficient to show that, for $j = k - i + 2, \ldots, k$, the sum of the integral in (4.2) and the integral corresponding to j in the second part of (4.1) is nonnegative. It is now easy to see that, in turn, it suffices to show that $e^{-x - D(\theta_{k-i+1}, \theta_j)} - e^{-x - D(\theta_j, \theta_{k-i+1})} \geq 0$, which is verified very easily.

Theorem 4.1

If $0 < p < 1$ and $0 < \Delta < \delta^*$, then $E[P(N) \mid (\mu, \sigma)] \geq p$ for all $(\mu, \sigma) \in \Omega$ provided that h in the stopping rule (2.3) is given by

$$\Pr\{U_s \le U_k + \{h(\delta^* + \Delta) - d\}W/\nu, \ s = 1, \ldots, k-1\}$$

$$+(k-1)\Pr\left\{\begin{array}{c} U_s \le U_1 + (h\delta^* - d)W/\nu, \ s = 2, \ldots, k-1; \\ U_k \le U_1 + \{h(\delta^* - \Delta) - d\}W/\nu \end{array}\right\}$$

$$= kp. \tag{4.3}$$

Proof of Theorem 4.1

For a fixed i, $1 \le i \le k-1$, consider $(\mu, \sigma) \in \Omega_i(\Delta)$. By Lemmas 4.1 and 4.2 and by the symmetry of $E[S(N) \mid (\mu, \sigma)]$ in $(\theta_{k-i+1}, \ldots, \theta_k)$, we see that this expectation is nonincreasing in θ_s, $s = 1, \ldots, k-i$, and nondecreasing in θ_s, $s = k-i+1, \ldots, k-1$. Thus the infimum of $E[S(N)]$ over $\Omega_i(\Delta)$ occurs when $\theta_1 = \ldots = \theta_{k-1} = \theta_k - \Delta$. Since this configuration is independent of the particular i chosen,

$$\inf_\Omega E[P(N)] = \min_{1 \le i \le k} \inf_{\Omega_i(\Delta)} \frac{E[S(N)]}{i} = \frac{1}{k} \inf_{\Omega_k^0(\Delta)} E[S(N)].$$

Note that $E[S(N)]$ is constant over $\Omega_k^0(\Delta)$ and hence this value, denoted by Q^*, can be evaluated for $\theta = (0, \ldots, 0, \Delta)$. Thus

$$Q^* = \sum_{j=1}^{k-1} \Pr\left[\begin{array}{c} U_s \le U_j + (N\delta^* - d\hat\sigma)/\sigma, \ s = 1, \ldots, k-1; \ s \ne j, \\ U_k \le U_j + (-N\Delta + N\delta^* - d\hat\sigma)/\sigma \end{array}\right]$$

$$\quad + \Pr\{U_s \le U_k + (N\Delta + N\delta^* - d\hat\sigma)/\sigma, \ s = 1, \ldots, k-1\}$$

$$\ge \sum_{j=1}^{k-1} \Pr\left[\begin{array}{c} U_s \le U_j + (h\delta^* - d)\hat\sigma/\sigma, \ s = 1, \ldots, k-1; \ s \ne j, \\ U_k \le U_j + \{h(\delta - \Delta) - d\}\hat\sigma/\sigma \end{array}\right]$$

$$\quad + \Pr\{U_s \le U_k + \{h(\delta^* + \Delta) - d\}\hat\sigma/\sigma, \ s = 1, \ldots, k-1\}$$

$$= (k-1)\Pr\left[\begin{array}{c} U_s \le U_1 + \frac{(h\delta^* - d)}{\nu}W, \ s = 2, \ldots, k-1; \\ U_k \le U_1 + \frac{\{h(\delta^* - \Delta) - d\}}{\nu}W \end{array}\right]$$

$$\quad + \Pr\{U_s \le U_k + \{h(\delta^* + \Delta) - d\}W/\nu, s = 1, \ldots, k-1\}.$$

The statement of the theorem now follows.

Now, let Q_L^* denote the left-hand side of (4.3). Given k, n, δ^*, Δ, P^*, and d obtained as described in Section 3, we need to evaluate h satisfying (4.3), namely, $Q_L^* = kp$. Towards this end, we obtain in the next section expressions for Q_L^*.

5 Expressions for Q_L^*

Let $a_1 = h(\delta^* + \Delta) - d$, $a_2 = h\delta^* - d$, and $a_3 = h(\delta^* - \Delta) - d$ so that $a_1 > a_2 > a_3$. Then

$$Q_L^* = I_k(a_1) + (k-1)J_k(a_2, a_3)$$

where

$$I_k(a_1) = \Pr\{U_s \le U_k + \frac{a_1 W}{\nu}, \ s = 1, \ldots, k-1\}$$

and

$$J_k(a_2, a_3) \ = \ \Pr\left\{U_s \le U_1 + \frac{a_2 W}{\nu}, \ s = 2, \ldots, k-1 \ and \right.$$

$$\left. U_k \le U_1 + \frac{a_3 W}{\nu}\right\}. \tag{5.1}$$

Also, let $z^+ = \max(0, z)$.

First, consider

$$I_k(a_1) \ = \ \int_0^\infty \int_0^\infty \left[1 - e^{-(u + a_1 w/\nu)^+}\right]^{k-1} e^{-u} g_\nu(w) du\, dw$$

$$= \ \sum_{\alpha=0}^{k-1} (-1)^\alpha \binom{k-1}{\alpha} \int_0^\infty \int_0^\infty e^{-u - \alpha(u + a_1 w/\nu)^+} g_\nu(w) du\, dw.$$

Considering the two cases of $a_1 \ge 0$ and $a_1 < 0$, the double integral in the preceding expression for $I(a)$ can be evaluated in a straightforward manner and we get after some routine algebraic manipulations

$$I_k(a_1) = \sum_{\alpha=0}^{k-1} (-1)^\alpha \binom{k-1}{\alpha} I'_\alpha(a_1) \tag{5.2}$$

where

$$I'_\alpha(a_1) = \begin{cases} \left(1 + \frac{2\alpha a_1}{\nu}\right)^{-\nu/2} - \frac{\alpha}{\alpha+1}\left(1 + \frac{2\alpha a_1}{\nu}\right)^{-\nu/2}, & a_1 \ge 0 \\ 1 - \frac{\alpha}{\alpha+1}\left(1 - \frac{2a_1}{\nu}\right)^{-\nu/2}, & a_1 < 0. \end{cases}$$

The two cases in $I'_\alpha(a_1)$ can now be combined to yield

$$I'_\alpha(a_1) \ = \ \left[1 + \frac{\alpha}{\nu}(a_1 + |a_1|)\right]^{-\nu/2}$$

$$- \frac{\alpha}{\alpha+1}\left[1 + \frac{\alpha+1}{\nu}|a_1| + \frac{\alpha-1}{\nu}a_1\right]^{-\nu/2}.$$

Next, $J_k(a_2, a_3)$ in (5.1) can be written as

$$J_k(a_2, a_3) = \sum_{\alpha=0}^{k-2} (-1)^\alpha \binom{k-2}{\alpha} J'_\alpha(a_2, a_3) \tag{5.3}$$

where

$$J'_\alpha(a_2, a_3) = \int_0^\infty \int_0^\infty e^{-u-\alpha(u+\frac{a_2 w}{\nu})^+}[1 - e^{-(u+\frac{a_3 w}{\nu})^+}]g_\nu(w)du\,dw. \quad (5.4)$$

By splitting (5.4) into two double integrals, we can rewrite (5.3) as

$$J_k(a_2, a_3) = I_{k-1}(a_2) - \sum_{\alpha=0}^{k-2}(-1)^\alpha \binom{k-2}{\alpha} J^*_\alpha(a_2, a_3)$$

where

$$J^*_\alpha(a_2, a_3) = \int_0^\infty \int_0^\infty e^{-u-\alpha(u+\frac{a_2 w}{\nu})^+-(u+\frac{a_3 w}{\nu})^+} g_\nu(w)du\,dw.$$

Noting that $0 < a_3 < a_2$, for evaluating $J^*_\alpha(a_2, a_3)$, we need to consider three cases, namely, (1) $0 < a_3 < a_2$, (ii) $a_3 < 0 < a_2$, and (iii) $a_3 < a_2 < 0$. Carrying out the evaluation of $J^*_\alpha(a_2, , a_3)$ in these cases, and combining them into a unified form, we get

$$\begin{aligned}
J^*_\alpha(a_2, a_3) = & \left[1 + \frac{\alpha}{\nu}(a_2 + \mid a_2 \mid) + \frac{1}{\nu}(a_3 + \mid a_3 \mid)\right]^{-\nu/2} \\
& - \frac{\alpha}{\alpha+1}\left[1 + \frac{(\alpha-1)a_2 + (\alpha+1) \mid a_2 \mid}{\nu} + \frac{a_3 + \mid a_3 \mid}{\nu}\right]^{-\nu/2} \\
& - \frac{1}{(\alpha+1)(\alpha+2)}\left[1 + \frac{2\alpha a_2 + (2+\alpha) \mid a_3 \mid - \alpha \mid a_3 \mid}{\nu}\right]^{-\nu/2}. \quad (5.5)
\end{aligned}$$

Summarizing all the preceding results, we have

$$Q^*_L = I_k(a_1) + (k-1)[I_{k-1}(a_2) - H_k(a_2, a_3)]$$

where

$$H_k(a_2, a_3) = \sum_{\alpha=0}^{k-2}(-1)^\alpha \binom{k-2}{\alpha} J^*_\alpha(a_2, a_3)$$

with $I_k(a_1)$ and $J^*_\alpha(a_2, a_3)$ given by (5.2) and (5.5), respectively.
    For the case of $k = 2$, it can be directly seen that

$$Q^*_L = I_2(a_1) + I_2(a_3),$$

where

$$I_2(a) = 1 - \{1 + \frac{1}{\nu}(a + \mid a \mid)\}^{-\nu/2} + \frac{1}{2}\{1 + \frac{2}{\nu} \mid a \mid\}^{-\nu/2}.$$

    The expressions for Q^*_L obtained in this section and the fact that Q^*_L is monotonically increasing in h can now be used to compute h satisfying $Q^*_L = kp$.

Acknowledgment

The authors are thankful to Professor Woo-Chul Kim of Seoul National University for a helpful discussion.

References

Bechhofer, R.E. (1954). A single-sample multiple decision procedure for ranking means of normal populations with known variances. *Ann. Math. Statist.* **25** 16 - 39.

Bechhofer, R.E., Dunnett, C.W. and Sobel, M. (1954). A two-sample multiple-decision procedure for ranking means of normal populations with a common unknown variance. *Biometrika* **41** 170 - 176.

Bristol, D.R., Chen, H.J. and Mithongtae, J.S. (1992). Comparing exponential guarantee times with a control. In *Frontiers of Modern Statistical Inference Problems–II* (E.J. Dudewicz and E. Bofinger, eds.), American Sciences Press, Syracuse, New York, To appear.

Desu, M.M., Narula, S.C. and Villarreal, B. (1977). A two-stage procedure for selecting the best of k exponential distributions. *Comm. Statist. A–Theory Methods* **6** 1223 - 1230.

Desu, M.M. and Sobel, M. (1968). A fixed-subset size approach to a selection problem. *Biometrika* **55** 401 - 410. Corrections and amendments: **63** (1976), 685.

Gupta, S.S. (1956). On a decision rule for a problem in ranking means. Mimeo. Series No. 150, Institute of Statistics, University of North Carolina, Chapel Hill, North Carolina.

Gupta,, S.S. and Leu, L.-Y. (1986). Isotonic procedures for selecting populations better than a standard: two-parameter exponential distributions. In *Reliability and Quality Control* (A.P. Basu, ed.), Elsevier Science Publishers B.V., Amsterdam, 167 - 183.

Kim, W.-C. and Lee, S.-H. (1985). An elimination type two-stage selection procedure for exponential distributions. *Comm. Statist.–Theor. Meth.* **14** 2563 - 2571.

Leu, L.-Y. and Liang, T. (1990). Selection of the best with a preliminary test for two-parameter exponential distributions. *Comm. Statist.– Theor. Meth.* **19** 1443 - 1455.

Raghavachari, M. and Starr, N. (1970). Selection problems for some terminal distributions. *Metron* **28** 185 - 197.

Santner, T.J. (1976). A two-stage procedure for selecting δ^*-optimal means in the normal model. *Comm. Statist. A-Theory Methods* **5** 283 - 292.

Chapter 21

A Note on Selection Procedures and Selection Constants

STEFAN DRIESSEN   Medical Research and Development Unit, Organon International B.V., Oss, The Netherlands

Abstract   This paper will start off with a renewed discussion about the origin of Subset Selection Procedures (SSPs) and the strong resemblance with Multiple Comparison Procedures (MCPs). At first sight contradictory, it will be shown that MCPs can be identical to SSPs as well as are less appropriate than SSPs with respect to the goal of selecting the best treatment of a set of different treatments. Extensions of existing SSPs for the one-way layout to more general designs will be indicated, so that also for such designs the advantages of SSPs can be exploited. These SSPs use selection constants (critical values) that need to be defined for each treatment separately. The determination as well as some interesting properties of these selection constants in relation to experimental designs will be discussed. It appears that for some designs these selection constants can still be all the same. It will be shown that a specific kind of symmetry in a design, autocongruency called, is a sufficient condition for this property. In more detail the concept autocongruency is explained and examples are given.

1   Introduction

Suppose we have $t(\geq 2)$ treatments $\P_1, \ldots, \P_t$ in an experiment and from each treatment n observations (a so-called balanced one-way layout). Let Y_{ij} denote the jth observation from treatment $\P_i$, $i = 1, \ldots, t$, $j = 1, \ldots, n$. We assume as model for the observations

$$Y_{ij} = \tau_i + \sigma\varepsilon_{ij}, \ i = 1, \ldots, t, \ j = 1, \ldots, n, \tag{1.1}$$

with $\tau_1, \cdots, \tau_t$ the unknown treatment means, σ the common unknown standard deviation and the $\varepsilon'_{ij}s$ mutually independent standard normal variables, representing the random errors of the observations. Let

$$\tau_{(1)} \leq \cdots \leq \tau_{(t)}$$

denote the ordered treatment means and let $\P_{(i)}$ be the treatment with treatment mean $\tau_{(i)}$, $i = 1, \cdots, t$. If treatments with larger means are interpreted as "better", then treatment $\P_{(t)}$ is called the *best treatment* for it has the largest treatment mean $\tau_{(t)}$. Our goal is to select this best treatment.

The classical approach to tackle this problem has been to start off with testing the so-called homogeneity hypothesis that the parameter values are all the same. But the outcome of this test does not inform the experimenter properly with respect to the goal of selecting the best. In the first place, the hypothesis that there is no difference between treatments is not realistic (as was already pointed out by Cochran and Cox 1957, p. 5) and an experiment with large samples of each treatment will surely bring this to light at any prespecified significance level. In the second place, a significant result from the homogeneity test only tells one that there exist differences between the treatments but not which treatment is best.

Therefore the homogeneity test is often followed by techniques of multiple comparisons or simultaneous inference that try to determine which treatments differ from which others and in what direction, in order to try to identify the best treatment. The problem with this indirect approach is the obligatory translation of interim results of significant differences between treatments towards an interpretable inference with respect to the selection of the best. So although additional information is obtained about the relative merits of the treatments, this approach gives no satisfactory answer to the selection problem.

The inadequacy of tests of homogeneity and multiple comparison procedures has led to the development of the Ranking and Selection theory which tries to give more direct and easy-to-interpret inferences regarding the identification of the best treatment. In the next section we will throw a bit different light on this reasoning, yet come to the same conclusion.

2 Subset Selection, Multiple Comparisons

For the goal of selecting the best treatment Gupta (1956) proposed a SSP that selects a (minimal) subset of treatments such that the probability of a correct selection (CS), i.e. the probability that the unknown best treatment is included in the subset, is at least P^* (the P^*-condition). We recall this rule. Let $\bar{Y}_i$, $i = 1, \ldots, t$, be the mean of the observations from treatment $\P_i$, which we will call for sake of convenience the sample mean of treatment $\P_i$, and let S^2 be the pooled sample variance. Further, let $\bar{Y}_{[1]} \leq \ldots \leq \bar{Y}_{[t]}$ denote the ordered sample means and $\P_{[1]}, \ldots, \P_{[t]}$ the corresponding treatments. Gupta's rule selects a treatment $\P_i$, $i = i, \ldots, t$, if and only if

$$\bar{Y}_i \geq \bar{Y}_{[t]} - \delta_{t,\nu,P^*} \cdot \frac{S}{\sqrt{n}}, \tag{2.1}$$

where $\delta = \delta_{t,\nu,P^*}$ is Gupta's selection constant satisfying the equation

$$\int_0^\infty \int_{-\infty}^\infty \Phi^{t-1}(x + \delta y)\phi(x)q_\nu(y)dxdy = P^*, \tag{2.2}$$

where $\Phi(\cdot)$ and $\phi(\cdot)$ are the Normal distribution and density function, respectively, and $q_\nu(\cdot)$ is the density function of a $\sqrt{\chi_\nu^2/\nu}$ variable where χ_ν^2 is a chi-squared distribution with ν degrees of freedom.

Now, let us consider now MCPs that can test the significance of any pairwise difference between treatments, control the experimentwise error rate at a value α and possess the transitivity property, i.e., larger "distances" are more "significant". Then we can easily define a SSP that is adapted from such a MCP by:

select each treatment that does *not significantly differ* from treatment $\P_{[t]}$ according to the MCP at simultaneous level α.

As an example, consider the well known all-pairwise comparison procedure by Tukey (1953) that tests the significance of any pairwise difference between treatment means at simultaneous level α as follows:

any pair of treatments $\P_i$ and $\P_j$, $i, j = 1, \cdots, t$, $i \neq j$, is judged significantly different if and only if

$$|\bar{Y}_i - \bar{Y}_j| \geq q_{t,\nu,1-\alpha}\frac{S}{\sqrt{n}},$$

where $q_{t,\nu,1-\alpha}$ is the upper α-point of the Studentized range distribution with parameters t and $\nu = t(n - 1)$ (degrees of freedom).

According to the definition, the SSP adapted from Tukey's MCP selects a treatment $\P_i$, $i = 1, \ldots, t$, if and only if

$$\bar{Y}_i > \bar{Y}_{[t]} - q_{t,\nu,1-\alpha} \frac{S}{\sqrt{n}}. \tag{2.3}$$

Notice that the SSPs (2.1) and (2.3) are equivalent except for the values

$$\delta_{t,\nu,P^*} \text{ and } q_{t,\nu,1-\alpha}.$$

This strong resemblance calls for comparisons. The first way to do such is a kind of comparison found in the literature when a MCP is set against a SSP with respect to their performances. In such comparisons the SSP and MCP are chosen so that $\alpha = 1 - P^*$ (e.g. $P^* = 0.95$ and $\alpha = 0.05$). Theorem 2.1 tells us that this always turns out to the advantage of Gupta's SSP with respect to the size of the selected subset.

Theorem 2.1   (Driessen 1992)

For all $t \geq 2$, $\nu \in \mathbf{N} \cup \{\infty\}$ and $P^* \in (1/t, 1)$:

$$\delta_{t,\nu,P^*} \leq q_{t,\nu,P^*}.$$

Proof of Theorem 2.1

Let $t \geq 2$, $\nu \in \mathbf{N}$ and $P^* \in (1/t, 1)$. For sake of convenience we abbreviate δ_{t,ν,P^*} and q_{t,ν,P^*} with δ and q, respectively. The simultaneous confidence intervals for all pairwise differences between the treatment means (Tukey 1953) are based on the pivotal events

$$A_d \;=\; [\bar{Y}_i - \bar{Y}_j - d\frac{S}{\sqrt{n}} \leq \tau_i - \tau_j$$

$$\leq \bar{Y}_i - \bar{Y}_j + d\frac{S}{\sqrt{n}} \quad \text{for all } i, j, i \neq j], \qquad d > 0.$$

From the definition of q it follows that $P[A_q] = P^*$. The constant δ satisfies

$$P[B_\delta^{(t)}] = P^*,$$

where

$$B_d^{(t)} = [\bar{Y}_{(t)} - \tau_{(t)} \geq \bar{Y}_{(i)} - \tau_{(i)} - d\frac{S}{\sqrt{N}} \text{ for all } i \neq t], \qquad d > 0.$$

Notice that the event $B_d^{(t)}$ "increases" in d, i.e., if $d_1 \leq d_2$ then $B_{d_1}^{(t)} \subseteq B_{d_2}^{(t)}$. From $P[A_q] = P[B_\delta^{(t)}] = P^*$ and the observation that for each value $d > 0$

we have $A_d \subseteq B_d^{(t)}$, it readily follows that $\delta \leq q$. The case $\nu = \infty$ ($S = \sigma$) can be treated similarly. ■

There seems to be no other reason to choose $\alpha = 1 - P^*$ than a practical one. Yet such a comparison is not on an even basis. If one wants to compare two procedures as SSPs, then one should also evaluate them as such. So each has to be determined on the base of satisfying the P^*-property. Thus should hold:

$$\delta_{t,\nu,P^*} = q_{t,\nu,1-\alpha}. \tag{2.4}$$

But if (2.4) holds, both SSPs are simply identical and further comparisons, e.g. with respect to the subset size, seem useless. Yet (2.4) offers us a second way of comparing MCPs to SSPs. Namely, determine for any fixed value of P^* the value of α such that (2.4) holds. The latter is then the experimentwise error rate of Tukey's original MCP with which we select a subset that contains the best treatment with minimal probability P^* by using the SSP adapted from the MCP. Table 1 shows values of α as function of t in case of $P^* = 0.95$ and $\nu = \infty$ (the variance known case). From Table 1 we clearly see the conservatism of MCPs with respect to

Table 1: Relation between α and P^*.

	$P^* = 0.95$							
t	2	3	.	20	.	50	.	100
α	0.10	0.14	.	0.53	.	0.74	.	0.88

selection of the best as compared to SSPs if we would think that using α would lead to a minimum probability of a correct selection of $1 - \alpha$.

As a consequence we could try to consider other MCPs, adapt them to SSPs and compare them to already existing SSPs. Driessen (1992) has done such for the well known Gabriel-Scheffé MCP but could only establish the minimum probability for $t = 3$. The same problems arise with the SSPs that are proposed by Somerville (1984, 1985), which can be seen as adapted forms of MCPs like Newman-Keuls' and Einot-Gabriel's (see Hochberg and Tamhane 1987 for descriptions of these MCPs).

3 Extensions

It will be clear that SSPs are very appropriate to addressing the problem of selecting the best treatment. Yet, as often mentioned in the literature, practical applications seem to be scarce. One reason for this may be that the theory of Ranking and Selection is mainly restricted to the one-way layout and then also to the balanced case. In practice however, one often uses other designs like (incomplete) block designs or designs with a covariate for which no (exact) selection procedures were available. But lately (Driessen 1992 and Chang and Hsu 1992) also for these more complicated designs than the one-way layout, SSPs have been given that satisfy the P^*-condition. Since in the next section we will discuss selection constants appearing in rules for block designs we give a SSP for a general block design that satisfies the P^*-condition.

Let $Y_{ij\ell}$ be the ℓth observation of treatment $\P_i$ in block j, $i = 1, \cdots, t$, $j = 1, \cdots, b$, $\ell = 1, \cdots, n_{ij}$. As a model for the observations we assume that

$$Y_{ij\ell} = \mu + \tau_i + \beta_j + \sigma \varepsilon_{ij\ell}, \tag{3.1}$$

with μ the general mean, τ_i the effect of treatment $\P_i$, β_j the effect of block j, σ the common unknown standard deviation and the $\varepsilon_{ij\ell}$'s mutually independent standard normal variables. Let S^2 denote the usual estimator of σ^2. We assume that the design is connected, i.e., all elementary contrasts $\tau_i - \tau_j$ are estimable. Least squares estimates of the treatment effects are given by the vector

$$\hat{\tau} = (\hat{\tau}_1, \cdots, \hat{\tau}_t)' = C^- Q,$$

with C^- a generalized inverse of $C = R - NK^{-1}N'$, $Q = T - NK^{-1}B$ (the corrected treatment totals), N the incidence matrix and R and K both diagonal matrices with on the diagonal the number of replicates and block sizes, respectively. For sake of convenience we define

$$\nu_{ij} = \text{var} \frac{(\hat{\tau}_i - \hat{\tau}_j)}{\sigma}, \qquad i, j = 1, \cdots, t.$$

The rule selects a treatment $\P_i$, $i = 1, \cdots, t$, if and only if

$$\hat{\tau}_i \geq \hat{\tau}_j - \Delta_i \sqrt{\nu_{ij}} S \qquad \text{for all } j \neq i, \tag{3.2}$$

where each selection constant Δ_i, $i = 1, \cdots, t$, is such that

$$P[\hat{\tau}_i - \tau_i \geq \hat{\tau}_j - \tau_j - \Delta_i \sqrt{\nu_{ij}} S \qquad \text{for all } j \neq i] = P^*. \tag{3.3}$$

4   Selection Constants

The selection constants that appear in SSP (3.2) are defined for each treatment separately by (3.3). This equation can be rewritten as

$$P^* = P[T^i \leq (\Delta_i, \cdots, \Delta_i)'], \tag{4.1}$$

where $T^i = Z^i/(S/\sigma)$ with S/σ and Z^i, $i = 1, \cdots, t$, independent random variables, distributed as a $\sqrt{\chi^2_\nu/\nu}$ variable and a multinormal vector with zero means and correlation matrix V_i of order $t - 1$, respectively. This correlation matrix is described for notational convenience by $\{\rho^i_{jj'}\}$ although the indices j and j' attain only the values in the set $\{1, ..., i-1, i+1, ..., t\}$. The entries $\{\rho^i_{jj'}\}$ of V_i, $i = 1, \cdots, t$, $j \neq i$, $j' \neq i$, are given by

$$\begin{cases} 1 & \text{if } j = j' \\ \frac{\nu_{ji}+\nu_{j'i}-\nu_{jj'}}{2\sqrt{\nu_{ji}}\sqrt{\nu_{j'i}}} & \text{if } j \neq j'. \end{cases}$$

Notice that in case of a pairwise balanced design each correlation matrix V_i, $i = 1, \cdots, t$, attains the simple form $\frac{1}{2}(\mathbf{I}_{t-1} + \mathbf{1}_{t-1}\mathbf{1}'_{t-1})$ which implies (Gupta 1963) that for all $i = 1, \cdots, t$, (4.1) can be rewritten into the well-known integral form:

$$\int_0^\infty \int_{-\infty}^\infty \Phi^{t-1}(x + \sqrt{2}\Delta_i y)\phi(x)q_\nu(y)dxdy = P^*.$$

Clearly, this leads to equal selection constants for all treatments. Except for this and some other special cases (Dunnett 1989), the evaluation of a multivariate t (or normal) probability is a problem, for no simple numerical procedures are available. Yet it is reasonable and feasible to approximate the selection constants by simulating for each $i = 1, \cdots, t$, the distribution of the vector T^i and thus solve (4.1). So the practical determination of the selection constants itself is possible and we will let this matter rest in the sequel.

Having seen that for pairwise balanced designs the selection constants are all equal it is interesting to investigate whether or not more (classes of) designs have this special property. The following lemma, which follows directly from (4.1), gives an introductory view on this aspect.

Lemma 4.1

For any $i = 1, \cdots, t$, if the coordinates of T^i can be permuted such that the resulting vector is distributed as T^1, then

$$\Delta_i = \Delta_1.$$

We rephrase Lemma 4.1 towards the correlation matrices corresponding to the vectors T^i, $i = 1, \ldots, t$, by using concepts from matrix theory. A permutation matrix is a square matrix with exactly one 1 per row and column, and all other entries zero. Two square real matrices A and B of order t are called *strongly congruent* if there exists a permutation matrix, say P, such that $PAP' = B$. It is easily verified that strong congruency is an associative and transitive property of square real matrices of the same order.

Corollary 4.1

For any $i = 1, \cdots, t$, if the correlation matrices V_i and V_1 are strongly congruent, then

$$\Delta_i = \Delta_1.$$

Some more definitions first. We call a design *autocongruent*, if for all $i = 1, \cdots, t$, the matrices V_i and V_1 are strongly congruent. So for an autocongruent design it holds that all selection constants are equal.

We call a square real matrix A autocongruent if and only if for each $i = 1, \cdots, t$, there exists a permutation matrix E_i such that

$$\left\{ \begin{array}{ll} E_i(e_i) & = e_1 \\ E_i A E_i' & = A, \end{array} \right. \tag{4.2}$$

where e_i, $i = 1, \cdots, t$, is the elementary vector with t coordinates all equal to zero except for the ith, being equal to 1.

We will restrict our discussion of selection constants in the following to block designs. The following theorem directly links a characteristic of a block design, in case the C matrix, to equality of selection constants.

Theorem 4.1 (Driessen 1991)

Let $C = R - NK^{-1}N'$ be given for a block design. If C is autocongruent, then the design is autocongruent and therefore has equal selection constants.

Corollary 4.2

If a block design is equireplicate (all treatments appear an equal number of times), proper (equal block sizes) and binary (entries of N are equal to 0 or 1), then:

   if the concurrence matrix NN' is autocongruent, then the design
   is autocongruent.

Corollary 4.2 gives lead to a nice interpretation with respect to the concept of autocongruency in (experimental) design:

   the way a treatment concurs with the rest and this rest mutually,
   is the same for each treatment.

Example In a simple lattice for $t = 4$ treatments, the treatments have 2 replicates each, and there are $b = 4$ blocks, containing 2 plots each. The incidence and concurrence matrix of this design, respectively, are

$$N = \begin{bmatrix} 1 & 0 & 1 & 0 \\ 1 & 0 & 0 & 1 \\ 0 & 1 & 1 & 0 \\ 0 & 1 & 0 & 1 \end{bmatrix} \quad \text{and} \quad NN' = \begin{bmatrix} 2 & 1 & 1 & 0 \\ 1 & 2 & 0 & 1 \\ 1 & 0 & 2 & 1 \\ 0 & 1 & 1 & 2 \end{bmatrix}.$$

The following three equations make clear that the design is autocongruent:

$$\begin{aligned} E_2 N N' E_2' &= N N', \\ E_3 N N' E_3' &= N N', \\ E_4 N N' E_4' &= N N', \end{aligned}$$

with

$$E_2 = \begin{bmatrix} 0 & 1 & 0 & 0 \\ 1 & 0 & 0 & 0 \\ 0 & 0 & 0 & 1 \\ 0 & 0 & 1 & 0 \end{bmatrix}, \quad E_3 = \begin{bmatrix} 0 & 0 & 1 & 0 \\ 0 & 0 & 0 & 1 \\ 1 & 0 & 0 & 0 \\ 0 & 1 & 0 & 0 \end{bmatrix} \quad \text{and}$$

$$E_4 = \begin{bmatrix} 0 & 0 & 0 & 1 \\ 0 & 0 & 1 & 0 \\ 0 & 1 & 0 & 0 \\ 1 & 0 & 0 & 0 \end{bmatrix}.$$

The simple lattice discussed in the example is a member of the class of so-called partially balanced incomplete block designs with two associate classes (PBIB(2)s). For this class of equireplicate, proper and binary designs it holds that any two treatments $\P_i$ and $\P_j$ are first associates or second associates and as such occur λ_1 or λ_2 times together in blocks of the design. Notice that these latter numbers are irrespective of the choice of treatments and that they only depend upon the association. This relationship between the treatments on which the design is based is called the association scheme. For PBIB(2)s we can write the concurrence matrix always as

$$NN' = r\mathbf{I}_t + \lambda_1 B_1 + \lambda_2 B_2,$$

with $\mathbf{I}_t$ the identity matrix of order t, and B_1 and B_2 as the so-called first and second association matrix, respectively, where an entry (i, j) of $B_1(B_2)$ is equal to 1, if i and j are first (second) associates.

Corollary 4.3

The first (or second) association matrix of a PBIBD(2) is autocongruent if and only if the design is autocongruent.

Proof of Corollary 4.3

The corollary directly follows from the observation that for a PBIBD(2)

$$\mathbf{I}_t + B_1 + B_2 = \mathbf{1}_t\mathbf{1}'_t. \quad \blacksquare$$

The class of *Latin Square* designs is a class of PBIBD(2)s based on an association scheme, which is defined by a generating array. Let a set of $t = n^2$ ($n \geq 2$) treatments be arranged in an $n \times n$ array onto which a set of $i - 2$ ($2 \leq i \leq n - 1$) mutually orthogonal Latin Squares of side n has been superimposed, where a Latin Square of side n is also an $n \times n$ array but based on a set of only n symbols, such that each row and each column contains each symbol exactly once. The first associates of a treatment are all the treatments in the same row or column of the array or associated with the same symbols in one of the Latin Squares. This is called an L_i-type association scheme. The well known lattice designs are L_2 designs.

Theorem 4.2

L_2 designs are autocongruent.

Proof of Theorem 4.2

In a L_2 design the first associates of any treatment are just the treatments in the same row or column in the generating array. Clearly, by permuting rows and columns of the generating array appropriately, any treatment can be transposed to the position of treatment 1 in the generating array and these permutations leave the association scheme invariant. This implies that the design is autocongruent. $\blacksquare$

One can show that all PBIB(2)s that are based on so-called group divisible and triangular association schemes are also autocongruent. But despite the strong symmetry of PBIBD(2)s they are not all autocongruent. A counter example is given by a latin square L_3 scheme for $t = 36$ treatments. It is defined by the following two matrices:

$$\begin{pmatrix} 1 & 2 & 3 & 4 & 5 & 6 \\ 7 & 8 & 9 & 10 & 11 & 12 \\ 13 & 14 & 15 & 16 & 17 & 18 \\ 19 & 20 & 21 & 22 & 23 & 24 \\ 25 & 26 & 27 & 28 & 29 & 30 \\ 31 & 32 & 33 & 34 & 35 & 36 \end{pmatrix} \text{ and } \begin{pmatrix} a & b & c & d & e & f \\ b & a & f & c & d & e \\ c & f & a & e & b & d \\ d & c & e & a & f & b \\ e & d & b & f & a & c \\ f & e & d & b & c & a \end{pmatrix}.$$

Treatments are first associates if and only if:
      they are in the same row in the left matrix; or
      they are in the same column in the left matrix; or
      they have the same symbols in the right matrix.

$$B_1 = \begin{pmatrix}
0 & 1 & 1 & 1 & 1 & 1 & 1 & 1 & 0 & 0 & 0 & 0 & 1 & 0 & 1 & 0 & 0 & 0 & 1 & 0 & 0 & 1 & 0 & 0 & 1 & 0 & 0 & 1 & 0 & 0 & 1 & 0 & 1 & 0 & 0 & 1 \\
1 & 0 & 1 & 1 & 1 & 1 & 1 & 1 & 0 & 0 & 0 & 0 & 0 & 1 & 0 & 0 & 1 & 0 & 0 & 1 & 0 & 0 & 0 & 1 & 0 & 1 & 1 & 0 & 0 & 0 & 0 & 1 & 0 & 1 & 0 & 0 \\
1 & 1 & 0 & 1 & 1 & 1 & 0 & 0 & 1 & 1 & 0 & 0 & 1 & 0 & 1 & 0 & 0 & 0 & 0 & 1 & 1 & 0 & 0 & 0 & 0 & 1 & 0 & 0 & 1 & 0 & 0 & 1 & 0 & 1 & 0 & 1 \\
1 & 1 & 1 & 0 & 1 & 1 & 0 & 0 & 0 & 1 & 1 & 0 & 0 & 0 & 0 & 1 & 0 & 1 & 1 & 0 & 0 & 1 & 0 & 0 & 0 & 1 & 0 & 1 & 0 & 0 & 0 & 0 & 1 & 1 & 0 & 0 \\
1 & 1 & 1 & 1 & 0 & 1 & 0 & 0 & 0 & 0 & 1 & 1 & 0 & 0 & 0 & 1 & 1 & 0 & 0 & 0 & 1 & 0 & 1 & 0 & 1 & 0 & 0 & 0 & 1 & 0 & 0 & 1 & 0 & 0 & 1 & 0 \\
1 & 1 & 1 & 1 & 1 & 0 & 0 & 0 & 1 & 0 & 0 & 1 & 0 & 1 & 0 & 0 & 0 & 1 & 0 & 0 & 0 & 0 & 1 & 1 & 0 & 0 & 0 & 1 & 0 & 1 & 1 & 0 & 0 & 0 & 0 & 1 \\
1 & 1 & 0 & 0 & 0 & 0 & 0 & 1 & 1 & 1 & 1 & 1 & 1 & 0 & 0 & 0 & 1 & 0 & 1 & 0 & 0 & 0 & 0 & 1 & 1 & 0 & 1 & 0 & 0 & 0 & 1 & 0 & 0 & 1 & 0 & 0 \\
1 & 1 & 0 & 0 & 0 & 0 & 1 & 0 & 1 & 1 & 1 & 1 & 1 & 0 & 1 & 1 & 0 & 0 & 1 & 0 & 0 & 1 & 0 & 0 & 1 & 0 & 0 & 1 & 0 & 0 & 1 & 0 & 0 & 0 & 0 & 1 \\
0 & 0 & 1 & 0 & 0 & 1 & 1 & 1 & 0 & 1 & 1 & 1 & 0 & 1 & 1 & 0 & 0 & 0 & 0 & 1 & 0 & 1 & 0 & 0 & 0 & 1 & 1 & 0 & 0 & 1 & 0 & 1 & 0 & 0 & 0 & 0 \\
0 & 0 & 1 & 1 & 0 & 0 & 1 & 1 & 1 & 0 & 1 & 1 & 1 & 0 & 0 & 1 & 0 & 0 & 0 & 1 & 0 & 1 & 0 & 0 & 0 & 0 & 1 & 0 & 1 & 0 & 0 & 0 & 1 & 1 & 0 & 0 \\
0 & 0 & 0 & 1 & 1 & 0 & 1 & 1 & 1 & 1 & 0 & 1 & 0 & 0 & 0 & 0 & 1 & 1 & 1 & 0 & 0 & 0 & 1 & 0 & 0 & 1 & 0 & 0 & 1 & 0 & 0 & 0 & 1 & 0 & 1 & 0 \\
0 & 0 & 0 & 0 & 1 & 1 & 1 & 1 & 1 & 1 & 1 & 0 & 0 & 0 & 0 & 1 & 0 & 1 & 0 & 0 & 1 & 0 & 0 & 1 & 1 & 0 & 0 & 0 & 0 & 1 & 0 & 1 & 0 & 0 & 0 & 1 \\
1 & 0 & 1 & 0 & 0 & 0 & 1 & 0 & 0 & 1 & 0 & 0 & 0 & 1 & 1 & 1 & 1 & 1 & 1 & 0 & 0 & 0 & 0 & 1 & 0 & 0 & 0 & 1 & 0 & 0 & 0 & 0 & 1 & 0 & 0 & 1 \\
0 & 1 & 0 & 0 & 0 & 1 & 0 & 1 & 1 & 0 & 0 & 0 & 1 & 0 & 1 & 1 & 1 & 1 & 1 & 0 & 1 & 0 & 0 & 1 & 0 & 0 & 1 & 0 & 1 & 0 & 0 & 1 & 1 & 0 & 0 & 0 \\
1 & 0 & 1 & 0 & 0 & 0 & 0 & 1 & 1 & 0 & 0 & 0 & 1 & 1 & 0 & 1 & 1 & 1 & 0 & 0 & 1 & 1 & 0 & 0 & 0 & 0 & 1 & 0 & 1 & 0 & 0 & 0 & 1 & 0 & 0 & 1 \\
0 & 0 & 0 & 1 & 1 & 0 & 0 & 0 & 0 & 1 & 0 & 1 & 1 & 1 & 1 & 0 & 1 & 1 & 0 & 0 & 1 & 1 & 0 & 0 & 1 & 0 & 0 & 1 & 0 & 0 & 0 & 1 & 0 & 1 & 0 & 0 \\
0 & 1 & 0 & 0 & 1 & 0 & 1 & 0 & 0 & 0 & 1 & 0 & 1 & 1 & 1 & 1 & 0 & 1 & 0 & 0 & 0 & 0 & 1 & 1 & 0 & 0 & 1 & 0 & 1 & 0 & 0 & 0 & 0 & 1 & 1 & 0 \\
0 & 0 & 0 & 1 & 0 & 1 & 0 & 0 & 0 & 0 & 1 & 1 & 1 & 1 & 1 & 1 & 1 & 0 & 1 & 0 & 0 & 0 & 1 & 0 & 1 & 0 & 0 & 0 & 1 & 0 & 0 & 1 & 0 & 0 & 0 & 1 \\
1 & 0 & 0 & 1 & 0 & 0 & 1 & 0 & 0 & 0 & 1 & 0 & 1 & 0 & 0 & 0 & 1 & 0 & 1 & 1 & 1 & 1 & 1 & 1 & 1 & 0 & 0 & 0 & 0 & 1 & 0 & 1 & 0 & 0 & 0 & 0 \\
0 & 1 & 1 & 0 & 0 & 0 & 0 & 1 & 0 & 1 & 0 & 0 & 1 & 1 & 0 & 0 & 0 & 1 & 0 & 1 & 1 & 1 & 1 & 1 & 0 & 1 & 0 & 0 & 0 & 1 & 0 & 1 & 0 & 0 & 1 & 0 \\
0 & 0 & 1 & 0 & 1 & 0 & 0 & 0 & 1 & 0 & 0 & 1 & 0 & 0 & 1 & 1 & 0 & 0 & 1 & 1 & 0 & 1 & 1 & 0 & 1 & 0 & 0 & 1 & 0 & 0 & 1 & 0 & 1 & 0 & 0 & 1 \\
1 & 0 & 0 & 1 & 0 & 0 & 0 & 1 & 0 & 1 & 0 & 0 & 0 & 0 & 1 & 1 & 0 & 0 & 1 & 1 & 1 & 0 & 1 & 1 & 0 & 0 & 0 & 1 & 1 & 0 & 0 & 0 & 0 & 1 & 0 & 1 \\
0 & 0 & 0 & 0 & 1 & 1 & 0 & 0 & 1 & 0 & 1 & 0 & 0 & 1 & 0 & 0 & 1 & 0 & 0 & 1 & 1 & 1 & 0 & 1 & 0 & 0 & 0 & 1 & 0 & 1 & 0 & 0 & 0 & 1 & 0 & 1 \\
1 & 0 & 0 & 0 & 0 & 1 & 1 & 0 & 0 & 0 & 0 & 1 & 0 & 0 & 1 & 0 & 0 & 0 & 1 & 1 & 0 & 0 & 0 & 0 & 1 & 1 & 1 & 1 & 1 & 1 & 1 & 0 & 0 & 0 & 0 & 0 \\
0 & 1 & 0 & 1 & 0 & 0 & 0 & 1 & 0 & 1 & 0 & 0 & 1 & 0 & 0 & 1 & 0 & 0 & 1 & 0 & 0 & 1 & 0 & 0 & 1 & 0 & 1 & 1 & 1 & 1 & 0 & 1 & 0 & 0 & 0 & 1 \\
0 & 1 & 0 & 0 & 0 & 1 & 1 & 0 & 0 & 0 & 0 & 1 & 0 & 0 & 0 & 0 & 1 & 0 & 0 & 0 & 1 & 1 & 1 & 1 & 1 & 1 & 1 & 0 & 0 & 0 & 1 & 0 & 1 & 0 & 0 & 1 \\
1 & 0 & 0 & 0 & 1 & 0 & 0 & 0 & 0 & 1 & 1 & 0 & 0 & 0 & 1 & 1 & 0 & 0 & 1 & 0 & 1 & 0 & 0 & 1 & 0 & 0 & 0 & 1 & 1 & 1 & 1 & 1 & 1 & 0 & 0 & 0 \\
0 & 1 & 0 & 1 & 0 & 0 & 0 & 1 & 0 & 0 & 1 & 0 & 0 & 1 & 0 & 0 & 1 & 0 & 0 & 0 & 1 & 1 & 1 & 0 & 0 & 0 & 1 & 0 & 1 & 1 & 1 & 1 & 0 & 1 & 1 & 0 \\
0 & 1 & 1 & 0 & 0 & 0 & 1 & 0 & 1 & 0 & 0 & 0 & 0 & 1 & 0 & 1 & 0 & 0 & 0 & 1 & 0 & 0 & 0 & 1 & 0 & 1 & 0 & 0 & 1 & 1 & 1 & 0 & 1 & 1 & 1 & 0 \\
0 & 0 & 0 & 1 & 0 & 1 & 0 & 0 & 1 & 1 & 0 & 0 & 0 & 1 & 0 & 1 & 0 & 0 & 0 & 0 & 1 & 1 & 0 & 1 & 1 & 1 & 0 & 1 & 1 & 0 & 1 & 1 & 1 & 0 & 1 & 0 \\
1 & 0 & 0 & 0 & 1 & 0 & 0 & 1 & 0 & 0 & 1 & 0 & 0 & 0 & 1 & 0 & 1 & 0 & 0 & 0 & 0 & 1 & 1 & 0 & 1 & 1 & 1 & 1 & 0 & 1 & 0 & 0 & 0 & 0 & 1 & 1 \\
0 & 0 & 1 & 0 & 0 & 1 & 0 & 0 & 0 & 1 & 0 & 1 & 1 & 0 & 0 & 0 & 0 & 1 & 0 & 1 & 1 & 0 & 0 & 0 & 1 & 0 & 1 & 0 & 0 & 0 & 1 & 1 & 1 & 1 & 1 & 1 \\
1 & 0 & 0 & 0 & 0 & 1 & 1 & 0 & 1 & 0 & 0 & 0 & 1 & 1 & 0 & 0 & 0 & 1 & 0 & 0 & 0 & 1 & 0 & 0 & 0 & 1 & 0 & 1 & 0 & 0 & 1 & 0 & 0 & 1 & 1 & 1 \\
0 & 1 & 0 & 0 & 1 & 0 & 0 & 1 & 0 & 0 & 0 & 1 & 0 & 1 & 0 & 1 & 0 & 0 & 0 & 1 & 1 & 0 & 0 & 0 & 1 & 1 & 0 & 0 & 0 & 0 & 1 & 0 & 1 & 1 & 1 & 1 \\
0 & 0 & 1 & 1 & 0 & 0 & 0 & 0 & 1 & 0 & 1 & 0 & 0 & 0 & 1 & 0 & 0 & 1 & 1 & 0 & 1 & 0 & 0 & 0 & 0 & 1 & 1 & 0 & 0 & 0 & 1 & 1 & 0 & 1 & 1 & 1 \\
1 & 0 & 0 & 0 & 0 & 1 & 0 & 1 & 0 & 0 & 0 & 1 & 0 & 0 & 1 & 0 & 0 & 1 & 0 & 0 & 1 & 0 & 0 & 1 & 0 & 0 & 1 & 0 & 1 & 0 & 0 & 0 & 1 & 1 & 1 & 0
\end{pmatrix}$$

Figure 1: First association matrix of the L_3 scheme for 36 treatments.

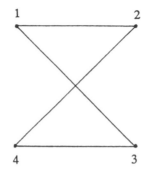

Figure 2: Graph of the simple lattice.

So, for instance, treatment 1 has as first associates the treatments contained in the set $\{2, 3, 4, 5, 6, 7, 8, 13, 15, 19, 22, 25, 29, 31, 36\}$. The first association matrix B_1 is given in Figure 1. Notice that $B_1^2 = 9\mathbf{I}_{36} + 6\mathbf{1}_{36}\mathbf{1}'_{36}$. It can be shown (Driessen 1992) that B_1 is *not* autocongruent (for example, for treatment $\P_{10}$ no permutation matrix exists that leaves the first association matrix invariant as required by (4.2)).

Since association matrices are square (0,1)-matrices we can represent them by graphs by simply connecting vertices i and j, that represent treatments $\P_i$ and $\P_j$, if and only if the entry (i, j) of the matrix is equal to 1. Vertices that are connected by edges in this way are called adjacent. This leads, for the association matrix B_1 of the simple lattice, to the graph in Figure 2. We call a graph autocongruent whenever for each number $i \neq 1$ there exists a permutation of the numbers that maps i on 1 and leaves the original graph structure invariant. It is easy to see that an association matrix is autocongruent if and only if its corresponding graph is autocongruent. So, in other words, we have to look for permutations of the numbers corresponding to the vertices that do not change original connections between vertices. In Figure 3 it is shown that the graph of the simple lattice is autocongruent. So such graphs can be made up for all first and second association matrices of PBIBD(2)s. We then get a class of graphs, the so-called strongly regular graphs (designated by the parameters t, d, a and c) that have properties:

- all t vertices have the same degree (d), that is, from each vertex originate d edges to other vertices.

- the number of vertices adjacent to i and j is equal to

$$\begin{cases} a & \text{if } i \text{ and } j \text{ are adjacent} \\ c & \text{if } i \text{ and } j \text{ are not adjacent}. \end{cases}$$

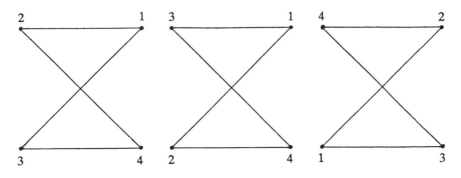

Figure 3: Autocongruency of the simple lattice.

Example The well known Petersen graph, shown in Figure 4, is a strongly regular graph with parameters $t = 10$, $d = 3$, $a = 0$ and $c = 1$. It is easily verified that the Petersen graph is indeed autocongruent. The graph corresponds to the second association matrix of the triangular design for 10 treatments.

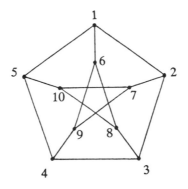

Figure 4: Petersen graph.

References

Chang, Y-J. and Hsu, J.C. (1992). Optimal designs for multiple comparisons with the best. *Jour. Statist. Plan. Infer.* **30** 45-62.

Cochran, W.G. and Cox, G.M. (1957). *Experimental Designs.* John Wiley and Sons, Inc., New York.

Driessen, S.G.A.J. (1991). Multiple comparisons with and selection of the best treatment in (incomplete) block designs. *Commun. Statist. - Theory and Methods* **20** 179 - 217.

Driessen, S.G.A.J. (1992). *Statistical Selection: Multiple Comparison Approach.* Ph.D. thesis, Dept. of Mathematics, Agricultural University Wageningen, The Netherlands.

Dunnett, C.W. (1989). Multivariate normal probability integrals with product correlation structure. *Appl. Statist.* **38** 564 - 579.

Gupta, S.S. (1956). *On a Decision Rule for a Problem in Ranking Means.* Ph.D. thesis, Dept. of Statistics, University of North Carolina, Chapel Hill, N.C.

Gupta, S.S. (1963). Probability integrals of multivariate normal and multivariate *t*. *Ann. Math. Statist.* **12** 1136 - 1144.

Hochberg, Y. and Tamhane, A.C. (1987). *Multiple Comparison Procedures.* John Wiley and Sons, Inc., New York.

Somerville, P.N. (1984). A multiple range subset selection procedure. *Jour. Statist. Comput. Simul.* **19** 215 - 226.

Somerville, P.N. (1985). A new subset selection method for normal populations. *Jour. Statist. Comput. Simul.* **22** 27 - 50.

Tukey, J.W. (1953). *The Problem of Multiple Comparisons.* Mimeographed monograph, Princeton University.

Chapter 22

Sharpening Subset Selection of Treatments Better than a Control

MANFRED HORN   Biometrical Unit, Hans Knöll Institute
of Natural Product Research, Jena, Germany

RÜDIGER VOLLANDT   Department of Medical Informatics and
Biomathematics, Friedrich Schiller University, Jena, Germany

Abstract   The selection of treatments that are better than a control can
be performed by multiple comparison procedures. If such a procedure controls the experimentwise error rate, one gets some kind of confidence estimate of the set of treatments that are better than the control. The higher
the power of the multiple comparison procedure is, the sharper is the selection and the better is the confidence estimate.

1   Introduction

The method of Dunnett (1955) for multiple comparisons of k treatments
with a control and the methods of Gupta and Sobel (1958) for selecting a
subset of treatments that are better than a control are closely related. Both
works assume normal distributions for the control and the k treatments with
expectations μ_0 and $\mu_i (i = 1, \cdots, k)$ and common variance σ^2.

In the sequel we assume, for simplicity, that the sample sizes of the
treatments and the control are all n, the sample means are denoted by

$\bar{x}_i(i = 0, 1, \cdots, k)$, and s denotes the estimate of σ having $f = (k+1)(n-1)$ degrees of freedom.

A multiple comparison method can be used to test the one-sided hypotheses $H_i : \mu_i \leq \mu_0$ or $H_i : \mu_i \geq \mu_0$ or the two-sided hypotheses $H_i : \mu_i = \mu_0 (i = 1, \cdots, k)$. With Dunnett's test $H_i : \mu_i \leq \mu_0$ will be rejected if

$$t_i = (\bar{x}_i - \bar{x}_0)\sqrt{n}/(s\sqrt{2}) > d$$

or equivalently

$$\bar{x}_i > \bar{x}_0 + ds\sqrt{2}/\sqrt{n},$$

and $H_i : \mu_i \geq \mu_0$ will be rejected if

$$t_i < -d$$

or equivalently

$$\bar{x}_i < \bar{x}_0 - ds\sqrt{2}/\sqrt{n}.$$

Here d denotes the one-sided $(1 - \alpha)$-quantile of the k variate t distribution with f degrees of freedom and correlation coefficient 0.5 (Dunnett 1955, Bechhofer and Dunnett 1988, Krishnaiah and Armitage 1966) or the distribution of the studentized maximum of k equally correlated normal random variables (Gupta et al. 1983). The use of this quantile guarantees that the experimentwise (Type 1) error rate does not exceed α.

Let us denote by B^+ the set of treatments with $\mu_i \geq \mu_0$, i.e. the set of treatments being not worse than the control. Subset selection has the aim to select a subset S_2 that contains B^+ with probability $1 - \alpha$. The corresponding selection rule ensuring $P(B^+ \subseteq S_2) \geq 1 - \alpha$ was formulated in Gupta and Sobel (1958). It demands to select all treatments for which

$$\bar{x}_i \geq \bar{x}_0 - ds\sqrt{2}/\sqrt{n}.$$

Evidently, there is some equivalence between this selection rule and the decision rule of the method of Dunnett. S_2 is the set of treatments for which the hypotheses $H_i : \mu_i \geq \mu_0$ cannot be rejected.

Instead of looking for a subset S_2 containing all treatments with $\mu_i \geq \mu_0$, one can aspire to get a subset S_1 containing, with probability $1 - \alpha$, no treatment with $\mu_i \leq \mu_0$ but only treatments with $\mu_i > \mu_0$. If B is the set of all treatments with $\mu_i > \mu_0$, the selection rule ensuring $P(S_1 \subseteq B) \geq 1 - \alpha$ demands to select all treatments for which

$$\bar{x}_i > \bar{x}_0 + ds\sqrt{2}/\sqrt{n},$$

that is, for which $H_i : \mu_i \leq \mu_0$ must be rejected by Dunnett's method. What we finally obtain is the relation

$$S_1 \subseteq B \subseteq B^+ \subseteq S_2$$

with $P(S_1 \subseteq B) \geq 1 - \alpha$ and $P(B^+ \subseteq S_2) \geq 1 - \alpha$. Thus S_1 can be regarded as lower $(1 - \alpha)$-confidence limit set of B, and S_2 can be regarded as upper $(1 - \alpha)$-confidence limit set of B^+. One can easily show that $P(S_1 \subseteq B \subseteq B^+ \subseteq S_2) \geq 1 - 2\alpha$. This inequality provides a two-sided $(1 - 2\alpha)$-confidence estimate of B and B^+, and one can obtain a two-sided $(1 - \alpha)$-confidence estimate when testing the one-sided hypotheses at level $\alpha/2$. (Besides, one can derive a similar inequality by using the two-sided Dunnett test. Then the use of the corresponding two-sided $(1-\alpha)$-quantiles provides a two-sided $(1 - \alpha)$-confidence estimate of B and B^+. But the two-sided $(1 - \alpha)$-quantiles are only a little smaller than the one-sided $(1 - \alpha/2)$-quantiles so that the gain can be neglected.)

What we intend is to sharpen the inequality $S_1 \subseteq B \subseteq B^+ \subseteq S_2$, replacing Dunnett's method by a more powerful multiple comparison procedure which rejects more and accepts fewer hypotheses so that the lower limit set S_1 becomes larger and the upper limit set S_2 becomes smaller. Additionally, we will investigate whether the way to get such confidence subset relations by multiple comparisons can be generalized.

2 Subset Selection by Multiple Comparisons in Case of k Treatments and One Control

Let us examine the one-sided hypotheses $H_i : \mu_i \leq \mu_0$ and $H_i : \mu_i \geq \mu_0 (i = 1, \cdots, k)$ by some multiple test procedure. Again, let B denote the set of treatments with $\mu_i > \mu_0$, B^+ the set of treatments with $\mu_i \geq \mu_0$, S_1 the set of treatments for which $H_i : \mu_i \leq \mu_0$ must be rejected by the multiple test procedure, and S_2 the set of treatments for which $H_i : \mu_i \geq \mu_0$ cannot be rejected. Generally, with the inequality $S_1 \subseteq B \subseteq B^+ \subseteq S_2$ one obtains some estimate of B and B^+. The subsets S_1 and S_2 can be regarded as lower and upper confidence limit sets if the probabilities $P(S_1 \subseteq B)$ and $P(B^+ \subseteq S_2)$ are given. The following result is important.

Theorem

The relations

$$P(S_1 \subseteq B) \geq 1 - \alpha$$

and

$$P(B^+ \subseteq S_2) \geq 1 - \alpha$$

are true if and only if the experimentwise (Type 1) error rates in testing the hypotheses $H_i : \mu_i \leq \mu_0$ and $H_i : \mu_i \geq \mu_0$ do not exceed α.

Proof of Theorem

The proof immediately follows from the definition of the experimentwise error rate $\alpha_{\exp} = P(\text{at least one true } H_i \text{ rejected})$. Now $\alpha_{\exp} \leq \alpha$, or equivalently, $1 - \alpha_{\exp} = P (\text{no true } H_i \text{ rejected}) \geq 1 - \alpha$ means, concerning the hypotheses $H_i : \mu_i \leq \mu_0 (i = 1, \cdots, k)$, that at least with probability $1 - \alpha$ no true H_i will be rejected, i.e. only some of the false H_i will be rejected, i.e. some of the treatments of B are selected, i.e. $S_1 \subseteq B$. Concerning the hypotheses $H_i : \mu_i \geq \mu_0$ this means that at least with probability $1 - \alpha$ all true H_i are accepted (and some false ones, too), i.e. at least all treatments of B^+ are selected, i.e. $B^+ \subseteq S_2$. ∎

Consequently, one should use only multiple comparison procedures that control the experimentwise error rate if one aspires to obtain confidence estimates of the sets B and B^+ with $P(S_1 \subseteq B) \geq 1 - \alpha$, $P(B^+ \subseteq S_2) \geq 1 - \alpha$ and $P(S_1 \subseteq B \subseteq B^+ \subseteq S_2) \geq 1 - 2\alpha$.

3   Sharper Subset Selection and Improved Confidence Estimates

Obviously, the procedure for obtaining subsets S_1 and S_2 with $S_1 \subseteq B \subseteq B^+ \subseteq S_2$ is not restricted to normal distributions. It can be used with any multiple comparison method. One will aspire to get S_1 and S_2 as close as possible to B and B^+, i.e. to get S_1 as large as possible and S_2 as small as possible.

A multiple test procedure will be said to have high power if it is able to reject false hypotheses H_i with relatively high probability. Thus high power means many rejected false hypotheses $H_i : \mu_i \leq \mu_0$, and few accepted false hypotheses $H_i : \mu_i \geq \mu_0$, i.e. large S_1 and small S_2.

One can find several methods that control the experimentwise error rate and have higher power than the multiple test of Dunnett (1955). One of them is the closed procedure of Marcus et al. (1976). This method is testing in a stepwise manner. The first step is identical with Dunnett's test: all H_i are tested by comparing the test statistics $t_i (i = 1, \cdots, k)$ defined in Section 1 with the same one-sided $(1-\alpha)$-quantile of the k variate t distribution. If no t_i exceeds this quantile no H_i can be rejected. If ℓ_1 statistics t_i exceed this quantile, the corresponding H_i must be rejected. The $k - \ell_1$ remaining hypotheses will be tested in the second step by comparing the corresponding t_1 against the $(1 - \alpha)$-quantile of the $(k - \ell_1)$ variate t distribution (with f degrees of freedom). This quantile is smaller than that of the k variate distribution. Therefore it is possible that some H_i are rejected in the second step that could not be rejected in the first step, i.e., by the original test of

Dunnett. If no hypothesis is rejected, the procedure stops. If ℓ_2 hypotheses are rejected in the second step, the third step demands to compare the remaining $k - \ell_1 - \ell_2$ statistics t_i with the $(1 - \alpha)$-quantile of the $(k - \ell_1 - \ell_2)$ variate t distribution. This is smaller than the $(1 - \alpha)$-quantile of the $(k - \ell_1)$ variate t distribution, so possibly one rejects further hypotheses. The last step may be the one-sided t test of one remaining hypothesis H_i by comparing t_i with the $(1 - \alpha)$-quantile of Student's t distribution (with f degrees of freedom). This quantile is smaller than the one-sided $(1 - \alpha)$-quantile of the j variate t distribution (with f degrees of freedom) for $j > 1$. So at least each H_i that is rejected by Dunnett's method will be rejected by the method of Marcus et al., i.e. the number of significances with the method of Marcus et al. cannot be smaller.

The idea in Marcus et al. (1976) is a general one and is not restricted to normal distributions and to stepwise uses of Dunnett's method. Any test (parametric or nonparametric) for the intersection and single hypotheses can be used.

There exist other general closed multiple comparison methods: one example is the method of Holm (1979) which can be recommended because of its simplicity. However, the procedure of Marcus et al. (1976) is the most powerful method.

4 Multiple Comparisons with a Control and Subset Selection: Normal Distributions, Known or Unknown μ_0 and σ

With normal distributions, the method of Dunnett is constructed for the case of unknown μ_0 and unknown σ. If one of these two parameters is known, or both are known, the decision rules of $H_i : \mu_i \leq \mu_0$ or $H_i : \mu_i \geq \mu_0$ are similar to those of the method of Dunnett as can be seen in Table 1.

All the decision rules in Table 1 can be used as rules to select subsets S_1 or S_2, containing, with probability $1 - \alpha$, no treatment with $\mu_i \leq \mu_0$ or all treatments with $\mu_i \geq \mu_0$, respectively. The rules of selecting S_2 for these different constellations of known or unknown μ_0 or σ have already been given in Gupta and Sobel (1958), and the definitions of the d_i have been given there, too. d_1 denotes the $(1 - \alpha)^{1/k}$-quantile of the standard normal distribution; d_2 is the $(1 - \alpha)$-quantile of the k variate normal distribution with correlation coefficient 0.5; and d_3 was only defined by some integral. In Rausch and Horn (1988) it was shown that d_3 is the $(1 - \alpha)$-quantile of the k variate t distribution with correlation coefficient 0 and $k(n - 1)$ degrees of freedom. d_4 is the one-sided $(1 - \alpha)$-quantile of the k variate t

Table 1: Rules for multiple comparisons and subset selection.

given values	subset S_1 reject $H_i : \mu_i \leq \mu_0$ if	subset S_2 accept $H_i : \mu_i \geq \mu_0$ if
μ_0, σ	$\bar{x}_i > \mu_0 + d_1 \frac{\sigma}{\sqrt{n}}$	$\bar{x}_i \geq \mu_0 - d_1 \frac{\sigma}{\sqrt{n}}$
$\bar{x}_0, \sigma$	$\bar{x}_i > \bar{x}_0 + d_2 \sqrt{2} \frac{\sigma}{\sqrt{n}}$	$\bar{x}_i \geq \bar{x}_0 - d_2 \sqrt{2} \frac{\sigma}{\sqrt{n}}$
μ_0, s	$\bar{x}_i > \mu_0 + d_3 \frac{s}{\sqrt{n}}$	$\bar{x}_i \geq \mu_0 - d_3 \frac{s}{\sqrt{n}}$
$\bar{x}_0, s$	$\bar{x}_i > \bar{x}_0 + d_4 \sqrt{2} \frac{s}{\sqrt{n}}$	$\bar{x}_i \geq \bar{x}_0 - d_4 \sqrt{2} \frac{s}{\sqrt{n}}$

distribution with f degrees of freedom and correlation coefficient 0.5, which was denoted simply by d in Section 1.

The selection of S_1 or S_2 according to the decision rules in Table 1 ensures that $P(S_1 \subseteq B) \geq 1 - \alpha$, $P(B^+ \subseteq S_2) \geq 1 - \alpha$ and $P(S_1 \subseteq B \subseteq B^+ \subseteq S_2) \geq 1 - 2\alpha$.

Again, the inequality $S_1 \subseteq B \subseteq B^+ \subseteq S_2$ can be sharpened when replacing the test given in Table 1 by a method like that in Marcus et al. (1976).

5 Example

The following example is adapted from one given in Dunnett (1955). The data are measurements on the breaking strength of fabric treated by $k = 4$ different chemical processes compared with a standard method of manufacture. The task now is to find two-sided 90%-confidence estimates for the sets B and B^+ of processes that are better or not worse, respectively than the standard process.

All sample sizes are $n = 3$. The average variance estimate is $s^2 = 19$, the standard deviation is $s = 4.36$. The degree of freedom is $f = (k+1)(n-1) =$

10. The sample means of the standard group and of the treated groups are $\bar{x}_0 = 50$, $\bar{x}_1 = 61$, $\bar{x}_2 = 58.5$, $\bar{x}_3 = 41.5$, $\bar{x}_4 = 41$. What we need in all tests is $s\sqrt{2}/\sqrt{n} = 4.36\sqrt{2}/\sqrt{3} = 3.56$. In Bechhofer and Dunnett (1988) one can find the one-sided percentage points d_4. For $\alpha = 0.05$, rho $= 0.5$, $f = 10$, $k = 4$ we get $d_4 = 2.46$, and for $k = 3$ we get $d_4 = 2.34$. Let μ_i $(i = 0, 1, \ldots, 4)$ denote the expected breaking strength under the standard treatment and under the 4 different processes.

When testing the hypotheses $H_i : \mu_i \leq \mu_0$ or $H_i : \mu_i \geq \mu_0$ $(i = 1, \ldots, 4)$ with Dunnett's method we have to compare $\bar{x}_i$ $(i = 1, \ldots, 4)$ with $\bar{x}_0 + d_4 s\sqrt{2}/\sqrt{n} = 50 + 2.46 \times 3.56 = 50 + 8.76 = 58.76$ or with $\bar{x}_0 - d_4 s\sqrt{2}/\sqrt{n} = 50 - 8.76 = 41.24$, respectively. The hypothesis $H_1 : \mu_1 \leq \mu_0$ must be rejected because $\bar{x}_1 > 58.76$. The other hypotheses $H_i : \mu_i \leq \mu_0$ $(i = 2, 3, 4)$ cannot be rejected. Thus the set S_1 we get with Dunnett's method contains process 1, only.

The hypothesis $H_4 : \mu_4 \geq \mu_0$ must be rejected because $\bar{x} < 41.24$. The other hypotheses $H_i : \mu_i \geq \mu_0$ $(i = 1, 2, 3)$ cannot be rejected. Thus the set S_2 we get with Dunnett's method contains processes 1, 2 and 3.

When testing the hypotheses $H_i : \mu_i \leq \mu_0$ and $H_i : \mu_i \geq \mu_0$ $(i = 1, \ldots, 4)$ with the procedure of Marcus et al. (1976) the first step is identical with Dunnett's method just used. Because $H_1 : \mu_1 \leq \mu_0$ could be rejected with Dunnett's test, H_2, H_3 and H_4 must be tested in a second step. Now we have $k = 3$ and must use $d_4 = 2.34$. $\bar{x}_i$ $(i = 2, 3, 4)$ must be compared with $\bar{x}_0 + d_4 \sqrt{2}/\sqrt{n} = 50 + 2.34 \times 3.56 = 58.33$. Then the hypothesis $H_2 : \mu_2 \leq \mu_0$ must be rejected because $\bar{x}_2 > 58.33$. The remaining hypotheses $H_3 : \mu_3 \leq \mu_0$ and $H_4 : \mu_4 \leq \mu_0$ cannot be rejected because $\bar{x}_3 < \bar{x}_0$ and $\bar{x}_4 < \bar{x}_0$. Therefore further steps of this procedure cannot bring more significances. The set S_1 we get with this procedure contains processes 1 and 2.

Because the hypothesis $H_4 : \mu_4 \geq \mu_0$ could be rejected by Dunnett's method, the hypotheses H_1, H_2 and H_3 will be examined in a second step. Now we have to compare $\bar{x}_i$ $(i = 1, 2, 3)$ with $50 - 2.34 \times 3.56 = 41.67$. The hypothesis $H_3 : \mu_3 \geq \mu_0$ must be rejected because $\bar{x}_3 < 41.67$. The remaining hypotheses $H_1 : \mu_1 \geq \mu_0$ and $H_2 : \mu_2 \geq \mu_0$ cannot be rejected because $\bar{x}_1 > \bar{x}_0$ and $\bar{x}_2 > \bar{x}_0$. Therefore further steps will not bring more significances. Thus S_2 now contains processes 1 and 2.

The advantage of the procedure of Marcus et al. over Dunnett's method can be seen when comparing the confidence limits we get, i.e., $S_1 = \{$ process 1$\}$ and $S_2 = \{$processes 1, 2, 3$\}$ with Dunnett's method and $S_1 = S_2 = \{$processes 1 and 2$\}$ with the closed procedure. With $\alpha = 0.05$ the relation $P(S_1 \subseteq B \subseteq B^+ \subseteq S_2) \leq 1 - 2\alpha = 0.9$ now provides by Dunnett's method the two-sided 90%-confidence estimates

$$\{\text{process 1}\} \subseteq \{\text{processes with } \mu_i > \mu_0\} \subseteq \{\text{processes 1, 2, 3}\}$$
$$\{\text{process 1}\} \subseteq \{\text{processes with } \mu_i \geq \mu_0\} \subseteq \{\text{processes 1, 2 ,3}\}$$

and by the method of Marcus et al. the two-sided 90%-confidence estimates

$$\{\text{process 1, 2}\} \subseteq \{\text{processes with } \mu_i > \mu_0\} \subseteq \{\text{processes 1, 2}\}$$
$$\{\text{process 1, 2}\} \subseteq \{\text{processes with } \mu_i \geq \mu_0\} \subseteq \{\text{processes 1, 2}\}$$

The upper and lower confidence limit sets coincide with the second method.

According Dunnett's method at least process 1 is better (not worse) and at most processes 1, 2 or 3 are better (not worse) than the standard process. According the closed procedure at least and at most processes 1 and 2 are better (not worse), i.e., these and only these two processes are better (not worse) than the standard.

6 Summary

Multiple comparisons of k treatments with a control are closely related to the selection of a subset of treatments that are better than a control. Let B denote the set of treatments with $\mu_i > \mu_0$, B^+ the set of treatments with $\mu_i \geq \mu_0$, S_1 the set of treatments for which $H_i : \mu_i \leq \mu_0$ must be rejected by the multiple test procedure, and S_2 the set of treatments for which $H_i : \mu_i \geq \mu_0$ cannot be rejected. If the multiple comparison method guarantees that the experimentwise error rate does not exceed α, the relations $P(S_1 \subseteq B) \geq 1 - \alpha$ and $P(B^+ \subseteq S_2) \geq 1 - \alpha$ are true, and with them $P(S_1 \subseteq B \subseteq B^+ \subseteq S_2) \geq 1 - 2\alpha$. They can be regarded as confidence estimates of B and B^+. In cases of normal distributions with common variance, the multiple test of Dunnett (1955) provides the subset S_2 given in Gupta and Sobel (1958). If Dunnett's method is replaced by a more powerful procedure, e.g. by the closed procedure of Marcus et al. (1976), S_1 will be enlarged and S_2 will be diminished. Thus the inequality $S_1 \subseteq B \subseteq B^+ \subseteq S_2$ will be sharpened and the confidence estimate of B and B^+ improved.

References

Bechhofer, R.E. and Dunnett, C.W. (1988). Percentage points of multivariate Student t distributions. In *Selected Tables in Mathematical Statistics* Vol. 11. American Mathematical Society, Providence, Rhode Island.

Dunnett, C.W. (1955). A multiple comparison procedure for comparing several treatments with a control. *Jour. Amer. Statist. Assoc.* **50** 1096 - 1121.

Dunnett, C.W. (1964). New tables for multiple comparisons with a control. *Biometrics* **20** 482 - 491.

Gupta, S.S. and Sobel, M. (1958). On selecting a subset which contains all populations better than a standard. *Ann. Math. Statist.* **29** 235 - 244.

Gupta, S.S., Panchapakesan, S. and Sohn, J.K. (1985). On the distribution of the studentized maximum of equally correlated normal random variables. *Commun. Statist. Simula. Comput.* **14** 103 - 135.

Holm, S. (1979). A simple sequentially rejective multiple test procedure. *Scand. Jour. Statist.* **6** 65 - 70.

Hommel, G. (1986). Multiple test procedures for arbitrary dependence structures. *Metrika* **33** 321 - 336.

Krishnaiah, P.R. and Armitage, J.V. (1966). Tables for multivariate distributions. *Sankhya, Ser. B* **28** 31 - 56.

Marcus, R., Peritz, E. and Gabriel, K.R. (1976). On closed testing procedures with special reference to ordered analysis of variance. *Biometrika* **63** 655 - 660.

Rausch, W. and Horn, M. (1988). Application and tabulation of the multivariate t distribution with $rho = 0$. *Biom. Jour.* **30** 595 - 605.

Chapter 23

Interim Analyses in Clinical Trials

PETER ARMITAGE   Department of Statistics, University of Oxford, Oxford, United Kingdom

1   Introduction

It is a particular pleasure to take part in this celebration of Charles Dunnett's career. His principal contributions are being extensively surveyed by others at this conference, but I could perhaps add an observation that goes back to my time as Editor of *Biometrics*. Many requests are received for permission to reproduce tables and other material published in the journal. By far the most common requests were for Dunnett (1964), a paper containing new tables for multiple comparisons with a control. I think this is still true, since I continue to receive requests which I duly pass on.

The topic of my paper may seem at first sight to be a little remote from Charles Dunnett's main fields of activity. There are, though, at least two important links with his work. Charles was one of the first people to present and study a decision-theoretic model for drug screening programmes (Dunnett 1961), some of the most suitable designs for which are two- or three-stage sequential procedures. More recently, Sylvester (1988) has followed a very similar approach in his proposals for the design of Phase II trials, a topic very close to my present theme. The second point is that the recurrent question of how, if at all, to allow for the effects of interim analyses is a particular case of multiplicity (Tukey 1977), and the most familiar other special case is that of multiple comparisons. So, in two senses we are firmly in Dunnett territory.

We are concerned in this paper with clinical trials, but it is worth noting that interim analyses of experimental data are common in many branches of scientific research; indeed it would often seem the natural way to proceed. The classical theory of experimental design deals predominantly with experiments of predetermined size, presumably because the pioneers of the subject, particularly R. A. Fisher, worked in agricultural research, where the outcome of a field trial is not available until long after the experiment has been designed and started. It is interesting to speculate how differently statistical theory might have evolved if Fisher had been employed in medical or industrial research.

I shall largely avoid technical detail, as I want to discuss some of the general aspects of interim analyses in clinical trials: the ethical constraints, the administrative context, the statistical objectives and the problems of interpretation.

Sequential methods for use in clinical trials have been discussed widely for almost 40 years. Much of the early discussion concerned the possibility of continuous monitoring of accumulating data, and a useful change of emphasis occurred in the 1970s, when attention was drawn to the advantages of discontinuous monitoring by group sequential methods. This approach fitted neatly into the growing practice of submitting interim results to a data monitoring committee (DMC), which would naturally meet at intervals rather than sit in continuous session. The distinction is often not as great as might be supposed. A DMC which normally meets twice a year may well arrange that when the data are closely approaching a previously agreed stopping boundary they are kept under effectively continuous scrutiny. The effect will be much the same as if a totally continuous scheme had been used. Nevertheless, it simplifies matters to assume discrete inspections of data, and I shall do so.

Rather than quote a long bibliography on the subject, I will mention a useful general review by Jennison and Turnbull (1990).

2 The Ethical Constraints

The main reason for performing interim analyses is ethical, namely, a desire not to prolong random assignment beyond a point at which clinically important differences in efficacy can safely be asserted. The concerns are essentially the same as those which the investigators must face before they start the trial. Random assignment is ethical if the investigator is sufficiently uncertain about the relative merits of rival treatments that he would be prepared to administer any of them to each patient in the trial. The continuation of a trial is unethical if this ceases to be true.

Much has been written over the decades about the ethics of clinical trials, but it is doubtful whether any advance in understanding has occurred since the writings of Bradford Hill, who died in April of 1992 at the age of 93. Some writers describe the necessary state of agnosticism as one of 'equipoise', where the arguments in favour of different treatments are neatly balanced. There is, however, an ambiguity as to whether one is considering opinion amongst the medical community (which is often divided), or the views of the individual investigator. My own view is that each doctor must make up his own mind, and that no-one should be surprised if different investigators, all well-informed, come to different views about the propriety of a proposed trial.

The word 'equipoise' tends to conjure up the image the investigator sitting on a knife-edge, his discomfort being relieved by even a small piece of information pushing him one way or the other. I think that a better image is that of an 'uncertainty plateau', from which escape is possible only after the accumulation of a good deal of data. My reason for saying this is that the choice of therapy is often a multidimensional matter. There may be many different measures of efficacy to consider, different time points at which to measure them, adverse effects of various sorts, difficulties of administration revealed only after a good deal of experience, and so on. All these have to be put in the balance, and it will often be true that a confident view cannot be reached until a substantial amount of information has accrued.

Royall (1991), in a paper followed by contributed discussions, appears to argue that most randomized trials are unethical. The essence of his case is that so-called 'demonstration' trials, in which the investigator seeks to demonstrate to the medical community an effect of which he is already convinced, are unethical; whereas 'experimental' trials, where the investigator is genuinely uncertain, are permissible. Royall believes that most trials are 'demonstration' trials, and therefore that most trials are unethical. I can accept the contrast in ethical standing between the two categories, but I believe strongly that most trials now undertaken are in the 'experimental' category. In my experience medical research investigators are well aware of the ethical problems, and of course they are subject to the guidance of local ethical committees. So-called *demonstration trials* would not normally be allowed.

A quite different reason for the possible termination of a trial as a result of interim analyses is essentially economic rather than ethical. If it becomes clear at an interim stage that, whatever happens in the remaining part of the trial, the final results will not indicate a clinically important difference, the investigators may decide to cut their losses, i.e. to stop the study and switch their resources to some other research effort. I shall comment further on this idea later in the paper.

3   The Monitoring Process

It has become common, at least in large multicenter studies in chronic diseases, for the investigators to appoint an independent monitoring committee (DMC or some similar name). In most cases this has the function of advising the investigators (perhaps through a steering committee) whether the trial should be terminated, or the protocol changed, at an interim stage. That is, the DMC does not itself have the power to take this decision. In most trials, treatments are masked to the investigators, insofar as this is possible, and the investigators will be unable to analyze the data themselves. They could certainly be provided with summaries by the trial statistician, but experience suggests that in many cases they would prefer to avoid the anxiety caused by scrutiny of suggestive trends — at least until the DMC draws the data to their attention. The DMC thus acts as a surrogate for the investigators, and should be informed at the outset of the sort of circumstances under which the investigators would not wish to proceed. They should remember that the mere demonstration of a significant difference in one outcome variable may not be enough to cause concern, for the reasons I have already outlined.

The DMC will usually contain at least one statistician and one subject-matter specialist, and may include lay persons such as philosophers or priests. It should be a matter of careful consideration whether or not the investigators themselves are represented (apart, of course, from the trial statistician who needs to present the data). There are good precedents both ways. Another question is whether the sponsor of the trial should be represented. When the sponsor is a pharmaceutical firm, the best plan, even from their own interests, is for them to remain strictly aloof.

The DMC will frequently have to deal with the multidimensional type of data I referred to earlier. That is, they will probably have to consider possible effects in several response variables, for both efficacy and safety, they will be interested in rates of accrual and compliance, and they will need to be informed about the progress of other research in the field. It follows that a decision to recommend termination will rarely be an automatic consequence of a predetermined statistical stopping rule, although such rules will often be major determinants in the decision.

It is not uncommon to find a difference emerging in the early response to treatment, when there is good reason to suspect that longer follow-up periods may reveal a different picture. In these circumstances the DMC will be unwise to jump too readily to the conclusion that one treatment is clearly preferable to another: it may be better to wait for the more comprehensive pattern to emerge. I will mention two examples of this phenomenon.

The first arises in trials for the treatment of diabetic retinopathy by photocoagulation with laser beams. Success is measured by the slowing down of deterioration in visual acuity, which may lead to blindness in many patients. However, the initial effect of photocoagulation may be to worsen visual acuity, and it will be important to ignore early effects and to compare responses at longer intervals after first treatment, perhaps after several years' time. The second example concerns the treatment of patients who are HIV-positive but are asymptomatic for AIDS. An important US trial, ACTG 019 (Volberding et al. 1990), showed that treatment by zidovudine (AZT) reduced the initial rate at which patients became symptomatic. The trial was terminated at an early stage, when some 10-15% of the patients had developed symptoms, after an average follow-up period of a little over a year. Although there is no reason to doubt the validity of this finding, it is a matter of judgement whether the difference would persist after longer periods of follow-up, when the patients converting would tend to be those who initially were less severely affected, and when some patients may have developed drug resistance. The investigators of an Anglo-French trial, Cocorde I, with much the same protocol, decided that the findings in ACTG 019 were not sufficiently conclusive to make them regard their own trial as unethical, and this trial is continuing.

The early termination of a trial may need to be handled with tact and diplomacy, for premature leakage of positive findings may lead to inaccurate press coverage and damage the relations between physicians and patients. The procedures for implementing new treatment schedules, to replace abandoned regimens, need to be worked out carefully although perhaps speedily. There is an interesting account of the various steps taken in the closure of a trial on cryotherapy for retinopathy of prematurity, in a recent paper by Palmer et al. (1991).

4   Criteria for Efficacy

In a trial to compare a new treatment, T, with a control treatment, C, suppose there is a single predominant efficacy variable, and that the difference between treatments is represented by the parameter θ. A common view is that one might recommend termination if the null hypothesis that $\theta = 0$ is in some sense significantly contradicted. But other considerations, such as difficulty of administration or concern about side effects, may make it more reasonably to postulate an indifference zone (θ_L, θ_U), within which the investigators would be prepared to use either treatment. Perhaps the lack of symmetry between T and C might require that $0 = \theta_L < \theta_U$. One might then favour stopping if it became clear that $\theta < \theta_L$ or $\theta > \theta_U$.

Some trials, particularly in-house pharmaceutical trials, are in the nature of equivalence studies. That is, the investigator would like to be able to assert that

$$\theta_L < \theta < \theta_U,$$

where the two bounds lie on either side of zero. Again one might consider significance tests against the two critical values, but in opposite directions to the previous case; that is, we might stop if $\theta > \theta_L$ and $\theta < \theta_U$.

These assessments could be regarded either as significance tests or be expressed in terms of upper and lower confidence limits. The question of exactly how one would perform the tests or calculate the limits is nontrivial, and I shall come to that in a moment.

Of course, this formulation is simplistic, partly because of the complexity of the possible reasons for stopping, and partly because of the difficulty in deciding on critical bounds for the indifference region. Freedman and Spiegelhalter, in a series of papers (see, for instance, their 1983 paper), have described their experience in eliciting from physicians in a cooperative group the doctor's own assessments of the critical bounds. These are likely to vary considerably between physicians, who may choose either to reach a consensus or to maintain different targets. The implication of the latter course is that the investigators might differ considerably as to when the trial ought to stop. There should be no shame attached to a decision of a particular investigator to quit earlier than the others.

5   Adjustments for Interim Analyses

All the assessments referred to in the previous section can be regarded as based on repeated significance tests on accumulating data. This immediately raises a fundamental point which has underlain most work on sequential testing: namely, that repeated testing of any null hypothesis increases the probability of rejection at some stage, and therefore increases both the Type I error probability and the power. Much work on sequential methods for clinical trials, whether for continuous or discrete monitoring, has sought to control this effect by the provision of plans that achieve specific Type I and Type II error probabilities. I shall discuss later the question whether this concern is or is not well-placed.

I should emphasize again the general point that the termination decision is unlikely to follow precisely any given sequential plan, although it might be strongly guided by such a plan. The probabilities associated with a sequential plan are therefore not always exact, and it will often be impossible to state precisely how inexact they are. Nevertheless, sequential plans with known frequency properties are very useful as approximations to the true situation, and give the investigators a good idea of the probabilistic effects of repeated testing.

Many recent publications, such as Jennison and Turnbull (1990), give details of, and make comparisons between, alternative group sequential plans for interim analyses. Discussions usually assume two-sided tests of a single null hypothesis, *e.g.* that $\theta = 0$, but there is no great difficulty in adapting the schemes to test two different values of θ, or for one-sided tests. Three commonly discussed approaches are associated with the names of (a) Pocock, based on repeated significance tests at a constant nominal level; (b) Haybittle and Peto, based on a very stringent constant nominal level until the last inspection, which is at a level close to the Type I error probability; and (c) O'Brien and Fleming, based on nominal levels which are initially very stringent but gradually become less so. Other workers have suggested using a sequence of tests in which the nominal levels are chosen flexibly subject to the condition that the Type I error probability is controlled. The choice between these schemes, and perhaps others having similar frequency properties, is best made in the light of prior judgement as to the plausibility of effects of different sizes. If, for instance, one was strongly inclined to the belief that differences, if they existed at all, would be small, one should be less attracted towards a scheme, like Pocock's, that encouraged early stopping, and more attracted towards the other schemes which require much stronger evidence for stopping at an early stage. This comment of course invites the criticism that one should in any case be using Bayesian methods, and I shall return to that theme later.

I have assumed so far in this section that termination would be prompted by evidence of differences in treatment efficacy. That is, I have assumed that the ethical motivation is uppermost. Earlier I mentioned the economic motivation, which might lead one to stop early if it was clear that the final results would almost certainly be, in some sense, conclusive. The final conclusion might be that there is no strong evidence of a difference, and it is in this context that the approach is usually discussed. Ware et al. (1985) speak of a 'futility factor', implying that in certain circumstances it is futile to continue a trial because the final conclusion is already very clear.

The techniques used for this form of early stopping are called 'stochastic curtailment'. Since the argument involves a prediction of future, unobserved, data, it is necessary to be clear on what hypotheses the prediction are made. Some early work used extremely conservative assumptions. For example, for the 'futility' type of stopping, future results might be predicted on the unlikely hypothesis that θ is equal to the nonzero value against which the plan has high power, even though the current data strongly contradict this value. More recent proponents have suggested that prediction should be made on a range of values more consistent with current data, and the most satisfactory approach seems to me to be Bayesian prediction, using the current posterior distribution of θ. I have previously pointed out that, under simplified assumptions (normality and diffuse prior) this is equivalent to a non-Bayesian 'pivotal' argument, in which the known past results are compared with the unknown future results.

I have one or two reservations about stochastic curtailment, which make me hesitate to advocate its general use. Firstly, for negative results for which the 'futility' argument is brought to bear, there is normally no ethical reason for early stopping. Moreover, all information has positive value, and the additional data which would become available if the trial were continued might contribute usefully to a subsequent overview even though this particular trial might be inconclusive. Secondly, for positive results for which it is claimed that the final result would be significant, it will usually be true that the current results are already impressive and might justify stopping on ethical grounds. In the cryotherapy trial mentioned earlier, conditional power arguments (Hardy et al. 1991) showed that the final result would almost certainly be significant; but at that stage the current results gave a $\chi^2_{(1)}$ value of about 20, which would seem to be sufficient evidence for stopping on ethical grounds alone. Thirdly, the question whether the final difference will be significant seems a little too stark, in view of the current emphasis on estimation as well as significance testing, and the likely relevance of more than one response variable.

In any clinical trial, attention must be given to the estimation of treatment effects, and not merely to the results of significance tests, important though these may be. If a sequential plan is followed, estimates calculated by familiar methods, such as differences between means or ratios of odds, are in general biased, and familiar formulae for confidence intervals may not have the intended coverage properties. Methods of adjustment are often available (see, for example, Whitehead 1983). The effects of adjustment are often small in comparison with random error, and it is perhaps debatable whether they should be used in situations where the sequential plan does not fully represent the stopping criteria.

A final point about estimation is that it might be desirable to calculate confidence intervals repeatedly in the course of interim analyses, even before the trial has stopped. Jennison and Turnbull, in a series of papers (see, e.g. the 1990 paper) have developed a system which guarantees a high probability of coverage for all the intervals so calculated. The main problem is that, to provide such a strong guarantee, the intervals have to be a good deal wider than would be obtained from standard formulae — another familiar link with multiple comparisons. It is perhaps too early to judge the value of this approach in practical situations.

6   The Bayesian Approach

I have touched once or twice on the possible use of Bayesian methods at various stages of the analysis of data from clinical trials. To many statisticians this would seem a natural, indeed perhaps the only proper, way to approach the various problems we have been discussing. I do not intend to embark on a serious discussion of the foundations of statistical inference, although I should perhaps explain that I tend to take a pluralist view, wishing to see the coexistence of different approaches to statistical inference. Many authors have written persuasive accounts of the application of Bayesian methods for specific aspects of trial data analysis. For example, Freedman and Spiegelhalter (1983), in their collaborative work with a group of investigators, got each physician to describe his own prior distribution for the difference parameter, with a view to clarifying whether they thought that clinically important effects were likely to exist.

My concern at present is with the use of Bayesian methods in the interim analyses. This in fact has been one of the main philosophical battlegrounds. The essential point is that Bayesian methods seem to remove completely any problems concerned with the effects of repeated analyses and optional stopping. The likelihood function, and therefore also the posterior probability distribution, depend only on the current data and not in any way on the number of previous or future inspections of the data. Considerations of Type I and Type II error probability, of bias of estimators and of coverage of confidence intervals are all regarded as irrelevant. That is fine, and wholly comforting, if these concepts are seen to be easily dispensable, but otherwise not.

In particular, many statisticians continue to be concerned about the effect of multiple inspections in enhancing the Type I error probability, and, contrary to the view of Berry (1987), I find that trial physicians also are sensitive to this issue. With reasonably diffuse priors, there is often a close connection between a nominal P-value and a posterior probability; *e.g.* with normality and a diffuse prior, when the estimate is 1.96 standard errors above zero the one-sided P-value for a test against zero, and the posterior probability of a negative parameter value, are both 0.025. The phenomenon whereby the probability of occurrence of a small P-value is enhanced by repeated testing could equally well be expressed in terms of the occurrence of a small posterior probability. The counter-argument is sometimes advanced that if we are concerned about the situation when the null hypothesis is true, we should assume a more concentrated prior distribution. But this seems to confuse a legitimate interest in the situation conditional on the null hypothesis (a 'what if' question) with the degree of belief that that hypothesis is true.

These concerns may seem inexplicable to a fully committed Bayesian. He will tend to favour the use of unadjusted P-values or their Bayesian equivalents. Even so, he may be willing to adjust his attitude to a P-value of a particular size, to take account of repeated testing. Dupont (1983), in a paper advocating non-adjustment and favouring a Bayesian approach, writes:

> "The time to be concerned about premature termination of a trial is when we are actually faced with the decision of stopping a trial before its scheduled end. In this case we may be justified in demanding much stronger strength of evidence than, say $P = 0.05$, since the probability of obtaining a spurious P-value of this size at some time during the course of a long trial is all too real."

These conflicting attitudes towards the assessment of evidence give rise to uncertainty about the way to present the analysis of sequentially monitored data in a final publication. Problems may arise for both frequentists and Bayesians. A frequentist may hesitate to use only adjusted probabilities, if only because of the complexity of the stopping criteria, to which I have already alluded. Such a statistician might give nominal P-values etc., and add adjusted values to give an indication of the likely effect of repeated analyses. A Bayesian might give an unadjusted summary together with standard errors to give an approximate indication of the spread of the like-lihood function, and then discuss the effect of different priors (Freedman and Spiegelhalter 1991). He might or might not want to comment on the effect of interim analyses along the lines of Dupont's comment quoted ear-lier.

I hope we can regard these differences of approach as reflecting legiti-mate differences in the way we choose to look at data, rather than as exem-plifying moral rectitude or terpitude, as sometimes seems to be implied by our discussions. Our continued wrangling does little to enhance our stand-ing and influence with the medical research community. The adoption of a more tolerant and flexible attitude will do us no harm. More importantly, by facilitating the promotion of well-planned and well-executed trials, it may even help the patients.

References

Berry, D.A. (1987). Interim analyses in clinical research. *Cancer Invest.* **5** 469 - 477.

Dunnett, C.W. (1961). The statistical theory of drug screening. In *Quantitative Methods in Pharmacology* (H. de Jonge, ed.), North-Holland, Amsterdam.

Dunnett, C.W. (1964). New tables for multiple comparisons with a control. *Biometrics* **20** 482 - 491.

Dupont, W.D. (1983). Sequential stopping rules and sequentially adjusted P values: does one require the other? *Control. Clin. Trials* **4** 3 - 10.

Freedman, L.S. and Spiegelhalter, D.J. (1983). The assessment of subjective opinion and its use in relation to stopping rules for clinical trials. *Statistician* **32** 153 - 160.

Freedman, L.S. and Spiegelhalter, D.J. (1991). Response to letter to the Editor. *Control. Clin. Trials* **12** 346 - 350.

Hardy, R.J., Davis, B.R., Palmer, E.A. and Tung, B., On behalf of the Cryotherapy for Retinopathy of Prematurity Cooperative Group (1991). Statistical considerations in terminating randomization in The Multicenter Trial of Cryotherapy for Retinopathy of Prematurity. *Control. Clin. Trials* **12** 293 - 303.

Jennison, C. and Turnbull, B.W. (1990). Interim monitoring of clinical trials. *Statist. Science* **5** 299 - 317.

Palmer, E.A., Hardy, R.J., Davis, B.R., Stein, J A., Mowery, R.L., Tung, B., Phelps, D.L., Schaffer, D.B., Flynn, J.T. and Phillips, C.L., On behalf of the Cryotherapy for Retinopathy of Prematurity Cooperative Group (1991) Operational aspects of terminating randomization in The Multicenter Trial of Cryotherapy for Retinopathy of Prematurity. *Control. Clin. Trials* **12** 277 - 292.

Royall, R.M. (1991). Ethics and statistics in randomized clinical trials (with discussion)., *Statist. Science* **6** 52 - 88.

Sylvester, R.J. (1988). A Bayesian approach to the design of Phase II clinical trials. *Biometrics* **44** 823 - 836.

Tukey, J. (1977). Some thoughts on clinical trials, especially problems of multiplicity. *Science* **198** 697 - 684.

Volberding, P.A., Lagakos, S.W., Koch, M.A., and 17 others, and the AIDS Clinical Trials Group of the National Institute of Allergy and Infectious Diseases (1990). Zidovudine in asymptomatic human immunodeficiency virus infection. *New Engl. Jour. Med.* **322** 941 - 949.

Ware, J.H., Muller, J E. and Braunwald, E. (1985). The futility index: an approach to the cost-effective termination of randomized clinical trials. *Amer. Jour. Med.* **78** 635 - 643.

Whitehead, J. (1983). *The Design and Analysis of Sequential Clinical Trials.* Horwood, Chichester.

Chapter 24

A New Procedure for Assessing Acute Oral Toxicity

ROBERT N. CURNOW Department of Applied Statistics, University of Reading, Reading, United Kingdom

ANNE WHITEHEAD Department of Applied Statistics, University of Reading, Reading, United Kingdom

1 Introduction

Concern has been expressed for a number of years now about the emphasis on the estimated median lethal dose (LD_{50}) when classifying the acute oral toxicity of chemical substances. We need to know not the dose likely to kill 50% of those receiving it but rather the level of morbidity and mortality that would result from doses likely to occur in intentional or in unintentional exposure to the substance. An estimate of the slope of the dose response curve at the LD_{50} can be used with the estimate of the LD_{50} to predict the consequences of exposure to other doses but this requires a strong reliance on assumptions about the distribution of tolerance in the population or, equivalently, the slope of the response curve. Even with these assumptions the accuracy of prediction is often low. Additionally, accurate estimation of the LD_{50} requires the use of doses spanning the LD_{50} and therefore the subjection of animals to mortality rates at and above the 50% level.

Because of these problems, a new method, called the Fixed Dose Procedure (FDP) for classifying substances according to their toxicity has been proposed by the British Toxicology Society (1984). Two principles are involved. First, observed morbidity, defined as "signs of toxicity", as well as mortality is used in the classification. Second, the procedure is sequential in that a fixed number of animals is subjected to a particular dose and the substance either classified on the basis of the results at that dose or a further set of animals is subjected to either a lower or a higher dose until a classification can be made. The FDP is now being considered by the European Community as an alternative, at least initially a concurrent alternative, to the LD_{50} test. Discussions are taking place with the regulatory authorities in other countries about its acceptability. The FDP was devised by toxicologists and, strangely, has not been referred to or, until now, examined by statisticians. The procedure has been tested internationally by a number of laboratories and comparisons made between its classifications of a number of substances and those that would have resulted from the standard LD_{50} test (van den Heuvel et al. 1990).

The purpose of this paper is to outline the derivation of the statistical properties of the fixed dose procedure for comparison with the known properties of the LD_{50} test. This will be done in relation to the level of agreement between the two methods of classification and of the number of animals exposed to experimentation and to resulting morbidity and mortality. A more detailed study, including a discussion of the relations between our results and those of the international validation study has been accepted for publication (Whitehead and Curnow 1992).

2   The LD_{50} Test

The international agreed guideline for LD_{50} tests published by the Organization for Economic Co-operation and Development (OECD 1981) recommended that ten rats (five male and five female) be tested at each of a minimum of three dose levels. The highest permitted test dose was reduced in 1987 (OECD 1987) from 5000 mg/kg body weight to 2000 mg/kg body weight. From such a study it should be possible to estimate the LD_{50} and calculate a confidence interval, provided that mortality is seen at or below the highest test dose.

In the European Community the classification of substances according to their toxic hazard is based entirely on the LD_{50} estimate, ignoring the slope or any other features of the dose response curve. Table 1 gives the classification according to the European Commission Directive (83/467/EEC 1983).

Table 1: Dose-Toxicity classification (European Commission Directive).

LD_{50} (mg/kg body wt.)	Classification
< 25	VERY TOXIC
25 − 200	TOXIC
200 − 2000	HARMFUL
> 2000	UNCLASSIFIED

3 The Fixed Dose Procedure

The version of the procedure described by van den Heuvel et al. (1990) is that ten animals (five female and five male) be tested at each dose level which is investigated. The dose levels available to be used in the test are 5, 50, 500 or 2000 mg/kg body weight. The initial dose level chosen should be that which is judged likely to produce evident toxicity but no mortality. Where no information is available upon which to make such a judgement, an initial 'sighting' study should normally be carried out. Where evident toxicity does not result from administration of the chosen dose level, the substance should be retested at the next higher dose level. If a severe toxic reaction occurs requiring a humane kill or if mortality occurs the substance should be retested at the next lower dose level. Evident toxicity is a general term describing clear signs of compound-related toxicity, but not such as to cause very severe pain, distress or mortality.

Figure 1 shows, as an example, the possible routes which would be taken by a toxicologist from a stating dose of 50 mg/kg based on the results of the tests at the dose levels used. Included are the toxic classifications based on these results. For each of the dose levels 5, 50 and 500 mg/kg body weight there are three outcomes, which result in different actions regarding the substance: one or more deaths; no deaths but one or more animals suffering toxicity; no deaths and no toxicity. However, if the substance is tested at 2000 mg/kg, the three outcomes resulting in different classifications are: 50% or more deaths; at least one death but less than 50% mortality; no mortality. This leads to one inconsistency for substances tested at 500 mg/kg. If the result is no deaths and some toxicity, the substance would be unclassified but if at 500 mg/kg there were no deaths or toxicity the substance may be classified as harmful when tested at 2000 mg/kg, albeit only in the unlikely event that 50% or more of the animals die at that dose.

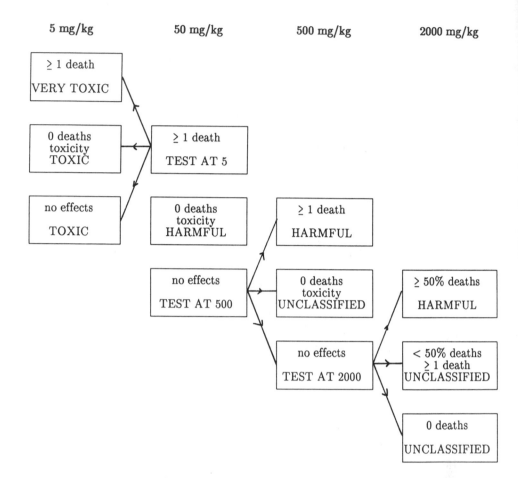

Figure 1: Fixed dose procedure pathways (starting dose of 50 mg/kg).

At each dose level tested results will be available from ten animals. Providing that there is reasonable agreement about what constitutes a toxic sign, each animal can be categorized into one of three groups — dead, suffering toxicity, or unaffected. In order to model the fixed dose procedure we need dose response relationships for death and for death or toxicity. If death can be considered as an extreme form of toxicity it is perhaps not unreasonable to propose parallel sigmoid shaped curves for toxicity and for toxicity or death as illustrated in Figure 2.

Let $p_1(x)$, $p_2(x)$ and $p_3(x)$ denote the respective probabilities that an animal dies, remains alive but suffers toxic signs, and is unaffected by dose x of a substance. Assuming that logit relationships are appropriate,

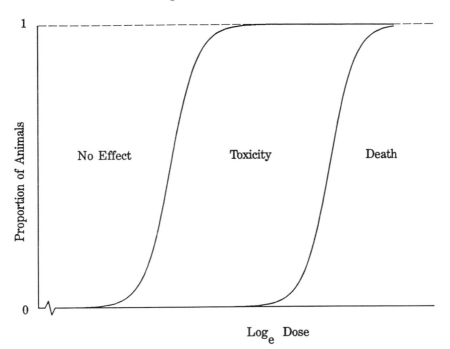

Figure 2: Dose-response relationships for death and for toxicity or death.

the dose-response relationships for death and for death or toxicity can then
be expressed in the form

$$\text{logit}(p_1(x)) = \alpha + \beta \log_e x \qquad (3.1)$$

$$\text{logit}(p_1(x) + p_2(x)) = \alpha' + \beta \log_e x \qquad (3.2)$$

where

$$\text{logit}(p_1(x)) = \log_e \left(\frac{p_1(x)}{1 - p_1(x)} \right)$$

and

$$\begin{aligned} \text{logit}(p_1(x) + p_2(x)) &= \log_e \left(\frac{p_1(x) + p_2(x)}{1 - p_1(x) - p_2(x)} \right) \\ &= \log_e \left(\frac{1 - p_3(x)}{p_3(x)} \right). \end{aligned}$$

Each particular substance has three parameters associated with it, name-
ly α, α' and β. We can, however, reparameterise the model in terms of the
$LD_{50}(\ell)$, β, and the factor (R) by which the dose needs to be increased in

order to equate (3.1) and (3.2). R would be a relative potency if (3.1) and (3.2) represented two different substances. We note that

$$\log_e \ell = -\frac{\alpha}{\beta} \qquad \text{and} \qquad \log_e R = \frac{\alpha' - \alpha}{\beta}.$$

For a substance with given values of ℓ, β and R, the probability of getting a particular outcome from a test of ten animals at a fixed dose, x, can be calculated. There are five such probabilities which need to be calculated.

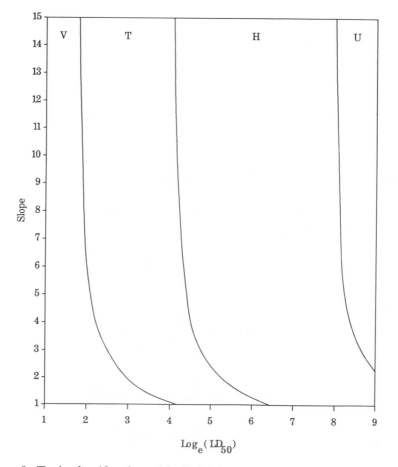

Figure 3: Toxic classification with the highest probability ($R = 50$, starting dose $= 50$ mg/kg). V = very toxic, T = toxic, H = harmful, and U = unclassified.

1. $P(\geq 1 \text{ death}) = 1 - (1 - p_1(x))^{10} = d(x)$

2. $P(\geq 50\% \text{ deaths}) = \sum_{j=5}^{10} \binom{10}{j} p_1(x)^j (1 - p_1(x))^{10-j} = d_1(x)$

3. $P(< 50\% \text{ deaths, but} \geq 1 \text{ death}) = \sum_{j=1}^{4} \binom{10}{j} p_1(x)^j (1-p_1(x))^{10-j}$
   $= d_2(x)$

4. $P(\text{no effects}) = (p_3(x))^{10} = n(x)$

5. $P(\text{some toxicity, no deaths}) = 1 - d(x) - n(x) = t(x)$.

The probabilities of following particular routes through the procedure (see Figure 1 when the starting dose is 50 mg/kg) can be calculated in terms of $d(x)$, $d_1(x)$, $d_2(x)$, $n(x)$ and $t(x)$. These lead to the probabilities of the substance being allotted to each of the four classes of toxicity, namely very toxic, toxic, harmful and unclassified.

Figure 3 shows the most likely classification for substances with values of ℓ and β varying from 4 to 8100 mg/kg and from 1 to 15 respectively when R is 50 and the starting dose is 50 mg/kg. The values of the parameters ℓ and β have been chosen to represent the range of values found in a number of previous studies. The sensitivity of the results to the value of R will be mentioned later.

4   The Starting Dose

Generally speaking, results show that the higher the starting dose used the more likely is the substance to be classified as less toxic. Companies may therefore choose the highest dose expected to give toxicity with only a small chance of mortality. We have tried to incorporate into our model the prior information a toxicologist will have available because the substance is similar to other tested components or from earlier work with the substance. We have assumed that the experimenter will have information about a substance equivalent to having carried out the sighting study recommended for use in the international validation study of the fixed dose procedure (van den Heuvel et al. 1990).

The sighting study consisted of three groups of one male and one female rat. One pair received 5, one pair 50 and one pair 500 mg/kg of the test substance. The dose level which resulted in evident toxicity but no mortality was identified. If evident toxicity was not seen at 500 mg/kg body weight the starting dose to be used in the main test was 2000 mg/kg.

There are six possible results from the two animals at one particular dose level giving a total of $6^3 = 216$ different sets of results taking all three

There are six possible results from the two animals at one particular dose level giving a total of $6^3 = 216$ different sets of results taking all three doses into consideration. We can associate every possible outcome from the sighting study with a starting dose for the fixed dose procedure and this is shown in Table 2. From this table, the probability that the starting dose is 5, 50, 500 or 2000 mg/kg can be calculated in terms of the values of ℓ, β and R for the substance.

Assuming that the sighting study is not an integral part of the fixed dose procedure but just provides the stating dose for the procedure we can calculate the probability that a substance is classified as very toxic by the fixed dose procedure as follows:

P (classify very toxic)

$$= \sum_{x=5,50,500,2000} P \text{ (classify very toxic given starting dose } x)P_x$$

where $P_x = P$ (choose starting dose x). The probabilities that a substance is classified as toxic, harmful and unclassified are calculated similarly.

Table 2: Choice of starting dose for the fixed dose procedure based on results from a sighting study (units throughout are mg/kg body weight).

RESULTS FROM SIGHTING STUDY	DOSE
Any mortality at 5 No mortality at 5, but mortality at 50	5
No mortality at 5 or 50, but mortality at 500 Toxicity at 5, no effects at 50 or 500 No mortality at 5, toxicity at 50, no effects at 500	50
No mortality at 5 or 50, toxicity at 500	500
No effects at 5, 50 and 500	2000

5 Toxic Classification of Substances

Figure 4 shows the most likely classification following a sighting study when $R = 50$ and for the same values of ℓ and β as in Figure 3 when the starting dose was fixed at 50 mg/kg. We have found that the classification with the highest probability does not alter greatly as R changes from 5 to 100 and so Figure 4 represents more generally how the classification depends on the LD_{50} and slope.

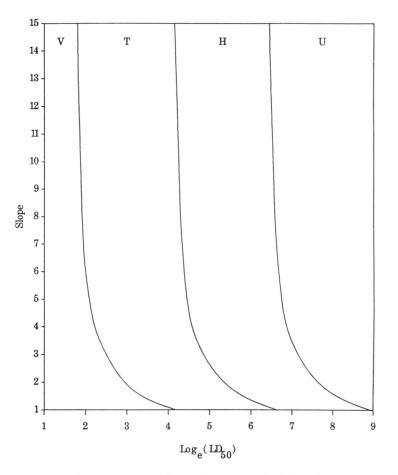

Figure 4: Toxic classification with the highest probability ($R = 50$, starting dose chosen from sighting study). V = very toxic, T = toxic, H = harmful, and U = unclassified.

For slopes greater than about 6 the most likely classification changes from very toxic to toxic, from toxic to harmful and from harmful to unclassified for LD_{50} values of approximately 7.5, 75, 750 mg/kg body weight respectively. The boundaries for the toxic classification based on the LD_{50} estimate are at 25, 200 and 2000 mg/kg body weight and so the fixed dose procedure is likely to classify substances as the same or less toxic than the LD_{50} classification. As the slope becomes shallower the substance is more likely to be given a higher toxic classification for a given LD_{50}. The possibility of changing the fixed dose levels to produce classifications more similar to those from the LD_{50} test should be considered (Whitehead and Curnow 1992).

6 Expected Numbers of Animals Required, Found Dead, Suffering Toxicity

For each starting dose, the expected number of animals required in the fixed dose procedure and also the expected number of deaths and number of animals suffering toxicity can be calculated. These expected numbers depend on the starting dose and in a similar way to the toxic classification probabilities we can calculate the expected numbers assuming that a sighting study has been carried out. The expected number of animals to be tested, assuming that a sighting study has been carried out but excluding those animals from the calculation, was calculated for a range of $LD_{50}s$, slopes and values of R. As R changes from 10 to 100 the expected numbers do not change very much. More animals will obviously be required for substances with an LD_{50} close to the boundaries of classification toxic/harmful and harmful/unclassified and for substances with shallow slopes. For the remaining substances the expected number lies between 10 and 11 animals compared with the fixed value of 30 for the LD_{50} test. As R decreases below 10 the expected number of animals required around the boundaries of classification and the size of these boundary regions increases. The absolute maximum number of animals required by the FDP will be 40, corresponding to the unlikely use of all four dose levels.

The expected number of deaths is also relatively insensitive to values of R between 10 and 100. Apart from substances that will be classified very toxic, the expected number lies between 0 and 2. The LD_{50} test would generally result in about 50% mortality, i.e., 15 deaths. As R decreases from 10 the expected numbers increase close to the boundaries of classification toxic/harmful and harmful/unclassified. The expected number of animals showing signs of toxicity are very dependent on the value of R and consequently no one choice of R can represent the general picture.

7   Discussion

The fixed dose procedure is likely to classify substances as the same or less toxic than the LD_{50} classification. For slopes greater than about 6 the most likely classification changes from very toxic to toxic, from toxic to harmful and from harmful to unclassified for LD_{50} values of approximately 7.5, 75 and 750 mg/kg body weight respectively. These values compare with values of 25, 200 and 2000 mg/kg associated with the LD_{50} tests. For lower slopes the FDP provides a more conservative classification, which is entirely appropriate when the unpredictability inherent in a low slope is considered.

Calculation of the expected numbers of animals to be tested and found dead showed that, in general, the fixed dose procedure would require considerbly fewer animals and cause fewer deaths than the LD_{50} test.

Acknowledgments

We are grateful to Mr. Michael Russell-Yarde whose work for his M.Sc. dissertation formed the starting point for this study, and to Mr. Keith Freeman for his assistance with the programming and computing work involved in preparing this paper.

References

British Toxicology Society (1984). Special report: A new approach to the classification of substances and preparations on the basis of their acute toxicity. *Human Toxic.* **3** 85 - 92.

Commission of the European Communities (1983). Council Directive 83/467/EEC amending for fifth time Council Directive 67/548/EEC on the approximation of laws, regulations and administrative provisions relating to the classification, packaging and labelling of dangerous substances; Annex III. *Off. J. Eur. Commun.* No. **L257** 13 - 33.

OECD. (1981). Guidelines for the testing of chemical substances - No. 401. *Acute oral toxicity*, OECD, Paris.

OECD. (1987). Guidelines for the testing of chemical substances - No. 401 (1987). *Acute oral toxicity*, OECD, Paris.

van den Heuvel, M.J., Clark, D.G., Fielder, R.J., Koundakjian, P.P., Oliver, G.J.A., Pelling, D., Tomlinson, N.J., and Walker, A.P. (1990). The international validation of a fixed-dose procedure as an alternative to the classical LD_{50} test. *Fd. Chem. Toxic.* **28** 469 - 482.

Whitehead, A. and Curnow, R.N. (1992). Statistical evaluation of the fixed dose procedure. To be published, *Fd. Chem. Toxic.*

Chapter 25

Estimating Risks of Progression to AIDS Using Serially Measured Immunologic Markers

JANET M. RABOUD   Department of Health Care and Epidemiology, University of British Columbia, Vancouver, British Columbia, Canada

RANDALL A. COATES   Department of Preventive Medicine and Biostatistics, University of Toronto, Toronto, Ontario, Canada

VERN T. FAREWELL   Departments of Health Studies and Statistics and Actuarial Science, University of Waterloo, Waterloo, Ontario, Canada

Abstract   A cohort of homosexual and bisexual men who have been infected with the Human Immunodeficiency Virus (HIV) are followed at quarterly intervals for up to 5 years. We compare risks of progression to Acquired Immunodeficiency Syndrome (AIDS) associated with changes in immunologic markers among individuals infected with HIV. Risks of progression are estimated after adjusting for treatment with AZT and the occurrence of opportunistic infections. The effects of differing measurement intervals and rates of change of the markers between individuals diagnosed with AIDS and AIDS-free individuals are investigated.

415

1   Introduction

The Toronto Sexual Contact Study (Coates et al. 1986) is a longitudinal study of healthy homosexual and bisexual males who have been exposed to HIV. Between July 1984 and July 1985, 249 men were recruited into the cohort. To be eligible for recruitment the men had to have had at least one sexual contact with a man (defined as a primary case) with either AIDS or an AIDS-related condition within one year of the onset of the primary case's disease. All diagnoses in primary cases were confirmed by the project physician, either by a personal evaluation or review of relevant medical documents. All primary cases with an AIDS-related condition had persistent generalized lymphadenopathy and had antibodies to HIV.

On enrollment and at each follow-up visit, cohort members underwent a complete physical examination, completed interviewer-administered questionnaires and had a battery of immunologic tests performed. The immunologic markers measured can be grouped into four categories: T cell counts (absolute T4 count, absolute T8 count, and the $T4/T8$ ratio), immunoglobulins (IgA, IgG and IgM), hematologic variables (white blood cells, red blood cells, hemoglobin, lymphocyte and platelet counts), and blastogenic responses ((PHA) phytohemagglutinin, (PWM) pokeweed mitogen, and (Con-A) concanavalin-A).

There was a total of 159 seropositive cohort members; 143 were seropositive at recruitment and 16 others seroconverted while under study. Only the seropositive men were included in the analysis. During the nearly 5 years of follow-up, 40 men developed AIDS. Of these, 39 cases of AIDS developed in seroprevalent cohort members and one case of AIDS arose in a seroconverter 30 months following the date of seroconversion.

There is considerable current interest in the study of immunologic markers in HIV disease. With data from the Toronto Sexual Contact Study we illustrate the use of relative risk regression models (Cox 1972) to represent the relationship between the risk of the development of AIDS and serially monitored immunologic variables. Special features are the choice of a time scale, the choice among correlated variables and the effect of intermittent and irregular monitoring. Incorporation of treatment and infection effects is also considered.

2   Methods

Relative risk regression models were used to examine the association between series of immunologic marker measurements and the risk of AIDS. For a single covariate $Z(t)$, the Cox model assumes that the hazard, or instantaneous risk of failure, at time t for an individual with covariate value $z(t)$ is related to an arbitrary underlying hazard, $\lambda_0(t)$, as follows:

$$\lambda(t, z(t)) = \lambda_0(t)e^{\beta z(t)},$$

where the covariate $Z(t)$ is allowed to vary with time t. The relative risk of failure associated with an increase of one unit in $Z(t)$ is thus e^{β}. The coefficient β is estimated using iterative procedures by maximizing the partial likelihood (Cox 1975), defined as

$$L(\beta, Z) = \prod_{i=1}^{n} \frac{e^{\beta z_i(t)}}{\sum_{j \in R_i} e^{\beta z_j(t)}}, \tag{2.1}$$

where n is the number of observed failures and R_i is the set of individuals at risk of failure at the time of the ith failure. The extension to multiple covariates follows easily. This model deals with time-dependent covariates in analyses of failure time data in a natural way since the likelihood is the product of a series of comparisons of the covariate values at each failure time. Prentice (1982) used this method to study the relationship between serial blood pressure measurements and the risk of cardiovascular disease.

In the analysis of the Toronto Sexual Contact Study data, survival time was calculated from the estimated date of HIV infection. For seroconverters, this date was estimated to be 90 days prior to the date of the first positive test result. For seroprevalent cohort members, the estimated date of infection was the date of the first sexual contact with the primary case which was considered to be the earliest possible date of infection. However, since the first sexual contact in some of the relationships occurred well before the first reported cases of AIDS in the United States, we modified some dates according to the following criteria: when the primary case had AIDS, the earliest possible date of infection could not antedate January 1978, resulting in a change of date for five cohort members; when the primary case had persistent generalized lymphadenopathy, the earliest possible date of infection in the corresponding contact could not be earlier than three years prior to the onset of persistent generalized lymphadenopathy in the primary case and could never be before January 1978. This latter criterion resulted in a change of date for four cohort members. Individual observations were left-truncated at the time of enrollment.

In many studies of seroprevalent individuals in which the dates of infection are unknown and there are no means to estimate them, survival time is calculated from the date of enrolment into the study. This choice of survival time leads to biased estimates of relative risk, as described by Brookmeyer and Gail (1987), due to phenomena known as onset confounding and frailty selection. For this reason, we felt that it was preferable to use the estimated date of infection as the starting point of follow-up rather than the date of enrolment into the study even though there was some uncertainty in estimating the dates of infection.

The immunologic markers were modelled with time-dependent covariates. In our analyses, a time-dependent covariate vector $Z(t)$ was updated just after the time of each visit for an individual and remained constant between visits. Let X_i be the value of the marker measured at the ith visit and T_i be the time of the ith visit, $i = 1, \ldots, p$. The time-dependent covariate representing a marker would be defined as follows:

$$Z(t) = \begin{cases} X_1 & T_1 < t < T_2 \\ X_2 & T_2 < t < T_3 \\ & \cdots \\ X_{p-1} & T_{p-1} < t < T_p \\ X_p & T_p < t \end{cases}$$

The covariate $Z(t)$ can be "lagged" by $\triangle$ time units by defining $Z(t)$ to be the value of the covariate measured at the visit closest to but not following $\triangle$ time units before time t. A lagged covariate would be defined as follows:

$$Z(t) = \begin{cases} X_1 & T_1 + \triangle < t < T_2 + \triangle \\ X_2 & T_2 + \triangle < t < T_3 + \triangle \\ & \cdots \\ X_{p-1} & T_{p-1} + \triangle < t < T_p + \triangle \\ X_p & T_p + \triangle < t \end{cases}$$

Lagged covariates were used in this analysis because the values of the immunologic markers were so extreme close to the time of AIDS diagnosis and because longer term predictions were of interest. The use of lagged covariates avoids the extreme values at the failure times and provides more valuable estimates of risks of progression to AIDS. Only failures which occur $\triangle$ time units after the start of follow-up contribute to the estimation of relative risks associated with covariates lagged by that amount. For example, with a lag period of one year, only 35 of the 40 failures were included in the analysis since 5 failures occurred within the first year of follow-up. In addition, individuals were excluded from the risk sets of failures which occurred within $\triangle$ time units of the individual's start of follow-up. A reduction in power results from lagging covariates longer periods of time, so there is a trade-off to be made between power and the examination of relative risks of developing AIDS further into the future. A lag period of one year was chosen because this resulted in an adequate sample size and because it was of interest to estimate relative risks of developing AIDS in one year's time.

As well as estimating risks of progression to AIDS associated with the immunologic markers, we estimated the changes in risks of AIDS associated with the occurrence of an opportunistic infection and treatment with AZT, also known as zidovudine or ZDV. Both of the events were modelled with binary time-dependent covariates, which were set equal to zero before the start of the treatment or the occurrence of the infection and set equal to one thereafter.

3   Preliminary Analysis

Table 1 gives results from some preliminary univariate analyses. Significance levels were calculated based on the asymptotic normality of the partial likelihood estimates. Of the immunologic markers, the levels of the T4/T8 ratio, the absolute T4 counts, IgA and the blastogenic responses were found to be the most strongly related to progression to AIDS. When examined in isolation, patients who were treated with AZT were found not to have significantly different rates of progression to AIDS from untreated patients while patients who had an opportunistic infection were found to have increased risk of AIDS compared with those who did not. Because of high correlation within categories of laboratory markers, only one marker from each category was chosen to be included in further analyses. With the

Table 1: Comparison of univariate models.

Covariate	Unit	Est. Coeff.	P-value
T4	100	−.51	.0000
T8	100	.05	.41
T4/T8	.1	−.43	.0000
IgA	100	.45	.0000
IgM	100	.17	.16
IgG	1000	.33	.30
WBC	1000	−.22	.10
RBC	1	−.27	.22
Hemoglobin	10	.03	.83
Lymphocyte	1000	−.17	.54
Platelet	10	−.05	.17
PHA	10000	−.17	.0002
PWM	10000	−.77	.0002
Con-A	10000	−.30	.0002
AZT	Yes/No	−1.43	.17
OI	Yes/No	1.04	.02

Table 2: Multivariate models.

Model	Covariate	Unit	Est. Coeff.	P-value
1	T4/T8	.1	$-.40$	.0000
	PHA	10000	$-.09$	.04
	IgA	100	.31	.01
	AZT	Yes/No	-2.58	.03
2	T4/T8	.1	$-.32$	.0002
	PHA	10000	$-.09$	.04
	IgA	100	.54	.01
	OI	Yes/No	.54	.20
3	T4/T8	.1	$-.41$	.0000
	PHA	10000	$-.09$	.07
	IgA	100	.29	.02
	AZT	Yes/No	-2.81	.01
	OI	Yes/No	.99	.02

use of pairwise comparisons (not shown), the T4/T8 ratio, IgA and PHA were chosen from the categories of T cells, immunoglobulins and blastogenic responses respectively. The nature of the associations between progression to AIDS and the various immunologic markers was further examined with multivariate models. In addition, because treatment with AZT was not randomly assigned but was available only to individuals meeting specific criteria of illness, we wanted to estimate the decrease in risk associated with AZT treatment after controlling for the state of health of the patients at the time they received the treatment. When the effect of treatment with AZT was examined in a model which included the levels of the three immunologic markers, patients treated with AZT were found to have significantly lower rates of progression to AIDS than untreated patients (Table 2, Model 1). Similarly, since individuals who have opportunistic infections are often sicker than those who do not, it was desired to quantify the effects of the infection over and above the general level of health of the individual. With the inclusion of the levels of the three immunologic markers, it was found that progression rates of patients who experienced an opportunistic infection were not significantly different from those of patients without infections (Table 2, Model 2). Interestingly, when all covariates were entered into the model simultaneously (Table 2, Model 3), both treatment with AZT and the occurrence of an opportunistic infection demonstrated significant associations with the risk of AIDS. Patients who were treated with AZT had significantly lower risks of AIDS than untreated patients while patients who had an opportunistic infection had higher risks of AIDS than those who did not.

4 Frequency of Measurement

The relative risk regression model assumes that time-dependent covariates are measured continuously and that the value of the covariate entered into the partial likelihood is the value of the covariate at time of the failure. In actuality, as in the Toronto study, the covariate is often not measured continuously, but periodically during the course of study. Thus, relative to what might be considered the true covariates of interest, the defined covariate is then out-of-date by some amount which depends on both the frequency of measurement and the rate of change of the covariate. For illustration, we assume that covariates change linearly in time between observation points and, simplistically, that only the next individual to fail is allowed a different slope from other individuals under study. Then we can define an offset term, δ, to be the difference in the amount that the covariate is out-of-date between the failing individual and the other members of the risk set:

$$\delta = (\gamma_1 \triangle_1 - \gamma_2 \triangle_2)/2, \qquad (4.1)$$

where γ_1 and γ_2 are the rates of change of the covariate for the failing individual and for all other members of the risk set respectively and $\triangle_1$ and $\triangle_2$ are the corresponding intervals of measurement. If the covariate is out-of-date by the same amount for all members of the risk set, then the estimate of the relative risk will not be biased; if this is not the case, then there may be bias introduced into the relative risk estimate. The estimated coefficient, β^*, was found to be related to the true coefficient, β, according to

$$\beta^* = \beta + k\delta/i(\beta), \qquad (4.2)$$

where k is the number of observed failures and $i(\beta)$ is the negative of the

Table 3: Bias in estimated coefficients due to differing rates of change and measurement intervals between failing and surviving individuals.

| Covariate | Rate of Change | | Offset | Est. Coeff. | | Adj. Coeff. |
	AIDS γ_1	non-AIDS γ_2	δ	β^*	$i(\beta)$	β
T4/T8	$-.00049$	$-.00030$	$-.018$	$-.43$	139	$-.44$
IgA	$-.00098$	$.00009$	$-.073$	$.45$	90	$.42$
PHA	$-.01086$	$-.00061$	$.722$	$-.17$	500	$-.22$

Table 4: Effect of frequency of measurement on estimates of coefficients (p denotes P-value).

Covariate	Every Visit		Second Visit		Third Visit		Enrolment	
	$\hat{\beta}$	p	$\hat{\beta}$	p	$\hat{\beta}$	p	$\hat{\beta}$	p
T4/T8	$-.431$	.0000	$-.347$	.0000	$-.259$	.0002	$-.214$	.0002
IgA	.445	.0000	.432	.0000	.365	.002	.262	.04
PHA	$-.174$	.0002	$-.166$	.0002	$-.122$	.0026	$-.122$	.001

second derivative of the log likelihood (Raboud 1991). In the Toronto Sexual Contact Study, it was found that the average number of days between visits for all cohort members was 107 (std. dev. $= 52$) but that the average number of days between the last two visits for AIDS patients was 139 (std. dev. $= 118$). A simple estimate of the difference in the amount that the covariates are out-of-date between the failing individuals and other members of the risk set is $\delta = [139\gamma_1 - 107\gamma_2]/2$. In Table 3 we show estimates of γ_1 and γ_2 for the T4/T8 ratio, IgA and PHA, as well as the offset term and the estimated and adjusted coefficients. The rate of change of the marker for all members of the risk set, γ_2, was estimated with the average rate of change of the marker over all pairs of visits. The rate of change of the marker for failing individuals, γ_1, was estimated with the average rate of change of the marker between the last two visits of failing individuals. The estimated coefficients, β^*, were obtained from Table 1. The values of the adjusted coefficients were found by solving for β in equation (4.2). We can see that the bias in the estimated coefficient due to the differences in the frequency of measurement and the rate of change of the marker is likely small for the T4/T8 ratio and IgA but may be moderately large for PHA.

The effect of increasing the interval of measurement was investigated by updating the time-dependent covariates at several different frequencies. Estimated coefficients were obtained when covariates were updated at each visit, at every second visit, at every third visit and when they were held constant at enrolment values. The results are shown in Table 4. The estimated coefficients of all three covariates studied decreased in magnitude as less frequent measurements were used. This is as one might expect, since if the rates of change for failing and surviving individuals remain constant and the differences in visit frequency are multiplied by a factor of two or three, then the amount of bias in the estimated coefficient should increase, according to (4.1) and (4.2). Thus, the more frequent updating of covariates leads to a reduction in the bias in the estimated coefficient when the rates of change are different between the failing individual and the other members of the risk set.

5   Discussion

Despite some limitations, the Toronto Sexual Contact Study provided a unique opportunity to study the natural history of HIV disease. One limitation of the study was that the dates of infection were not known previously for most members of the cohort, however, we were able to estimate these dates using information collected regarding the sexual relationship between the cohort members and the primary cases. The repeated measures of the immunologic markers over time allowed us to study the effect of infection with HIV on the markers. Due to the unique recruitment procedures, there exists the possibility of selection bias. Because the study population consisted entirely of homosexual and bisexual men, the results may not be generalizable outside this population.

In the analysis, the levels of the T4/T8 ratio, IgA and PHA were shown to be the most related to progression to AIDS, with the T4/T8 ratio having the strongest association. Patients who experienced an opportunistic infection were found to have increased risk of AIDS while patients treated with AZT were shown to have lower rates of progression to AIDS. Since only patients meeting specific criteria of illness had access to treatment with AZT, it is necessary to control for the general health of the patients who are receiving AZT when trying to evaluate the effectiveness of the treatment. We attempted to control for the patients' state of health by adjusting for the current levels of the T4/T8 ratio, IgA and PHA, however, caution must still be used when interpreting these results.

Differing frequencies of measurement or rates of change of the marker between the individual diagnosed with AIDS and the individuals who are AIDS-free at the time of diagnosis can introduce bias into estimates of relative risk. Estimates of relative risk were found to decrease in magnitude when time-dependent covariates were updated less frequently.

It is well known that the immunologic markers are measured with considerable variability. The variation in the markers is due to minor fluctuations in health and a variety of sources of laboratory measurement error, including variation among technicians, changes in equipment, brands of laboratory kits and repetition error. This issue is discussed in Raboud (1991). Error in the measurement of covariates has been found to bias estimates of relative risks obtained from the proportional hazards model. If an estimate of the variance of the error in measurement is available, it is possible to adjust estimates of relative risk of covariates which have been measured with error. When covariates are measured repeatedly over time, smoothing sequences of observations within individuals can help to reduce the variance of the error in measurement and subsequently the amount of bias in the estimated relative risk.

In this paper, we have focussed on relative risk regression models which provide a means to study the conditional distribution of a failure time, T, given a covariate $Z(t)$. The use of immunologic measurements, which are highly variable and intermittently monitored, to define $Z(t)$ creates methodologic concerns. These concerns, and the obvious desire to use the models for prediction will require models for the stochastic structure of a marker process. Lawless and Yan (1992), in a paper in this volume, consider Markov models for this purpose and reference other approaches.

Acknowledgments

J. Raboud was supported by a Ph.D. Fellowship, NHRDP AIDS Grant Program, Health and Welfare Canada, and a British Columbia Health Research Foundation Fellowship. Support is also acknowledged from Grant DA04722 U.S. National Institute on Drug Abuse

References

Brookmeyer, R., Gail and M.H. (1987). Biases in prevalent cohorts. *Biometrics* **43** 739 - 749.

Coates, R.A., et al. (1986). A prospective study of male sexual contacts of men with AIDS-related conditions (ARC) or AIDS: HTLV-III antibody, clinical, and immune function status at induction. *Canad. Jour. Pub. Health* **77** 1 26 - 32.

Cox, D.R. (1972). Regression models and life tables (with discussion). *Jour. Roy. Statist. Soc. B* **34** 187 - 220.

Cox, D.R. (1975). Partial likelihood. *Biometrika* **62** 269-276.

Lawless, J.F. and Yan. P. (1992). Some statistical methods for followup studies of disease with intermittent monitoring. [This volume, 427 - 446].

Prentice, R.L. (1982). Serial blood pressure measurements and cardiovascular disease in a Japanese cohort. *Amer. Jour. Epidem.* **116** 1 - 28.

Raboud, J.M. (1991). *The Effects of Errors in Measurement in Survival Analysis.* Doctoral Dissertation, University of Toronto.

Chapter 26

Some Statistical Methods for Followup Studies of Disease with Intermittent Monitoring

JERRY F. LAWLESS Department of Statistics and Actuarial Science, University of Waterloo, Waterloo, Ontario, Canada

PING YAN Department of Statistics and Actuarial Science, University of Waterloo, Waterloo, Ontario, Canada

Abstract We consider followup studies on disease progression for a group of subjects when individuals are seen only sporadically, and with the additional complication that the precise time of disease onset is unknown for many subjects. Problems of modelling and estimating disease duration and progression are discussed. Some methods based on survival distributions and disease state models are described and illustrated on a followup study of males infected with the human immunodeficiency virus.

1 Introduction

Studies of disease progression often involve the assembly of a cohort of individuals who are then observed ("followed up") over a period of time. Each study subject is typically seen sporadically, at which time clinical information about disease status is obtained, as well as laboratory measurements and information on other variables. Such intermittent observation makes detailed analysis of disease histories more difficult, particularly if the followup visits are widely spaced (e.g. see Kalbfleisch and Lawless 1985, 1988). A frequent problem is that for many subjects disease onset has occurred prior to enrolment in the study, so that information about key variables between the time of onset and enrolment may be missing. In some instances the time of disease onset is not even known, or known only to lie in some time interval; since time from onset is a crucial time variable for modelling and analysis of disease progression, this is a major difficulty. A final problem is that studies are often of such limited duration as to provide observation of only a portion of each subject's disease history, thus making inferences about the entire disease history process more tenuous.

The purpose of this paper is to consider some statistical methods for these types of situations. We assume that the objectives of the followup study include estimation of the duration of disease or stages of disease, and the relationship between fixed covariates (e.g. age, sex, treatment) or time-dependent covariates (e.g. laboratory measures) and disease progression. The discussion will be kept rather general and we will not directly consider important issues such as the assessment of treatments in clinical trials. As an example we will discuss a longitudinal study of individuals infected with the human immunodeficiency virus (HIV) that is thought to cause acquired immune deficiency syndrome (AIDS).

In Section 2 we consider the common "survival" problem where we analyze the time from disease onset until some clinical endpoint such as death. Section 3 deals with multi-state models for disease progression, and Section 4 discusses a followup study of HIV disease in a cohort of males. Section 5 deals with the joint analysis of clinical disease states and time-varying covariates, and Section 6 concludes the paper. In keeping with a broad orientation we will consider both parametric and semi-parametric methods, and the emphasis will be on laying out feasible approaches rather than on their detailed examination.

2   Survival Analysis

Suppose that the i'th subject $(i = 1, \cdots, n)$ has disease onset at calendar time x_i and is enroled in the study at time E_i. We allow for unknown time of onset by assuming that x_i is known to lie in the time interval $[L_i, R_i]$, where $-\infty < L_i \leq R_i$. Suppose that the time S_i between onset and some endpoint event is of interest; the endpoint event thus occurs at calendar time $T_i = x_i + S_i$. We wish to estimate the distribution of the S_i's.

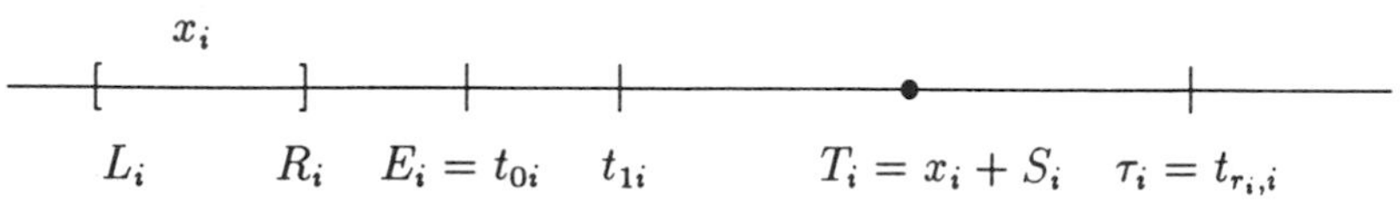

x_i

L_i R_i $E_i = t_{0i}$ t_{1i} $T_i = x_i + S_i$ $\tau_i = t_{r_i,i}$

Figure 1: A typical subject history.

We assume that it is a condition of enrolment in the study that $T_i \geq E_i$, but otherwise the selection of subjects is independent of their disease histories. Subject i is observed at followup times $E_i = t_{0i} < t_{1i} < \cdots < t_{r_i,i} = T_i$, where τ_i is the date of last followup. Figure 1 shows the situation for a subject who experienced the endpoint event prior to last followup. We suppose that S_i has density function $f_i(s)$ and corresponding cumulative distribution function (c.d.f.) $F_i(s)$ and survivor function (s.f.) $\bar{F}_i(s)$. The distribution of S_i may depend on covariates and we allow for either parametric or nonparametric estimation of the $F_i(s)$'s.

To deal with the possibly unknown time of disease onset we define $h_i(x)$ to be the density function for the time of onset for subject i. For now we assume that all covariates are fixed (i.e. not time-varying) and that if the endpoint event occurs during the period of observation for subject i, its time is known exactly. The data for subject i thus consist of the covariates z_i and

$$x_i \epsilon [L_i, R_i], \quad T_i^*, \quad \delta_i \tag{2.1}$$

where $T_i^* = \min(t_i, \tau_i)$ and $\delta_i = I(T_i^* = t_i)$. Assuming that the distribution of S_i is independent of x_i and conditioning on the knowledge that $T_i \geq E_i$, we obtain a likelihood contribution from the probability distribution of (2.1) as

$$L_{1i} = \frac{\int_{L_i}^{R_i} f_i(t_i - x)^{\delta_i} \bar{F}_i(\tau_i - x)^{1-\delta_i} h_i(x) dx}{\int_{-\infty}^{E_i} \bar{F}_i(E_i - x) h_i(x) dx}. \tag{2.2}$$

An alternative likelihood can be based on the distribution of T_i^*, given that $x_i \epsilon [L_i, R_i]$ and $T_i \geq E_i$. This gives

$$L_{2i} = \frac{\int_{L_i}^{R_i} f_i(t_i - x)^{\delta_i} \bar{F}_i(\tau_i - x)^{1-\delta_i} h_i^*(x) dx}{\int_{L_i}^{R_i} \bar{F}_i(E_i - x) h_i^*(x) dx}, \tag{2.3}$$

where

$$h_i^*(x) = \frac{h_i(x)}{\int_{L_i}^{R_i} h_i(x) dx} \qquad L_i \leq x \leq R_i$$

is the density for x_i, conditional on $x_i \epsilon [L_i, R_i]$.

In the case where x_i is known exactly we have $L_i = R_i = x_i$ and (2.2) and (2.3) both correspond to the usual likelihood contribution for a survival or censoring time, here with left-truncation at $E_i - x_i$ (e.g. Lawless 1982, Section 1.4). More generally, $h_i^*(x)$ or $h_i(x)$ would have to be estimated along with $f_i(s)$, or perhaps assumed known: see Section 2.1 below. We remark that it is tempting to use the fact that $x_i \epsilon [L_i, R_i]$, $T_i = t_i$ implies $t_i - R_i \leq S_i \leq t_i - L_i$ to write down an interval-censored likelihood contribution $F_i(t_i - L_i) - F_i(t_i - R_i)$ for subject i. This is in general incorrect, although (see (2.3)) in the case where $h_i^*(x)$ is uniform and $E_i \leq L_i$ it is correct. Jewell (1990, Section 4) considers likelihoods similar to (2.2) and (2.3) and makes some additional useful remarks.

Even with intermittent followup we usually know the exact time t_i of the endpoint event, if it precedes τ_i. If, however, we know only that $t_i \epsilon [A_i, B_i]$, say, then (2.2) and (2.3) are modified by replacing $f_i(t_i - x)$ with $F_i(B_i - x) - F_i(A_i - x)$.

We next discuss broad issues of estimation, looking first at parametric and then at non- or semi-parametric methods. Covariates are considered fixed; time-dependent covariates are discussed in Sections 3 to 5.

2.1   Parametric Estimation of Survival

Suppose that the distribution of S_i is specified up to a finite-dimensional parameter θ, so that $f_i(s) = f_i(s; \theta)$, and so on. If the $h_i^*(x)$'s are known or estimated separately, then θ can be estimated from

$$L_2(\theta) = \prod_{i=1}^{n} L_{2i}(\theta), \qquad (2.4)$$

assuming that observations on the n subjects are independent. Two possible assumptions are i) $h_i^*(x) = (R_i - L_i)^{-1}$, i.e. uniform over $[L_i, R_i]$ (e.g. Struthers and Farewell 1989), and ii) $x_i = (L_i + R_i)/2$. The effects of misspecifying the $h_i^*(x)$'s should of course be considered; it is helpful to compare analyses with different plausible choices of $h_i^*(x)$'s.

Sometimes one can estimate the $h_i(x)$'s by assuming all individuals are subject to a common disease onset process, as in De Gruttola and Lagakos (1989) who assume for a cohort of hemophiliacs exposed to HIV infection via blood transfusions that each faced similar risks of infection over time. If the marginal density for infection time is $h(x)$, then $h_i(x) = h(x)/H(\tau_i)$, where $H(t) = \int_0^t h(x)dx$. Brookmeyer and Goedert (1989) make a similar assumption, but allow $h_i(x)$ to depend on individual-specific covariates. The likelihood (2.2) can then be maximized jointly with respect to f and h.

2.2 Non- and Semi-Parametric Methods

It is often useful to estimate $F(s)$ non- or semi-parametrically. Reasons include a desire for more robustness than parametric models may enjoy, the use of nonparametric estimates to suggest or provide checks on parametric models and, in the case of relative risk or proportional hazards regression models (Cox 1972), the availability of simple methods for dealing with time-dependent covariates.

When there are no covariates, De Gruttola and Lagakos (1989) obtain nonparametric estimates of $F(s)$ and $H(x)$ in the case where $h_i(x) = h(x)$ is the same for all subjects, and there is no left-truncation. Gómez (1990) also considers this, as well as estimation of $F(s)$ when $h(x)$ is known. We will outline a method of nonparametric estimation for the general situation, assuming that the $h_i^*(x)$'s in (2.3) are known. To develop the estimates we assume that $f_i(s) = f(s)$, where $s = 0, 1, 2, \cdots$ is measured in discrete time units, and that the x_i's and t_i's are measured in the same units. Then (2.3) becomes

$$L_{2i} = \frac{\sum_{x=L_i}^{R_i} f(t_i - x)^{\delta_i} \bar{F}(\tau_i - x)^{1-\delta_i} h_i^*(x)}{\sum_{x=L_i}^{R_i} \bar{F}(E_i - x) h_i^*(x)}, \qquad (2.5)$$

where $\bar{F}(s) = \sum_{u=s+1}^{\infty} f(u) = 1 - F(s)$. We maximize the log likelihood $l(f) = \sum_{i=1}^{n} \log L_{2i}$ with respect to $f = (f(0), f(1), \cdots)$ to obtain the nonparametric estimate $\hat{f}$ and corresponding estimate of $F(s)$, $s = 0, 1, \cdots$. The log likelihood may be maximized by solving likelihood equations $\partial l / \partial f_j = 0$, but a more convenient approach is to use an $E - M$ algorithm. This involves an extension of the approaches in De Gruttola and Lagakos (1989) and Gómez (1990), and is discussed by Bacchetti and Jewell (1991) and Yan and Lawless (1991).

Semi-parametric regression methods are much harder to develop, except for the obvious extension of the above to deal with groups of subjects with different survival distributions. Even when times of onset are known, the extension of the Cox (1972) relative risk analysis to deal with intermittent monitoring schedules that vary across subjects is difficult (Finkelstein 1986). It is equally hard to develop methods when times of disease onset are uncertain. When the $h_i(x)$'s are equal, Kim et al. (1990) suggest an approach based on a discrete analogue of the proportional hazards model. However, the properties of this and similar procedures are difficult to study. For example, the regression parameters are sometimes inestimable but it is hard to characterize when this happens, even in simple situations. Further work in this area is needed. In the meantime, when times of endpoint events are observable, the main approach has been to carry out semi-parametric analyses based on relative risk models (Cox 1972) by using an estimated

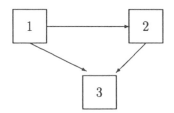

Figure 2: A three state disease progression model.

time of disease onset (e.g. Coates et al. 1990); analyses may be performed with different estimated times of onset. It should be noted that the precision of estimates tends to be overstated since uncertainty in the x_i's is not accounted for. Brookmeyer and Gail (1987) discuss biases in estimation that occur in relative risk analysis when the time origin (i.e. assumed date of disease onset) is after the actual time of onset.

When there are time-varying covariates such as laboratory measurements or treatment interventions, the approach just mentioned still applies. However, the results may be sensitive to the choice of x_i values, especially when the study cohort is fairly small and the $R_i - L_i$'s are large. Time-dependent covariates are discussed further in Section 5.

3   Models with Multiple Disease States

For many disease processes it is useful to model the passage of subjects through disease stages or states. For example, with the HIV process individuals are often classified clinically into stages according to whether they have certain sets of symptoms (e.g. see Longini et al. 1989 and Section 4 below). Similar models are used with cancer (e.g. Kay 1986) and other diseases. One reason to construct multi-state models is to provide a more comprehensive picture of disease than that given by onset and a single clinical endpoint. Another reason is to more effectively use information on fairly short portions of individual disease histories. In addition, multi-state models can sometimes help to deal with unknown times of onset. As we discuss below, when there is substantial information missing about subjects' disease histories, many multi-state models are hard to deal with. However, time-homogeneous Markov models can prove useful, as we indicate here and in Section 4.

Consider a model consisting of k states $\{1, \cdots, k\}$, with an individual being unequivocally in some state at any time t after disease onset. We let $Y_i(t)$ denote the state occupied by subject i at time t after onset. Figure 2, for example, portrays a three state model in which state 1 represents mild disease, stage 2 severe disease, and stage 3 death. The arrows indicate the transitions between states that are possible. (In this example we have supposed that transitions from state 2 back to state 1 are impossible.) The state $Y_i(0)$ occupied at onset would often, but not always, be state 1.

Let z_t denote the value of a vector of covariates at time t after onset (this does not preclude all covariates being fixed), and let H_t denote the individual's disease and covariate history up to time t. Transitions between states are governed by transition intensity functions

$$\lambda_{ij}(t \mid H_t, z_t) = \lim_{\Delta t \to 0} \frac{1}{\Delta t} Pr\{Y(t + \Delta t) = j \mid Y(t) = i, H_t, z_t\}, \ i \neq j. \quad (3.1)$$

Features of such models that may be of interest include baseline or average transition rates, sojourn times in various states, time until entry to certain states, and the relationship of covariates to such quantities.

We consider the observational schemes described in Sections 1 and 2 (e.g. see Figure 1); Kalbfleisch and Lawless (1988) comment on a broad range of observational plans. When subjects are followed up at points closely spaced in time and when the time of disease onset is known, we may effectively use continuous-time methods. Markov, semi-Markov and other types of models are easily fitted via parametric or semi-parametric means (e.g. see Andersen and Borgan 1985, Kalbfleisch and Lawless 1988, 1989). When followup is intermittent or when the time of onset is unknown, difficulties arise, as we now discuss.

For time-homogeneous Markov models, the difficulties created by intermittent monitoring or unknown time of onset are not too severe, and such models should be exploited where possible. The reason is that (ignoring covariates for convenience) the transition intensities (3.1) are in this case constants q_{ij} $(i \neq j)$, and the transition probabilities are stationary: $P_{ij}(t) = Pr\{Y(s + t) = j \mid Y(s) = i\}$ for $s \geq 0$, $t \geq 0$. The $P_{ij}(t)$'s are easily computed from the q_{ij}'s in matrix form (e.g. Cox and Miller 1965, ch. 4). If $q_{ii} = -\sum_{j \neq i} q_{ij}$ and $Q = (q_{ij})$ is the $k \times k$ intensity matrix, then the $k \times k$ transition probability matrix $P(t) = (P_{ij}(t))$ is given by

$$P(t) = \exp(Qt) = \sum_{r=0}^{\infty} \frac{(Qt)^r}{r!}. \quad (3.2)$$

Kalbfleisch and Lawless (1985, 1989) present methods for fitting and assessing such models when followup is intermittent; here the data on subject i consist of $Y_i(t_{i0}), Y_i(t_{i1}), \cdots, Y_i(t_{ir_i})$ where $E_i = t_{i0}$ and $\tau_i = t_{ir_i}$, as well as fixed covariates. Time-varying covariates may also be handled, provided that the transition intensities depend only on observable aspects of them.

In particular, if $Y_i(t_{ij}) = y_{ij}$ and z_{ij} is a covariate vector assumed to be fixed over $t_{i,j-1}$ to t_{ij}, then the likelihood contribution for subject i is

$$\prod_{j=1}^{r_i} P_{y_{i,j-1}y_{ij}}(t_{ij} - t_{i,j-1} \mid z_{ij}). \qquad (3.3)$$

This conditions on the fact that $Y_i(t_{i0}) = y_{i0}$, i.e. the state occupied at the time of enrolment. Note that (3.3) does not require knowledge of the time of onset since it depends only on the time differences $t_{ij} - t_{i,j-1}$.

By conditioning on $Y_i(t_{i0})$ we do not need to consider the time of onset. If, on the other hand, we wish to include an additional term for $Pr\{Y_i(t_{i0})\}$ in the likelihood, then we would need to use an approach like that in Section 2. Similarly, the fact that an individual must not have progressed to an endpoint state before enrolment does not alter (3.3), but would affect $Pr\{Y_i(t_{i0})\}$.

We remark that Markov models provide a way to deal with time-dependent covariates such as laboratory markers, by defining states according to current values of the markers. We illustrate this in Section 4. Kay (1986) uses a similar approach with cancer disease markers. It should be recognized in any application of these models that the time-homogeneous Markov assumption is a strong one, implying exponentially distributed sojourn times in each state. Careful definition of states is usually required to make this reasonable. The adequacy of the models should of course be assessed in any analysis.

More complicated models such as Markov models with time-dependent transition intensities or semi-Markov models in which sojourn times in states are independent random variables are unwieldy when followup is intermittent, and more so if the time of onset is unknown. It is feasible to fit models for a very simple setup like that in Figure 2, but we will not pursue this here.

We next discuss a specific longitudinal study.

4   A Followup Study of HIV Disease

We illustrate points discussed earlier on a followup study of 159 males infected with the human immunodeficiency virus (HIV). The data were kindly provided by Dr. Randy Coates and his colleagues. The study is described in some detail in Coates et al. (1990) and other references cited therein. Briefly, the study cohort consists of homosexual or bisexual males in Toronto who had at least one sexual contact with men diagnosed with AIDS or an AIDS-related condition. Subjects were healthy when recruited;

recruitment was in 1984-85. Of 249 men enroled in the study, 143 were HIV positive (and assumed infected with the HIV) at their time of enrolment, and 16 others became HIV positive (HIV+) during the study. It is these 159 HIV+ subjects whom we discuss here.

The subjects were seen approximately every three months after enrolment, at which times a variety of clinical observations and laboratory measurements were made. For the 16 subjects who became HIV+ after enrolment, the time of seroconversion is assumed known; we equate seroconversion with HIV infection and disease onset, although the date of actual HIV infection is a little before this. For the 143 subjects who were HIV+ at enrolment, it is assumed that their time of disease onset (seroconversion) lies in an interval $[L, R]$ that corresponds to the dates of their first and last sexual contacts with the man from whom they are presumed to have contracted the virus.

We will focus mainly on the problem of estimating the so-called incubation distribution, i.e. the distribution of time from HIV infection to AIDS. For convenience we take the calendar time origin for subject i to be $L_i = 0$, the left-hand endpoint of the interval $[L_i, R_i]$ within which their time of infection x_i is presumed to lie: thus, $0 \leq x_i \leq R_i$. For subjects who seroconverted after enrolment, we set $E_i = R_i = x_i = 0$, since for the purposes of disease followup, we are concerned only with the period after seroconversion. For subjects who were HIV+ at enrolment, we have $E_i > R_i \geq 0$. The time of AIDS and last followup are denoted by T_i and τ_i, as in Section 2 and Figure 1.

To estimate the incubation distribution, we first fitted a Weibull model with survivor function $\bar{F}(s; \lambda, \beta) = \exp\{-(\lambda s)^\beta\}$. The Weibull distribution has provided good fits to a variety of censored or truncated incubation time data (e.g. see Brookmeyer and Goedert 1989, and references therein). Earlier analyses of the Toronto cohort did not find a strong association between incubation time and fixed baseline covariates such as age and behavioural characteristics. Thus we ignored covariates and used the likelihood (2.4) to estimate λ and β, under different assumptions about $h_i^*(x)$. We remark that the data are heavily censored, as only 41 of the 159 subjects had AIDS by the date of last followup. Thirty-nine subjects have $R_i = 0$, but there are many large R_i values; 36 subjects have $R_i > 1000$ days, and the largest R_i is 2953 days. The largest τ_i value is 4481 days, but the vast majority are well under 3000 days, or about 8.2 years.

Results under four choices for $h_i^*(x)$ are as follows; figures in brackets after estimates represent approximate .95 confidence intervals, and m represents the median. The time units used in the analysis and figures below are days, but for convenience the median incubation time estimates are in years.

(i) $x_i = 0$: $\hat{\beta} = 1.65$, $\hat{\lambda} = .000257$, $\hat{m} = 8.5$ years (7.1-10.6)

(ii) $x_i = R_i/2$: $\hat{\beta} = 1.8$, $\hat{\lambda} = .000288$, $\hat{m} = 7.8$ years (6.6-9.6)

(iii) $h_i^*(x) = R_i^{-1}$ $(0 \le x \le R_i)$: $\hat{\beta} = 1.8$, $\hat{\lambda} = .000290$, $\hat{m} = 7.7$ years (6.6-9.6)

(iv) $h_i^*(x) = \frac{\alpha e^{-\alpha x}}{1 - e^{-\alpha R_i}}$ $(0 \le x \le R_i)$, with $\alpha = .0027$.
(This is based on assumptions about constant rates of sexual contact and infection transmission and makes earlier infection more probable.)
$\hat{\beta} = 1.75$, $\hat{\lambda} = .000270$, $\hat{m} = 8.2$ years (6.9-10.4)

The results from (i) - (iv) are in good agreement. As one would expect, the results from (i) and (iv) agree closely, as do those from (ii) and (iii); it does not matter much here whether a distribution $h_i^*(x)$ is used or whether a fixed x_i value roughly equal to the mean is used. The model (i), which assumes that infection occurred at the first sexual contact, provides conservative estimates for $F(s)$ that are on the low side, but differs little from the other models.

Nonparametric estimates of $F(s)$ as in (2.5) are more robust, and provide checks on parametric models. To illustrate, we consider model (ii). The estimate of $\bar{F}(s)$ obtained from (2.5) is then merely the Kaplan-Meier estimate adjusted for left-truncation; it gives an estimated median of 7.4 years, in good agreement with the fully parametric model (ii). In addition, a plot of $\log\{-\log \hat{\bar{F}}(s)\}$ vs. $\log s$ is roughly linear, consonant with a Weibull distribution for $\bar{F}(s)$. Figure 3 shows the Kaplan-Meier estimate and the Weibull estimate $F(s; \hat{\beta}, \hat{\lambda})$ of the incubation distribution function for (ii). Note that there is relatively little information about the right half of the distribution, since only 41 subjects have progressed to AIDS, and most followup times are considerably less than 10 years.

Coates et al. (1990) found evidence that laboratory markers such as T4 lymphocyte count, T4/T8 ratio and blastogenic responses to certain mitogens were prognostic regarding the transition to AIDS; see also Raboud et al. (1992). The evidence was based on a relative risk analysis (Cox 1972) that treated markers as time-dependent covariates, and assumed that $x_i = 0$ for all subjects. Such analyses do not provide predictions regarding progression to AIDS; to do this it is necessary to model the covariate processes. To attack this problem we consider a time-homogeneous Markov model with an absorbing AIDS state and with transient states defined according to marker values. For illustration we consider only T4 counts, and the six state model shown in Figure 4; a more searching analysis of various models will be given elsewhere. It is well known that in the course of HIV

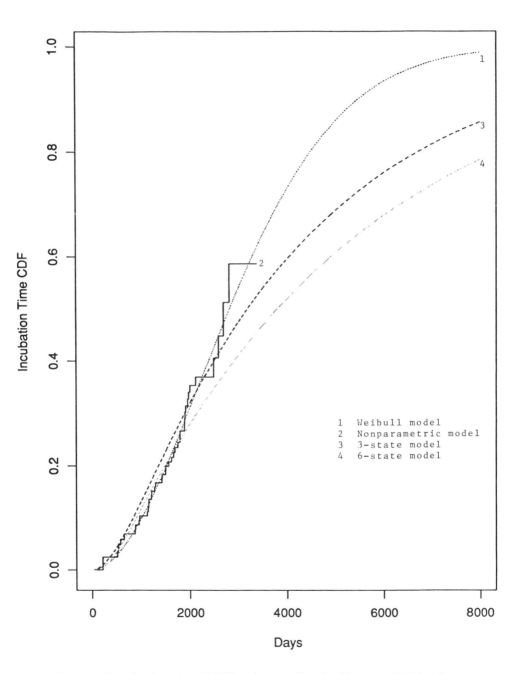

Figure 3: Incubation time CDF estimates for the Toronto AIDS cohort.

T4:   ≤ 900   700–899   500–699   200–499   < 200

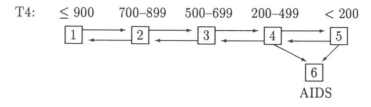

AIDS

Figure 4: A model with states defined by laboratory marker values.

disease there is a decline in T4 count, and that low T4 counts are associated with progression to AIDS. The definition of states is similar to that in Longini (1990), except that he views the decline in T4 counts as progressive and does not allow transitions from higher to lower states. This is undesirable since T4 counts and other laboratory markers are highly variable within individuals. Figure 5, for example, shows successive T4 counts for six subjects who seroconverted during the study.

The transition intensity matrix Q for the above model may be estimated using the likelihood (3.3) and methods in Kalbfleisch and Lawless (1985). The m.l.e. is (the time units re Q are days)

$$\hat{Q} = 10^{-1} \begin{pmatrix} -.111 & .111 & 0 & 0 & 0 & 0 \\ .060 & -.305 & .245 & 0 & 0 & 0 \\ 0 & .099 & -.197 & .098 & 0 & 0 \\ 0 & 0 & .038 & -.053 & .013 & .001 \\ 0 & 0 & 0 & .047 & -.060 & .013 \\ 0 & 0 & 0 & 0 & 0 & 0 \end{pmatrix}.$$

Transition probabilities can be computed from (3.2). For example, the one-year (365 day) transition probability matrix is

$$\hat{P}(365) = \begin{pmatrix} .10 & .13 & .26 & .44 & .06 & .02 \\ .07 & .11 & .24 & .48 & .08 & .03 \\ .06 & .09 & .22 & .50 & .09 & .04 \\ .04 & .07 & .20 & .52 & .11 & .06 \\ .02 & .05 & .13 & .41 & .18 & .22 \\ 0 & 0 & 0 & 0 & 0 & 1 \end{pmatrix}.$$

The transition intensities and probabilities reflect the fact that T4 counts are very variable within subjects. Thus, although T4 count is prognostic of AIDS, particularly in the sense that people do not usually convert to AIDS unless their T4 counts are low, the estimated probability that someone with a low T4 count (< 200) converts to AIDS within the next year is only .22.

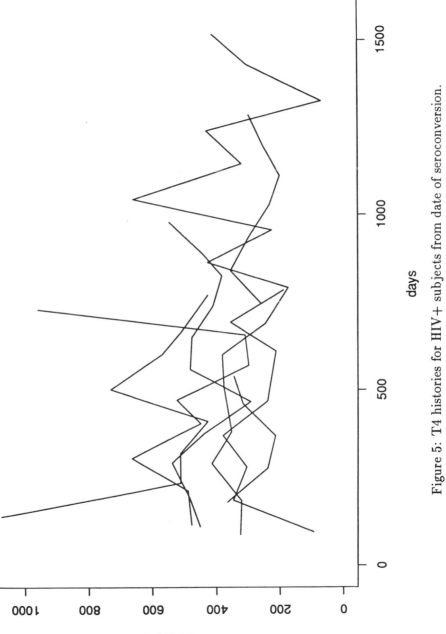

Figure 5: T4 histories for HIV+ subjects from date of seroconversion.

It is very common for a subject's T4 count to be below 200 at a followup visit and over 200 at the next visit.

We can also estimate the incubation distribution from this model. Assuming that at the time of infection subjects are in state 1, we have $F(t) = Pr\{Y(t) = 6 \mid Y(0) = 1\} = P_{16}(t)$. This gives the estimate $\hat{F}(t)$ shown in Figure 3. The estimated median incubation time is 10.5 years (.95 confidence interval 8.0 - 14.5 years), which is substantially larger than the earlier Weibull analyses. We remark that even if a subject is in one of states 2, 3 or 4 at the time of infection, the estimated waiting time distribution until AIDS does not change much. For example, from state 4 the estimated median waiting time until AIDS is still 9.5 years.

Two oversimplifications in the model above are the use of discrete states for the T4 counts and the time homogeneous Markov assumption. These facilitate analysis with the type of incomplete data seen here but are likely rather crude approximations to the actual T4-HIV disease process. We did check the fitted model by splitting the data into groups of successive T4 counts at pairs of times separated by approximately the same amount. (Subjects were followed up approximately every 90 days. Thus, for example, across the 159 subjects there are 273 occasions where two successive observations were approximately 85 days apart, and 752 which were approximately 95 days apart.) A comparison of total observed transitions from state i to j $(i, j = 1, \cdots, 6)$ for each group with expected numbers of transitions obtained from $\hat{Q}$ and (3.2) showed good agreement and did not provide evidence against the model. More sensitive model checks are needed; the lack of information about time of onset makes this more difficult, but techniques in Kalbfleisch and Lawless (1985, 1989) and De Stavola (1988) may be applied by using imputed times of onset.

We also fitted time-homogeneous Markov models with states defined according to clinical disease stages; this has also been done by Longini et al. (1989). In particular, we considered a progressive three state model where State 1 was "HIV+ without AIDS and with normal immune function", State 2 was "HIV+ without AIDS but abnormal immune function" and State 3 was "AIDS". Estimated mean sojourn times in States 1 and 2 are 1.3 and 10.5 years, respectively, and the estimated median incubation time is 8.7 years. The estimated incubation time c.d.f. is shown in Figure 3. There was mild evidence of lack of fit of the model; alternative state definitions are being considered.

Figure 3 portrays four estimates of the incubation time c.d.f. $F(s)$; it is interesting to compare them and note their limitations. The two survival analysis estimates (the Weibull estimate and the nonparametric estimate) are in close agreement. If the assumptions about infection times are correct (see below) these are reliable, although it must be emphasized that the data are informative only about the part of the distribution up to about 3000 days (8.2 years). There is no direct evidence about the right tail of the distribution, and the Weibull distribution may not be reasonable in that region. Also, the Weibull and nonparametric estimates corresponding to case (i) above ($x_i = 0$) would lie a little below the ones shown.

The c.d.f.'s estimated from the multi-state models agree quite well with the survival analysis estimates up to about 1800 days (for the six state model) and 2500 days (for the three state model),after which they lie well below the survival analysis estimates. One is inclined to attribute more validity to the survival estimates, but a caution is in order: the survival analysis estimates are more dependent on the correctness of the data and assumptions regarding the infection times x_i. If many subjects were in fact infected outside the presumed intervals $[L_i, R_i]$ the results could be biased. We remark in addition that changes in the nonparametric estimate of $F(s)$ after about 2200 days is determined by a small number of subjects. The multi-state models are less susceptible to assumptions about the x_i's but, on the other hand, make stronger assumptions about the disease process. They are also more affected by marker measurement errors (re the six state model) and variability in disease staging (re the three state model). The confidence limits for all four c.d.f.'s are of course fairly wide, as the confidence limits for the median incubation times suggest. We remark that Pawitan and Self (1991) have estimated the median incubation time from this study as 7.2 years, using rather different methods to those here.

Finally, we remark that the estimates of $F(s)$ up to about seven years are here somewhat larger than those in some other studies, and estimates of median incubation time are smaller. For example, a cohort of hemophiliacs and one of homosexual males both give a median of about 10 years (Brookmeyer and Goedert 1989). Further followup of the Toronto cohort is desirable, to see whether longer incubation times are more in line with the Weibull model or the multi-state estimates, which agree more closely with other studies.

5   Time-Dependent Covariates

We have alluded to the difficulty of dealing with time-dependent covariates. Here we study the problem in a little more detail and consider an approach.

Suppose that the survival time S (time from onset to endpoint, as before) has hazard function $\lambda(s; H_s, z_s)$, where H_s represents the history of the disease process and covariates up to time s after onset. We will restrict attention to models for which λ depends on H_s only through the current covariate values z_s, and consider relative risk models with

$$\lambda(s; H_s, z_s) = \lambda_0(s; \alpha) g(z_s; \beta), \qquad (5.1)$$

where α and β are vectors of parameters, λ_0 is a baseline hazard function, and g is a positive-valued function. The use of a parametric model for λ_0, as opposed to the usual nonparametric approach (Cox 1972), leads to relatively simple methods when times of onset are known, but monitoring is intermittent. We now consider how one might deal with unknown onset times.

Conditional on the covariate process, the survivor function and density for S are respectively

$$\bar{F}_i(s) = \exp\{-\int_0^s \lambda_0(u; \alpha) g(z_{i,u}; \beta) du\} \qquad (5.2)$$

$$f_i(s) = \lambda_0(s; \alpha) g(z_{i,s}; \beta) \bar{F}_i(s). \qquad (5.3)$$

Consider now the likelihood contribution L_{2i} of (2.3), assuming that $h_i^*(x)$ is specified. The difficulty in utilizing this is that it depends on subject i's covariate history $z_{i,s}$ for $0 \le s \le E_i - x$, i.e. over the calendar time interval from onset to enrolment. This information is unavailable. We mention two situations, however, where the problem can be circumvented.

1. If $\lambda(s; H_s, z_s)$ depends only on the covariate values and not s, then in (5.1) we can take $\lambda_0(s; \alpha) = 1$. In this case (2.3) gives

$$L_{2i} = g(z_{i,t_i-x}; \beta)^{\delta_i} \exp\{-\int_{E_i-x}^{t_i^*-x} g(z_{i,s}; \beta) ds\}, \qquad (5.4)$$

where $t_i^* = \min(t_i, \tau_i)$. (To obtain (5.4) from (2.3) we have to note that the expressions in (5.4) do not depend on the time of onset x, but only on covariate values between onset and final followup.)

2. If $g(z_s; \beta)$ is constant for $0 \le s \le S^*$, i.e. if the covariate effect is not time-varying up to time S^* after onset, then if it should be the case that $E_i - x \le S^*$ for all x in $[L_i, R_i]$, (5.2) gives

$$\bar{F}_i(E_i - x) = \bar{F}_0(E_i - x; \alpha)^{g(z_i^c; \beta)}$$

where $F_0(s; \alpha) = \int_0^s \lambda_0(t; \alpha) dt$ and $z_{i,s} = z_i^c$ for $0 \le s \le S^*$. Hence L_{2i} is computable, assuming that z_i^c is measured at enrolment.

With either approaches 1 or 2 it is necessary because of the intermittent monitoring to impute values for time-varying covariates between followup times. It is expedient to use a model for $\lambda(s; \alpha)$ which yields closed form expressions for (5.2) and (5.3). A Weibull hazard $\lambda_0(s; \alpha) = \alpha s^{\alpha-1}$ and an assumption that z_s is constant or linear between followup times is convenient and quite flexible.

We will briefly illustrate approach 1; a more detailed discussion of both approaches will be given elsewhere. As an example, suppose there is a single time-dependent covariate z_s and that in (5.4) we have $g(z_s) = \exp(\alpha + \beta z_s)$. Then the likelihood obtained by multiplying contributions of the form (5.4) may be written as

$$\exp\{ \sum_{i:\delta_i=1} (\alpha + \beta z_i^*) - \alpha \sum_{i=1}^{n}(t_i^* - E_i) - \beta \sum_{i=1}^{n} Z_i^* \}, \qquad (5.5)$$

where $z_i^* = z_{i, t_i - x_i}$ is the covariate value for subject i at the endpoint time, $t_i^* = \min(t_i, \tau_i)$ and

$$Z_i^* = \int_{E_i - x_i}^{t_i^* - x_i} z_{i,s} \, ds$$

is the integral of the covariate process over the observation period for subject i. Maximization of (5.5) is direct, assuming we can impute the Z_i^*'s.

This approach can be extended to multi-state models. Nielsen (1991) and Fusaro et al. (1991) have considered a similar approach and nonparametric estimation of $g(z_s)$. The assumption of a constant baseline hazard or intensity is of course strong, and needs to be checked. We note also that there is a close connection between this approach and that used with the six state model in Section 4.

6   Concluding Remarks

As we have indicated, statistical analysis of disease histories is more difficult when the time of onset is unknown, and the amount of information is also limited when the onset intervals $[L_i, R_i]$ are long. It is thus important to determine onset times as precisely as possible; with the x_i's known fairly closely, standard survival and multi-state model analysis can be brought to bear. We have outlined some procedures here for the case when some $R_i - L_i$'s are large, and further study is warranted.

More research is also needed to model better the process for laboratory markers and to utilize markers for predicting disease progression. Standard relative risk analysis (Cox 1972, Andersen and Borgan 1985) estimates the effect of markers on transition intensities among disease stages and gives predictions if a marker process history over the course of the disease is specified. For better prediction, joint models of markers and disease stages are needed. Berman (1990) and De Gruttola et al. (1991) model markers using Gaussian processes, for example, but relate markers to the incubation hazard function in a naive way. Measurement error and the entire measurement systems underlying laboratory markers also deserve study (Raboud et al. 1992).

Some authors (e.g. Munoz et al. 1989) have attempted to use marker process histories to estimate unknown times of onset. This is clearly feasible in principle, but needs a good deal of further study. With highly variable markers such as T4 lymphocyte counts, the degree of uncertainty in estimates of x_i obtained in this way is so large as to provide useful additional information only when $R_i - L_i$ is very large. (That is, the uncertainty in the estimated x_i is often larger than $R_i - L_i$.)

Finally, we remark that the distinction between disease stages, laboratory markers and other clinical variables is somewhat artificial, and what is sought are relevant ways of measuring disease progression and of predicting subjects' prognoses in functional terms. It is important that statistical disease process modelling keep this in mind.

Acknowledgments

We thank Randy Coates, Janet Raboud and their colleagues in the Toronto AIDS Study for supplying the data in Section 4, and Randy Coates, Vern Farewell, Robert Gentleman and Janet Raboud for helpful discussions. This research was supported in part by grant number 1-R01-DA 04722 from the National Institute on Drug Abuse, coordinated by the Societal Institute of the Mathematical Sciences, and by a grant to J.F. Lawless from the Natural Sciences and Engineering Research Council of Canada.

References

Andersen, P.K. and Borgan, O. (1985). Counting process models for life history data: a review (with discussion). *Scand. Jour. Statist.* **12** 97 - 158.

Bacchetti, P. and Jewell, N.P. (1991). Nonparametric estimation of the incubation distribution of AIDS based on a prevalent cohort with unknown infection times. *Biometrics* **47** 947 - 960.

Berman, S.M. (1990). A stochastic model for the distribution of HIV latency times based on T4 counts. *Biometrika* **77** 733 - 741.

Brookmeyer, R. and Gail, M.H. (1987). Biases in prevalent cohorts. *Biometrics* **43** 739 - 749.

Brookmeyer, R. and Goedert, J. (1989). Censoring in an epidemic with an application to hemophilia-associated AIDS. *Biometrics* 45 325 - 335.

Coates, R.A. et al (1990). Cofactors of progression to AIDS in a cohort of male sexual contacts with men with HIV disease. *Amer. Jour. Epidem.* 132 717 - 722.

Cox, D.R. (1972). Regression models and life tables (with Discussion). *Jour. Roy. Statist. Soc. (B)* 34 187 - 220.

Cox, D.R. and Miller, H.D. (1965). *The Theory of Stochastic Processes.* Methuen, London.

De Gruttola, V. and Lagakos, S. (1989). Analysis of doubly-censored survival data, with application to AIDS. *Biometrics* 45 1 - 11.

De Gruttola, V., Lange, N. and Dafni, U. (1991). Modeling the progression of HIV infection. *Jour. Amer. Statist. Assoc.* 86 569 - 577.

de Stavola, B.L. (1988). Testing departures from time homogeneity in multistate Markov processes. *Appl.Statist.* 37 242 - 250.

Finkelstein, D.M. (1986). A proportional hazards model for interval-censored failure time data. *Biometrics* 42 845 - 854.

Fusaro, R., Nielsen, P. and Scheike, T. (1991). Marker-dependent hazard estimation: an application to AIDS. University of California, Berkeley, Biostatistics Technical Report #21.

Gomez, G. (1990). Estimation of induction distribution with doubly censored data and application to AIDS. To appear in Proceedings of the IMS Meeting (Uppsala, Sweden).

Jewell, N.P. (1990). Some statistical issues in studies of the epidemiology of AIDS. *Statist. Med.* 9 1387 - 1416.

Kalbfleisch, J.D. and Lawless, J.F. (1985). The analysis of panel data under a Markov assumption. *Jour. Amer. Statist. Assoc.* 80 863 - 871.

Kalbfleisch, J.D. and Lawless (1988). Likelihood analysis of multi state models for disease incidence and mortality. *Statist. Med.* 7 149 - 160.

Kalbfleisch, J.D. and Lawless, J.F. (1989). Some statistical methods for panel life history data. In *Analysis of Data in Time* (A.C. Singh and P. Whitridge, eds.), 185 - 192. Statistics Canada, Ottawa.

Kay, R. (1986). A Markov model for analyzing cancer markers and disease states in survival studies. *Biometrics* **42** 855 - 865.

Kim, M.Y., De Gruttola, V. and Lagakos, S.W. (1990). Analyzing doubly censored data with covariates, with application to AIDS. Unpublished manuscript.

Lawless, J.F. (1982). *Statistical Models and Methods for Lifetime Data.* John Wiley and Sons, Inc., New York.

Longini, I. (1990). Modelling the decline of CD4 T-Lymphocyte counts in HIV-infected Individuals. *Jour. Acq. Immune Def. Synd.* **3** 930 - 931.

Longini, I. et al. (1989). Statistical analysis of the stages of HIV infection using a Markov model. *Statist. Med.* **8** 831 - 843.

Munoz, A. et al. (1990). Estimating the distribution of times from HIV seroconversion to AIDS using multiple imputation. *Statist. Med.* **9** 505 - 514.

Nielsen, J. (1991). Marker-dependent hazard estimation. Unpublished manuscript.

Pawitan, Y. and Self, S. (1991). Modelling disease marker processes in AIDS. Unpublished manuscript.

Raboud, J.M., Coates, R.A., and Farewell, V.T. (1992) Estimating risks of progression to AIDS using serially measured immunologic markers. [This volume, 415 - 425].

Struthers, C.A. and Farewell, V.T. (1989). A mixture model for time to AIDS data with left truncation and an uncertain origin. *Biometrika* **76** 814 - 817.

Yan, P. and Lawless, J.F. (1991). Nonparametric estimation of a survival distribution when time of onset is uncertain. Unpublished manuscript.

Chapter 27

Exact Solutions to the Behrens-Fisher Problem

BALDEO K. TANEJA [1] Center for Drug Evaluation and Research, U.S. Food and Drug Administration, Rockville, Maryland

EDWARD J. DUDEWICZ Department of Mathematics, Syracuse University, Syracuse, New York

Abstract The Behrens-Fisher problem is the problem of hypothesis testing of equality of means of two normal sets with unknown and unequal variances. In the statistical literature, a number of solutions to this problem have been proposed but they are either non-exact or involve discarding information by throwing away some of the observations. While these non-exact solutions are well-known and often compared to each other (see Best and Rayner 1987), the exact solutions are lesser known. This paper describes and compares two exact solutions: Chapman's solution and the Prokof'yev-Shishkin solution. Recommendations are provided on which method to choose under a given set of specifications.

1 Introduction

"The most frequently occurring problem in applied statistics is $\cdots$ [that of] the comparison of the means of two populations $\cdots$" according

---

[1] This work was performed prior to the first author's joining the Food and Drug Administration. The views expressed in this paper are those of the authors and not necessarily those of the Food and Drug Administration

to H. Scheffé (1970, p. 1501). For many practical problems which can be solved by statistics (especially problems of processing signals in noise), assumption of a normal model for observations is appropriate and adequate. In a number of instances, these problems are formulated as problems of hypothesis-testing of equality of means of two normal sets (for example: in two channels of a receiver; in two time, space, or frequency intervals; in two "rays"; etc.) with respective variances which are arbitrary and unknown (hence, which may not be the same). In this case we are dealing with what is known as the Behrens-Fisher problem. This problem is an inherently heteroscedastic one in many cases, and is the subject of a myth that there is no (exact) solution ⋯ perhaps due to a misapprehension of a result on the Behrens-Fisher problem version in the no-design-control case where one cannot control the sample sizes. (However, even in the no-design-control case, practical solutions are available; see Scheffé 1970.) In the design-control case (e.g. in the problem of processing sonar or radio signals in noise, where data are acquired sequentially, or in medical studies designed for termination based on sequential analysis of the data), statistical control (e.g. of probabilities of a false report and of correct detection of signal independent of the noise level) is possible with an *exact* solution. Thus, while in the statistical literature a number of *non-exact* "solutions" have been proposed (e.g., those of Neyman-Bartlett, Scheffé, Hsu and Welch, etc.; see Dudewicz 1976), and are often compared to each other (e.g., see Best and Rayner 1987), these often involve discarding information by throwing away some of the observations. In this paper we consider two lesser known exact solutions: one attempted by the Soviets V.N. Prokof'yev and A.D. Shishkin (1974) (denoted as PS below) which its authors say (p. 142) involves "⋯ some loss of information ⋯"; and, Chapman's Procedure, which is what the Heteroscedastic Method yields in the present case. (Dudewicz and Bishop 1979 have given the rudiments of development of a Heteroscedastic Method in the framework of statistical decision theory and have studied its properties. This work also provides exact solutions for more complex problems than that of this paper, and details including tables needed for application have been given elsewhere.)

The exact solutions are considered in Section 2, their efficiency compared and selected tables provided in Section 3, and recommendations on which method to choose (under a set of specifications) provided in Section 4.

A referee notes that "Although Scheffé's claim that the general problem of comparing two normal means is 'the most frequently occurring problem in [applied] statistics' may have been accurate 25 or 30 years ago, the development of new probability models and techniques in applied statistics coupled with the fantastic computing resources now available make Scheffé's statement invalid in the eyes of many statisticians today." We tend to agree with the referee, and certainly one should not confuse "most frequently occurring" with "most important". Nevertheless, many studies still seem to have such a comparison as their basic goal. A referee also noted that procedures which (like the exact procedures studied in this paper) involve sequential sampling are "perhaps one of the least frequent instances of normal means comparison." We agree, but feel that this may well be due to the fact that the sequential procedures (and their exact properties and much weaker assumptions) are, as noted earlier in this paper, much lesser known among practitioners, and not available yet in most texts. Since (for ethical as well as cost reasons) sequential analysis is experiencing a resurgence of activity, we expect that the exact properties and weak assumptions of the current methods of this paper will become more widely used in the years ahead.

2 Two Exact Solutions to the Behrens-Fisher Problem

Let X_1, $X_2, \ldots$ and Y_1, $Y_2, \ldots$ denote two sequential samples from normal distributions $N(\mu_1, \sigma_1^2)$ and $N(\mu_2, \sigma_2^2)$ respectively where μ_1, μ_2, σ_1^2 and σ_2^2 are all unknown. We wish to perform an α-level test of the null-hypothesis $H_0 : \mu_1 = \mu_2$ against the alternative $H_1 : \mu_1 \neq \mu_2$ (where σ_1^2 and σ_2^2 are unequal) such that when $\triangle = \triangle_1$ then $\beta = \beta_1$, where $\triangle$ is the absolute difference between the two means, β is the power of the test and $\triangle_1 > 0$ and $0 < \beta_1 < 1$ are specified numbers.

We now describe the two procedures that yield exact solutions.

Prokof'yev-Shishkin (1974) Procedure

First, on the basis of the two sequences X_1, $X_2, \ldots$ and Y_1, $Y_2, \ldots$ form a single sequence Z_1, $Z_2, \cdots$ where $Z_i = X_i - Y_i$, $i = 1, 2, \cdots$. Then Z_1, $Z_2, \cdots$ is a sequential sample from a normal distribution $N(\mu_1 - \mu_2, \sigma_1^2 + \sigma_2^2)$.

Take an initial sample Z_1, $Z_2, \cdots, Z_{n_0}$ of size $n_0 \geq 2$ from this population and calculate

$$\overline{Z}(n_0) = \frac{1}{n_0} \sum_{i=1}^{n_0} Z_i,$$

$$S^2 = \frac{1}{n_0 - 1} \sum_{i=1}^{n_0} (Z_i - \overline{Z}(n_0))^2, \qquad \text{and}$$

$$n = \max(n_0, [S^2/d])$$

where $[x]$ denotes the smallest integer $\geq x$, and $d > 0$ is chosen in such a way that, when $\Delta = \Delta_1$, $\beta = \beta_1$. (The choice of d will be addressed below.)

Take $n - n_0$ additional observations $Z_{n_0+1}, Z_{n_0+2}, \cdots, Z_n$ and calculate

$$\overline{Z} = \frac{1}{n}\sum_{i=1}^{n} Z_i,$$

and

$$T = \frac{\sqrt{n}\overline{Z}}{S}.$$

Then reject $H_0 : \mu_1 = \mu_2$ in favor of $H_1 : \mu_1 \neq \mu_2$ if and only if $|T| > L$ where L is obtained from the condition

$$\frac{\alpha}{2} = \int_L^\infty f_{n_0}(x)dx,$$

where $f_{n_0}(x)$ denotes the probability density function of a Student's t random variable with $n_0 - 1$ degrees of freedom.

Note that total number of observations here (let us denote it by M) is $2n$, which is of course a random variable.

The Choice of d.

The power function of the test is obtained as

$$\beta\left(\frac{\Delta}{\sqrt{d}}\right) = \int_{-\infty}^{L_1} f_{n_0}(x)dx + \int_{L_2}^\infty f_{n_0}(x)dx$$

where $r = n_0 - 1$, $L_1 = -L - \frac{\Delta}{\sqrt{d}}$, $L_2 = L - \frac{\Delta}{\sqrt{d}}$. Once the desired α is chosen, L is obtained from

$$\frac{\alpha}{2} = \int_L^\infty f_{n_0}(x)dx.$$

Then for a desired value β_1 for $\beta\left(\frac{\Delta}{\sqrt{d}}\right)$, the integral equation

$$\int_{-\infty}^{-L-\Delta/\sqrt{d}} f_{n_0}(x)dx + \int_{L-\Delta/\sqrt{d}}^\infty f_{n_0}(x)dx = \beta_1$$

is solved for $\Delta/\sqrt{d}$. Finally, by using the value Δ_1 for Δ in $\Delta/\sqrt{d}$, we obtain d.

Chapman (1950) Procedure

Take initial samples $X_1, X_2, \cdots, X_{n_0}$ and $Y_1, Y_2, \cdots, Y_{n_0}$ (both of size $n_0 \geq 2$) from the normal populations $N(\mu_1, \sigma_1^2)$ and $N(\mu_2, \sigma_2^2)$ respectively, and calculate

$$\overline{X}(n_0) = \frac{1}{n_0} \sum_{i=1}^{n_0} X_i,$$

$$\overline{Y}(n_0) = \frac{1}{n_0} \sum_{i=1}^{n_0} Y_i,$$

$$S_1^2 = \frac{1}{n_0 - 1} \sum_{i=1}^{n_0} (X_i - \overline{X}(n_0))^2$$

$$S_2^2 = \frac{1}{n_0 - 1} \sum_{i=1}^{n_0} (Y_i - \overline{Y}(n_0))^2$$

$$n_1 = \max\left(n_0 + 1, [S_1^2/h^2]\right), \qquad \text{and}$$

$$n_2 = \max\left(n_0 + 1, [S_2^2/h^2]\right),$$

where $[x]$ denotes the smallest integer $\geq x$ and $h > 0$ is chosen in such a way that when $\Delta = \Delta_1$ then $\beta = \beta_1$. (The choice of h will be discussed in detail below.)

Take $n_1 - n_0$ additional observations $X_{n_0+1}, X_{n_0+2}, \cdots, X_{n_1}$ from $N(\mu_1, \sigma_1^2)$ and take $n_2 - n_0$ additional observations $Y_{n_0+1}, Y_{n_0+2}, \cdots, Y_{n_2}$ from $N(\mu_2, \sigma_2^2)$ and then calculate

$$\overline{X}(n_1 - n_0) = \frac{1}{n_1 - n_0} \sum_{i=n_0+1}^{n_1} X_i,$$

$$\overline{Y}(n_2 - n_0) = \frac{1}{n_2 - n_0} \sum_{i=n_0+1}^{n_2} Y_i,$$

$$\tilde{X} = b_1 \overline{X}(n_0) + b_2 \overline{X}(n_1 - n_0),$$

$$\tilde{Y} = c_1 \overline{Y}(n_0) + c_2 \overline{Y}(n_2 - n_0)$$

where

$$b_1 = \frac{n_0}{n_1}\left(1 + \left(1 - \frac{n_1}{n_0}\left(1 - \frac{n_1 - n_0}{S_1^2/h^2}\right)\right)^{\frac{1}{2}}\right),$$

$$b_2 = 1 - b_1,$$

$$c_1 = \frac{n_0}{n_2}\left(1 + \left(1 - \frac{n_2}{n_0}\left(1 - \frac{n_2 - n_0}{S_2^2/h^2}\right)\right)^{\frac{1}{2}}\right),$$

and

$$c_2 = 1 - c_1.$$

Then reject $H_0 : \mu_1 = \mu_2$ in favor of $H_1 : \mu_1 \neq \mu_2$ if and only if

$$|\tilde{X} - \tilde{Y}| > \frac{c_{1-\frac{\alpha}{2}}(n_0)}{1/h}$$

where $c_{1-\gamma}(n_0)$ is the value of c such that

$$\int_{-\infty}^{\infty} F_{n_0}^{k-1}(z + c)f_{n_0}(z)dz = 1 - \gamma, \qquad k = 2$$

where $F_{n_0}(\cdot)$ and $f_{n_0}(\cdot)$ are respectively the cumulative distribution function (c.d.f.) and probability density function (p.d.f.) of a Student's t random variable with $n_0 - 1$ degrees of freedom, and $h > 0$ is chosen in such a way that when $\triangle = \triangle_1$ then $\beta = \beta_1$.

Note that the total number of observations here (let us denote it by N) is $n_1 + n_2$, which is also a random variable.

The Choice of h

Denoting the integral $\int_{-\infty}^{\infty} F_{n_0}(z + h)f_{n_0}(z)$ by $P_{n_0}(h)$, the power function of the test is

$$\beta(\frac{\triangle}{h}) = P_{n_0}\left(-c_{1-\frac{\alpha}{2}}(n_0) - \frac{\triangle}{h}\right) + P_{n_0}\left(-c_{1-\frac{\alpha}{2}}(n_0) + \frac{\triangle}{h}\right).$$

By choosing a suitable level α, $c_{1-\frac{\alpha}{2}}(n_0)$ is obtained from the integral equation

$$\int_{-\infty}^{\infty} F_{n_0}(z + c)f_{n_0}(z)dz = 1 - \frac{\alpha}{2}.$$

Then for a desired value β_1 for $\beta(\frac{\triangle}{h})$, the equation

$$P_{n_0}\left(-c_{1-\frac{\alpha}{2}}(n_0) - \frac{\triangle}{h}\right) + P_{n_0}\left(-c_{1-\frac{\alpha}{2}}(n_0) + \frac{\triangle}{h}\right) = \beta_1$$

is solved for $\frac{\triangle}{h}$. Finally, by using $\triangle_1$ for $\triangle$ in $\frac{\triangle}{h}$, we obtain h.

For $x < 0$, we use the relation

$$P_{n_0}(-x) = 1 - P_{n_0}(x)$$

to find $P_{n_0}(x)$.

3 Efficiency Comparison, Selected Tables

Since both the procedures take the same number of observations in the initial samples and both satisfy the same requirements on the level and power, it is reasonable to compare them via their expected total sample sizes.

N and M are the total numbers of observations for the Chapman (1950) and Prokof'yev-Shishkin (1974) procedures, respectively, in order to attain power β for testing $H_0 : \mu_1 = \mu_2$ against $H_1 : \mu_1 \neq \mu_2$ (when σ_1^2 and σ_2^2 are unknown and unequal) for specified $\triangle$. To compare $\mu_N \equiv E(N)$ and $\mu_M \equiv E(M)$, we estimate the ratio

$$R \equiv \frac{\mu_N}{\mu_M}$$

on the basis of a simple random sample (N_i, M_i) of (N, M), $i = 1, 2, \cdots, s$. (See Appendix on ratio estimation.)

From the definition of R, it is clear that if $R = 1$, then on the average both procedures perform equally well; if $R < 1$, then on the average Chapman's procedure is better than that of Prokof'yev-Shishkin; and if $R > 1$, then the Prokof'yev-Shishkin procedure is better. In order to be able to estimate R (using the classical ratio estimator $\hat{R}$), we first need values of d and h for a specified $\triangle$ for the two procedures.

For selected values of n_0, α and β ($n_0 = 5, 10, 30; \alpha = 0.01, 0.05; \beta = 0.50, 0.75, 0.90, 0.95, 0.99$), we have computed values of $\frac{\triangle}{\sqrt{d}}$ and $\frac{\triangle}{h}$ (Table 1). A typical entry in Table 1 looks like: 4.60306394 (5), which means that the value reported is accurate up to ± 5 units in the last digit reported. For a specified value of $\triangle$, values of d (for use in the Prokof'yev-Shishkin procedure) and h (for use in Chapman's procedure) can be obtained from this table, accurate up to at least 5 decimal places.

For $\alpha = 0.01, 0.05$; $n_0 = 5, 10$; $\triangle = 0.25, 0.50, 1.00, 1.50, 2.00$; $\beta = 0.50, 0.75, 0.90, 0.95, 0.99$; $\sigma_1^2 = 0.10, 1.0, 10.0$; and $\sigma_2^2/\sigma_1^2 = 1.0, 1.5, 2.0, 5.0, 10.0, 20.0, 100.0$, and ∞, we have computed values of $\hat{R}$ and have reported them in Table 2. The last column of Table 2 corresponding to $\sigma_2^2/\sigma_1^2 = \infty$ gives approximate values of R on the basis of the relationship

$$R = \frac{E(N)}{E(M)} = \frac{E(n_1 + n_2)}{E(2n)} \doteq \frac{\frac{\sigma_1^2}{h^2} + \frac{\sigma_2^2}{h^2}}{2\frac{\sigma_1^2 + \sigma_2^2}{d}} = \frac{d}{2h^2}.$$

Table 2 was computed using random number generator UNI to generate a stream of uniform random variables on (0,1), the Box-Muller (1958) transformation to obtain two independent sequences of normal random variables from $N(\mu_1, \sigma_1^2)$ and $N(\mu_2, \sigma_2^2)$, 64-point Gauss-Legendre quadrature (see

Stroud and Secrest 1966), and the scheme of Dudewicz et al. (1975) for evaluating the integrals involved. Entries in Table 2 are given as "*" if there is no significant difference between the two procedures (i.e., the ratio is 1), in order to make it easier to see, at a glance, where the breaks occur between values smaller-than/greater-than 1 in that table.

4 Recommendations: Which Method to Use

Recall (see Section 3) that an entry greater than 1 in Table 2 indicates the Prokof'yev-Shishkin procedure should be recommended (as it has the same power and level guarantees, with a lower expected sample size total); an entry less than 1 indicates the Chapman procedure should be recommended; and an entry * indicates that there is no significant difference between the two procedures (so one may use whichever is simpler in the application setting without suffering a larger sample size due to that choice).

A careful inspection of Table 2 shows that Chapman's procedure improves in comparison to Prokof'yev-Shishkin's as $\triangle \downarrow$, $\sigma_2^2/\sigma_1^2 \uparrow$, $\sigma_1^2 \uparrow$, $\beta \uparrow$, $\alpha \downarrow$, $n_0 \downarrow$. Also, if one has fixed values of n_0, α, $\triangle$ and β, then, in some cases, one of the two procedures can be clearly recommended. For example, if $n_0 = 5$, $\alpha = .01$, $\triangle = 1.00$ and $\beta = 0.99$, then Chapman's procedure performs slightly better than that of Prokof'yev-Shishkin (no matter what the variance-ratio is); whereas, if $n_0 = 10, \alpha = .05, \triangle = .05$, and $\beta = .90$, then the Prokof'yev-Shishkin procedure performs slightly better than that of Chapman (again no matter what the variance-ratio).

We performed a regression analysis for the understanding of the relationship between R and n_0, σ_1^2, α, $\triangle$, β and σ_2^2/σ_1^2; and also for predicting R so that a procedure can be recommended. The fitted model was

$$\begin{aligned} \hat{R} = \quad & 0.93778 + 0.00136n_0 - 0.00529\sigma_1^2 + 1.91247\alpha \\ & +0.03897 \,\triangle -0.02469\beta - 0.00025\sigma_2^2/\sigma_1^2 \end{aligned}$$

and it explained 47.10% of the total variation along with 0.0556 standard error of estimate. Examination of residuals revealed normality of errors but non-constancy of error variance. Use of $1/R$ as the dependent variable fixed this problem, and normality of errors was also retained. The prediction equation was

$$\begin{aligned} \widehat{1/R} = \quad & 1.07584 - 0.00169n_0 + 0.00485\sigma_1^2 - 1.93546\alpha \\ & -0.03536 \,\triangle +0.01646\beta + 0.00023\sigma_2^2/\sigma_1^2 \end{aligned}$$

and it explained 50.05% of the total variation along with 0.0507 standard error of estimate. At 0.05 level of significance, all the variables (namely

n_0, σ_1^2, α, $\triangle$, β and σ_2^2/σ_1^2) are significant. Note that the patterns we saw above in the parameters are reflected in the regression equation: positive signs for the three noted for increase and negative signs for the three noted for decrease.

To allow for interactions among pairs of variables, we added crossproducts of those variables. Use of R as dependent variable did not satisfy the assumption of constancy of error variance, but $1/R$ did. The prediction equation was

$$
\begin{aligned}
\widehat{1/R} \;=\; & 1.24118 - 0.01242n_0 + 0.00314\sigma_1^2 - 5.58503\alpha - 0.09026\,\triangle \\
& -0.08231\beta + 0.00043\sigma_2^2/\sigma_1^2 + 0.00036(n_0 * \sigma_1^2) \\
& +0.24814(n_0 * \alpha) - 0.00106(n_0 * \triangle) + 0.00395(n_0 * \beta) \\
& -0.000008(n_0 * \sigma_2^2/\sigma_1^2) - 0.00616(\sigma_1^2 * \alpha) + 0.00607(\sigma_1^2 * \triangle) \\
& -0.00784(\sigma_1^2 * \beta) - 0.000039(\sigma_1^2 * \sigma_2^2/\sigma_1^2) + 0.15335(\alpha * \triangle) \\
& +2.07799(\alpha * \beta) - 0.00248(\alpha * \sigma_2^2/\sigma_1^2) + 0.03807(\triangle * \beta) \\
& +0.00023(\triangle * \sigma_2^2/\sigma_1^2) - 0.00021(\beta * \sigma_2^2/\sigma_1^2)
\end{aligned}
$$

and it explained 62.89% of the total variation along with 0.0438 standard error of estimate. Using all possible interactions in the model improved the coefficient of determination by only 1.26%. Note that while the % of variability explained is not close to 100% (it is 62.89% as noted above), the model is deemed to fit well since $1/R = \mu_M/\mu_N$ is predicted with a standard deviation of only .0438. (Thus, e.g., a 95% confidence interval for the ratio has a proportional closeness of less than 10%, which is quite good prediction.)

If $\widehat{1/R}$ is less than 1 for given values of n_0, σ_1^2, α, $\triangle$, β and σ_2^2/σ_1^2, then Prokof'yev-Shishkin's procedure is better; and if $\widehat{1/R}$ is greater than 1, then Chapman's procedure is better; and if $\widehat{1/R}$ is 1, then both the procedures are equivalent.

The average of all $1/R$ values is 1.00393 with standard deviation 0.07134. Extreme values of $1/R$ are 0.77328 and 1.13379. A referee "was somewhat surprised to find out that Prokof'yev-Shishkin's test [ever] performed better than Chapman's test". Indeed, when we initially undertook this work we expected to be able to show uniform superiority of Chapman's test. If one looks to the case of known variances and asks: is it better to take n data pairs, or n_1 from source 1 and n_2 from source 2, one obtains mild insight: if the level and power are to be the same and the unpaired case allocates samples so that $\sigma_1^2/n_1 = \sigma_2^2/n_2$, then one finds $n_1 + n_2 = 2n$, or $R = 1$. However (see below) Chapman's test is to be recommended when high power is needed for detecting a small difference.

While choosing a method is a function of the true (and usually unknown) parameters, we emphasize that both the methods are exact and the choice between them involves only how much sample size one will need to pay for an exact attainment of desired level and power (while with the approximate procedures one does not know what level and power one has).

For a recommendation, we recommend that Chapman's procedure be used. The reason for this is that R is the ratio $E(N)/E(M)$ where N is the sample size needed by the Chapman procedure. From Table 2 we see that R is typically less than 1 when power in high and $\triangle$ is low (experimenter's requirements). In virtually all cases the expected sample sizes are within 10% of each other, however, so the loss of an incorrect choice is small; but, if variance ratio is very far from 1, then Chapman's procedure is preferred.

One can take more data-driven approach also. Once the specifications are set, both procedures use n_0 as first-stage sample size. So, take n_0 samples, estimate σ_1^2 and σ_2^2, use Table 2 and decide which procedure to choose. Then take an appropriate number of additional observations and make a decision.

Appendix on Ratio-Estimation in Sampling Theory

Since both the procedures take the same number of observations in the initial samples and both satisfy the same requirements on the level and power, it is reasonable to compare them via their expected total sample sizes.

N and M are the total numbers of observations for the Chapman (1950) and Prokof'yev-Shishkin (1974) procedures, respectively, in order to attain power β_1 for testing $H_0 : \mu_1 = \mu_2$ against $H_1 : \mu_1 \neq \mu_2$ (when σ_1^2 and σ_2^2 are unknown and unequal) for specified $\triangle_1$. To compare $\mu_N \equiv E(N)$ and $\mu_M \equiv E(M)$, we estimate the ratio

$$R \equiv \frac{\mu_N}{\mu_M}$$

on the basis of a simple random sample (N_i, M_i) of (N, M), $i = 1, 2, \cdots, s$. From the literature (e.g. Cochran 1977, Des Raj 1968, and Murthy 1967) on ratio-estimation in sampling theory, for this case the following two ratio-estimators can be considered:

(1) The Classical Ratio-Estimator $\hat{R}$;

(2) The Ratio-Type Estimator $\overline{R}$.

Let $\hat{\mu}_N$ and $\hat{\mu}_M$ be unbiased estimators for μ_N and μ_M respectively and let

$$e_N = \frac{\hat{\mu}_N - \mu_N}{\mu_N}, e_M = \frac{\hat{\mu}_M - \mu_M}{\mu_M}.$$

For deriving expressions for bias and variance of estimators $\hat{R}$ and $\overline{R}$ we need the two assumptions:

(a) $\left| \frac{\hat{\mu}_M - \mu_M}{\mu_M} \right| < 1$ i.e. $\hat{\mu}_M$ lies between 0 and 2 μ_M;

and

(b) terms of the degree greater than 2 in (e_N, e_M) in the expansion for $(1 + e_N)(1 + e_M)^{-1}$ can be neglected.

According to Murthy (1967), both these assumptions are likely to be valid only for large samples unless the population is fairly homogeneous (in which case even with a smaller sample size the two assumptions may be valid).

The Classical Ratio-Estimator $\hat{R}$ is defined as $\hat{R} = \overline{N}/\overline{M}$ where $\overline{N} = \sum_{i=1}^{s} N_i/s$ and $\overline{M} = \sum_{i=1}^{s} M_i/s$. Since N and M vary from sample to sample, Cochran (1977) has noted that the distribution of $\hat{R}$ is annoyingly intractable. For practical applications, few results are available. The principal available results about $\hat{R}$ are:

(i) it is consistent; and

(ii) it is biased.

Under the above-mentioned assumptions, the approximate bias and variance of $\hat{R}$ are

$$B(\hat{R}) = \frac{R}{s} \left[C_M^2 - \rho \, C_M C_N \right]$$

$$V(\hat{R}) = \frac{R^2}{s} \left[C_M^2 - 2\rho \, C_M C_N + C_N^2 \right]$$

where ρ is the correlation between N and M; and C_M and C_N are the coefficients of variation of M and N, respectively. From these expressions, it is clear that the approximate bias and the variance decrease with increasing s. It may also be seen that the ratio of the squared bias to the variance,

$$\frac{B^2(\hat{R})}{V(\hat{R})} = \frac{1}{s} \cdot \frac{(C_M^2 - \rho \, C_M C_N)^2}{(C_M^2 - 2\rho \, C_M C_N + C_N^2)},$$

decreases with increasing s, hence for large samples the bias is negligible compared to the variance.

(iii) Since unbiased estimators of σ_M^2, σ_N^2, and cov (M, N) are given by

$$s_M^2 = \frac{1}{s-1} \sum_{i=1}^{s} (M_i - \overline{M})^2$$

$$s_N^2 = \frac{1}{s-1} \sum_{i=1}^{s} (N_i - \overline{N})^2, \text{ and}$$

$$S_{M,N} = \frac{1}{s-1} \sum_{i=1}^{s} (M_i - \overline{M})(N_i - \overline{N})$$

respectively, we obtain estimators of the bias and variance of $\hat{R}$ as

$$b(\hat{R}) = \frac{1}{\overline{M}^2} \sum_{i=1}^{s} \frac{M_i(\hat{R}M_i - N_i)}{s(s-1)}$$

and

$$v(\hat{R}) = \frac{1}{\overline{M}^2} \sum_{i=1}^{s} \frac{(N_i - \hat{R}M_i)^2}{s(s-1)}.$$

(iv) The limiting distribution as $s \to \infty$ of $\hat{R}$ is normal, subject to "mild" restrictions on the type of population from which sampling is being done. As a working rule, Cochran suggests the large sample results may be used if the sample size exceeds 30 and is also large enough so that the coefficients of variation of $\overline{M}$ and $\overline{N}$ are both less than 10%.

(v) If the sample is large enough that the normal approximation is valid, then confidence limits for R may be obtained as

$$\hat{R} \pm z\sqrt{v(\hat{R})}$$

where z is the normal deviate corresponding to the chosen confidence coefficient.

The Ratio-Type Estimator $\overline{R}$ is defined as $\overline{R} = \frac{1}{s} \sum_{i=1}^{s} \frac{N_i}{M_i}$. We observe that:

(i) It is not consistent;
(ii) It is biased. The bias of $\overline{R}$ is

$$B(\overline{R}) = -\frac{\text{cov}\left(\frac{N}{M}, M\right)}{E(M)}.$$

It is not desirable to use $\overline{R}$ because the bias of $\overline{R}$ does not depend on the sample size and hence does *not* decrease with increasing s (unlike the bias of the estimator $\hat{R}$). (Although Goodman and Hartley 1958 have corrected $\overline{R}$ for bias to obtain an unbiased estimator, they require knowledge of the denominator of R. In our case, this denominator is not known.)

For our purposes, before recommending the use of $\hat{R}$ or $\overline{R}$, let us consider some applications.

Suppose we have two normal populations $N(\mu_1, \sigma_1^2)$ and $N(\mu_2, \sigma_2^2)$, and we wish to have an $\alpha = 0.01$ level test of $H_0 : \mu_1 = \mu_2$ against $H_1 : \mu_1 \neq \mu_2$

(where σ_1^2 and σ_2^2 are assumed unknown and unequal) such that, when $\Delta_1 = 0.25$, $\beta_1 = 0.50$. Suppose we agree to take initial samples of size $n_0 = 5$ from the two populations. Then

$$\frac{\Delta_1}{\sqrt{d}} = 4.60306394 \quad \text{and} \quad \frac{\Delta_1}{h} = 6.14878057,$$

hence

$$d = 0.0029497 \quad \text{and} \quad h = 0.0406584.$$

To help in the choice of an appropriate estimator, note that

$$R \;=\; \frac{E(N)}{E(M)} = \frac{E(n_1 + n_2)}{E(2n)}$$

$$\doteq \; \frac{\frac{\sigma_1^2}{h^2} + \frac{\sigma_2^2}{h^2}}{2\frac{\sigma_1^2 + \sigma_2^2}{d}}$$

$$= \; \frac{d}{2h^2} = \frac{1}{2}\frac{\left(\frac{\Delta_1}{h}\right)^2}{\left(\frac{\Delta_1}{\sqrt{d}}\right)^2} = 0.8922.$$

Applications

For the applications described here we used random number generator UNI (see Dudewicz and Ralley 1981) to generate uniform $(0,1)$ random variables. Various results are summarized in Table 3, where we have taken $\alpha = 0.01$, $\Delta_1 = 0.25$, $\beta_1 = 0.50$, $n_0 = 5$, $\mu_1 = 0$, $\mu_2 = 0.25$, and $\sigma_1^2 = 0.10$. The seven columns in Table 3 correspond to seven values of σ_2^2 : 0.10, 0.15, 0.20, 0.50, 1.00, 2.00, 10.00; so that the variance-ratio $\frac{\sigma_2^2}{\sigma_1^2}$ takes the values: $1.0, 1.5, 2.0, 5.0, 10.0, 20.0, 100.0$.

For the first column in Table 3, we generated 10,000 sets of 10 uniform $(0,1)$ random numbers each. Within each set of 10 uniform $(0,1)$ random numbers, we applied the Box-Muller (1958) transformation to obtain two independent initial samples (each of size 5) from $N(\mu_1, \sigma_1^2)$ and $N(\mu_2, \sigma_2^2)$. For the first set, the 10 uniform $(0,1)$ random numbers generated were:

$$0.40461671$$
$$0.69982326$$
$$0.38076901$$
$$0.33324194$$
$$0.48070574$$
$$0.57326061$$

0.96209621
0.10630631
0.11367542
0.26607591

with the following corresponding independent initial samples.

$N(\mu_1, \sigma_1^2)$	$N(\mu_2, \sigma_2^2)$
−0.40443082	0.11809584
0.38069714	0.03049569
−0.17002891	−0.09291420
0.05444895	0.31901789
0.65609427	0.18350285

On the basis of these two samples, the total sample sizes required by the two procedures to satisfy the same requirements on the level and the power are $N = 118$ and $M = 124$ (where we have used $h = 0.0406584$ and $d = 0.0029497$). So, we have $(N_1, M_1) = (118,124)$ for the first set of 10 uniform (0,1) random numbers. In this way, the 10,000 sets produce a random sample (N_i, M_i) of size $s = 10,000$ for (N, M). So, for the construction of first column in Table 3, we have available 10,000 values of (N, M). Similarly, we have 10,000 values of (N, M) for every column in Table 3, and compute various statistics and confidence intervals; these results are summarized in Table 3.

Recommendations

We observe that $\hat{R}$ is closer to $\frac{d}{2h^2}$ than is $\overline{R}$, and recommend the use of $\hat{R}$ for our case. (Also, $\overline{R}$ is not desirable for reasons noted in the previous section.) Since large sample results apply, we recommend the use of $\hat{R}$ along with a 95% confidence interval for R based on $\hat{R}$.

Table 1: Values of $\Delta/\sqrt{d}$ and Δ/h for selected n_0, α and β.

n_0	α		β 0.50	0.75	0.90
5	0.01	$\Delta/\sqrt{d}$	4.60306394(5)	5.34373785(7)	6.1354985(1)
		Δ/h	6.14878057(8)	7.28664517(11)	8.4392067(2)
5	0.05	$\Delta/\sqrt{d}$	2.76955137(5)	3.51114784(7)	4.3007897(1)
		Δ/h	3.93235615(8)	5.07194398(11)	6.2233639(2)
10	0.01	$\Delta/\sqrt{d}$	3.24969183(5)	3.95247233(6)	4.6327802(1)
		Δ/h	4.40990381(7)	5.44030230(10)	6.4085158(1)
10	0.05	$\Delta/\sqrt{d}$	2.26029947(5)	2.96396283(6)	3.6444189(1)
		Δ/h	3.17854034(7)	4.20981891(10)	5.1782189(1)
30	0.01	$\Delta/\sqrt{d}$	2.75637820(5)	3.43942829(6)	4.0678191(1)
		Δ/h	3.84999659(7)	4.82670647(9)	5.7160507(1)
30	0.05	$\Delta/\sqrt{d}$	2.04483383(5)	2.72819705(6)	3.3566384(1)
		Δ/h	2.87965245(7)	3.85666190(9)	4.7460413(1)

n_0	α		β 0.95	0.99
5	0.01	$\Delta/\sqrt{d}$	6.7328783(3)	8.3392776(22)
		Δ/h	9.2541947(4)	11.2755190(28)
5	0.05	$\Delta/\sqrt{d}$	4.8946023(3)	6.4807478(22)
		Δ/h	7.0360317(4)	9.0435714(28)
10	0.01	$\Delta/\sqrt{d}$	5.0828480(2)	6.0710773(12)
		Δ/h	7.0240694(3)	8.2972840(15)
10	0.05	$\Delta/\sqrt{d}$	4.0944406(2)	5.0822634(12)
		Δ/h	5.7938056(3)	7.0670078(15)
30	0.01	$\Delta/\sqrt{d}$	4.4555126(2)	5.2184071(8)
		Δ/h	6.2569812(2)	7.2979426(11)
30	0.05	$\Delta/\sqrt{d}$	3.7443418(2)	4.5072433(8)
		Δ/h	5.2869770(2)	6.3279414(11)

Taneja and Dudewicz

Table 2a: Comparison of Prokof'yev-Shishkin and Chapman procedures for selected n_0, σ_1^2, α, Δ, β and σ_2^2/σ_1^2: $n_0 = 5$, $\sigma_1^2 = 0.10$, $\alpha = .01$.

		σ_2^2/σ_1^2							
Δ	β	1.0	1.5	2.0	5.0	10.0	20.0	100.0	∞
0.25	0.99	0.9116	0.9182	0.9148	0.9097	0.9133	0.9104	0.9143	0.9141
	0.95	0.9464	0.9443	0.9426	0.9496	0.9423	0.9463	0.9433	0.9446
	0.90	0.9479	0.9463	0.9505	0.9451	0.9429	0.9431	0.9462	0.9460
	0.75	0.9248	0.9297	0.9373	0.9254	0.9299	0.9286	0.9282	0.9297
	0.50	0.9015	0.8828	0.8867	0.8916	0.8909	0.8945	0.8923	0.8922
0.50	0.99	0.9058	0.9179	0.9126	0.9179	0.9144	0.9154	0.9131	0.9141
	0.95	0.9443	0.9449	0.9475	0.9476	0.9464	0.9459	0.9444	0.9446
	0.90	0.9419	0.9445	0.9506	0.9473	0.9433	0.9453	0.9452	0.9460
	0.75	0.9439	0.9472	0.9266	0.9326	0.9364	0.9285	0.9295	0.9297
	0.50	0.9013	0.9028	0.9017	0.8971	0.8930	0.8963	0.8922	0.8922
1.00	0.99	0.9415	0.9250	0.9241	0.9164	0.9153	0.9140	0.9163	0.9141
	0.95	0.9865	0.9725	0.9712	0.9563	0.9580	0.9505	0.9447	0.9446
	0.90	*	*	0.9758	0.9692	0.9686	0.9539	0.9479	0.9460
	0.75	1.0344	1.0086	*	0.9727	0.9501	0.9435	0.9329	0.9297
	0.50	1.0799	1.0363	1.0226	0.9691	0.9393	0.9161	0.8973	0.8922
1.50	0.99	1.0105	0.9825	0.9775	0.9553	0.9384	0.9253	0.9148	0.9141
	0.95	1.1034	1.0780	1.0601	1.0204	0.9932	0.9730	0.9489	0.9446
	0.90	1.1324	1.1097	1.0978	1.0515	1.0111	0.9831	0.9539	0.9460
	0.75	1.1693	1.1463	1.1309	1.0836	1.0316	0.9855	0.9423	0.9297
	0.50	1.1908	1.1761	1.1653	1.1113	1.0473	0.9859	0.9116	0.8922
2.00	0.99	1.1242	1.0814	1.0670	1.0231	0.9787	0.9502	0.9215	0.9141
	0.95	1.1787	1.1617	1.1490	1.1164	1.0679	1.0124	0.9599	0.9446
	0.90	1.1905	1.1789	1.1675	1.1477	1.0963	1.0359	0.9651	0.9460
	0.75	1.1977	1.1940	1.1890	1.1756	1.1302	1.0569	0.9596	0.9297
	0.50	1.1997	1.1992	1.1979	1.1906	1.1564	1.0729	0.9354	0.8922

Table 2b: Comparison of Prokof'yev-Shishkin and Chapman procedures for selected n_0, σ_1^2, α, Δ, β and σ_2^2/σ_1^2: $n_0 = 5$, $\sigma_1^2 = 0.10$, $\alpha = .05$.

		σ_2^2/σ_1^2							
Δ	β	1.0	1.5	2.0	5.0	10.0	20.0	100.0	∞
0.25	0.99	0.9767	0.9716	0.9763	0.9756	0.9747	0.9705	0.9730	0.9736
	0.95	1.0383	1.0220	1.0336	1.0326	1.0346	1.0328	1.0341	1.0332
	0.90	1.0393	1.0495	1.0495	1.0459	1.0519	1.0485	1.0458	1.0469
	0.75	1.0503	1.0367	1.0466	1.0424	1.0385	1.0398	1.0430	1.0433
	0.50	1.0518	1.0131	*	1.0106	1.0067	1.0118	1.0088	1.0080
0.50	0.99	0.9787	0.9720	0.9722	0.9689	0.9743	0.9756	0.9737	0.9736
	0.95	1.0371	1.0272	1.0310	1.0394	1.0340	1.0376	1.0337	1.0332
	0.90	1.0585	1.0352	1.0495	1.0479	1.0557	1.0444	1.0464	1.0469
	0.75	1.0643	1.0596	1.0586	1.0559	1.0481	1.0462	1.0432	1.0433
	0.50	1.0685	1.0592	1.0482	1.0359	1.0289	1.0147	1.0130	1.0080
1.00	0.99	1.0189	1.0092	*	0.9880	0.9825	0.9797	0.9749	0.9736
	0.95	1.1125	1.0956	1.0891	1.0799	1.0637	1.0509	1.0378	1.0332
	0.90	1.1468	1.1284	1.1192	1.1231	1.0926	1.0763	1.0526	1.0469
	0.75	1.1793	1.1728	1.1717	1.1640	1.1353	1.1009	1.0562	1.0433
	0.50	1.1971	1.1941	1.1922	1.2095	1.1839	1.1165	1.0346	1.0080
1.50	0.99	1.1263	1.0957	1.0863	1.0530	1.0240	*	0.9802	0.9736
	0.95	1.1862	1.1806	1.1806	1.1859	1.1465	1.0999	1.0475	1.0332
	0.90	1.1967	1.1928	1.1930	1.2281	1.2059	1.1468	1.0689	1.0469
	0.75	1.1997	1.1996	1.2007	1.2426	1.2538	1.2019	1.0796	1.0433
	0.50	1.2000	1.2000	1.2001	1.2157	1.2610	1.2420	1.0759	1.0080
2.00	0.99	1.1864	1.1723	1.1698	1.1480	1.0999	1.0451	0.9916	0.9736
	0.95	1.1996	1.1993	1.2001	1.2314	1.2342	1.1764	1.0677	1.0332
	0.90	1.2000	1.1997	1.2005	1.2375	1.2763	1.2290	1.0935	1.0469
	0.75	1.2000	1.2000	1.2002	1.2188	1.2741	1.2794	1.1195	1.0433
	0.50	1.2000	1.2000	1.2000	1.2017	1.2317	1.2847	1.1293	1.0080

Table 2c: Comparison of Prokof'yev-Shishkin and Chapman procedures for selected n_0, σ_1^2, α, Δ, β and σ_2^2/σ_1^2: $n_0 = 5$, $\sigma_1^2 = 1.00$, $\alpha = .01$

		σ_2^2/σ_1^2							
Δ	β	1.0	1.5	2.0	5.0	10.0	20.0	100.0	∞
0.25	0.99	0.9114	0.9181	0.9147	0.9096	0.9133	0.9104	0.9143	0.9141
	0.95	0.9462	0.9442	0.9425	0.9495	0.9423	0.9463	0.9433	0.9446
	0.90	0.9477	0.9461	0.9504	0.9450	0.9429	0.9431	0.9462	0.9460
	0.75	0.9243	0.9294	0.9371	0.9253	0.9298	0.9285	0.9282	0.9297
	0.50	0.9006	0.8821	0.8861	0.8913	0.8907	0.8944	0.8923	0.8922
0.50	0.99	0.9047	0.9172	0.9120	0.9175	0.9142	0.9153	0.9131	0.9141
	0.95	0.9427	0.9437	0.9466	0.9470	0.9461	0.9457	0.9444	0.9446
	0.90	0.9391	0.9427	0.9492	0.9465	0.9429	0.9450	0.9452	0.9460
	0.75	0.9384	0.9431	0.9229	0.9310	0.9354	0.9279	0.9294	0.9297
	0.50	0.8876	0.8931	0.8937	0.8928	0.8907	0.8950	0.8919	0.8922
1.00	0.99	0.9224	0.9113	0.9137	0.9104	0.9119	0.9122	0.9159	0.9141
	0.95	0.9443	0.9435	0.9468	0.9419	0.9497	0.9459	0.9438	0.9446
	0.90	0.9430	0.9501	0.9396	0.9471	0.9507	0.9471	0.9464	0.9460
	0.75	0.9305	0.9305	0.9305	0.9318	0.9267	0.9303	0.9302	0.9297
	0.50	0.9023	0.8897	0.8992	0.8912	0.8930	0.8906	0.8919	0.8922
1.50	0.99	0.9097	0.9112	0.9178	0.9203	0.9171	0.9138	0.9123	0.9141
	0.95	0.9460	0.9487	0.9496	0.9449	0.9467	0.9473	0.9435	0.9446
	0.90	0.9443	0.9520	0.9552	0.9502	0.9468	0.9471	0.9461	0.9460
	0.75	0.9375	0.9375	0.9376	0.9316	0.9296	0.9276	0.9298	0.9297
	0.50	0.9033	0.8970	0.9022	0.8898	0.8905	0.8930	0.8914	0.8922
2.00	0.99	0.9270	0.9126	0.9161	0.9210	0.9150	0.9144	0.9140	0.9141
	0.95	0.9535	0.9434	0.9476	0.9441	0.9511	0.9450	0.9452	0.9446
	0.90	0.9521	0.9553	0.9441	0.9516	0.9480	0.9475	0.9455	0.9460
	0.75	0.9439	0.9545	0.9424	0.9331	0.9303	0.9299	0.9308	0.9297
	0.50	0.9308	0.9176	0.9107	0.9011	0.8979	0.8968	0.8933	0.8922

Table 2d: Comparison of Prokof'yev-Shishkin and Chapman procedures for selected n_0, σ_1^2, α, Δ, β and σ_2^2/σ_1^2: $n_0 = 5$, $\sigma_1^2 = 1.00$, $\alpha = .05$.

		σ_2^2/σ_1^2							
Δ	β	1.0	1.5	2.0	5.0	10.0	20.0	100.0	∞
0.25	0.99	0.9766	0.9716	0.9763	0.9755	0.9747	0.9705	0.9730	0.9736
	0.95	1.0384	1.0220	1.0337	1.0326	1.0346	1.0328	1.0341	1.0332
	0.90	1.0394	1.0497	1.0496	1.0459	1.0520	1.0486	1.0458	1.0469
	0.75	1.0501	1.0367	1.0467	1.0424	1.0384	1.0398	1.0430	1.0433
	0.50	1.0137	1.0117	1.0072	1.0097	1.0063	1.0115	1.0087	1.0080
0.50	0.99	0.9773	0.9711	0.9711	0.9683	0.9740	0.9754	0.9737	0.9736
	0.95	1.0331	1.0248	1.0289	1.0380	1.0331	1.0371	1.0336	1.0332
	0.90	1.0533	1.0309	1.0461	1.0454	1.0541	1.0435	1.0462	1.0469
	0.75	1.0468	1.0453	1.0474	1.0475	1.0430	1.0434	1.0426	1.0433
	0.50	*	1.0106	*	*	1.0114	1.0048	1.0109	1.0080
1.00	0.99	0.9783	0.9787	0.9770	0.9719	0.9729	0.9746	0.9739	0.9736
	0.95	1.0358	1.0383	1.0315	1.0351	1.0345	1.0350	1.0344	1.0332
	0.90	1.0508	1.0457	1.0384	1.0544	1.0462	1.0498	1.0471	1.0469
	0.75	1.0434	1.0497	1.0465	1.0410	1.0443	1.0466	1.0442	1.0433
	0.50	1.0168	1.0175	1.0157	1.0167	1.0118	1.0066	1.0096	1.0080
1.50	0.99	0.9821	0.9732	0.9753	0.9714	0.9739	0.9751	0.9741	0.9736
	0.95	1.0262	1.0374	1.0396	1.0342	1.0325	1.0323	1.0324	1.0332
	0.90	1.0573	1.0431	1.0485	1.0526	1.0526	1.0505	1.0469	1.0469
	0.75	1.0544	1.0615	1.0435	1.0625	1.0445	1.0489	1.0416	1.0433
	0.50	1.0610	1.0452	1.0448	1.0373	1.0222	1.0168	1.0093	1.0080
2.00	0.99	0.9710	0.9711	0.9792	0.9765	0.9745	0.9713	0.9753	0.9736
	0.95	1.0445	1.0402	1.0509	1.0367	1.0350	1.0339	1.0334	1.0332
	0.90	1.0657	1.0655	1.0632	1.0629	1.0569	1.0515	1.0476	1.0469
	0.75	1.0884	1.0763	1.0762	1.0672	1.0588	1.0550	1.0465	1.0433
	0.50	1.1261	1.1075	1.1024	1.0771	1.0519	1.0340	1.0132	1.0080

Table 2e: Comparison of Prokof'yev-Shishkin and Chapman procedures for selected n_0, σ_1^2, α, Δ, β and σ_2^2/σ_1^2: $n_0 = 5$, $\sigma_1^2 = 10.0$, $\alpha = .01$.

		σ_2^2/σ_1^2							
Δ	β	1.0	1.5	2.0	5.0	10.0	20.0	100.0	∞
0.25	0.99	0.9114	0.9181	0.9147	0.9096	0.9133	0.9104	0.9143	0.9141
	0.95	0.9462	0.9442	0.9425	0.9495	0.9423	0.9463	0.9433	0.9446
	0.90	0.9477	0.9461	0.9503	0.9450	0.9429	0.9431	0.9462	0.9460
	0.75	0.9243	0.9293	0.9370	0.9253	0.9298	0.9285	0.9282	0.9297
	0.50	0.9006	0.8820	0.8860	0.8913	0.8907	0.8944	0.8923	0.8922
0.50	0.99	0.9046	0.9172	0.9119	0.9175	0.9142	0.9153	0.9131	0.9141
	0.95	0.9426	0.9436	0.9466	0.9470	0.9461	0.9457	0.9444	0.9446
	0.90	0.9390	0.9426	0.9491	0.9464	0.9429	0.9450	0.9452	0.9460
	0.75	0.9382	0.9430	0.9228	0.9309	0.9354	0.9279	0.9294	0.9297
	0.50	0.8873	0.8929	0.8935	0.8927	0.8907	0.8950	0.8919	0.8922
1.00	0.99	0.9221	0.9111	0.9135	0.9103	0.9119	0.9121	0.9159	0.9141
	0.95	0.9439	0.9432	0.9466	0.9418	0.9496	0.9459	0.9438	0.9446
	0.90	0.9425	0.9498	0.9393	0.9469	0.9506	0.9471	0.9464	0.9460
	0.75	0.9298	0.9299	0.9299	0.9315	0.9265	0.9302	0.9301	0.9297
	0.50	0.9006	0.8882	0.8980	0.8906	0.8926	0.8904	0.8918	0.8922
1.50	0.99	0.9087	0.9104	0.9173	0.9201	0.9170	0.9137	0.9123	0.9141
	0.95	0.9445	0.9476	0.9487	0.9444	0.9465	0.9472	0.9434	0.9446
	0.90	0.9422	0.9504	0.9540	0.9494	0.9464	0.9469	0.9461	0.9460
	0.75	0.9331	0.9344	0.9351	0.9302	0.9288	0.9272	0.9297	0.9297
	0.50	0.8921	0.8897	0.8963	0.8864	0.8887	0.8920	0.8912	0.8922
2.00	0.99	0.9246	0.9109	0.9147	0.9202	0.9146	0.9141	0.9140	0.9141
	0.95	0.9488	0.9393	0.9447	0.9424	0.9501	0.9445	0.9451	0.9446
	0.90	0.9438	0.9495	0.9399	0.9488	0.9466	0.9468	0.9453	0.9460
	0.75	0.9271	0.9432	0.9338	0.9280	0.9274	0.9283	0.9305	0.9297
	0.50	0.8934	0.8908	0.8889	0.8892	0.8913	0.8933	0.8925	0.8922

Table 2f: Comparison of Prokof'yev-Shishkin and Chapman procedures for selected n_0, σ_1^2, α, Δ, β and σ_2^2/σ_1^2: $n_0 = 5$, $\sigma_1^2 = 10.0$, $\alpha = .05$.

		σ_2^2/σ_1^2							
Δ	β	1.0	1.5	2.0	5.0	10.0	20.0	100.0	∞
0.25	0.99	0.9766	0.9716	0.9763	0.9755	0.9747	0.9705	0.9730	0.9736
	0.95	1.0384	1.0220	1.0337	1.0326	1.0346	1.0328	1.0341	1.0332
	0.90	1.0394	1.0497	1.0497	1.0459	1.0520	1.0486	1.0458	1.0469
	0.75	1.0502	1.0368	1.0467	1.0424	1.0385	1.0398	1.0430	1.0433
	0.50	1.0137	1.0117	*	1.0097	1.0063	1.0115	1.0087	1.0080
0.50	0.99	0.9772	0.9711	0.9711	0.9683	0.9740	0.9754	0.9737	0.9736
	0.95	1.0332	1.0249	1.0289	1.0380	1.0332	1.0371	1.0336	1.0332
	0.90	1.0534	1.0309	1.0461	1.0454	1.0542	1.0435	1.0462	1.0469
	0.75	1.0470	1.0454	1.0475	1.0475	1.0431	1.0434	1.0426	1.0433
	0.50	*	1.0106	*	*	1.0113	1.0047	1.0109	1.0080
1.00	0.99	0.9781	0.9785	0.9769	0.9718	0.9729	0.9746	0.9739	0.9736
	0.95	1.0359	1.0384	1.0315	1.0351	1.0345	1.0350	1.0344	1.0333
	0.90	1.0507	1.0458	1.0384	1.0544	1.0461	1.0498	1.0471	1.0469
	0.75	1.0421	1.0486	1.0457	1.0405	1.0440	1.0464	1.0441	1.0433
	0.50	*	1.0120	1.0106	1.0139	1.0100	1.0056	1.0094	1.0080
1.50	0.99	0.9808	0.9724	0.9747	0.9710	0.9737	0.9749	0.9741	0.9736
	0.95	1.0236	1.0361	1.0381	1.0333	1.0319	1.0320	1.0323	1.0332
	0.90	1.0531	1.0395	1.0466	1.0504	1.0512	1.0499	1.0467	1.0469
	0.75	1.0402	1.0523	1.0340	1.0565	1.0407r	1.0469	1.0411	1.0433
	0.50	*	*	1.0111	1.0150	1.0079	1.0094	1.0077	1.0080
2.00	0.99	0.9658	0.9679	0.9763	0.9749	0.9735	0.9708	0.9751	0.9736
	0.95	1.0328	1.0310	1.0441	1.0315	1.0319	1.0322	1.0330	1.0332
	0.90	1.0439	1.0524	1.0498	1.0528	1.0508	1.0483	1.0470	1.0469
	0.75	1.0380	1.0382	1.0420	1.0414	1.0428	1.0407	1.0446	1.0433
	0.50	*	*	1.0093	*	*	1.0082	1.0077	1.0080

Table 2g: Comparison of Prokof'yev-Shishkin and Chapman procedures for selected n_0, σ_1^2, α, Δ, β and σ_2^2/σ_1^2: $n_0 = 10$, $\sigma_1^2 = 0.10$, $\alpha = .01$.

		σ_2^2/σ_1^2							
Δ	β	1.0	1.5	2.0	5.0	10.0	20.0	100.0	∞
0.25	0.99	0.9323	0.9359	0.9354	0.9314	0.9349	0.9343	0.9338	0.9339
	0.95	0.9649	0.9523	0.9590	0.9525	0.9546	0.9529	0.9544	0.9548
	0.90	0.9579	0.9560	0.9541	0.9559	0.9569	0.9570	0.9576	0.9568
	0.75	0.9487	0.9479	0.9400	0.9491	0.9468	0.9476	0.9472	0.9473
	0.50	0.9213	0.9197	0.9230	0.9202	0.9235	0.9203	0.9198	0.9208
0.50	0.99	0.9379	0.9374	0.9386	0.9303	0.9342	0.9374	0.9331	0.9339
	0.95	0.9670	0.9634	0.9584	0.9561	0.9597	0.9558	0.9552	0.9548
	0.90	0.9757	0.9657	0.9629	0.9667	0.9615	0.9557	0.9569	0.9568
	0.75	*	0.9780	0.9829	0.9680	0.9575	0.9544	0.9483	0.9473
	0.50	1.0382	1.0166	1.0104	0.9819	0.9557	0.9394	0.9248	0.9208
1.00	0.99	1.0641	1.0441	1.0382	1.0176	0.9826	0.9599	0.9386	0.9339
	0.95	1.0932	1.0876	1.0867	1.0978	1.0507	1.0063	0.9658	0.9548
	0.90	1.0979	1.0957	1.1018	1.1395	1.0876	1.0277	0.9726	0.9568
	0.75	1.0998	1.0988	1.1015	1.1634	1.1387	1.0608	0.9709	0.9473
	0.50	1.1000	1.1000	1.1000	1.1380	1.1879	1.1016	0.9611	0.9208
1.50	0.99	1.0999	1.0994	1.1005	1.1576	1.1215	1.0402	0.9572	0.9339
	0.95	1.1000	1.1000	1.1005	1.1561	1.1984	1.1197	0.9907	0.9548
	0.90	1.1000	1.1000	1.1001	1.1388	1.2190	1.1581	1.0020	0.9568
	0.75	1.1001	1.1000	1.1000	1.1102	1.1947	1.2096	1.0142	0.9473
	0.50	1.1000	1.1000	1.1000	1.1006	1.1359	1.2241	1.0230	0.9208
2.00	0.99	1.1000	1.1000	1.1000	1.1297	1.2027	1.1425	0.9808	0.9339
	0.95	1.1000	1.1000	1.1000	1.1074	1.1930	1.2306	1.0255	0.9548
	0.90	1.1000	1.1000	1.1000	1.1020	1.1639	1.2446	1.0430	0.9568
	0.75	1.1000	1.1000	1.1000	1.1001	1.1205	1.2339	1.0673	0.9473
	0.50	1.1000	1.1000	1.1000	1.1000	1.1016	1.1560	1.1025	0.9208

Table 2h: Comparison of Prokof'yev-Shishkin and Chapman procedures for selected n_0, σ_1^2, α, Δ, β and σ_2^2/σ_1^2: $n_0 = 10$, $\sigma_1^2 = 0.10$, $\alpha = .05$.

Δ	β	σ_2^2/σ_1^2							
		1.0	1.5	2.0	5.0	10.0	20.0	100.0	∞
0.25	0.99	0.9639	0.9690	0.9600	0.9654	0.9647	0.9682	0.9656	0.9668
	0.95	*	*	*	*	1.0040	*	*	1.0012
	0.90	1.0088	1.0119	1.0115	1.0097	1.0073	1.0074	1.0087	1.0094
	0.75	1.0095	1.0088	1.0149	1.0115	1.0112	1.0068	1.0090	1.0087
	0.50	1.0097	1.0060	*	0.9951	0.9935	0.9916	0.9897	0.9888
0.50	0.99	0.9741	0.9754	0.9768	0.9672	0.9694	0.9650	0.9672	0.9668
	0.95	1.0245	1.0194	1.0212	1.0117	1.0091	1.0057	*	1.0012
	0.90	1.0459	1.0406	1.0434	1.0386	1.0255	1.0167	1.0103	1.0094
	0.75	1.0814	1.0791	1.0869	1.0931	1.0591	1.0370	1.0149	1.0087
	0.50	1.0988	1.0977	1.1063	1.1645	1.1226	1.0661	1.0036	0.9888
1.00	0.99	1.0929	1.0892	1.0946	1.1109	1.0606	1.0165	0.9780	0.9668
	0.95	1.0997	1.1001	1.1052	1.1938	1.1714	1.1020	1.0217	1.0012
	0.90	1.1000	1.1000	1.1023	1.1935	1.2256	1.1450	1.0382	1.0094
	0.75	1.1000	1.1000	1.1000	1.1397	1.2550	1.2188	1.0573	1.0087
	0.50	1.1000	1.1000	1.1000	1.1023	1.1660	1.2740	1.0979	0.9888
1.50	0.99	1.1000	1.1000	1.1004	1.1604	1.2091	1.1295	1.0016	0.9668
	0.95	1.1000	1.1000	1.1000	1.1211	1.2379	1.2423	1.0625	1.0012
	0.90	1.1000	1.1000	1.1000	1.1069	1.2011	1.2846	1.0853	1.0094
	0.75	1.1000	1.1000	1.1000	1.1002	1.1278	1.2696	1.1296	1.0087
	0.50	1.1000	1.1000	1.1000	1.1000	1.1007	1.1477	1.1970	0.9888
2.00	0.99	1.1000	1.1000	1.1000	1.1076	1.1992	1.2397	1.0371	0.9668
	0.95	1.1000	1.1000	1.1000	1.1006	1.1380	1.2741	1.1143	1.0012
	0.90	1.1000	1.1000	1.1000	1.1000	1.1135	1.2383	1.1521	1.0094
	0.75	1.1000	1.1000	1.1000	1.1000	1.1006	1.1466	1.2214	1.0087
	0.50	1.1000	1.1000	1.1000	1.1000	1.1000	1.1020	1.2932	0.9888

Table 2i: Comparison of Prokof'yev-Shishkin and Chapman procedures for selected n_0, σ_1^2, α, Δ, β and σ_2^2/σ_1^2: $n_0 = 10$, $\sigma_1^2 = 1.00$, $\alpha = .01$.

		σ_2^2/σ_1^2							
Δ	β	1.0	1.5	2.0	5.0	10.0	20.0	100.0	∞
0.25	0.99	0.9321	0.9357	0.9352	0.9312	0.9348	0.9343	0.9338	0.9339
	0.95	0.9647	0.9521	0.9589	0.9524	0.9545	0.9528	0.9544	0.9548
	0.90	0.9578	0.9557	0.9538	0.9557	0.9568	0.9569	0.9575	0.9568
	0.75	0.9480	0.9474	0.9395	0.9489	0.9467	0.9476	0.9472	0.9473
	0.50	0.9189	0.9181	0.9216	0.9195	0.9231	0.9201	0.9198	0.9208
0.50	0.99	0.9346	0.9351	0.9366	0.9293	0.9335	0.9371	0.9330	0.9339
	0.95	0.9580	0.9578	0.9526	0.9530	0.9579	0.9549	0.9550	0.9548
	0.90	0.9595	0.9540	0.9521	0.9603	0.9578	0.9538	0.9565	0.9568
	0.75	0.9482	0.9453	0.9524	0.9484	0.9464	0.9486	0.9470	0.9473
	0.50	0.9217	0.9210	0.9212	0.9206	0.9196	0.9207	0.9209	0.9208
1.00	0.99	0.9319	0.9337	0.9342	0.9359	0.9347	0.9345	0.9332	0.9339
	0.95	0.9575	0.9567	0.9476	0.9520	0.9561	0.9558	0.9550	0.9548
	0.90	0.9612	0.9620	0.9566	0.9592	0.9579	0.9571	0.9577	0.9568
	0.75	0.9517	0.9469	0.9524	0.9490	0.9461	0.9479	0.9472	0.9473
	0.50	0.9320	0.9253	0.9277	0.9249	0.9270	0.9202	0.9210	0.9208
1.50	0.99	0.9412	0.9353	0.9381	0.9382	0.9364	0.9338	0.9346	0.9339
	0.95	0.9636	0.9593	0.9612	0.9590	0.9545	0.9566	0.9551	0.9548
	0.90	0.9674	0.9599	0.9654	0.9623	0.9598	0.9592	0.9572	0.9568
	0.75	0.9776	0.9698	0.9712	0.9599	0.9512	0.9529	0.9492	0.9473
	0.50	1.0209	*	0.9912	0.9746	0.9518	0.9361	0.9236	0.9208
2.00	0.99	0.9484	0.9505	0.9422	0.9367	0.9356	0.9354	0.9340	0.9339
	0.95	0.9902	0.9847	0.9788	0.9730	0.9675	0.9625	0.9562	0.9548
	0.90	1.0205	*	*	0.9866	0.9731	0.9638	0.9585	0.9568
	0.75	1.0560	1.0395	1.0408	1.0126	0.9914	0.9682	0.9508	0.9473
	0.50	1.0915	1.0823	1.0813	1.0688	1.0169	0.9744	0.9312	0.9208

Table 2j: Comparison of Prokof'yev-Shishkin and Chapman procedures for selected n_0, σ_1^2, α, Δ, β and σ_2^2/σ_1^2: $n_0 = 10$, $\sigma_1^2 = 1.00$, $\alpha = .05$.

		σ_2^2/σ_1^2							
Δ	β	1.0	1.5	2.0	5.0	10.0	20.0	100.0	∞
0.25	0.99	0.9637	0.9689	0.9599	0.9653	0.9647	0.9682	0.9656	0.9668
	0.95	*	*	*	*	1.0039	*	*	1.0012
	0.90	1.0088	1.0118	1.0114	1.0096	1.0073	1.0074	1.0087	1.0094
	0.75	1.0083	1.0076	1.0140	1.0108	1.0108	1.0067	1.0089	1.0087
	0.50	*	*	0.9904	0.9882	0.9898	0.9896	0.9893	0.9888
0.50	0.99	0.9667	0.9695	0.9716	0.9642	0.9676	0.9641	0.9670	0.9668
	0.95	*	*	*	*	*	*	*	1.0012
	0.90	1.0109	1.0101	1.0089	1.0104	1.0102	1.0086	1.0086	1.0094
	0.75	1.0119	1.0114	1.0093	1.0151	1.0115	1.0115	1.0097	1.0087
	0.50	0.9838	0.9834	0.9891	0.9866	0.9896	0.9914	0.9878	0.9888
1.00	0.99	0.9657	0.9683	0.9669	0.9670	0.9667	0.9657	0.9673	0.9668
	0.95	*	*	*	*	*	1.0033	*	1.0012
	0.90	*	1.0115	1.0129	1.0113	1.0124	1.0101	1.0091	1.0094
	0.75	1.0252	1.0091	1.0164	1.0139	1.0112	1.0097	1.0087	1.0087
	0.50	1.0411	1.0317	1.0296	1.0180	1.0095	*	0.9908	0.9888
1.50	0.99	0.9691	0.9744	0.9719	0.9684	0.9660	0.9674	0.9663	0.9668
	0.95	1.0111	1.0097	1.0115	1.0111	1.0081	1.0033	1.0030	1.0012
	0.90	1.0403	1.0312	1.0351	1.0305	1.0201	1.0167	1.0096	1.0094
	0.75	1.0711	1.0683	1.0783	1.0688	1.0478	1.0292	1.0122	1.0087
	0.50	1.0976	1.0957	1.1045	1.1473	1.1036	1.0539	1.0021	0.9888
2.00	0.99	1.0057	*	0.9925	0.9808	0.9773	0.9749	0.9681	0.9668
	0.95	1.0643	1.0604	1.0606	1.0515	1.0328	1.0146	1.0046	1.0012
	0.90	1.0855	1.0787	1.0912	1.0958	1.0605	1.0386	1.0152	1.0094
	0.75	1.0981	1.0980	1.1099	1.1693	1.1265	1.0752	1.0224	1.0087
	0.50	1.1000	1.1001	1.1014	1.1786	1.2108	1.1327	1.0187	0.9888

Table 2k: Comparison of Prokof'yev-Shishkin and Chapman procedures for selected n_0, σ_1^2, α, Δ, β and σ_2^2/σ_1^2: $n_0 = 10$, $\sigma_1^2 = 10.0$, $\alpha = .01$.

		σ_2^2/σ_1^2							
Δ	β	1.0	1.5	2.0	5.0	10.0	20.0	100.0	∞
0.25	0.99	0.9320	0.9357	0.9352	0.9312	0.9348	0.9343	0.9338	0.9339
	0.95	0.9647	0.9520	0.9589	0.9524	0.9545	0.9528	0.9544	0.9548
	0.90	0.9577	0.9557	0.9538	0.9557	0.9568	0.9569	0.9575	0.9568
	0.75	0.9480	0.9474	0.9395	0.9489	0.9466	0.9476	0.9472	0.9473
	0.50	0.9187	0.9180	0.9215	0.9195	0.9231	0.9201	0.9198	0.9208
0.50	0.99	0.9345	0.9350	0.9365	0.9293	0.9335	0.9371	0.9330	0.9339
	0.95	0.9579	0.9578	0.9525	0.9530	0.9578	0.9549	0.9550	0.9548
	0.90	0.9594	0.9540	0.9520	0.9603	0.9578	0.9538	0.9565	0.9568
	0.75	0.9481	0.9452	0.9522	0.9484	0.9463	0.9486	0.9470	0.9473
	0.50	0.9213	0.9207	0.9209	0.9205	0.9195	0.9206	0.9209	0.9208
1.00	0.99	0.9315	0.9335	0.9338	0.9357	0.9346	0.9345	0.9332	0.9339
	0.95	0.9571	0.9563	0.9473	0.9518	0.9560	0.9558	0.9550	0.9548
	0.90	0.9607	0.9616	0.9562	0.9589	0.9577	0.9570	0.9577	0.9568
	0.75	0.9493	0.9452	0.9509	0.9483	0.9457	0.9477	0.9472	0.9473
	0.50	0.9213	0.9179	0.9214	0.9212	0.9250	0.9191	0.9208	0.9208
1.50	0.99	0.9389	0.9336	0.9366	0.9375	0.9360	0.9336	0.9345	0.9339
	0.95	0.9569	0.9558	0.9575	0.9568	0.9533	0.9559	0.9550	0.9548
	0.90	0.9561	0.9517	0.9588	0.9582	0.9574	0.9579	0.9570	0.9568
	0.75	0.9430	0.9454	0.9488	0.9458	0.9434	0.9488	0.9483	0.9473
	0.50	0.9191	0.9211	0.9186	0.9264	0.9238	0.9216	0.9206	0.9208
2.00	0.99	0.9319	0.9398	0.9332	0.9309	0.9325	0.9337	0.9337	0.9339
	0.95	0.9497	0.9541	0.9512	0.9560	0.9578	0.9572	0.9551	0.9548
	0.90	0.9598	0.9563	0.9567	0.9559	0.9557	0.9545	0.9566	0.9568
	0.75	0.9474	0.9503	0.9529	0.9431	0.9496	0.9464	0.9463	0.9473
	0.50	0.9205	0.9282	0.9236	0.9194	0.9211	0.9230	0.9205	0.9208

Table 21: Comparison of Prokof'yev-Shishkin and Chapman procedures for selected n_0, σ_1^2, α, Δ, β and σ_2^2/σ_1^2: $n_0 = 10$, $\sigma_1^2 = 10.0$, $\alpha = .05$.

Δ	β	σ_2^2/σ_1^2							
		1.0	1.5	2.0	5.0	10.0	20.0	100.0	∞
0.25	0.99	0.9636	0.9689	0.9599	0.9653	0.9647	0.9682	0.9656	0.9668
	0.95	*	*	*	*	1.0039	*	*	1.0012
	0.90	1.0088	1.0118	1.0114	1.0096	1.0073	1.0074	1.0087	1.0094
	0.75	1.0083	1.0076	1.0140	1.0108	1.0108	1.0067	1.0089	1.0087
	0.50	*	*	0.9904	0.9882	0.9898	0.9895	0.9893	0.9888
0.50	0.99	0.9666	0.9694	0.9716	0.9642	0.9676	0.9641	0.9670	0.9668
	0.95	*	*	*	*	*	*	*	1.0012
	0.90	1.0109	1.0101	1.0089	1.0104	1.0102	1.0086	1.0086	1.0094
	0.75	1.0120	1.0114	1.0094	1.0152	1.0115	1.0116	1.0097	1.0087
	0.50	0.9832	0.9830	0.9888	0.9865	0.9895	0.9913	0.9878	0.9888
1.00	0.99	0.9654	0.9681	0.9666	0.9669	0.9666	0.9656	0.9673	0.9668
	0.95	*	*	*	*	*	1.0032	*	1.0012
	0.90	*	1.0101	1.0116	1.0106	1.0120	1.0098	1.0091	1.0094
	0.75	1.0167	*	1.0089	1.0092	1.0084	1.0084	1.0085	1.0087
	0.50	0.9883	0.9902	0.9872	0.9868	0.9910	0.9886	0.9888	0.9888
1.50	0.99	0.9638	0.9701	0.9684	0.9662	0.9649	0.9668	0.9662	0.9668
	0.95	*	*	*	*	*	*	1.0024	1.0012
	0.90	1.0109	1.0086	1.0096	1.0113	1.0089	1.0110	1.0084	1.0094
	0.75	*	1.0098	1.0124	1.0082	1.0108	1.0095	1.0081	1.0087
	0.50	0.9837	0.9860	0.9900	0.9843	0.9904	0.9912	0.9889	0.9888
2.00	0.99	0.9727	0.9700	0.9663	0.9645	0.9677	0.9700	0.9671	0.9668
	0.95	*	1.0076	*	*	*	*	*	1.0012
	0.90	1.0123	*	*	1.0066	1.0069	1.0099	1.0091	1.0094
	0.75	1.0084	1.0116	1.0085	1.0082	1.0091	1.0105	1.0086	1.0087
	0.50	0.9896	0.9904	0.9852	0.9874	0.9874	0.9891	0.9883	0.9888

Table 3: Simulation results for recommending an appropriate ratio estimator. $\alpha = 0.01$, $\Delta_1 = 0.25$, $\beta_1 = 0.50$, $n_0 = 5$, $\mu_1 = 0$, $\mu_2 = 0.25$, $\sigma_1^2 = 0.10$.

	$\sigma_2^2 = 0.10$	$\sigma_2^2 = 0.15$	$\sigma_2^2 = 0.20$	$\sigma_2^2 = 0.50$
	$\frac{\sigma_2^2}{\sigma_1^2} = 1.0$	$\frac{\sigma_2^2}{\sigma_1^2} = 1.5$	$\frac{\sigma_2^2}{\sigma_1^2} = 2.0$	$\frac{\sigma_2^2}{\sigma_1^2} = 5.0$
$\overline{N}$	122.6721	152.0915	183.1695	361.7294
$\overline{M}$	136.0702	172.2852	206.5824	405.6946
$cv(\overline{N})$ in %	0.4952	0.5020	0.5272	0.6008
$cv(\overline{M})$ in %	0.7010	0.7036	0.7038	0.7134
$\overline{R}$	1.2778	1.2259	1.2406	1.1338
$SD(\overline{R})$	.0124	.0113	.0126	.0099
1.96 $SD(\overline{R})$	.0244	.0222	.0247	.0194
95% C.I. for $E(\overline{R})$ based on $\overline{R}$				
Lower CL	1.2534	1.2037	1.2159	1.1144
Upper CL	1.3022	1.2481	1.2653	1.1532
$\hat{R}$	0.9015	0.8828	.8867	0.8916
$b(\hat{R})$	-0.0000	-0.0000	-0.0000	-0.0000
$v(\hat{R})$	0.0000	0.0000	0.0000	0.0000
95% C.I. for R based on $\hat{R}$				
Lower CL	0.8898	0.8707	0.8751	0.8818
Upper CL	0.9133	0.8948	0.8983	0.9015
$corr(N, M)$	0.7002	0.7312	0.7500	0.8485

Table 3: Simulation results for recommending an appropriate ratio estima-
tor. $\alpha = 0.01$, $\Delta_1 = 0.25$, $\beta_1 = 0.50$, $n_0 = 5$, $\mu_1 = 0$, $\mu_2 = 0.25$, $\sigma_1^2 = 0.10$
(continued).

	$\sigma_2^2 = 1.00$	$\sigma_2^2 = 2.00$	$\sigma_2^2 = 10.00$
	$\frac{\sigma_2^2}{\sigma_1^2} = 10.0$	$\frac{\sigma_2^2}{\sigma_1^2} = 20.0$	$\frac{\sigma_2^2}{\sigma_1^2} = 100.0$
$\overline{N}$	660.7329	1265.0819	6060.5338
$\overline{M}$	741.6462	1414.3544	6792.0242
$cv(\overline{N})$ in %	0.6430	0.6832	0.6978
$cv(\overline{M})$ in %	0.7098	0.7140	0.7027
$\overline{R}$	1.0267	0.9685	0.9101
$SD(\overline{R})$	.0057	.0037	.0015
$1.96\ SD(\overline{R})$	.0112	.0073	.0029
95% *C.I.* for $E(\overline{R})$ based on $\overline{R}$			
Lower *CL*	1.0155	0.9612	0.9072
Upper *CL*	1.0379	0.9758	0.9130
$\hat{R}$	0.8909	0.8945	.8923
$b(\hat{R})$	-0.0000	-0.0000	-0.0000
$v(\hat{R})$	0.0000	0.0000	0.0000
95% *C.I.* for R based on $\hat{R}$			
Lower *CL*	0.8825	0.8874	0.8865
Upper *CL*	0.8993	0.9015	0.8981
$corr(N, M)$	0.9133	0.9546	0.9902

References

Best, D.J. and Rayner, J.C.W. (1987). Welch's approximate solution for the Behrens-Fisher problem. *Technometrics* **29** 205 - 210.

Box, G.E.P and Muller, M.A. (1958). A note on the generation of random normal deviates. *Ann. Math. Statist.* **29** 610.

Chapman, D.G. (1950). Some two sample tests. *Ann. Math. Statist.* **21** 601-606.

Chen, H.J. (1978). On approximations to the inverse Student's *t* distribution function. *Jour. Statist. Comput. Simul.* **7** 167 - 180.

Cochran, W.G. (1977). *Sampling Techniques* (Third Edition). John Wiley and Sons, Inc., New York.

Dudewicz, E.J. (1972). Statistical Inference with unknown and unequal variances. *Transactions of the Annual Quality Control Conference of the Rochester Society for Quality Control* **28** 71 - 85.

Dudewicz, E.J. (1976). *Introduction to Statistics and Probability.* Holt, Rinehart and Winston, New York.

Dudewicz, E.J. and Bishop, T.A. (1979). The heteroscedastic method. In *Optimizing Methods in Statistics* (J.S. Rustagi, ed.), Academic Press, Inc., New York, 183 - 203.

Dudewicz, E.J. and Ralley, T.G. (1981). *The Handbook of Random Number Generation and Testing with TESTRAND Computer Code.* American Sciences Press, Inc., Columbus, Ohio.

Dudewicz, E.J., Ramberg, J.S., and Chen, H.J. (1975). New tables for multiple comparisons with a control (unknown variances). *Biometrische Zeitschrift* **17** 13 - 26.

Goodman, L.A. and Hartley, H.O. (1958). The precision of unbiased ratio-type estimators. *Jour. Amer. Statist. Assoc.* **53** 491 - 508.

Murthy, M.N. (1967). *Sampling Theory and Methods.* Statistical Publishing Society, Calcutta, India.

Prokof'yev, V.N. and Shishkin, A.D. (1974). Successive classification of normal sets with unknown variances. *Radio Engineering and Electronic Physics* **19** (Issue No. 2) 141-143.

Raj, D. (1968). *Sampling Theory.* McGraw-Hill Book Company, New York.

Scheffé, H. (1970). Practical solutions of the Behrens-Fisher problem. *Jour. Amer. Statist. Assoc.* **65** 1501 - 1508.

Stroud, A.H. and Secrest, D. (1966). *Gaussian Quadrature Formulas.* Prentice-Hall Inc., Englewood Cliffs, New Jersey.

Chapter 28

An Easy Way to Use Dunnett t with SAS Least Squares Means

SPENCER M. FREE JR. Private Consultant, Newtown Square, Pennsylvania

During my career in the pharmaceutical industry, I found many opportunities to use the Dunnett t procedure. A common application was a routine screening procedure in the laboratories. As part of the procedure, one could build in the decision process rules with specific t ratios or p-values. This paper presents Dunnett t applications where the decision process included several combinations of degrees of freedom in the error term and groups of interest to be compared to the control. The statistical analyses started with multiple regression models that included blocking restrictions and covariates. As a result, the means of interest were always least squares means. The familiar SAS format for least squares means usually includes p-values for all possible paired comparisons. These p-values are all two sided for the Student t ratio. This paper describes a way to use these p-values to carry out a Dunnett t procedure.

1 First Example

This application was developed for the pathology/toxicology department in the research and development division of a large chemical company. This department includes a group of scientists who carry out studies in laboratory animals. Most of the company's new compounds are studied in a series of experiments that investigate potential toxicity in a dose response relationship format. The study design is routine. However, the findings are often unique for each compound.

The study design calls for a control group and two or more groups of animals exposed to increasing concentrations of the compound of interest. Animals are individually caged. Equal sample sizes per group are studied for each sex. Individual animal body weights and feed consumption are frequently recorded. At the end of the study, all animals are sacrificed to provide weights for several organs. When there are obvious signals of toxicity, selected organs for all animals are observed and graded for degree of pathology. Very few animals are lost during the course of the experiment. Thus each experiment provides a large number of data sets all with the same sample sizes.

The biologists that carry out the bulk of the study are also responsible for the first data analyses. To help these biologists identify the signals of potential toxicity, a routine set of univariate analyses are in place. For each criterion, the data are analyzed by sex and then combined over sexes. Analyses that combine the sexes are preferred when qualifying assumptions are met. Many of the analyses include one or more covariates. All analyses are carried out with SAS. For each analysis, least squares means and p-values for all possible paired comparisons are printed out. We set up a data analysis procedure that deliberately avoids reviewing most of the tables of least squares means and p-values.

The data analysis procedure begins with a general screen for each subset of data. This includes a few obvious orthogonal contrasts. These are: control versus all concentrations of the compound; linear and quadratic trends across concentrations; and occasionally a cubic trend. Most often, the non linear trends are used to help identify results that are unusual for one concentration.

The biologists are trained to use the contrasts before they review any averages. They identify data sets with no measurable effect among the concentrations and stop that data review. And, they develop biological expectations for the data sets where some effect of the compound may have been measured.

Only positive signals from the screening are followed up with a review of the table of least squares means. Most often, these tables are for data that are combined over sexes. Given biological expectations for each data set, each data review starts with a comparison of the control mean with the means for each of the concentrations of the compound of interest. The statistical criterion used is the two sided Dunnett t.

The first column of p-values is the column of interest for the review. As

Table 1: Dunnett t (two sided $P \leq 0.05$). P-Values for associated Student t in least squares means table (to find the largest p that will be statistically significant).

Degrees Freedom	Compound Concentrations Tested 2	3	4	5
20	0.0274	0.0195	0.0153	0.0128
25	0.0273	0.0193	0.0152	0.0126
30	0.0273	0.0192	0.0151	0.0125
35	0.0272	0.0192	0.0150	0.0124
40	0.0272	0.0192	0.0150	0.0124
45	0.0272	0.0191	0.0149	0.0123
50	0.0272	0.0191	0.0149	0.0123
60	0.0271	0.0190	0.0149	0.0122
70	0.0271	0.0190	0.0148	0.0122
80	0.0271	0.0190	0.0148	0.0122
100	0.0270	0.0190	0.0148	0.0122

Table 2: Dunnett t ratios. Control versus compound comparisons.

Degrees Freedom	Concentrations Tested 2	3	4	5
20	2.379	2.540	2.651	2.735
25	2.344	2.500	2.607	2.688
30	2.321	2.474	2.578	2.657
35	2.305	2.455	2.558	2.635
40	2.293	2.441	2.543	2.619
45	2.283	2.431	2.531	2.607
50	2.276	2.422	2.522	2.597
60	2.265	2.410	2.508	2.582
70	2.258	2.401	2.499	2.572
80	2.252	2.394	2.491	2.564
100	2.244	2.385	2.481	2.544

you all know, the *p*-values in the whole table are for two sided Student *t*. To make the biologist's review quick and easy, we developed a table that shows the *p*-value (smaller *p*'s) that would be statistically significant ($P \leq 0.05$, two sided) for the Dunnett *t* criterion.

The table used by the biologists is attached as Table 1. I noted above that most data sets have the same, or nearly the same, sample sizes. So, the biologist finds one critical *p*-value in the table. This critical value can be used to make judgements for all the control versus compound comparisons of interest in the experiment. The biologists' experience and the biological expectations developed in the screen phase help to avoid false positive findings. A reference table of Dunnett *t* ratios is given in Table 2.

The senior scientists who discuss the results of the data review were very impressed with how well the biologists understood the full data set. They quickly found that following the same steps provided an efficient review of any findings that needed additional attention.

The efficiency associated with the continuous data analyses soon identified the complications that were introduced by the qualitative data sets. We found that we could use SAS's CATMOD for ranked categorical data to develop similar routine procedures.

The screening step was easy and obvious. Using the same CATMOD procedure to develop single degree of freedom comparisons for the control frequency distribution versus the appropriate frequency distribution for each of the concentrations of the compound was relatively easy too. As you know, the format provides a Chi Square and an associated *p*-value for each comparison. After considering several alternative decision processes, we decided to have the biologists use the print out *p*-values just like they used the table with least square means. To date, this has been more than satisfactory.

2 Second Example

This comes from a regulatory submission that I helped to prepare for a small company. The efficacy data were generated in two pivotal clinical trials. Each was a randomized, double blind clinical trial that included the same four treatments. The data set included more than one efficacy criterion. And for this application, the data were analyzed in actual measurement units and in percent change from baseline units.

The parallel clinical trials were designed to support the efficacy of a two compound combination product. Thus the four treatments of interest were negative control, each compound alone and the combination product. The analysis models included strata and some baseline covariates. Tables

Table 3: Calculations for Dunnett t comparisons.

Data Set			One Sided		Two Sided	
Study	Time	*df*	Dunnett	Ref *p*-value	Dunnett	Ref *p*-value
Percent of Pretreatment Baseline Comparisons						
100	Wk 3	44	2.118	0.0398	2.433	0.0191
	Wk 6	38	2.127	0.0400	2.447	0.0191
	Last	44	2.118	0.0398	2.433	0.0191
101	Wk 3	33	2.137	0.0401	2.463	0.0192
	Wk 6	27	2.148	0.0402	2.478	0.0193
	Last	33	2.137	0.0401	2.463	0.0192

Table 4: Summary of the Dunnett t comparisons: $+$ = statistically significant one sided only $(P < 0.05)$,* = statistically significant two sided $(P < 0.05$, at least).

Data Set		Combination Versus		
Study	Time	Placebo	RxA	RxB
Percent of Pretreatment Baseline Comparisons				
100	Wk 3	+	−	*
	Wk 6	*	*	*
	Last	*	*	*
101	Wk 3	+	−	−
	Wk 6	*	−	−
	Last	*	*	−

of least squares means provided the summary data needed to support the submission.

The regulatory format called for separate analyses for each pivotal clinical trial and an analysis that combined the data from both trials. Statistical support for the combination product was to be based upon comparisons within each clinical trial. I noted above that there was more than one efficacy criterion. So, there were several tables of least squares means and the associated Student t p-values for all possible paired comparisons.

When I was writing the statistical report for this project, I found no established regulatory guidelines for the decision process. I knew a conservative approach was expected. For my decision criteria, I defined the combination product as the "control" and I used the Dunnett t procedure for comparisons with each of the other three treatments. For this application, I presented decision criteria for one sided and two sided Dunnett t.

I did not compute Dunnett t values for all the comparisons that I needed to support the efficacy of the combination. I wrote a special appendix in which I developed a table somewhat like the one I used in my first example. There were just a few combinations of degrees of freedom so a relatively small table satisfied all of my needs. This table is attached as Table 3.

Since this was a regulatory submission, the appendix included some more specific tables, such as Table 4, that helped the reviewers follow my decision process.

I should add a postscript here. When the submission was reviewed, there was a decision process guideline in place. This called for two sided Student t for all comparisons with the combination in each pivotal clinical study.

I have presented two examples where all the details were in my files. I can recall several other examples where I used the same process to avoid calculating a large number of t ratios. Most of those applications were laboratory experiments with large numbers of data sets that had similar sample sizes within the data set.

3   Conclusion

When one wants to use the Dunnett t procedure with SAS's least squares means and associated p-values, there is a little more work do to arrive at each control versus "treatment" comparison that suggests a statistically significant difference decision. When a data set calls for calculating more than a few t ratios, I have found another approach to be useful.

I find the appropriate Dunnett t's for the situation. I find the associated Student t, p-value for these Dunnett t's. I develop a small table and a set of instructions that explain how to use the new p-values with the SAS print out.

The Dunnett t procedure would be much easier to carry out if SAS would print t ratios on one side of the diagonal and p-values on the other side in their least square means tables.

Chapter 29

Future Directions for National Health Information Systems

DAVID F. BRAY Health Policy Research and Evaluation Unit, Queen's University, Kingston, Ontario, Canada

ROBERT LUSSIER Canadian Centre for Health Information, Statistics Canada, Ottawa, Ontario, Canada

Abstract This paper outlines the current sources of health information in Canada, as well as the roles of the National Health Information Council and that of the Canadian Centre for Health Information. This material is presented in the context of the review just being completed by the National Task Force on Health Information.

1 The National Health Information Council

The Council was created in December of 1988 by the Conference of Deputy Ministers of Health together with the Chief Statistician of Canada. It is the vehicle of federal/provincial/territorial coordination in the planning and development of national health information systems. It has representation from each provincial/territorial/federal department of health and from Statistics Canada. It is one of several advisory committees to the Conference of Deputy Ministers of Health but is unique in having representation from most of these committees. Council has as its mandate: the

Table 1: Teams initiated by the National Task Force on Health Information.

1. *Coordinates for the Measurement of Health and Health Policy Issues*
   Information needed to measure the health of populations
   Health policy information requirements
   Information needed to support policy issues
   Identification of key actors, stake holders, their information needs

2. *Public Information*
   Requirements for marketing health information
   Health indicators
   Implications of privacy and confidentiality concerns

3. *Information Needed for Health Protection/Promotion*
   Information to support lifestyle modification
   Information to support community interactions and interventions

4. *Health Care Management Information*
   Comparability of health services information
   Information for the economic analysis of health
   Information to support health human resource management
   Health care quality assurance and outcomes

5. *Information for Health Science and Research*
   Information in support of epidemiology
   Information on health determinants — socio-economic, lifestyles,
     environment, genetic, birth conditions, inadvertent influences, etc.
   Health information analysis: Potentials and impediments

6. *Development of a Health Information Structural Model (Template)*
   Development of a structural model (template)
   Requirements for development/maintenance of databases, software,
     and access
   Mapping existing information programs onto the template,
     identification of data gaps and provincial systems directions
   Relationship to health information production processes
     (e.g. medical, resource, socio-economic data, population survey
     data, linkage, matching and integration)
   Relevant international data experiences

development and maintenance of a national consensus on national health information systems; the consideration of requirements, roles, responsibilities and plans; the provision of advice and direction to organizations which maintain national health information systems; as well as ensuring the availability of accurate, relevant and comparable health information in Canada in a timely manner.

2 Canadian Centre for Health Information

The Centre was established in the fall of 1989 by renaming the former Health Division of Statistics Canada. The Centre remains part of Statistics Canada but unlike its predecessor, takes its priority directions from the Council. This is accomplished by having the Council make recommendations to the Centre through the Chief Statistician. The Centre operates a series of health care and health status information programs, and maintains a series of relevant data bases detailed in Bray (1991). The key product is the quarterly journal, *Health Reports*, which contains timely articles, highlights of recently released files and a listing of all information available from the Centre. The Centre cooperatively develops changes to health information systems by encouraging the adoption of uniform data sets and ensuring the confidentiality of records. The Centre responds to requests for information and undertakes client driven research on a cost recovery basis.

3 The National Task Force

The Task Force was initiated by the Deputy Ministers of Health and the Chief Statistician in cooperation with the Council to undertake a major review and synthesis of health information in Canada. The Task Force is chaired by the immediate past Chief Statistician, Dr. Martin Wilk. It has representation from all sectors, government, non-government, and private.

The objective of the Task Force is to support the Council in improving health information in Canada by formulating planning recommendations for NHIC. The approach involves building consensus around information and systems needed to support policy issues, as well as the mechanisms and resources needed to modify and extend current collections or add new collections. A wide range of project teams have been initiated to undertake the detailed studies (Table 1).

The work has highlighted the reality that health is more than health care. To focus attention, a template (Figure 1) has been developed, which shows the individuals within an environmental milieu. The latter is divided

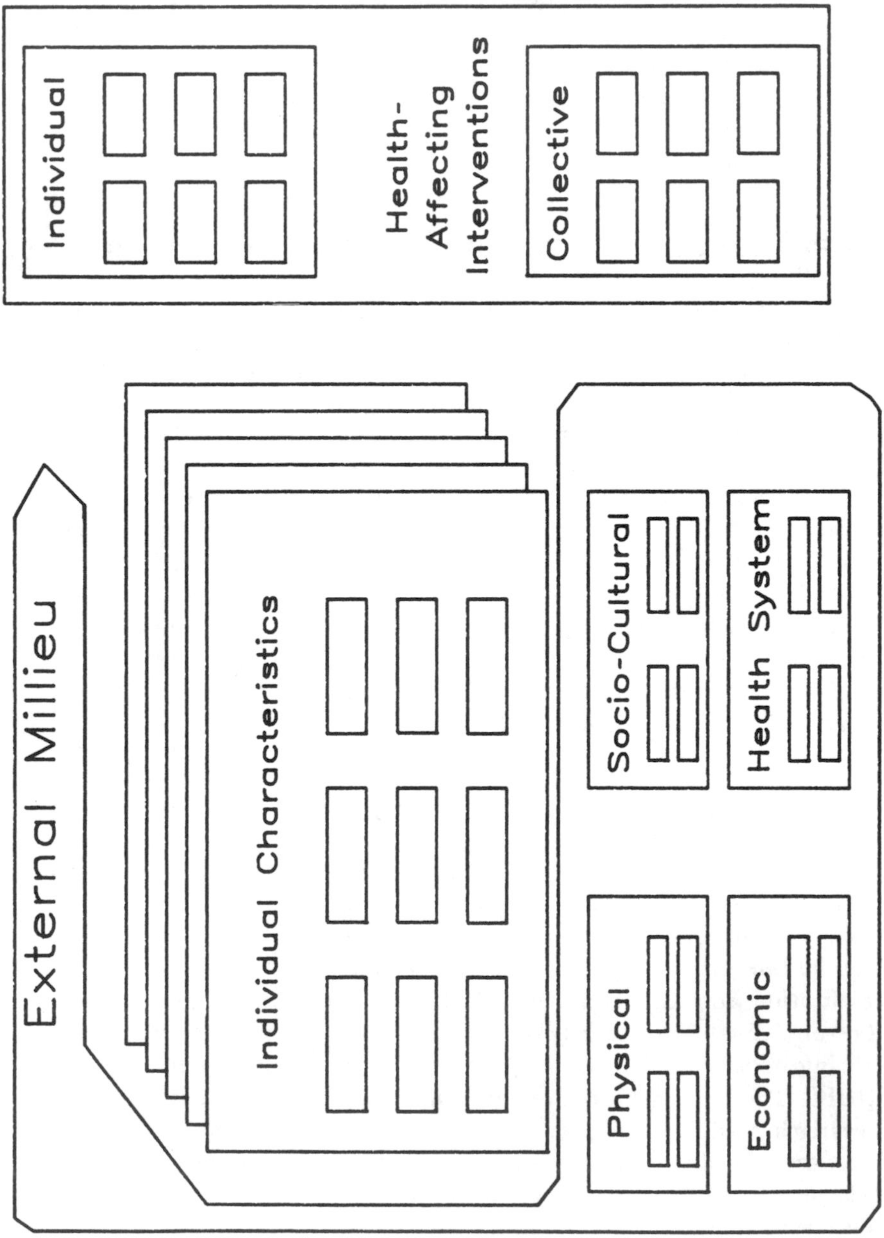

Figure 1: A template for health information: overview.

into four parts, Physical, Socio-Cultural, Economic, and the Health Care System. On the right of Figure 1, health affecting interventions are identified as those which affect either the individual directly as in the provision of a service, or collectively as when a society modifies its physical facilities or socio-economic support systems.

4   Health Status and Function

There is considerable information being collected in both administrative files (vital statistics, environmental contaminates and traffic injuries) and scheduled surveys (General Social Survey). This information has already provided Canada with a broad understanding of gross changes affecting the health of Canadians. For example, Figure 2 displays the increasing longevity of Canadians. In 1921 60% of Canadians lived to age 65 while in 1985 this percentage had risen to 80%. The leading causes of death (Figure 3) have also shifted dramatically, infection and respiratory diseases having become a much smaller proportion while cancer and cardiovascular diseases have become much larger proportions.

Ranking diseases by death certificates is admittedly a crude procedure. One way to at least recognize that those diseases which affect younger persons would have a greater impact on longevity and hopefully more quality years is to consider potential years of life lost (Figure 4). Here we see that cancer becomes more significant than cardiovascular disease.

5   Health Care

This is the area in which the greatest amount of information already exists including procedures performed and diagnoses on separation for those who stay overnight at hospitals. Information on the operation of hospitals and nursing homes, registration figures for those involved in the delivery of care, as well as inventories of drug and device products are available.

Expenditure patterns are always of interest and Figure 5 shows the trend from 1976 - 1987. Public Hospital Expenditures have not risen as much as popular discussion might lead one to expect (from 10 to 12 billion constant dollars). The changing age structure and hospital practice is clearly reflected in the shift from a relatively uniform age structure in 1961 to very marked age dependent relationship in 1988-9 (Figure 6).

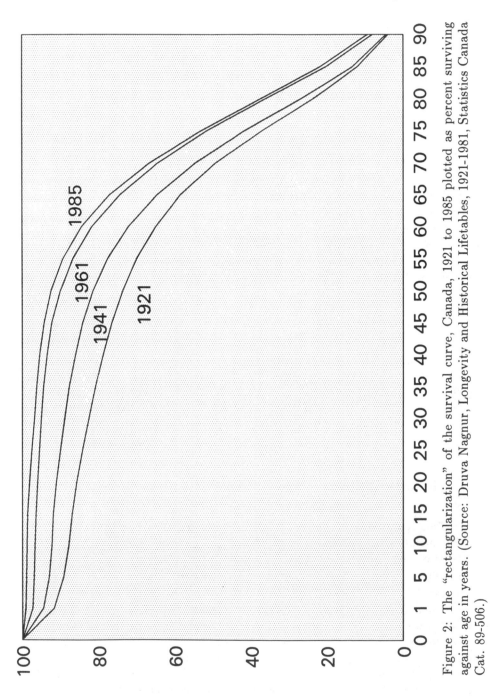

Figure 2: The "rectangularization" of the survival curve, Canada, 1921 to 1985 plotted as percent surviving against age in years. (Source: Druva Nagnur, Longevity and Historical Lifetables, 1921-1981, Statistics Canada Cat. 89-506.)

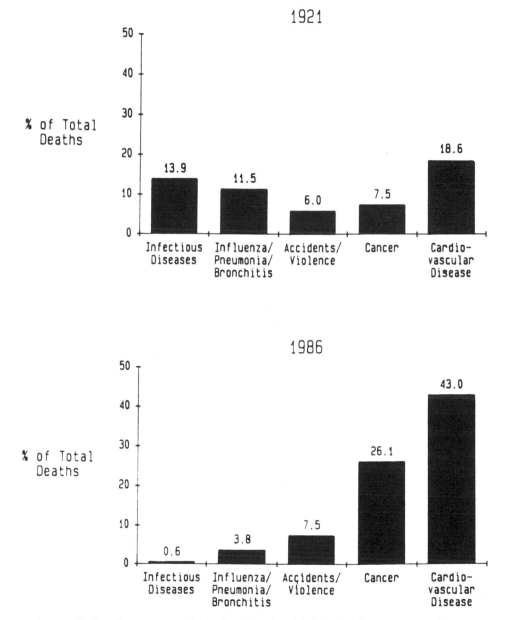

Figure 3: Leading causes of death, Canada, 1921 to 1986 as percent of total deaths. (Source: Vital Statistics and Disease Registries Section; Health Division, Statistics Canada.)

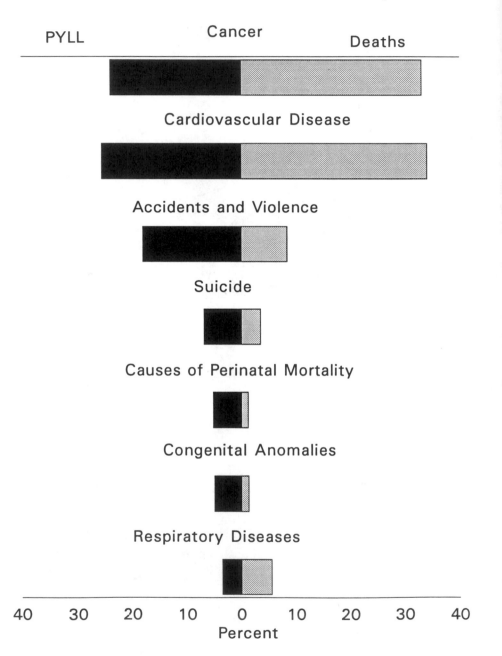

Figure 4: Potential years of life lost (PYLL) and deaths before age 75 for selected causes of death, Canada, 1989. (Source: Health Status Section, Canadian Center for Health Information, Statistics Canada.)

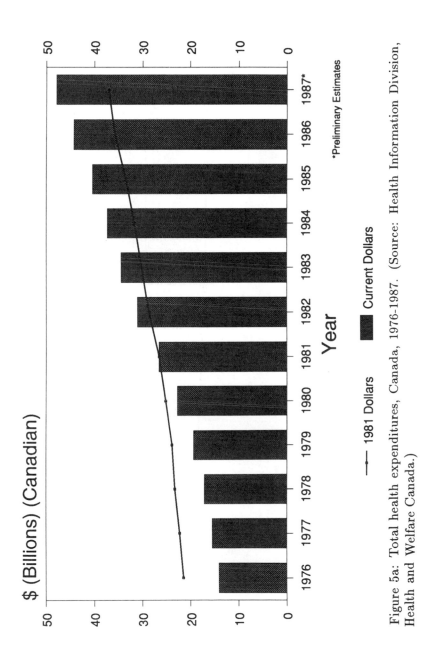

Figure 5a: Total health expenditures, Canada, 1976-1987. (Source: Health Information Division, Health and Welfare Canada.)

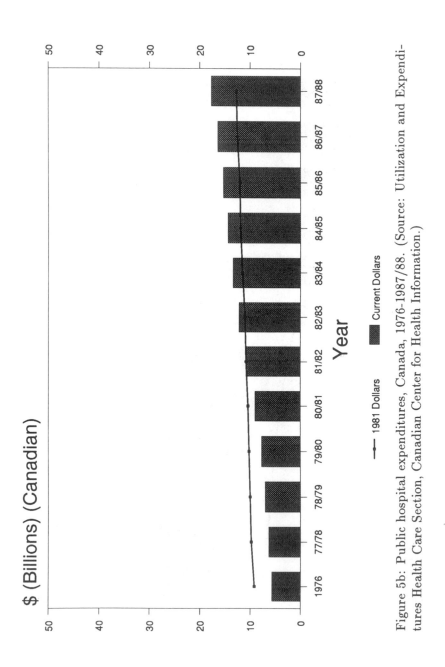

Figure 5b: Public hospital expenditures, Canada, 1976-1987/88. (Source: Utilization and Expenditures Health Care Section, Canadian Center for Health Information.)

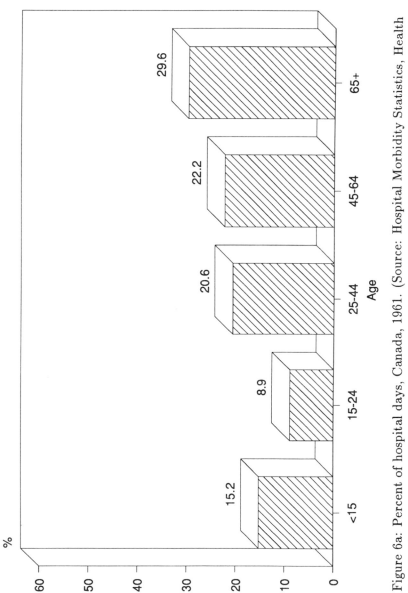

Figure 6a: Percent of hospital days, Canada, 1961. (Source: Hospital Morbidity Statistics, Health Division.)

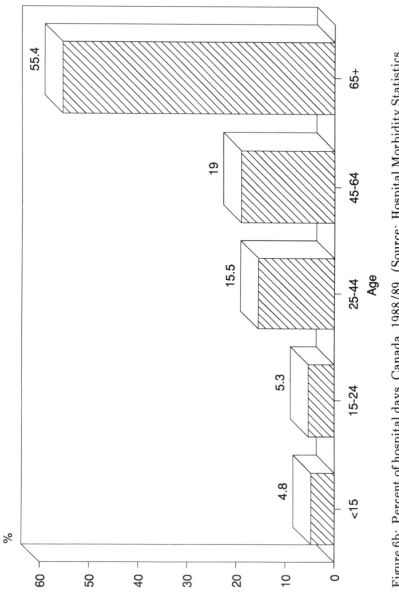

Figure 6b: Percent of hospital days, Canada, 1988/89. (Source: Hospital Morbidity Statistics, Health Division.)

6 Social Environment

The template separates the Socio-Cultural Environment from the Economic Environment. Nevertheless, for this paper, the data bases are combined. There are both administrative (Marriages and Divorces) and scheduled survey information (Survey of Consumer Finance). The broadest coverage, however, is provided by the Census. If Figure 2 displays the increasing longevity of Canadians, Figure 7 illustrates that the degree of improvement continues to vary within region of the country and income level of census tract. While both have shown much improvement, there remains 1.5 years of life expectancy difference between the best and worst regions and 3.5 years between lowest and highest income census tracts. A similar relationship has also been found for rates of infant mortality (Figure 8).

7 Physical Environment

A wide range of data collections currently record changes in ambient air, water and soil. In addition, the Geological Survey of Canada has characterized the mineral concentrations typically found in Canada and one system, the National Radiation Dose Register, monitors the exposure of those who are routinely exposed to radiation emitting devices.

Those collections which reflect geographical differences can and have been mapped by the Geographical Survey. Dr. R. Boyle (1991), of the Geological Survey has prepared Figure 9 which illustrates the complex interrelationships among the various aspects of the physical environment as they relate to humans especially as these concern ingestion and respiration. Statistics Canada in collaboration with National Health and Welfare has mapped indications of how mortality relates to location of dwelling. Of course, we are a mobile society and a full assessment would require an estimate of the specific exposure of individuals to specific insults deriving from the physical environment summed over residences and workplaces. Data from the 1986 census depicting the mobility patterns of Canadians since the previous census showed that:

56.3 % lived in the same residence;

24.2 % lived in a different residence but in the same municipality;

13.5 % lived in different municipalities in the same province, territory;

4.0 % lived in a different province or territory; and

2.0 % had entered from outside of Canada.

The Canadian Mortality Data Base, which includes the coded causes of all deaths which have occurred in Canada since 1950, is a national trea-

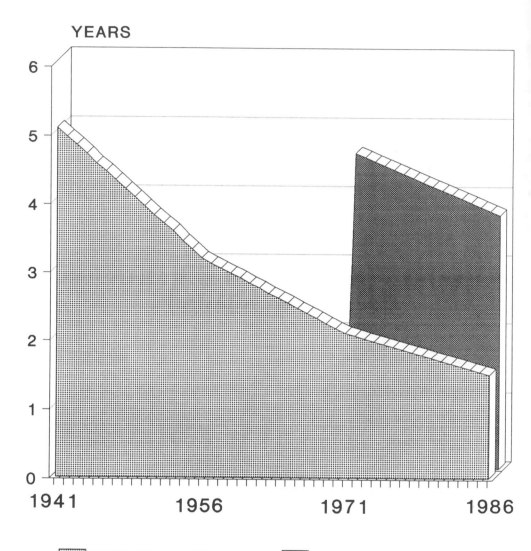

Figure 7: Disparities in life expectancy, Canada, 1941-1986. (Source: Statistics Canada, Health and Welfare Canada [Wilkins, Adams, and Brancker, 1989].)

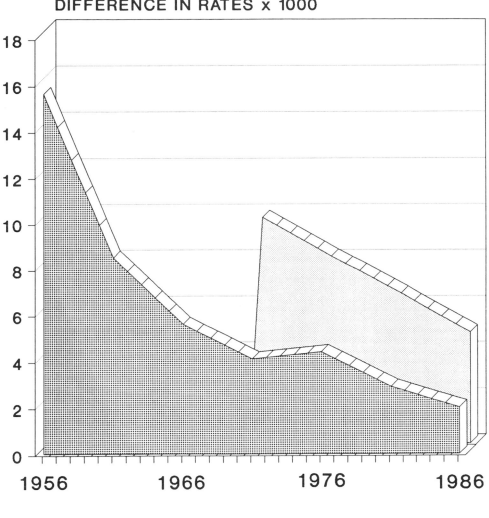

DIFFERENCE IN RATES x 1000

REGIONAL DISPARITY   RICH-POOR DISPARITY

Figure 8: Disparities in infant mortality, Canada, 1956-1986. (Source: Statistics Canada, Health and Welfare Canada [Wilkins, Adams, and Brancker, 1989].)

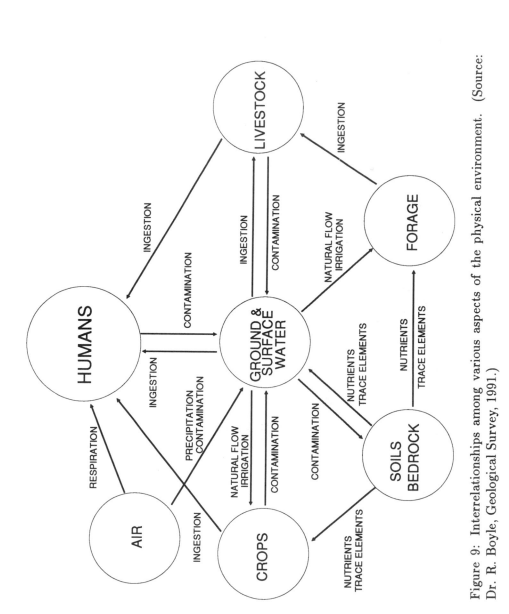

Figure 9: Interrelationships among various aspects of the physical environment. (Source: Dr. R. Boyle, Geological Survey, 1991.)

sure. It is being used to quantify the experience of employees in selected subject industries/settings relative to the general population (see, for example, Smith and Newcombe 1982).

8   Emerging Directions

Since the Lalonde Report (1974) there has been increasing interest in health promotion and prevention. More recently, community based programs have drawn increasing interest. Happily these two areas mesh well together.

Equal levels of interest have been expressed in the management of the personnel providing care as well as considering such provision in terms of the management of the outcome of providing care. Even these activities are being considered from the local perspective with the result that local decision making could now be addressed over the whole range of services which could be provided to a local population.

While research is going forward based on information from each of these topics (Table 2), much more could be done if the information were available in a more comprehensive manner (Table 3).

Table 2: Examples of current research and analysis.

1. *Health Promotion and Prevention* Levels and Correlates of Behavioral Risk Factors 2. *Community Based Programs* Community Health Information Systems Project 3. *Management of Health/Medical Personnel* Health Personnel in Canada 4. *Outcomes Management* Epidemiological Studies Based on the Canadian Mortality Data Base Coronary Artery By-pass Surgery in Canada 5. *Local Decision-Making* Health Practices of Edmontonians Toronto Community Health Survey

Table 3: Research opportunities with an enhanced database.

1. *Health Promotion and Prevention* Direct Relationship Between Risk Factors and Utilization and Disease 2. *Community Based Programs* Individual Rather than Ecological Studies 3. *Management of Health/Medical Personnel* Physician Practice Pattern 4. *Outcomes Management* Person-Based Utilization Records Disease Incidence 5. *Local Decision-Making* Balance Between Institutional and Community Programs

9   Enhancements and Statistical Methods

Specific desirable enhancements to the database and data collection required to bring about such comprehensiveness are:

> Continuing Health Survey
> Linkage of Survey to Administrative Data and to the Census
> Adoption of Personal Identifiers
> Development of Medical Care and Hospital Care Personal Records
> Development of Consistent Standards
> Quality Assurance
> Timeliness

Several of these are themselves areas of application for statistical methods. For example, data quality issues (recording and coding) are topics well known to statistics. While personal identifiers will certainly aid in the development of administrative files and in linking them to survey results, statistics is needed to develop the sampling procedures to make such files amenable to analysis. Given that the sample is drawn, we must consider what methods to use in describing trends and how cohort or panel studies may help. Finally, access to records is central. If we are unable to devise acceptable and persuasive methods to ensure access to analysts all the efforts in increasing the quantity and quality of useful data will have been in vain.

References

Boyle, D.R. (1991). Geochemical environment and its relationship to the development of health status indicators. In *Environmental Health Status Indicators*, (R.S. McColl, ed.), University of Waterloo Press, In press.

Bray, D.F. (1991) Health activities in Statistics Canada, an overview. *Health Reports* 3 No. 4, Supplement, 55 *pp.*

Lalonde, M. (1974). A New Perspective on the Health of Canadians. Department of Supply and Services Catalogue No. H31-1374, Ottawa, Canada.

Smith, M.E. and Newcombe, H.B. (1982). Use of Canadian mortality data base for epidemiology follow-up. *Can. Jour. Public Health* 73 39 - 46.

Wilkins, Adams, and Brancker (1989). Changes in mortality by income in urban Canada from 1971 to 1986. *Health Reports* 1 137 - 174.

Chapter 30

Concomitants of Order Statistics: Review and Recent Developments

H.A. DAVID Department of Statistics, Iowa State University, Ames, Iowa

Abstract Let (X_i, Y_i), $i = 1, \cdots, n$, be independent pairs of variates. If $X_{r:n}$ denotes the r-th ordered X-variate, then the Y-variate paired with $X_{r:n}$ is termed the *concomitant of the r-th order statistic* and denoted by $Y_{[r:n]}$. After a review of basic results, an outline will be given of progress since 1982 in both the theory and application of concomitants. These developments, due to various authors, include (a) Selection through an associated variable; (b) Estimation of the correlation coefficient for sensitive data; (c) Concomitants of extreme order statistics; (d) Dependence structure of concomitants.

1   Introduction

Let (X_i, Y_i), $i = 1, \cdots, n$, be independent pairs of variates from a bivariate distribution with cumulative distribution function (c.d.f.) $F(x, y)$. If the pairs are ordered by the X_i, then the Y-variate associated with the r-th order statistics $X_{r:n}$ will be denoted by $Y_{[r:n]}$ and termed the concomitant of the r-th order statistic (David 1973).

The most important use of concomitants arises in selection procedures when $k(< n)$ individuals are chosen on the basis of their X-values. Then the corresponding Y-values represent performance on an associated characteristic. For example, if the top k out of n bulls, as judged by their genetic make-up, are selected for breeding, then $Y_{[n-k+1:n]}, \cdots, Y_{[n:n]}$ might represent the average milk yields of their female offspring. Or X might be the score on a screening test and Y the score on a later test. There are related problems dealing with the estimation of parameters from data in which selection has taken place. The study of some aspects of these problems antedates the term concomitant of order statistics (e.g., Watterson 1959). However, the occurrence of concomitants in a variety of contexts independently prompted also another term, *induced order statistics* (Bhattacharya 1974). As pointed out by Sen (1981), linear functions of concomitants may also be viewed as mixed rank statistics (Ghosh and Sen 1971).

The resulting burst in the study of concomitants as a class of statistics has been reviewed in David (1981, 1982) and very thoroughly in Bhattacharya (1984). It is the purpose of the present paper to provide an overview of research on concomitants published since the earlier reviews and also to indicate some work in progress. We mention here a recent paper (Do and Hall 1992) in which it is shown that a simulation method suggested by Efron (1990) for more efficiently approximating bootstrap distributions is closely related to techniques based on concomitants of order statistics. The authors develop the asymptotic properties of the method from this viewpoint.

2   Basic Results

Suppose that X_i and Y_i $(i = 1, \cdots, n)$ have means μ_X, μ_Y, variances σ_X^2, σ_Y^2, and are linked by the linear regression model ($|\rho| < 1$)

$$Y_i = \mu_Y + \rho \frac{\sigma_Y}{\sigma_X}(X_i - \mu_X) + Z_i, \tag{2.1}$$

where the X_i and Z_i are mutually independent. Then from 2.1 it follows that $EZ_i = 0$, $\mathrm{var}\, Z_i = \sigma_Y^2(1 - \rho^2)$ and $\rho = \mathrm{corr}\,(X, Y)$. In the special case when the X_i and Z_i are normal, X_i and Y_i are bivariate normal. Ordering on the X_i, we have for $r = 1, \cdots, n$

$$Y_{[r:n]} = \mu_Y + \rho \frac{\sigma_Y}{\sigma_X}(X_{r:n} - \mu_X) + Z_{[r]}, \tag{2.2}$$

where $Z_{[r]}$ denotes the particular Z_i associated with $X_{r:n}$. In view of the independence of the X_i and the Z_i, we see that the set of $X_{r:n}$ is independent of the $Z_{[r]}$, the latter being mutually independent, each with the same distribution as Z_i.

Setting

$$\alpha_{r:n} = E\left(\frac{X_{r:n} - \mu_X}{\sigma_X}\right) \quad \text{and} \quad \beta_{rs:n} = \text{cov}\left(\frac{X_{r:n} - \mu_X}{\sigma_X}, \frac{X_{s:n} - \mu_X}{\sigma_X}\right)$$

$r, s = 1, \cdots, n$, we have from 2.2

$$
\begin{aligned}
E\, Y_{[r:n]} &= \mu_Y + \rho\, \sigma_Y \alpha_{r:n} \\
\text{var}\, Y_{[r:n]} &= \sigma_Y^2(\rho^2 \beta_{rr:n} + 1 - \rho^2) \\
\text{cov}\,(Y_{[r:n]}, Y_{[s:n]}) &= \rho^2 \sigma_Y^2 \beta_{rs:n} \quad r \neq s.
\end{aligned}
\tag{2.3}
$$

The distribution of $Y_{[r:n]}$ for finite n follows directly from 2.2. Now suppose that $n \to \infty$ with $r/n \to \lambda$, a constant. We must distinguish between the quantile case $(0 < \lambda < 1)$ and the extreme-value case $(\lambda = 0$ or $1)$. For ease of writing take $\mu_X = \mu_Y = 0$, $\sigma_X = \sigma_Y = 1$ in 2.2, so that

$$Y_{[r:n]} = \rho X_{r:n} + Z_{[r]}. \tag{2.4}$$

Since in the quantile case $X_{r:n}$ converges in probability to $F_X^{-1}(\lambda)$, we see at once that the asymptotic distribution of $Y_{[r:n]} - \rho F_X^{-1}(\lambda)$ coincides with the distribution of Z. The situation is more complicated in the extreme-value case (see David 1981, p. 283). In the bivariate normal situation one finds that for $\lambda = 0, 1$

$$Y_{[r:n]} \pm \rho(2\log n)^{1/2} \sim N(0, 1 - \rho^2). \tag{2.5}$$

This is, of course, in sharp contrast to the asymptotic behavior of $X_{r:n}$.

A generalization of 2.1 may be noted here. Let $Y_i = g(X_i, Z_i)$ represent a general regression model of Y on X, where neither the X_i nor the Z_i need be identically distributed (but are still independent). Then

$$Y_{[r:n]} = g(X_{r:n}, Z_{[r]}) \qquad r = 1, \cdots, n. \tag{2.6}$$

From the mutual independence of the X_i and the Z_i it follows that $Z_{[r]}$ has the same distribution as the Z_i accompanying $X_{r:n}$ and that the $Z_{[r]}$ are mutually independent (Kim and David 1990).

For completeness we mention also previously reviewed finite-sample and asymptotic results on the moments and distribution of concomitants, obtained without structural assumptions such as 2.6 (Bhattacharya 1974, Yang 1977, Galambos 1987, p. 316). The case when X is a continuous random variable but Y is discrete has been treated by Jha and Hossein (1986).

An important function of the concomitants is the *induced selection differential*

$$D_{[k,n]} = \frac{1}{k} \sum_{i=n-k+1}^{n} (Y_{[i:n]} - \mu_Y)/\sigma_Y$$

which measures the superiority on Y of the k individuals ranked highest on X. The asymptotic distribution of this statistic, suitably standardized, is investigated in both the extreme and the quantile cases by Nagaraja (1982). In the quantile case closely related results were obtained by Bhattacharya (1976); as might be expected, the limiting distribution is normal under mild assumptions.

The asymptotic distribution of general linear functions of concomitants is treated by Yang (1981 a,b). Specifically, he establishes the asymptotic normality under mild regularity conditions of statistics of the form

$$\frac{1}{n} \sum_{i=1}^{n} J\left(\frac{i}{n+1}\right) Y_{[i:n]}$$

and

$$\frac{1}{n} \sum_{i=1}^{n} J\left(\frac{i}{n+1}\right) H(X_{i:n}, Y_{[i:n]}),$$

where J is a bounded smooth function which may depend on n, and $H(x,y)$ is a real-valued function. The second statistic may also be written as

$$T(F_n) = \int_{-\infty}^{\infty} \int_{-\infty}^{\infty} J(F_n(x)) H(x,y) dF_n(x,y),$$

where F_n is the empirical distribution function. Whereas Yang obtained the asymptotic normality of

$$n^{1/2}[T(F_n) - E(T(F_n))],$$

Sandström (1987) proves the asymptotic normality of

$$n^{1/2}[T(F_n) - T(F)]$$

and also, under certain assumptions, of

$$n^{1/2}[T(F_n) - T(F_N)],$$

where F_N is the c.d.f. of a finite population of size N.

3   Selection Through an Associated Variable

Yeo and David (1984) consider the problem of choosing the best k objects out of n when, instead of measurements y_i of primary interest, only associated measurements x_i $(i = 1, \cdots, n)$ are available or feasible. For example, y_i could represent future performance of an individual, with current score x_i, or y_i might be an expensive measurement on the i-th object, perhaps destructive, and x_i an inexpensive measurement. It is assumed that the n pairs (x_i, y_i) are a random sample from a continuous population. The actual values of the x_i are not required, only their ranks. A general expression is developed for the probability π that the s objects with the largest X-values include the k objects $(k \leq s)$ with the largest Y-values. When X and Y are bivariate normal with correlation coefficient ρ, a table of $\pi =_n\pi_{s:k}(\rho)$, for selected values of the parameters, gives the smallest s for which $\pi \geq P^*$, where P^* is preassigned.

Example   From 10 objects it is desired to select a subset of size s that will contain the k best objects $(k = 1, 2, 3)$ with probability at least 0.9. We give a table of s for $\rho = 0.7, 0.8, 0.9$. Thus if we want to be at least 90%

ρ	k	1	2	3
0.7		5	7	na
0.8		4	6	7
0.9		3	5	6

certain that the object with the highest Y-value is in the chosen subset for $\rho = 0.8$, we need to select the four objects with the highest X-value. The full table gives the actual inclusion probability $_{10}\pi_{4:1}(0.8)$ as 0.9183 and also shows that the object with the highest X-value has probability 0.5176 of having the highest Y-value. Another table tells us that for the object with the highest X-value to have probability ≥ 0.90 of having the highest Y-value, would require $\rho \geq 0.993$! For a 50:50 chance $\rho = 0.783$.

With the help of a computer program it is also possible to base the selection of the best object on the actual values of the x_i rather than on their ranks (Yeo and David 1984).

Suppose now that the cost of each Y-measurement is c. Unaware of the preceding approach, Feinberg (1991) has pursued the same aim of using the ordering of the x_i to reduce the number, n_c, of objects for which y_i needs to be measured. He chooses n_c to maximize the difference in expected utility and expected cost:

$$E[\max(Y_{[n:n]}, \cdots, Y_{[n-n_c+1:n]})] - cn_c. \tag{3.1}$$

This is difficult to carry out but Feinberg (1991) includes some simulation results in the bivariate normal case. The asymptotic distribution of

$$V_{n,k} = \max(Y_{[n:n]}, \cdots, Y_{[n-k+1:n]}),$$

and of related statistics, is currently being investigated by H.N. Nagaraja and myself.

Pinhas (1983) studies the following question. Let (X_i, Y_i) be n independent pairs with common c.d.f. $F(x, y)$. Suppose that the stochastic utility of the i-th of n mutually exclusive choices can be represented by $u[\psi_i(Y_i)]$, where u and ψ_i are both increasing function. Then the utility of the information in $(Y_1, \cdots, Y_n)$ on the best choice is $\max Eu[\psi_i(Y_i)]$. If the ranks of the observations $x_1, \cdots, x_n$ are known, the utility is shown to be increased to the average under the $n!$ permutations τ of $(1, \cdots, n)$ of $\max Eu[\psi_{\tau(i)}(Y_{[i]})]$, where $\tau(i)$ is given by $X_{\tau(i)} = X_{(i)}$.

4 Parameter Estimation, Hypotheses Tests

Watterson (1959) treated the linear estimation of the parameters of a bivariate normal population under various forms of censoring. Harrel and Sen (1979) used the method of maximum likelihood in one of these situations, namely when $x_{1:n}, \cdots, x_{k:n}$ and $y_{[1:n]}, \cdots, y_{[k:n]}$ are available. They also give a test of independence of X and Y. Gill et al. (1990) use Tiku's simplified maximum likelihood estimators to deal with two-sided censoring in the same situation, but with possibly more than one set of concomitants.

A rather basic point is made by Lo and McKinlay (1990), namely that in some practical situations the effect of previous selection of objects by their X-values is ignored and the concomitants for the chosen objects are treated simply as random $Y's$. The authors examine the resulting effects on some standard tests of significance, with specific reference to financial asset pricing models.

Motivated by confidentiality considerations, Spruill and Gastwirth (1982) have made the following interesting use of concomitants. The aim is to estimate the correlation coefficient between two sensitive random variables X and Y, data on which is kept by agencies A and B, respectively. Agency A is asked to divide the $N = nm$ individuals into n groups of size m by ordering on x, and to provide the group identification of each individual as well as the group means and variances, viz., for group k, $k = 1, \cdots, n$,

$$\bar{x}_k = \sum x_{(i)}/m, \qquad s_{x,k}^2 = \sum (x_{(i)} - \bar{x}_k)^2/m,$$

where the sums extend over $i = (k-1)m+1, \cdots, km$. Given only the group identifications, agency B simply provides

$$\bar{y}_k = \sum y_{[i]}/m, \qquad s_{y,k}^2 = \sum (y_{[i]}) - \bar{y}_k)^2/m.$$

The least-squares estimate of ρ can now be obtained from $\hat{\rho} = \hat{\beta}\hat{\sigma}_x/\hat{\sigma}_Y$ as

$$\hat{\rho} = \frac{\sum_{k=1}^{n}(\bar{x}_k - \bar{x})(\bar{y}_k - \bar{y})[\sum_{i=1}^{N}(x_i - \bar{x})^2]^{1/2}}{\sum_{k=1}^{n}(\bar{x}_k - \bar{x})^2[\sum_{i=1}^{N}(y_i - \bar{y})^2)]^{1/2}},$$

where $\bar{x}$ and $\bar{y}$ are grand means. Note that only the overall sample variances of X and Y occur in $\hat{\rho}$; the $s_{x,k}^2$ and $s_{y,k}^2$ are needed for assessing the efficiency, E, of $\hat{\rho}$. In the bivariate normal case the authors find $\hat{\rho}$ to be nearly unbiased and E to exceed 0.8 for $n = 10$ in the cases studied ($N = 100, \ 1000; \rho = .25, \ .50, \ .75, \ .9$).

Guilbaud (1985) considers related questions of inference in a slightly more general setting. Let $F(x,y)$ be the c.d.f. of $(X, \ Y)$ and let $0 = \lambda_0 < \lambda_1 < \cdots < \lambda_{k-1} < \lambda_k = 1$. Also let $\xi_j = F_X^{-1}(\lambda_j)$, $j = 1, \cdots, k-1$, $\xi_0 = -\infty$, $\xi_k = \infty$, where $F_X(x)$, the marginal c.d.f. of X, has positive derivatives in the neighborhoods of $\lambda_1, \cdots, \lambda_{k-1}$.

Given a random sample $(x_i, \ y_i)$, $i = 1, \cdots, n$, from $F(x,y)$, natural estimates of the class means are

$$\bar{x}_j = \sum_i x_{(i)}/n_j, \ \bar{y}_j = \sum_i y_{[i]}/n_j, \quad j = 1, \cdots, k$$

where the sums now extend over the n_j observations with x-values in (ξ_{j-1}, ξ_j). With $\mu_{X,j} = E(\bar{X}_j)$, $\mu_{Y,j} = E(\bar{Y}_j)$, Guilbaud gives the 2k-variate asymptotic normal distribution of $n^{1/2}[(\bar{X}_j - \mu_{X,j}), \ (\bar{Y}_j - \mu_{Y,j})]$ with the help of which large-sample tests and confidence intervals can be constructed. He also treats stratified random sampling.

5   Concomitants of Extreme Order Statistics

Gomes (1981, 1984) considers the following situation. Suppose we have N sequences of observations each arranged in *descending* order of magnitude:

$$X'_{1i} \geq X'_{2i} \geq \ldots \geq X'_{n_i,i} \qquad i = 1, \cdots, N.$$

If these sequences are ordered by their maxima X'_{1i}, what is the asymptotic distribution $(n_i \to \infty)$ of the concomitant $X'_{2i}, \cdots, X'_{ki}$ variates, suitably normalized (k fixed)?

To answer this, suppose the sequence $X_1' \geq X_2' \geq \ldots \geq X_n'$ admits constants $a_n(> 0)$ and b_n as well as a nondegenerate extremal c.d.f. $G(x)$ such that

$$\lim_{n \to \infty} Pr(X_1' \leq a_n x + b_n) = G(x)$$

for all x in the set of continuity points of G. Then it is known that the asymptotic joint p.d.f. of $[(X_1' - b_n)/a_n, \cdots, (X_k' - b_n)/a_n]$ is given by

$$h(x_1, \cdots, x_k) = \prod_{i=1}^{k-1} [g(x_i)/G(x_i)]g(x_k) \qquad x_1 \geq \ldots \geq x_k \qquad (5.1)$$

where $g(x) = G'(x)$ (Dwass 1966, Weissman 1975).

We confine ourselves to the case $k = 2$ and write $(X_1' - b_n)/a_n = X$, $(X_2' - b_n)/a_n = Y$. Then from 5.1, as $n \to \infty$,

$$f_{X,Y}(x,y) = g(x)g(y)/G(x) \qquad x > y,$$

so that

$$f_X(x) = g(x) \text{ and } f_Y(y) = g(y)[-\log G(y)].$$

Consequently (Gomes 1984), $U_i = -\log G(X_i)$, $i = 1, \cdots, n$, are independent, identically distributed (i.i.d.) standard exponential random variables and $V_i = -\log G(Y_i)$ are gamma i.i.d. random variables with shape parameter 2. We thus have a linear model

$$V_i = U_i + W_i \qquad i = 1, \cdots, N,$$

where the W_i are i.i.d. standard exponentials independent of the U_i. Hence, ordering on the U_i, we have in our standard notation

$$-\log G(Y_{[N-i+1:n]}) = U_{i:N} + W_{[i]} \qquad i = 1, \cdots, N \qquad (5.2)$$

where the $W_{[i]}$ are i.i.d. standard exponentials, independent of the $U_{i:n}$. This gives (in principle) the distribution of $Y_{[N-i+1:N]}$. If i remains fixed as $N \to \infty$, then $U_{i:N}$ converges in probability to zero and hence from (5.2) $Y_{[N-i+1:N]}$ also has limiting c.d.f. $G(x)$.

So far we have not specified the particular form of $G(x)$. Gomes (1981) gives more explicit results for the distribution and moments of $Y_{[N-i+1:N]}$, for any i, when $G(x) = G_0(x) = e^{-e^{-x}} (-\infty < x < \infty)$. She also deals with the estimation of location and scale parameters when $G(x)$ is replaced by $G((x - \lambda)/\delta)$.

6 Dependence Structure of Concomitants

Definition The random variables $X_1, \cdots, X_n\ (= X)$ are said to be *associated* if $\mathrm{cov}\,[h_1(X), h_2(X)] \geq 0$ for all pairs of increasing functions h_1, h_2 for which the covariance exists.

Let X_i and Z_i $(i = 1, \cdots, n)$ be mutually independent random variables and $Y_i = g(X_i, Z_i)$, leading to 2.6. Then from results on associated random variables given in e.g., Barlow and Proschan (1975), it is easy to show that the concomitants are associated if g is monotone (Kim and David 1990). For $(X_{1:n}, \cdots, X_{n:n})$ and $(Z_{[1]}, \cdots, Z_{[n]})$ are independent sets of associated random variables, so that their union is also associated. Since any monotone functions of associated random variables are associated, $Y_{[1:n]}, \cdots, Y_{[n:n]}$ are associated.

Note that association implies positive quadrant dependence, so that for any y_1, y_2.

$$\mathrm{Pr}\{Y_{[r:n]} \leq y_1,\ Y_{[s:n]} \leq y_2\} \geq \mathrm{Pr}\{Y_{[r:n]} \leq y_1\}\mathrm{Pr}\{Y_{[s:n]} \leq y_2\}.$$

Kim and David (1990) also show that the concomitants satisfy a stronger form of dependence, multivariate total positivity of order two (MTP$_2$) (Karlin and Rinott 1980) if each $Z_{[r]}$ has a Pólya frequency function of order two (PF$_2$).

Acknowledgment

This research has been supported by the U. S. Army Research Office.

References

Barlow, R E. and Proschan, F. (1975). *Statistical Theory of Reliability and Life Testing.* Holt, Rinehart and Winston, New York.

Bhattacharya, P.K. (1974). Convergence of sample paths of normalized sums of induced order statistics. *Ann. Statist.* **2** 1034 - 1039.

Bhattacharya, P.K. (1976). An invariance principle in regression analysis. *Ann. Statist.* **4** 621 - 624.

Bhattacharya, P.K. (1984). Induced order statistics: Theory and applications. In *Handbook of Statistics* Vol. 4, (P.R. Krishnaiah and P.K. Sen, eds.), 383 - 403.

David, H.A. (1973). Concomitants of order statistics. *Bull. Inst. Internat. Statist.* **45** (1) 295 - 300.

David, H.A. (1981). *Order Statistics* (Second edition). John Wiley and Sons, Inc., New York.

David, H A. (1982). Concomitants of order statistics: theory and applications. In *Some Recent Advances in Statistics* (J. Tiago de Oliveira, ed.), 89 - 100. Academic Press, New York.

Do, K.-A. and Hall, P. (1992). Distribution estimation using concomitants of order statistics, with applications to Monte Carlo simulation for the bootstrap. *Jour. Roy. Statist. Soc.* **B 54** 595 - 607.

Dwass, M. (1966). Extremal processes II. *Ill. Jour. Math.* **10** 381 - 391.

Efron, B. (1990). More efficient bootstrap computations. *Jour. Amer. Statist. Assn.* **85** 79 - 89.

Feinberg, F.M. (1991). Modelling optimal cutoff severity: A comparison of discrete and continuous cutoff rules under imperfect information. Submitted for publication.

Galambos, J. (1987). *The Asymptotic Theory of Extreme Order Statistics* (Second edition). Krieger, Florida.

Ghosh, M. and Sen, P K. (1971). On a class of rank order tests for regression with partially informed stochastic predictors. *Ann. Math. Statist.* **42** 650 - 661.

Gill, P.S., Tiku, M.L., and Vaughan, D.C. (1990). Inference problems in life testing under multivariate normality. *Jour. Appl. Statist.* **17** 133 - 147.

Gomes, M.I. (1981). An i-dimensional limiting distribution function of largest values and its relevance to the statistical theory of extremes. In *Statistical Distributions in Scientific Work*, Vol. 6, (C. Taillie et al., eds.), 389 - 410, Reidel, Holland.

Gomes, M.I. (1984). Concomitants in a multidimensional extreme model. In *Statistical Extremes and Applications*, (J. Tiago de Oliveira, ed.), 353 - 364, Reidel, Holland.

Guilbaud, O. (1985). Statistical inference about quantile class means with simple and stratified random sampling. *Sankhyā* **B 47** 272 - 279.

Harrell, F.E. and Sen, P.K. (1979). Statistical inference for censored bivariate normal distributions based on induced order statistics. *Biometrika* **66** 293 - 298.

Jha, V.D. and Hossein, M.G. (1986). A note on concomitants of order statistics. *Jour. Ind. Soc. Agric. Statist.* **38** 417 - 420.

Karlin, S. and Rinott, Y. (1980). Classes of orderings of measures and related correlation inequalities. 1. Multivariate totally positive distributions. *Jour. Multivar. Anal.* **10** 467 - 498.

Kim, S.H. and David, H.A. (1990). On the dependence structure of order statistics and concomitants of order statistics. *Jour. Statist. Plan. Infer.* **24** 363 - 368.

Lo, A.W. and MacKinlay, A.C. (1990). Data-snooping biases in tests of financial asset pricing models. *Rev. Financial Studies* **3** 431 - 467.

Nagaraja, H.N. (1982). Some asymptotic results for the induced selection differential. *Jour. Appl. Prob.* **19** 253 - 261.

Pinhas, M. (1983). Variables concomitantes et information qualitative. *Metron* **41** 147 - 153.

Sandström, A. (1987). Asymptotic normality of linear functions of concomitants of order statistics. *Metrika* **34** 129 - 142.

Sen, P.K. (1981). Some invariance principles for mixed rank statistics and induced order statistics and some applications. *Commun. Statist. - Theory Meth.* **10** 1691 - 1718.

Spruill, N.L. and Gastwirth, J. (1982). On the estimation of the correlation coefficient from grouped data. *Jour. Amer. Statist. Assn.* **77** 614 - 620.

Watterson, G.A. (1959). Linear estimation in censored samples from multivariate normal populations. *Ann. Math. Statist.* **30** 814 - 824.

Weissman, I. (1975). Multivariate extremal processes generated by independent nonidentically distributed random variables. *Jour. Appl. Prob.* **12** 477 - 487.

Yang, S.S. (1977). General distribution theory of the concomitants of order statistics. *Ann. Statist.* **5** 996 - 1002.

Yang, S.S. (1981a). Linear functions of concomitants of order statistics with application to nonparametric estimation of a regression function. *J. Amer. Statist. Assn.* **76** 658 - 662.

Yang, S.S. (1981b). Linear combinations of concomitants of order statistics with application to testing and estimation. *Ann. Inst. Statist. Math.* **33** 463 - 470.

Yeo, W.B. and David, H.A. (1984). Selection through an associated characteristic with applications to the random effects model. *Jour. Amer. Statist. Assn.* **79** 399 - 405.

Chapter 31

Residual Plots for Minimal Resolution IV Designs

PETER W.M. JOHN   Department of Mathematics, University of Texas, Austin, Texas

Abstract   Some experimenters use a normal plot of the residuals to identify which observations in a 2^{n-k} factorial experiment are outliers. But this diagnostic procedure fails when a model with all the main effects is fitted to a foldover design for n factors in $2n$ runs with resolution IV, such as 2^{4-1} and 2^{8-4}. In this situation the residuals of observations in the same foldover pair are identical. The plot degenerates to n duplicate pairs of points and one cannot tell which observation in any pair is an outlier. This paper develops the theoretical background of this phenomenon and its consequences in detail, with an illustrative example from semiconductor technology.

1   Introduction

Many experimenters prefer not to rely upon significance tests with questionable error terms to decide which of the effects in a 2^n factorial, or in a 2^{n-k} fractional factorial, are important. Instead, they use a normal plot, or a half normal plot, of the contrasts as a diagnostic device. It is reasonable to apply the same idea to the question of identifying unusually

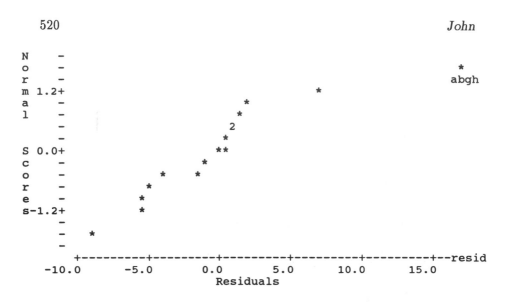

Figure 1: Normal plot of residuals.

large residuals. This conservative approach has been suggested, for example, by Box et al. (1978). If there are no outliers, the normal plot of the residuals will approximate a straight line. Possible outliers will stand out by being off the line to the upper right, or to the lower left.

However, a problem arises with the residual plots because the residuals are not independent. This difficulty is illustrated in Figures 1 and 2.

Figure 1 shows a typical normal plot of the residuals for a 2^{5-1} factorial experiment. It is clear from the plot that the point with the largest residual is an outlier. The experimenter is then faced with a problem. Should the point be repeated? If it cannot be repeated, should the observation be dropped from the data set, or should the model be modified to accommodate that exceptional value?

In Figure 2, on the other hand, the dependencies between the residuals are clear; the sixteen residuals occur in eight pairs of duplicates that cannot be separated. It is the residual plot when a model consisting of all the main effects is fitted to the data for a 2^{8-4} experiment of resolution IV. The data set is presented in Section 3. The degeneracy into eight pairs of duplicate residuals is not a coincidence of the data. It was caused by the combination of model and design and would have occurred no matter what the observations were. The purpose of this paper is to prove that this phenomenon of duplicate residuals always happens when a model consisting of all the main effects is fitted to a minimal resolution IV design and to investigate some of its consequences.

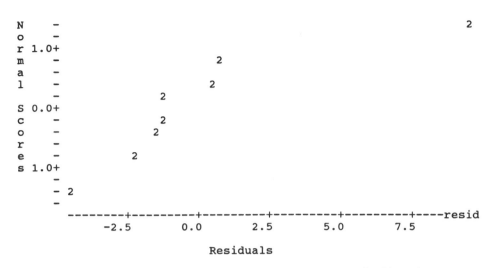

Figure 2: Normal plot of residuals. All residuals are double points.

2   Minimal Resolution IV Designs

In resolution IV fractions of 2^n factorials, the main effects are estimable clear of two factor interactions, but the two factor interactions are aliased with one another. With such a fraction, an experimenter may choose to fit a model that includes only the n main effects by regression, and then plot the residuals.

A minimal resolution IV fraction for n factors has $2n$ runs. It is the smallest resolution IV design for n factors; hence the name *minimal*. All minimal designs are foldover designs. These properties were proved by Webb (1968) and Margolin (1969).

Minimal resolution IV designs can be obtained by starting with a resolution III design for the first $n - 1$ factors in n points. This is run at the low level of the last factor. Then the resolution III design is folded over and run at the high level of the last factor. Folding over consists of repeating the points of the first design with the levels of all the factors changed from high to low or vice versa. The final design thus consists of n pairs of complementary points, called foldover pairs. The most popular examples are the 2^{4-1} fraction that consists of the following eight points, and the

	A	B	C	D
(1)	−	−	−	−
ac	+	−	+	−
bc	−	+	+	−
ab	+	+	−	−
abcd	+	+	+	+
bd	−	+	−	+
ad	+	−	−	+
cd	−	−	+	+

2^{8-4} fraction that is listed in Table 1.

A detailed account of the properties of foldover designs and their use is given by Box and Hunter (1961).

Minimum foldover designs do not have to be orthogonal fractions. The following example is a non-orthogonal minimal design for three factors in six runs (John 1962):

	A	B	C
(1)	−	−	−
a	+	−	−
ab	+	+	−
abc	+	+	+
bd	−	+	+
c	−	−	+

3   A Minimal Design with an Outlier

Table 1 shows the data for a minimal design for eight factors in sixteen runs. This data set comes from an experiment to investigate the sheet resistivity of a P-doped well in a silicon substrate. The response observed is a measure of the sheet resistivity.

The extreme right column of the table shows the residuals when the main effects model has been fitted to the data. The normal plot of them appeared earlier in Figure 2. The residuals of the points in foldover pairs are equal. They form the double points in Figure 2. One pair of residuals is clearly "off the line". They are 9.40 for *abgh* and its complement *cdef*. One, or both, of those observations may be an outlier, but we cannot separate them and tell which is the offender.

Table 1: Sheet resistivity.

	A	B	C	D	E	F	G	H	Y	res
(1)	−	−	−	−	−	−	−	−	567.2	−1.30
aefg	+	−	−	−	+	+	+	−	533.9	−1.35
befh	−	+	−	−	+	+	−	+	535.7	−1.45
abgh	+	+	−	−	−	−	+	+	522.4	9.40
cegh	−	−	+	−	+	−	+	+	557.6	0.50
acfh	+	−	+	−	−	+	−	+	522.3	−2.20
bcfg	−	+	+	−	−	+	+	−	590.7	−4.45
abce	+	+	+	−	+	−	−	−	568.4	0.85
abcdefgh	+	+	+	+	+	+	+	+	526.2	−1.30
bcdh	−	+	+	+	−	−	−	+	559.4	−1.35
acdg	+	−	+	+	−	−	+	−	557.4	−1.45
cdef	−	−	+	+	+	+	−	−	592.4	9.40
abdf	+	+	−	+	−	+	−	−	539.4	0.50
bdeg	−	+	−	+	+	−	+	−	569.3	−2.20
adeh	+	−	−	+	+	−	−	+	496.4	−4.45
dfgh	−	−	−	+	−	+	+	+	529.3	0.85

4   The Main Theorem

We now prove the main theoretical result of this paper.

Theorem

Suppose that a main effects model is fitted to a minimal resolution IV design for n factors. The residuals of two observations that form a foldover pair are equal.

Proof of Theorem

We denote the factor levels by the coordinates ± 1 and fit the regression model

$$y = \beta_0 + \beta_1 x_1 + \beta_2 x_2 + \ldots + \beta_n x_n + e.$$

We order the points in the fraction, as we have in the examples, so that the points in the second half are the foldover images of the points in the first half in the same order. The design matrix may then be written as

$$X = \begin{pmatrix} 1 & V \\ 1 & -V \end{pmatrix},$$

where V is a square matrix with n rows and columns, and 1 is a vector of ones. The points that correspond to the i-th and $(n+i)$-th rows for X are a foldover pair.

The vector of residuals is $[I - H]Y$, where H denotes the hat matrix

$$H = X(X'X)^{-1}X'.$$

In the present situation,

$$X'X = \begin{pmatrix} 2n & 0' \\ 0 & 2V'V \end{pmatrix},$$

$$2n(X'X)^{-1} = \begin{pmatrix} 1 & 0' \\ 0 & nV^{-1}(V')^{-1} \end{pmatrix},$$

whence

$$2nH = \begin{pmatrix} 1 & V \\ 1 & -V \end{pmatrix} \begin{pmatrix} 1 & 0 \\ 0 & nV^{-1}(V')^{-1} \end{pmatrix} \begin{pmatrix} 1' & 1' \\ V' & -V' \end{pmatrix}$$

$$= \begin{pmatrix} 1 & n(V')^{-1} \\ 1 & -n(V')^{-1} \end{pmatrix} \begin{pmatrix} 1' & 1' \\ V' & -V' \end{pmatrix}$$

$$= \begin{pmatrix} J + nI & J - nI \\ J - nI & J + nI \end{pmatrix},$$

where J is a square matrix of ones, so that

$$2n[I - H] = \begin{pmatrix} nI - J & nI - J \\ nI - J & nI - J \end{pmatrix}.$$

The theorem is proved when we notice that the i-th and $(n + i)$-th rows of $2n[I - H]$ are identical. ∎

5   Calculating the Residuals

In the case of four factors in eight runs, we have

$$8[I - H] = \begin{pmatrix} 4I - J & 4I - J \\ 4I - J & 4I - J \end{pmatrix}.$$

The vector of observations is

$$[(1), \ ab, \ ac, \ bc, \ abcd, \ cd, \ bd, \ ad]',$$

and so the residuals for (1) and for $abcd$ are each equal to

$$[4(1) + 4abcd - (1) - ab - ac - bc - abcd - cd - bd - ad]/8 = [(1) + abcd]/2 - \bar{y}.$$

This illustrates the general formula for minimal foldover designs with main effects models:

> The common residual for the members of a foldover pair is obtained by subtracting $\bar{y}$ from their average.

6   The Next Step

It is important to emphasize that the theorem in Section 4 depends upon the design being minimal. The derivation of the hat matrix depended upon V being a square, non-singular matrix. If a foldover design has $2n$ points but fewer than n factors, V has n rows but fewer than n columns. Consequently, $V'V^{-1}$ is no longer equal to $V^{-1}(V')^{-1}$ and the hat matrix does not have the form that it takes in the theorem. The difficulty with the normal plot of the residuals does not follow. This suggests that the experimenter should drop one or more of the weakest factors from the model and refit.

In the example of Table 1, the normal plot of the effects shows clearly that A, C and H are the only important effects. The residual plot that was shown in Figure 1 is the plot of the residuals for the data in Table 1 after the model with the main effects of only A, C, H is fitted. The problem of duplicate pairs of residuals has disappeared. It is obvious that $abgh$, rather than $cdef$, is the outlier.

7   Omitting an Observation

Another strategy would be to drop each member of the suspect pair in turn and see what is achieved. In this section we show what happens when we omit an observation, refit the main effects model to the remaining $N-1$ points, and predict the value of the missing observation from that fit. We shall see that this strategy has little to recommend it and that it may raise more questions than it answers.

Suppose that the regression model

$$Y = X\beta + e$$

is fitted to N data points, and that $\hat{\beta}$ is the estimate of the vector of coefficients. Let x_i' be the i-th row of the data matrix, i.e., the set of coordinates of the i-th point, let h_{ii} be the i-th diagonal element of H, and let e_i be the residual of the i-th point. If the i-th point is now omitted and the model is fitted to the remaining $N-1$ points, the new vector of coefficients, $\beta_{(i)}$ is given by

$$(1 - h_{ii})[\hat{\beta} - \hat{\beta}_{(i)}] = (X'X)^{-1}x_i e_i.$$

(See, for example, Weisberg 1985, p. 126.)

Denote the vector of residuals from the new fit by δ; in this case, the residual δ_j will be the difference between the observation y_j and its predicted value from the second fit. Let e be the vector of residuals in the original regression. Then

$$\delta = Y - X\hat{\beta}_{(i)} = e + X(X'X)^{-1}x_i e_i/(1 - h_{ii}),$$

and

$$
\begin{aligned}
\delta_j &= e_j + x_j'(X'X)^{-1}x_i e_i/(1 - h_{ii}) \\
&= e_j + h_{ij} e_i/(1 - h_{ii}).
\end{aligned}
$$

Applying this result to the present situation, in which

$$2nH = \left(\begin{array}{cc} J + nI & J - nI \\ J - nI & J + nI \end{array} \right),$$

we see that

(i) for the omitted point, $h_{ii} = (n + 1)/2n$ and

$$\delta_i = e_i/(1 - h_{ii}) = 2ne_i/(n - 1);$$

(ii) for the other point in the foldover pair,

$$
\begin{aligned}
h_{i,n+i} &= (1 - n)/2n \text{ and} \\
\delta_{n+i} &= e_{n+i} - e_i = e_i - e_i = 0;
\end{aligned}
$$

(iii) for the remaining points, $h_{ij} = 1/2n$ and $\delta_j = e_j + e_i/(n - 1)$.

This says that the other point of the pair fits the new regression perfectly. The residuals of the remaining points again occur in pairs, because they are obtained by adding the same quantity $e_i/(n - 1)$ to the old residuals; furthermore the new residuals, and, hence, the new sum of squares for error, are the same whichever of the two points we drop.

8   Omitting Points in the Example

We apply the results of Section 7 to the example. The original residuals were shown in Table 1. The residuals when *abgh* is omitted and the main effects model is fitted to the other fifteen points are shown in Table 2.

Table 2: Residuals when $abgh$ is omitted.

(1)	$aefg$	$befh$	$abgh$	$cegh$	$acfh$	$bcfg$	$abce$
0.04	−0.01	−0.11	(21.5)	1.84	−0.86	−3.11	2.19
$abcdefgh$	$bcdh$	$acdg$	$cdef$	$abdf$	$bdeg$	$adeh$	$dfgh$
0.04	−0.01	−0.11	0.00	1.84	−0.86	−3.11	2.19

Notice that:

(1) the new residual for $cdef$ is zero;

(2) the original residual for $abgh$ and for $cdef$ is $e_i = 9.40$; the new residuals for the points other than $abgh$ and $cdef$ are obtained by adding

$$e_i/(n-1) = 9.40/7 = 1.3429$$

to their earlier values;

(3) the new residual for $abgh$ is $16(9.40)/7 = 21.5$ and so the predicted value of $abgh$ from the other fifteen points is

$$522.4 - 21.5 = 500.9.$$

If we omit $cdef$ instead of $abgh$, the residual for $abgh$ will be zero, the predicted value of $cdef$ will be $592.4 - 21.5 = 570.9$, and the other residuals will be the same as they are in Table 2.

9 The New Estimates When a Predicted Value is Used

The result of omitting a point and substituting its predicted value is to subtract the quantity $\delta_i = 2ne_i/(n-1)$ from the observation. This, in turn, subtracts a quantity $\pm\delta_i$ from each of the n contrasts for the main effects and, in the case of orthogonal designs, subtracts δ_i/n from the estimate of each main effect. Since the regression coefficients are equal to half the estimates of the corresponding effects, they will be reduced by

$$\pm\delta_i/2n = e_i/(n-1).$$

In our example, $\delta_i = 21.5$. The predicted value for $abgh$, when it is omitted, is 500.9. Subtracting 21.5 from $abgh$ decreases the A, B, G and H contrasts by 21.5, and decreases the estimates of those main effects by 21.5/8. The change also increases the estimates of the main effects of C, D, E, F by

21.5/8. On the other hand, replacing $cdef$ by its predicted value decreases the value of $cdef$ by 21.5; it increases the A, B, G and H contrasts and decreases the C, D, E, and F contrasts by that amount. The residual sum of squares is the same in both cases. The regression coefficients are changed by ± 1.34.

The new regression coefficients in the two cases are shown in Table 3. It is interesting to note that when the actual outlier $abgh$ is omitted and replaced by its predicted value, there is only one change in the list of effects that would be declared significant if one were to use a t-test at $\alpha = 5\%$. The original three significant factors, A, C, H, have been joined by B. On the other hand, when $cdef$, which was not considered to be an outlier, is changed, all factors except G acquire a significant t (and F) value at the 5% level. This strategy has served only to muddy the waters further.

Table 3: Estimated regression coefficients.

	change $abgh$			change $cdef$		
	Coef	t		Coef	t	
Const	546.656	948.35		546.656	948.35	
A	−16.044	−27.83	A	−13.356	−23.17	A
B	2.094	3.63	B	4.781	8.29	B
C	12.644	21.93	C	9.956	17.27	C
D	−0.431	−0.75	D	−3.119	−5.41	D
E	0.831	1.44	E	−1.856	−3.22	E
F	−0.419	−0.73	F	−3.106	−5.39	F
G	−0.994	−1.72	G	1.694	2.94	G
H	−18.181	−31.54	H	−15.494	−26.88	H

10   Omitting Both Points

We have seen that dropping one of the points and predicting a replacement value is not satisfactory. The experimenter might consider omitting both points and using only the remaining fourteen. But this only worsens the situation.

At first glance, there would seem to be no problem. After all, we still have fourteen observations and there are only eight main effects plus the constant term to be estimated. That should leave five degrees of freedom. But there is a difficulty. When a foldover pair is lost from a minimal foldover design, the resolution of the design formed by the remaining points falls from four to two. We can no longer even estimate the main effects.

The reason is that the non-singularity of $X'X$ requires that the matrix, $V'V$, be nonsingular. When a foldover pair is dropped from the fraction, the general form of the design matrix remains the same, but corresponding rows have been dropped from V and $-V$; V now has n columns, and only $n-1$ rows. It follows that the rank of $V'V$ is at most $n-1$ and that $X'X$ is singular.

11 Summary

This paper began with a discussion of the use of normal plots of the residuals to identify outliers in 2^{n-k} factorials. For the popular minimal resolution IV fractions (such as four factors in eight runs and for eight factors in sixteen runs), these plots behave in a peculiar way. The minimal fractions are foldover designs, and the residuals of each member of any foldover pair are identical.

If, therefore, the experimenter insists on making a residual plot using all the factors in a minimal resolution IV experiment the plot will be useless for detecting outliers. On the other hand, if the experimenter waits until some of the unimportant factors have been detected and dropped from the model before making the residual plot, the mathematical difficulty will disappear. Then, as in Figure 1, the plot will give useful results. Another way of saying this is that the experimenter should get a good handle on the correct model before starting on residual analysis!

Acknowledgment

This research was supported by the Semiconductor Research Corporation Grant SRC-89-MJ-136.

References

Box, G.E.P. and Hunter, J S. (1961). The 2^{k-p} fractional factorial designs. *Technometrics* **3** Part I 311 - 352, Part II 449 - 458.

Box, G.E.P., Hunter, J.S. and Hunter, W.G. (1978). *Statistics for Experimenters*. John Wiley and Sons, Inc., New York.

Daniel, C. (1976). *Applications of Statistics to Industrial Experimentation*. John Wiley and Sons, Inc., New York.

John, P.W.M. (1962). Three-quarter replicates of 2^n designs. *Biometrics* **18** 172 - 184.

Margolin, B.H. (1969). Resolution IV fractional factorial designs. *Jour. Roy. Statist. Soc. B* **31** 514 - 523.

Webb, S.R. (1968). Non-orthogonal designs of even resolution. *Technometrics* **10** 291 - 300.

Weisberg, S. (1985). *Applied Linear Regression* (Second Edition). John Wiley and Sons, Inc., New York.

Chapter 32

An Optimal Stopping Rule for Multinomial Inverse Sampling Problems

MILTON SOBEL Department of Statistics, University of California, Santa Barbara, California

PINYUEN CHEN Department of Mathematics, Syracuse University, Syracuse, New York

Abstract Each cell in a given subset of the cells with a multinomial distribution has specified frequency quotas and the remaining cells are quota-free. Our goal is to get all of these quotas satisfied with a high specified probability. It is clear that an optimal solution for minimizing the expected waiting time includes stopping as soon as the goal is reached. However, we set two practical and realistic conditions on the waiting-time procedure. Although we still stop when the goal is reached, to specify the procedure explicitly we need a few more stopping points *outside the goal*. The problem is to add these stopping points so that the expected total time until stopping is minimized. In the simplest solution we use the concept of the length of the initial run of a combined quota-free cell to determine one of the new stopping points and allow all the others to be determined by one of the conditions. Randomization then yields the optimal solution to our waiting-time problem. The main tool used is the Dirichlet integral.

1   Introduction

A frequency quota for any one cell in a multinomial with k cells is defined as the minimum positive frequency for that cell which you are waiting to achieve. If we have frequency quotas $f_1, f_2, \cdots, f_c$ for a subset c cells $(c \leq k)$ then we way that the goal is reached when all the c quotas are satisfied. The cells without quotas are said to be quota-free. The use of the term 'quota' to denote the minimum desired frequency for a cell (rather than the maximum) also appeared in a related paper by Anderson et al. (1982) on quota fulfilment dealing with both multinomial and hypergeometric distributions. Waiting time and quota fulfilment problems have been of interest to statisticians for a long time (cf., the references in the paper cited above).

There are two related problems at hand. The first problem is simply to calculate the mean, variance etc. of the waiting-time for the unrestricted problem of waiting for the goal to be reached. The second problem is based on putting two restrictions on the waiting-time problem. Using "RG" for reaching the goal and "T" for the total number of observations required, these restrictions are

$$P\{RG\} \geq P^*, \qquad \text{with } P^* \text{ specified,} \qquad (1.1)$$

$$P\{\underline{S} \leq T \leq \overline{S}\} = 1 \qquad (1.2)$$

with $\underline{S} \leq \overline{S}$ both given and such that condition (1.1) can be satisfied.

It is intuitively clear that the optimal solution will be to stop as soon as the goal is reached. However, with these two above restrictions a few stopping points have to be added that are not in the goal and the problem is how to add these stopping points in an optimal manner, i.e., so as to minimize the expected waiting time for terminating the procedure.

The unrestricted problem is a simple application of the Dirichlet integrals which are defined, studied, and tabled in Sobel et al. (1977, 1985) (which we refer to below as SUF1 and SUF2, respectively) and these results are also used in the restricted problem. The restricted problem is solved by allowing all new stopping points, except for one, to be determined by the given upper bound $\overline{S}$. For the one exceptional stopping point, we use the concept of the length r of the initial run of observations from a combined quota-free cell. Thus the $k - c$ quota-free cells are first combined to form a single cell, say C_0, and the initial run is a run from C_0 before any observation from any of the c cells with positive quotas. We find two successive integers, $r - 1$ and r, such that the $P\{RG\} \geq P^*$ if we use as a stopping point an initial run from C_0 of length r and $< P^*$ if we use as a stopping point aninitial run from C_0 of length $r - 1$, respectively. Then randomization is used between these two procedures to get $P\{RG\} = P^*$ exactly and the same randomization of the $E\{T\}$ - values furnishes the desired optimal solution.

To avoid certain degenerate and semi-degenerate cases at the outset, it will be necessary to assume that P^* is not too small and that $\overline{S}$ is large enough for a solution to exist; these additional assumptions will be quantified later.

As an illustration of the use of this formulation we consider a few "baby problems" in which the parents specify in advance that their goal is to have b boys and g girls and that they want to have at least $\underline{S}$ children and at most $\overline{S}$ children with probability 1, and would like to have probability at least P^* of reaching their goal. Assuming only single births with $p_b = p_g = 1/2$ and that $b > 0$ and $g = 0$, then for $P^* > 1/2$ we use the "girls" as the quota-free cell and our solution for minimizing the expected total number $E\{T\}$ of children required then applies directly to this case. Some tables for selected values of $\overline{S}$, b, g and P^* (with $\underline{S} = b + g$) are given at the end of the paper; some cases with b and g both positive are also given.

The above binomial "baby problem" was also treated in Sobel and Ebneshahrashoob (1989) but with slightly different restrictions. There the assumption $P\{\underline{S} \leq T \leq \overline{S}\} = 1$ was replaced by the restriction that with probability one we would have either b boys or g girls. The optimal solution in that case has some unbounded features and hence is quite different from the present procedure. (However a bounded version of that procedure given in Sobel and Ebneshahrashoob (1989, Section 1) as an illustration (with the notation N in place of $\overline{S}$) turns out to be an initial run procedure within the framework of our present paper.) Hence, although that paper is related to the present one, the present paper has a different formulation and is more general in that it also deals with multinomial waiting-time problems. Other papers dealing with run quotas are also in the literature (e.g., Ebneshahrashoob and Sobel 1991) but these are not concerned with the initial run length of a specified quota-free cell or of a combined low quota cell consisting of a subset of specified cells.

2 Procedures for Different Cases

In the discussion below we make use of curtailment bounds for stopping points. For simplicity we assume a positive frequency quota b for boys only, in which case the curtailment bound $\overline{S} - b + 1$ appears on the girl axis only. If we have positive frequency quotas b and g for boys and girls respectively, then we use both curtailment bounds: $\overline{S} - b + 1$ on the girl-axis and $\overline{S} - g + 1$

on the boy axis.

We wish to distinguish a regular case in which the $P\{RG\}$ reaches the specified P^* for some value of $r(\underline{S} \leq r \leq \overline{S} - b + 1)$ from an irregular case in which we have to combine quota-free and low-quota cells before finding an optimal solution. To define some notation needed for this let $p_i > 0$ denote the cell probability associated with the frequency quota $f_i(i = 1, 2, \cdots, c)$. For convenience and simplicity we assume that the specified $\underline{S}$ is equal to the sum of the c given frequency quotas. In both of the two cases being defined, we first combine all quota-free cells into a single quota-free cell, denoted by C_0, with probability $p_0 = 1 - \sum_{i=1}^{c} p_i$. In the irregular case we further combine C_0 with one (or more) low-quota cells and denote the result by C_0^* with cell probability p_0^* and let c^* denote the number of quota cells not in C_0^*, i.e., $p_0^* = 1 - \sum_{i=1}^{c} p_i$.

In both of the above cases we wish to show that the optimal procedure for minimizing $E\{T\}$ always has the following properties:

1. Stop as soon as the goal is reached (provided you did not get through any stopping points).

2. All the stopping points that are not in the goal (with at most a small number of exceptions) are determined by the idea of curtailment. Thus, if the goal is b boys only and you reach (α boys, $\overline{S} - b + 1$ girls) for any $\alpha < b$ then you stop, since you cannot reach the goal in the remaining $b - \alpha - 1$ observations.

3. If $P\{RG\} > P^*$ for all values of r available to us then we put a star (*) before $\overline{S}$ in the table and reconsider the problem with $\overline{S}$ reduced to $\overline{S} - 1$; we continue this provided the new $\overline{S} \geq \underline{S}$ with $P\{RG\} < P^*$ for at least one value of r available to us. The structure of the stopping points is not affected by these steps.

4. In the usual case of one exceptional stopping point in (2), this point is determined by condition (1.1) and *the length r of an initial run of observations from C_0* (or from C_0^* in the degenerate case), i.e., we find the *smallest* positive integer r such that if we use an initial run of length r from C_0 as a stopping point then $P\{RG\} \geq P^*$, *and* if we use an initial run of length $r - 1 \geq \underline{S}$ from C_0 as a stopping point, we have $P\{RG\} < P^*$. In this case the final result is obtained by randomization between $r - 1$ and r to get P^* exactly; the same randomization on $E\{T\}$ yields the desired optimal result for $E\{T\}$.

5. In the unusual case (where we need two or more exceptional stopping points not dictated by curtailment), we find that uniformly in the value of r_1 considered, the value of $P\{RG\} > P^*$ for some $\overline{S}$ and

$P\{RG\} < P^*$ for $\overline{S} - 1$. In particular this is true for $r_1 = \underline{S}$ and hence the value $r_1 - 1$ is not available to us without contradicting condition (ii). In this case we set $r_1 = \underline{S}$ and reduce the value r_2 of the total number of girls before the second boy needed for stopping until we find a value for which $P\{RG\} < P^*$ or until we find a pair $(r_2 - 1, r_2)$ for which the $P\{RG\}$-values straddle P^*. If we find such a pair then we randomize between these two so that $P\{RG\} = P^*$ exactly and the same randomization on the associated $E\{T\}$-values give the desired optimal value (and strategy) for $E\{T\}$. If we do not find such a pair then we set $r_2 = r_1 = \underline{S}$ and reduce the value r_3 of the total number of girls before the third boy needed for stopping until we find a value r_3 for which $P\{RG\} < P^*$ or until we find a pair $(r_3 - 1, r_3)$ for which the $P\{RG\}$-values straddle P^*, etc.

As a result of the above procedure we may find several stopping points with abscissa $\underline{S}$, at most one in the open interval between $\underline{S}$ and $\overline{S} - b + 1$ (assuming still that the goal is b boys), and the remaining stopping points of $\overline{S} - b + 1$ all determined by curtailment, this characterizes the structure of the stopping points in the general case.

In the examples studied in Table 1 only one case was found where we had to go beyond the initial run r_1 to get an optimal solution. Namely for $p = .50$, $P^* = .75$ and $\overline{S} = 7, 8, 9$ or 10 for the Goal: 3 boys and 1 girl, it was necessary to reduce r_2 from $\overline{S} - b + 1 = 7 - 3 + 1 = 5$, to $4 = \underline{S}$ to get a $P\{RG\} = .72656$; for $r_2 = 5$ the value was $.75781$. Randomizing to get exactly $P^* = .75$ we obtain $E\{T\} = 5.34375$, which is smaller than any of the 4 entries marked with a # in Table 1.

The first of our two cases revolves around the existence of the r-value in property (4); in the second case there is no r-value (consistent with curtailment) which enables us to reach or exceed P^*. The "baby problem" with a goal of (say) 2 boys and 1 girl is an example of the irregular case since there is no quota-free cell; for $p_b = 1 - p_g \le 1/2$ the optimal solution uses the girl "cell" to form C_0^* and stops with a sufficiently long initial run of girls. Formally we have the two cases as follows:

Case 1: Regular Case

For some $r \ge \underline{S}$ (and consistent with curtailment at $\overline{S}$), we have $P\{RG\} \ge P^*$. Let r_1 denote the smallest such integer. If $r_1 > \underline{S}$ then for $r = r_1$ we have $P\{RG\} < P^*$ and we randomize between $r_1 - 1$ and r_1 to get P^* exactly.

Case 2: Irregular Case

There is no positive integer $r \ge \underline{S}$ consistent with curtailment such that Case 1 holds. This could happen because $\overline{S}$ is too small or because p_0 is too close to zero; in particular if the quota-free cell C_0 with one (or more) low-quota cells to form C_0^* and the exceptional point is determined by P^* and the initial run length from C_0^*. It should be noted that we may not get any solution if either $\overline{S}$ is unreasonably small or if no combinations are possible. It may also happen that after the combination we are again in the irregular case; we include all of these as belonging to Case 2.

To illustrate Case 2, consider the "baby problem" with the goal of 2 boys and 1 girl, where $p_b = p_g = 1/2$ and there is no quota-free cell. Here we use the girl category as the new cell C_0^* and stop with an initial run of girls of length r_1, where r_1 is determined by P^*. Thus for $P^* = .90$, $\underline{S} = 3$ and $\overline{S} = 7, 8$ and 9, we obtain, respectively, $r_1 = 5, 4$, and 4 as the initial run length of girls needed for stopping. The exact answers for $E\{T\}$ are 4.17917, 4.13667 and 4.11774, respectively, which are given in Table 1 below. For these cases the Dirichlet or unrestricted answer is 4.5 and the closer upper bound obtained by letting $\overline{S} \to \infty$ (for any P^*) is $.5 + 4P^*$, which is equal to 4.1 for $P^* = .90$; the latter result will be explained and illustrated after our main result below.

To illustrate the general structure (5) of the solution (which goes beyond the initial run) we point out that the goal: 3 boys and 1 girl (with only 2 cells) with $P^* = .75$, $p = .50$, $\underline{S} = 4$ and $\overline{S} = 7, 8, 9$, or 10, is included on Page 2 of Table 1. Since a * appears on the first line of this subgroup (in front of 7), we have to reduce the stopping point before the second boy from the curtailed value $\overline{S} - b + 1 = 7 - 3 + 1 = 5$ to $\underline{S} = 4$. The values of $P(RG)$ for stopping at $r_2 = 5$ and at $r_2 = 4$ are $.75781 > P^*$ and $.72656 < P^*$. Randomizing between these two to get exactly $P^* = .75$, we use the same randomization to obtain the desired optimal value $E\{T\} = 5.34375$ which is smaller than any of the four entries marked with a # in Table 1. Since this happened in only one case of Table 1, we claim the optimal solution will *usually* depend only on the initial run from the combined quota-free cell.

As another illustration of the general structure (5) which is not in Table 1 below, consider the goal of 7 boys and 0 girls with $p = .50$, $P^* = .95$ and $\overline{S} = 21$. For $\overline{S} \leq 20$ (resp., $\overline{S} \geq 21$) the value of $P\{RG\} < .95$ (resp., $> .95$, strictly) if we look for a solution based only on the initial run of girls with $r_1 = 7$ and the other stopping points all at 14 (resp., 15) (by virtue of curtailment). Hence we set $r_1 = \underline{S} = 7$ and move r_2 downward, keeping all the other stopping points at 15. For the pair $(r_2 - 1, r_2) = (8, 9)$ we obtain $P\{RG\} = .94772$ (resp., .95268) and randomizing between these two gives us exactly .95. The same randomization of $E\{T\}$ yields the result 13.65436, which is the desired optimal solution for $P^* = .95$. It appears that for larger P^* (i.e., as $P^* \to 1$) there will be a succession of $\overline{S}$-values for which the optimal solution simplifies (as on Page 2 of Table 1) and depends only on the initial run from the combined quota-free cell. In the illustration above with 7 boys and 0 girls for $P^* = .99$ this is true for $\overline{S} = 25$, 26 and 27. Since this result has not been shown it should be treated as a further conjecture. This could be a justification for referring to case (4) as the 'usual' case.

3   Derivation of $P\{RG\}$ and $E\{T\}$ Formulas

Consider any waiting-time procedure that stops as soon as the goal is reached except for a few isolated stopping points not in the goal. In our examples these (restriction) points are needed because of the presence of an upper bound $\overline{S}$ on the total number of observations given in Condition (1.2). Since the p_i for the quota cells are positive for each $i(i = 1, 2, \cdots, c)$, it follows that in the unrestricted (and hence unbounded) problem the value of $P\{RG\}$ is 1; the proof of this is omitted. For any closed set of stopping points the sum of the probabilities of reaching a stopping point without going through any other stopping point is 1, provided that we sum over all (and only those) stopping points that can be reached with positive probability. Let the set of stopping points not in the goal be denoted by $Z_1, Z_2, \cdots, Z_w$ and let N_j denote the number of paths from the origin to Z_j which do not go through any other Z-values or through any of the points in the goal. Then we have the following lemma for the procedure based on these Z-values.

Lemma 2.1

For the above described (binomial or multinomial) model

$$P\{RG\} = 1 - \sum_{j=1}^{w} N_j P\{Z_j\}, \qquad (3.1)$$

where $P\{Z_j\}$ is the probability of reaching Z_j with any specified order of observations.

Proof of Lemma 2.1

Since $N_j P\{Z_j\}$ is the total probability of reaching $Z_j(j = 1, 2, \cdots, w)$ and we have a closed set of stopping points, this is an immediate consequence of the fact that the total sum over all stopping points (inside and outside the goal) is 1. ∎

The result (3.1) is quite useful in all of our examples because the value of w is generally small relative to the total number of stopping points and because it leads to an analogous result for the expected total number T

of observations needed to reach any stopping point (i.e., to terminate the procedure with restrictions present).

For the unrestricted problem with c quota cells (i.e., with the ith cell having cell probability p_i and a positive quota f_i $(i = 1, 2, \cdots, c)$), we use the Dirichlet form of the expectation given in equation (5.8) of SUF2 with $\gamma = 1$ for the first moment and with b replaced by c; the unrestricted result (which we call the Dirichlet coefficient) is

$$E\{T_U | \underset{\sim}{f}, \underset{\sim}{p}\} = \sum_{i=1}^{c} \frac{f_i}{1/p_i} C_{p_{\sim i}/p_i}^{(c-1)}(\underset{\sim i}{f}, f_i + 1) \qquad (3.2)$$

where $\underset{\sim i}{f} = (f_1, f_2, ..., f_{i-1}, f_{i+1}, ..., f_c)$, $\underset{\sim i}{p} = (p_1, p_2, ..., p_{i-1}, p_{i+1}, ..., p_c)$ and $C_{\underset{\sim}{a}}^{(c-1)}(\underset{\sim}{r}, m)$ is a $(c-1)$-fold integral defined, studied and tabled in SUF2. For equal arguments in $\underset{\sim}{r}$ and equal components in $\underset{\sim}{a}$, we can easily read off the C-values from the tables in SUF2 and hence easily compute the value of (3.2). In the remainder of this paper we use the symbol r and r_1 (interchangeably), for the length of the initial run from the combined quota-free cell.

For each of the w stopping points not in the goal, there is a zero or a partial fulfilment of the goal requirements for the quota cells. Let the vector $\underset{\sim}{f}^{(j)}$ denote the remaining unfulfilled quotas for the jth stopping point outside the goal. Note that some of the c components may now be zero. Then the expected saving due to the use of the jth stopping points outside the goal is $E\{T_U | (\underset{\sim}{f}^{(j)}, \underset{\sim}{p})\}$. In analogy with Lemma 2.1, we now have a result for $E\{T\}$ using the same procedure based on Z_j $(j = 1, 2, \cdots, w)$.

Lemma 2.2

For the above described (binomial or multinomial) model

$$E\{T\} = E\{T_U | \underset{\sim}{f}, \underset{\sim}{p}\} - \sum_{j=1}^{w} N_j P\{Z_j\} E\{T_U | \underset{\sim}{f}^{(j)}, \underset{\sim}{p}\}, \qquad (3.3)$$

where $P\{Z_j\}$ is as before and all Dirichlet coefficients on the right side of (3.3) are easily obtained from SUF2 (eqn. 5.8) with $\gamma = 1$. Outside the sum in (3.3) the components of f are the original frequency quotas but inside the sum some of them (or all of them) may be reduced in value.

The simplicity of (3.3) cannot be appreciated without an illustration. In the "baby problem" with 2 boys and 1 girl as the goal, we can plot the procedure in 2 dimensions with girls (respectively, boys) on the x-axis (respectively, the y-axis). In this binomial problem with $p_b = p_g = 1/2$

there are $w = b + g = 3$ stopping points, not in the goal; these are $(0, \overline{S})$, $(r, 0)$ and $(\overline{S} - 1, 1)$. The corresponding last three Dirichlet coefficients for (3.3) are 2, 4, and 2, respectively, using the geometric distribution with $p_b = p_g = 1/2$. The first Dirichlet coefficient in (3.3), by SUF2 (eqn. 5.8) is

$$
\begin{aligned}
E\{T_U | \underset{\sim}{f}, \underset{\sim}{p}\} &= \frac{2}{(1/2)} C_1^{(1)}(1, 3) + \frac{1}{(1/2)} C_1^{(1)}(2, 2) \\
&= 4 \left(\frac{7}{8} \right) + 2 \left(\frac{1}{2} \right) = 4.5.
\end{aligned}
\tag{3.4}
$$

Hence, the value of $E\{T\}$ by (3.3) is

$$
E\{T\} = 4.5 - 2 \left(\frac{1}{2} \right)^{\overline{S}} - 4 \left(\frac{1}{2} \right)^r - 2r \left(\frac{1}{2} \right)^{\overline{S}-1},
\tag{3.5}
$$

since the number of paths for the above 3 stopping points in 1, 1 and r, respectively.

The corresponding value of $P\{RG\}$ from (3.1) using the same N_j-values is clearly

$$
P\{RG\} = 1 - \left(\frac{1}{2} \right)^{\overline{S}} - \left(\frac{1}{2} \right)^r - r \left(\frac{1}{2} \right)^{\overline{S}-1}.
\tag{3.6}
$$

For any given value of $\overline{S}$ (say, $\overline{S} = 10$) and given P^* (say, $P^* = .95$) and $\underline{S} = 3$, we set the right side of (3.6) $\geq .95$ and solve for the smallest integer r obtaining $r = 5$ and hence this is an example of a regular case (Case 1). After randomization between $r = 4$ and $r = 5$ to obtain $P^* = .95$ exactly, the same randomization between $r = 4$ and $r = 5$ yields the result $E\{T\} = 4.31089$, by using (3.5); this result is included in the tables at the end of this paper. In the next section we show why this procedure based on the length of the initial run (or on the total number of girls before the jth boy is obtained, $j = 2, 3, \cdots$) from the quota-free cell (regular or irregular cases) or from a combined low-quota cell (degenerate case) is optimal, i.e., produces the smallest value of $E\{T\}$.

4   Proof of Optimality

The proof is carried through for several special examples, using at least one example for each of the two cases mentioned in Section 2; after this we append a few remarks necessary to apply our method to any multinomial problem within our model. It should be borne in mind, although we stick to the more usual values of P^* (like .95) that the form of the optimality result does not depend on these values and holds for all values of these parameters, provided we stay in the same cases. Thus, if P^* decreases from (say) .95 toward zero, we enter Case 2 at some point; we claim that the corresponding procedures are optimal for both cases.

Consider as an example of the regular case (Case 1) the goal of 2 girls in the "baby problem." so that $\underline{S} = 2$. Since boys have no quota, they constitute, a quota-free cell and plotting boys on the x-axis (with girls on the y-axis), we add the stopping point $(r, 0)$ to the stopping point $(x, 1)$ with $r \leq x \leq \overline{S}$; here we take an arbitrary $p_g = p$ and $p_b = 1 - p$ but actually assume only that $p_g = p$ for the quota-free cell is sufficiently removed from 0 so that we remain in the regular case.

Before proceeding we explain why we wrote $x \geq r$ above. Suppose first that $x < r - 1$. Then the set of stopping points do not form a closed procedure since we can first get $r - 1$ boys then 1 girl and then more boys without satisfying the upper bound condition. Suppose now that $x = r - 1$. Then the point $(x, 0)$ now has the property that we take one more observation and stop without reaching the goal in either case then $P\{RG\}$ is unchanged and $E\{T\}$ is diminished if we replace the stopping point $(r, 0)$ by $(x, 0)$. Hence $x = r - 1$ also cannot lead to an optimal solution; thus we must have $x \geq r$.

Then from (3.1) we have to find values of x such that for some $r(\underline{S} \leq r \leq x \leq \overline{S})$

$$(1 - p)^r + r(1 - p)^x p \leq 1 - P^*. \tag{4.1}$$

We now treat r and x both as continuous variables so that we attain equality in (4.1) without randomization. For fixed P^* and $0 < p \leq 1$ we have from (3.3)

$$\begin{aligned} E\{T\} &= \frac{2}{p} - \frac{2}{p}(1 - p)^r - \frac{r}{p}(1 - p)^x p \\ &= \frac{2}{p} - \frac{(1 - p)^r}{p} - \frac{1}{p}(1 - P^*) = \frac{1 + P^* - (1 - p)^r}{p}. \end{aligned} \tag{4.2}$$

This is non-decreasing in r and if we can show from (4.1) with equality that r and x go in opposite directions then it follows that $E\{T\}$ is decreasing in x and we obtain the best result at $x = \overline{S}$. Recall that r is the length of an initial run of boys for which we stop. From equality in (4.1) for fixed P^* we obtain by differentiation

$$\frac{dr}{dx} = \frac{rp(1 - p)^{x - r} \ln(1 - p)}{\ln \frac{1}{1 - p} - p(1 - p)^{x - r}}. \tag{4.3}$$

Since $r \leq x$ and $\ln \left(\frac{1}{1-p}\right) \geq p$ for $0 \leq p \leq 1$, it follows that the denominator in (4.3) is positive and hence $\frac{dr}{dx}$ is negative. It follows that for this regular case $E\{T\}$ is minimized by setting $x = \overline{S}$. Since this is a typical regular case, the general proof for any such regular case will be similar.

The proof for the (irregular) case of two cells with equal quotas is somewhat different from the above. The special goal of 1 boy and 1 girl is easy to prove using (3.1) and (3.2) but is not typical; we therefore consider the goal of 2 boys and 2 girls, which is more typical of the general case. Let $p_g = p$, $p_b = 1 - p$ and assume $p \leq \frac{1}{2}$. Here there is no quota-free cell and we show by taking the exceptional stopping point as an initial run of boys $\left(p_b > \frac{1}{2}\right)$ of sufficient length r to satisfy condition (i) that we obtain an optimal result for $E\{T\}$. Plotting girls (resp. boys) on the x-axis (resp., y-axis), let $a \geq r$ denote the stopping point on the line $y = 1$, let b (resp., c) denote the stopping points on the line $x = 0$ (resp., line $x = 1$). For the same reason as expressed earlier, we assume that $c \geq b$. From (3.1) and (3.2) we obtain the two results

$$
\begin{aligned}
P\{RG\} &= 1 - (1 - p)^r - rp(1 - p)^a - p^b - b(1 - p)p^c \\
&\geq 1 - P^*,
\end{aligned} \tag{4.4}
$$

$$
\begin{aligned}
E\{T\} &= \frac{2}{p}C^{(1)}_{(1-p)/p}(2,3) + \frac{2}{1-p}C^{(1)}_{p/(1-p)}(2,3) \\
&\quad -\frac{2}{p}(1-p)^r - r(1-p)^a - \frac{2}{1-p}p^c - bp^c.
\end{aligned} \tag{4.5}
$$

Treating r, a, b and c as continuous parameters, we obtain an equality in (4.4) and, applying this to (4.5), by solving for $(1-p)^r$, we obtain

$$
\begin{aligned}
E\{T\} &= \frac{2}{p}C^{(1)}_{(1-p)/p}(2,3) + \frac{2}{1-p}C^{(1)}_{p/(1-p)}(2,3) - \frac{2}{p}(1-P^*) \\
&\quad +r(1-p)^a + 2p^b\left(\frac{1-2p}{p(1-p)}\right) + bp^c\left(\frac{2-3p}{p}\right).
\end{aligned} \tag{4.6}
$$

We now consider three steps, all with P^* fixed. In the first step we play r against a and show that $E\{T\}$ decreases with a and increases with r, so that for any values of b and c we get the best result by pushing a up to its maximum value $\overline{S} - 1$. This step is already clear from (4.6) without taking derivatives in (4.4). In the second step we play r against c with a and b fixed. For $p \leq 2/3$ it follows from (4.6) that $E\{T\}$ is decreasing in c and increasing in r and thus the second step in our proof also follows from (4.6) and we push b up to its maximum value $\overline{S} - 1$. For $p \leq 1/2$ (and hence also $\leq 2/3$) we wish in the third step of the proof to play r against b with a and c fixed. The result (4.6) is clearly increasing in r for fixed P^*. Hence, if we can show from (4.4) with equality that r and b are going in opposite directions, the proof will be complete. By differentiation of (4.4) with equality to P^* we obtain, letting $q = 1 - p$,

$$\frac{\partial r}{\partial c} = -\frac{p^b}{q^r}\left[\frac{\ln\frac{1}{p} - qp^{c-b}}{\ln\frac{1}{q} - pq^{a-r}}\right] \tag{4.7}$$

Since $a \geq r$, $c \geq b$ and $\ln\left(\frac{1}{x}\right) \geq 1 - x$ for all $x(0 \leq x \leq 1)$, we use this inequality twice and it follows that $\frac{\partial r}{\partial c}$ is negative. Hence we obtain the minimum $E\{T\}$ by pushing c up to its maximum value $\overline{S} - 1$.

Note that we assumed $p \leq 1/2$ in the above proof. If $p > 1/2$ then we define the exceptional point by an initial run of girls, rather than boys, and the same proof holds. Moreover, in a problem with equal quotas for both cells the proof will be similar.

The multinomial case with 3 or more cells will now be illustrated by a "baby problem" application with 3 cells denoted by B (for boy), G (for girl) and S (for stillborn) with cell probabilities p_b, p_g and $p_s = 1 - p_b - p_g$, respectively. Our goal is simply to get at least one boy *and* at least one girl, so that the S cell is quota-free. Assume that p_s is large enough so that for some $r \geq 1$ we have $P\{RG|r-1\} < P^* \leq P\{RG|r\}$; although r could be 1, it will more typically be 4 or 5. We can think of $\underline{S} = 2$ and $\overline{S} = 10$ to be definite, but these values are not utilized in the essential steps below.

The waiting-time procedure $R = R(L, M)$ will be such that if you have at least one girl (and no boys) then the stopping total, denoted by L, will depend only on the total number of children. Similarly, if you have at least one boy (and no girls) the stopping total, denoted by M, will depend only on the total number of children. We do not assume that $L = M$ (since p_b and p_g are not necessarily equal) but both are subject to the same upper bound $\overline{S}$. These assumptions are justified because the probability of getting at least one boy (resp., girl) in n tries depends only on n and not on the number of girls obtained. We have to play r against L (and then r against M) to show that both L and M are to be increased to $\overline{S}$ for the optimal result, i.e., for minimizing $E\{T\}$.

The probability of reaching the goal $P\{RG\}$ is given by

$$
\begin{aligned}
P\{RG\} = \ & 1 - p_s^r - \sum_{i=1}^{L}\left[\binom{L}{i} - \binom{L-r}{i}\right]p_g^i p_s^{L-i} \\
& - \sum_{j=1}^{M}\left[\binom{M}{j} - \binom{M-r}{j}\right]p_b^j p_s^{M-j}
\end{aligned}
$$

$$= 1 - p_s^r - \left[(p_g + p_s)^L - p_s^r(p_g + p_s)^{L-r}\right] \tag{4.8}$$
$$- \left[(p_b + p_s)^M - p_s^r(p_b + p_s)^{M-r}\right].$$

Using the same Dirichlet coefficients as in (3.2) and letting

$$
\begin{aligned}
W_{1,1} &= \frac{1}{p_b} C^{(1)}_{p_g/p_b}(1,2) + \frac{1}{p_g} C^{(1)}_{p_b/p_g}(1,2) \\
&= \frac{1}{p_b + p_g}\left[1 + \frac{p_b}{p_g} + \frac{p_g}{p_b}\right], \tag{4.9}
\end{aligned}
$$

the expected waiting time for procedure R is given by

$$
\begin{aligned}
E\{T\} &= W_{1,1}(1 - p_s^r) - W_{0,1}\left[(p_b + p_s)^M - p_s^r(p_b + p_s)^{M-r}\right] \\
&\quad - W_{1,0}\left[(p_g + p_s)^L - p_s^r(p_g + p_s)^{L-r}\right], \tag{4.10}
\end{aligned}
$$

where $W_{01} = 1/p_g$ (resp., $W_{10} = 1/p_b$) denotes the waiting time for a girl (resp., a boy). Setting (4.8) equal to P^* and eliminating L between (4.8) and (4.10) we obtain for fixed P^* and any fixed M

$$
\begin{aligned}
E(T) &= -p_s^r(W_{1,1} - W_{1,0}) + W_{1,1} - W_{1,0}(1 - P^*) \\
&\quad + (W_{1,0} - W_{0,1})\left[(p_b + p_s)^M - p_s^r(p_b + p_s)^{M-r}\right]. \tag{4.11}
\end{aligned}
$$

It is fairly straightforward to show (in all 3 cases described below) that for both $p_b \le p_g$ and $p_b \ge p_g$, the value in (4.11) is increasing in r; we omit these details. Hence we have to show in (4.8) for (say) L fixed as well as P^* that if r increases then M must decrease (or at least be non-increasing). Equating (4.8) to P^* we have, using an equivalent form,

$$
\begin{aligned}
P\{RG\} &= 1 - (p_g + p_s)^L - (p_b + p_s)^M \\
&\quad - p_s^r\left[1 - (p_g + p_s)^{L-r} - (p_b + p_s)^{M-r}\right] = P^*. \tag{4.12}
\end{aligned}
$$

Differentiation in (4.12) with respect to r, holding L fixed, yields

$$
\begin{aligned}
\frac{dM}{dr}\ln\left(\frac{1}{p_b + p_s}\right)&\left[\frac{(p_b + p_s)^M - p_s^r(p_b + p_s)^{M-r}}{p_s^r}\right] \\
&= \ln p_s + (p_g + p_s)^{L-r}\ln\left(\frac{p_g + p_s}{p_s}\right) \\
&\quad + (p_b + p_s)^{M-r}\ln\left(\frac{p_b + p_s}{p_s}\right). \tag{4.13}
\end{aligned}
$$

Case 1: Assume r is strictly less than both L and M.

Since the 2nd and 3rd term have positive logarithms as coefficients and since r is strictly less than both L and M, we can assume that $L - r = M - r = 1$ in order to show that the right side of (4.13) is non-positive, i.e., if either $L - r$ or $M - r > 1$ then the right side of (4.13) is reduced. Hence (using the fact that $p_g + p_b + p_s = 1$) it is sufficient to show that

$$(p_g + p_s)\ln(p_g + p_s) + (p_b + p_s)\ln(p_b + p_s) - p_s \ln p_s \leq 0 \qquad (4.14)$$

or equivalently, setting $p_b + p_s = 1 - p_g$ and $x = p_s$, that for $0 \leq x \leq 1 - p_g$ and any p_g $(0 \leq p_g \leq 1)$

$$(p_g + x)\ln(p_g + x) + (1 - p_g)\ln(1 - p_g) - x \ln x \leq 0. \qquad (4.15)$$

By differentiation with respect to x in (4.15) we find that the left side of (4.15) is increasing in x for any fixed p_g and takes its maximum value at $x = 1 - p_g$. Since this gives an equality in (4.15), the inequality in (4.15) holds. Hence $\partial M/\partial r$ is negative in (4.13) and hence M has to be pushed up to $\bar{S}$ in order to get the smallest value for $E\{T\}$. Since the same result holds for L, we have proved the result we wanted to show for Case 1.

Case 2: Assume $L = r_0$, $1 \leq r \leq r_0$ and $r < M$.

A similar proof to the above can be used to show that $\partial M/\partial r$ is negative and hence M has to be pushed up to $\bar{S}$ in order to get the smallest value for $E\{T\}$; we omit the details

Case 3: Assume that $L = M = r_0$ and $1 \leq r \leq r_0$.

It is obvious that if both L and M are fixed than $P\{RG\}$ and $E(T)$ will both be increasing in r and we omit this detail.

The optimal results for $P\{RG\}$ and $E\{T\}$ are then obtained by setting both L and M equal to $\bar{S}$ in (4.8) and (4.10) and using the smallest integer r that will make $P\{RG\}$ in (4.8) equal to at least the specified P^*.

In summary, we have shown for the binomial case and also for a trinomial case that the optimal solution has the properties stated earlier. We believe that the same properties hold for any similar multinomial problem. Since we have not given a general proof we state our result in the form of a conjecture.

Conjecture

The optimal solution for any such multinomial inverse sampling problem is to find a stopping rule with the three or four features given in Section 2. In all cases we first combine the quota-free cells (or assume there is only 1 quota-free cell C_0 with cell probability p_0). If p_0 is too small we have to combine the quote-free cell (a) with a low quota cell when the quotas are unequal and (b) with the cell having a higher cell probability when the quotas are equal. The fact that the variable stopping points (with only one exception) are pushed to their highest (*and/or* lowest) value makes it easy to compute the optimal waiting-time procedure and its properties for any particular goal and for any specified values of P^*, $\underline{S}$ and $\overline{S}$.

The Dirichlet result for the unrestricted waiting-time problem was given in (3.2) and appears as the first term in (3.3). As a result of the final statement in the theorem above, the value $\overline{S}$ appears as an exponent on the right side of both (3.1) and (3.3). Taking the limit as $\overline{S} \to \infty$ in the final result will generally give us useful upper bounds on both $P\{RG\}$ and $E\{T\}$. Thus for the example in Section 3 with $P^* = .95$, we obtain

$$P\{RG\} \leq 1 - \left(\frac{1}{2}\right)^5 = .96875 \text{ and}$$

$$E\{T\} \leq 4.5 - 4\left(\frac{1}{2}\right)^5 = 4.37500. \tag{4.16}$$

For the example in (4.1) and (4.2) with $P^* = .95$ and $p = .5$ and $\overline{S} = 8$, we find that $r = 6$ and the bounds are

$$P\{RG\} \leq (1-p)^r = .98438 \text{ and}$$

$$E\{T\} \leq \frac{2}{p} - \frac{2}{p}(1-p)^r = 3.93750; \tag{4.17}$$

the exact value, using the second expression in (4.2), is 3.85547.

Notes to Table 1

$\underline{S}$, $\overline{S}$ are the lower and upper bounds, resp., $p = $ cell probability for a boy, and $r = r_1 = $ stopping value for the length of the initial run from the quota-free cell C_0 (or from C_0^*).
$*\overline{S}$ indicates that $P\{RG\} > P^*$ (strictly) for any allowable value of r_1. Note that the next line up (at $\overline{S} - 1$) has a smaller $E\{T\}$-value and whenever possible, we switch to that line without violating any condition. In each subgroup the first value of $\overline{S}$ listed is the smallest consistent with the specified P^*.
If the top line in any subgroup has a $*$ in front of $\overline{S}$ as in only one case on Page 2 of Table 1 then we are in the unusual Case 5 and have to reduce the stopping value before the second boy; we use the total number of girls before the second boy.
\# These are not optimal results and the explanation is in the paper at the end of Section 2; they are the only such results in Table 1.

Table 1: Optimal results for various goals and for specified P^*, $\underline{S}$, and $\overline{S}$.

	$P^* = .75$			$P^* = .90$			$P^* = .95$			$P^* = .99$		
	$\overline{S}$	r	E{T}	$\overline{S}$	r	E{T}	$\overline{S}$	r	E{T}	$\overline{S}$	r	E{T}
					Goal: 1 boy, 0 girl (2 cells)							

Goal: 1 boy, 0 girl (2 cells)

	$\overline{S}$	r	E{T}	$\overline{S}$	r	E{T}	$\overline{S}$	r	E{T}	$\overline{S}$	r	E{T}
	2	2	1.5000	4	4	1.8000	5	5	1.9000	7	7	1.9800
	3	2	1.5000	5	4	1.8000	6	5	1.9000	8	7	1.9800
p = .50	4	2	1.5000	6	4	1.8000	7	5	1.9000	9	7	1.9800
	5	2	1.5000	7	4	1.8000	8	5	1.9000	10	7	1.9800
	UE		2.0000	UE		2.0000	UE		2.0000	UE		2.0000
	5	5	3.0000	9	9	3.6000	11	11	3.8000	17	17	3.9600
	6	5	3.0000	10	9	3.6000	12	11	3.8000	18	17	3.9600
p = .25	7	5	3.0000	11	9	3.6000	13	11	3.8000	19	17	3.9600
	8	5	3.0000	12	9	3.6000	14.11		3.8000	20	17	3.9600
	UE		4.0000	UE		4.0000	UE		4.0000	UE		4.0000

Goal: 2 boys, 0 girl (2 cells)

	$\overline{S}$	r	E{T}	$\overline{S}$	r	E{T}	$\overline{S}$	r	E{T}	$\overline{S}$	r	E{T}
	5	3	3.1667	7	4	3.6607	8	6	3.8396	11	8	3.9672
	6	3	3.0714	8	4	3.6283	9	5	3.8183	12	7	3.9634
p = .50	7	3	3.0333	9	4	3.6137	10	5	3.8089	13	7	3.9617
	8	3	3.0161	10	4	3.6068	11	5	3.8044	14	7	3.9608
	UE		4.0000	UE		4.0000	UE		4.0000	UE		4.0000
	10	9	6.6041	15	11	7.3798	18	13	7.6955	24	20	7.9465
	11	7	6.3609	16	10	7.3247	19	12	7.6668	25	19	7.9382
p = .25	12	6	6.2459	17	9	7.2892	20	12	7.6482	26	18	7.9331
	13	6	6.1750	18	9	7.2651	21	12	7.6352	27	16	7.9296
	UE		8.0000	UE		8.0000	UE		8.0000	UE		8.0000

Goal: 3 boys, 0 girl (2 cells)

	$\overline{S}$	r	E{T}	$\overline{S}$	r	E{T}	$\overline{S}$	r	E{T}	$\overline{S}$	r	E{T}
	7	4	5.0156	9	5	5.6338	11	6	5.7806	14	8	5.9582
	*8	3	5.0390	10	4	5.5154	12	5	5.7424	15	8	5.9495
p = .50	*9	3	5.1328	11	4	5.4620	13	5	5.7229	16	7	5.9451
	*10	3	5.1856	12	4	5.4336	14	6	5.7124	17	7	5.9427
	UE		6.0000	UE		6.0000	UE		6.0000	UE		6.0000
	15	9	10.1675	20	13	11.2979	23	18	11.7082	31	21	11.9316
	16	8	9.8123	21	11	11.1602	24	14	11.6147	32	19	11.9185
p =.25	17	7	9.6046	22	10	11.0709	25	13	11.7608	33	18	11.9093
	18	6	9.4579	23	10	11.0066	26	12	11.5223	34	18	11.9025
	UE		12.0000	UE		12.0000	UE		12.0000	UE		12.0000

Table 1: Optimal results for various goals and for specified $P^*, \underline{S}$, and $\overline{S}$ (continued).

	$P^* = .75$			$P^* = .90$			$P^* = .95$			$P^* = .99$		
						Goal: 1 boy, 1 girl (2 cells)						
	$\overline{S}$	r	E{T}	$\overline{S}$	r	E{T}	$\overline{S}$	r	E{T}	$\overline{S}$	r	E{T}
	4	3	2.5000	5	4	2.8000	6	5	2.9000	9	7	2.9800
	5	3	2.5000	6	4	2.8000	7	5	2.9000	10	7	2.9800
p=.50	6	3	2.5000	7	4	2.8000	8	5	2.9000	11	7	2.9800
	7	3	2.5000	8	4	2.8000	9	5	2.9000	12	7	2.9800
	UE		3.0000	UE		3.0000	UE		3.0000	UE		3.0000
	6	5	3.3340	10	9	3.9333	12	11	4.1333	18	17	4.2933
	7	5	3.3335	11	9	3.9333	13	11	4.1333	19	17	4.2933
p=.25	8	5	3.3334	12	9	3.9333	14	11	4.1333	20	17	4.2933
	9	5	3.3333	13	9	3.9333	15	11	4.1333	21	17	4.2933
	UE		4.3333	UE		4.3333	UE		4.3333	UE		4.3333
						Goal: 2 boys, 1 girl (2 cells)						
	5	4	3.7500	7	5	4,1792	8	6	4.3500	11	8	4.4683
	*6	3	3.8750	8	4	4.1367	9	5	4.3225	12	7	4.4639
p=.50	*7	3	3.9375	9	4	4.1177	10	5	4.3109	13	7	4.4619
	*8	3	3.9688	10	4	4.1087	11	5	4.3054	14	7	4.4610
	UE		4.5000	UE		4.5000	UE		4.5000	UE		4.5000
	10	9	6.6874	15	11	7.4631	18	13	7.7788	24	20	8.0298
	11	7	6.4443	16˙	10	7.4081	19	12	7.7501	25	19	8.0215
p=.25	12	6	6.3292	17	9	7.3725	20	12	7.7316	26	18	8.0165
	13	6	6.2583	18	9	7.3484	21	12	7.7184	27	17	8.0129
	UE		8.0333	UE		8.0333	UE		8.0333	UE		8.0333
						Goal: 3 boys, 1 girl (2 cells)						
	*7	4	5.3906#	9	6	5.8969	11	7	6.0329	14	8	6.2084
	*8	4	5.6016#	10	5	5.7697	12	5	5.9935	15	8	6.1996
p=.50	*9	4	5.7227#	*11	4	5.8291	13	5	5.9734	16	8	6.1951
	*10	4	5.7910#	*12	4	5.8501	14	5	5.9626	17	8	6.1928
	UE		6.2500	UE		6.2500	UE		6.2500	UE		6.2500
	15	9	10.1883	20	13	11.3187	23	18	11.7290	31	21	11.9524
	16	8	9.8331	21	11	11.1810	24	14	11.6355	32	19	10.1230
p=.25	17	7	9.6255	22	10	11.0917	25	13	11.5816	33	18	10.1137
	18	6	9.4787	23	10	11.0274	26	12	11.5432	34	18	10.1069
	UE		12.0208	UE		12.0208	UE		12.0208	UE		12.0208

References

Anderson, K., Sobel, M., and Uppuluri, V.R.R. (1982). Quota fulfilment times. *Can. Jour. Statist.* **10** 73 - 88.

Ebneshahrashoob, M. and Sobel, M. (1991). Personal communication.

Sobel, M., and Ebneshahrashoob, M. (1989). Optimal sequential stopping rules for certain binomial sampling problems. *Recent Developments in Statistics and Their Applications* (J. Klein and J. Lee, eds.), Freedom Academy, 333 - 362.

Sobel, M., Uppuluri, V.R R. and Frankowski, K. (1977). *Selected Tables in Mathematical Statistics*, Vol. 4. Published by AMS and IMS.

Sobel, M., Uppuluri, V.R.R. and Frankowsk, K. (1979). *Selected Tables in Mathematical Statistics*, Vol. 9. Published by AMS and IMS.

Publications of
Charles W. Dunnett

Dunnett, C.W. and Hopkins, J.W. (1951). Two stage acceptance sampling by attributes. *American Soc. Test. Mat.*, Spec. Tech. Pub. No. 114.

Campbell, J., McLaughlan, J., Clark, J. and Dunnett, C.W. (1953). The six-point design in the U.S.P. microbiological assay of vitamin B 12. *Jour. Amer. Pharm. Assoc.* **42** 276 - 283.

Dunnett, C.W. (1954). Statistical planning and field trials. In *Proc. 14th Annual Meeting, Animal Health Institute* **72**.

Dunnett, C.W. and Sobel, M. (1954). A bivariate generalization of Student's *t*-distribution, with tables for some special cases. *Biometrika* **41** 153 - 169.

Bechhofer, R.E., Dunnett, C.W., and Sobel, M. (1954). A two-sample multiple decision procedure for ranking means of normal populations with a common unknown variance. *Biometrika* **41** 170 - 176.

Dunnett, C.W. and Crisafio, R. (1955). The operating characteristics of some official weight variation tests for tablets. *Jour. Pharm. and Pharmacol.* **7** 314 - 327.

Dunnett, C.W. and Sobel, M. (1955). Approximations to the probability integral and certain percentage points of a multivariate analogue of Student's *t*-distribution. *Biometrika* **42** 258 - 260.

Dunnett, C.W. (1955). A multiple comparison procedure for comparing several treatments with a control. *Jour. Amer. Stat. Assoc.* **50** 1096 - 1121.

Dunnett, C.W. (1955). Multiple comparisons with a standard. *Trans. Amer. Soc. Qual. Cont.* **9** 485 - 491.

Dunnett, C.W. (1955). Statistical need for new tables involving the multivariate bormal distribution. In *18th Annual Meeting of the Institute of Mathematical Statistics*, New York City, December 27–30, 1 - 4.

Dearborn, E.H., Litchfield, J.T., Eisner, H.J., Corbett, J.J. and Dunnett, C.W. (1957). The effects of various substances on the absorption of Tetracycline in rats. *Antiobiotic Med. Clin. Ther.* 4 627 - 641.

Dunnett, C.W. (1957). Screening problems in the pharmaceutical industry. *Jour. Amer. Stat. Assoc.* 52 368.

Dunnett, C.W. (1960). Tables of the bivariate normal distribution with correlation $1/\sqrt{2}$. *Math. Comput.* 14 79.

Dunnett, C.W. and Lamm, R.A. (1960). Some tables of the multivariate normal probability integral with correlation coefficients 1/3. *Math. Comput.* 14 290.

Dunnett, C.W. (1960). On selecting the largest of k normal population means (with discussion). *Jour. Roy. Stat. Soc. B* 22 1 - 30.

Dunnett C.W (1961). Approaches to some problems in drug screening and selection. Presented at the *Gordon Research Conference in Chemistry and Chemical Engineering*, August 7.

Dunnett, C.W. (1961). Statistical theory of drug screening. In *Quantitative Methods in Pharmacology*, North Holland Publishing Company, 212 - 231.

Curnow, R.N. and Dunnett, C.W. (1962). The numerical evaluation of certain multivariate normal integrals. *Ann. Math. Statist.* 33 571 - 579.

Dunnett, C.W. and Lamm, R.A. (1963). Sequential procedures for drug screening. *Jour. Amer. Stat. Assoc.* 58 549.

Dunnett, C.W. (1964). New tables for multiple comparisons with a control. *Biometrics* 20 482 - 491.

Dunnett, C.W. (1968). Screening and selection. In *International Encyclopedia of the Social Sciences*, Macmillan and Free Press, 14 123 - 127.

Dunnett, C.W. (1968). Biostatistics in pharmacological testing. In *Medicinal Research: Biology and Chemistry III, Pharmacological Testing Methods*, Marcel Dekker, 7 - 49.

Dunnett, C.W. (1970). Multiple comparisons. In *Statistics in Endocrinology*, MIT Press, 79 - 103.

Dunnett, C.W. (1971). Optimum procedures for drug screening, In *38th Session of the International Statistical Institute*, Washington, 118 - 122.

Dunnett, C.W. (1971). Optimum procedures for drug screening, Spring Meeting of Eastern North American Region of Biometric Society, University Park, Pennsylvania, April 21–23, 1 - 10.

Dunnett, C.W. (1972). Drug screening: The never-ending search for new and better drugs. In *Statistics, A Guide to the Unknown*, Holden-Day, 23 - 33.

Dunnett, C.W. and Gent, M. (1977). Significance testing to establish equivalence between treatments, with special reference to data in the form of 2×2 tables. *Biometrics* **33** 593 - 602.

Dent, P., McCulloch, P., Wesley-James, O., MacLaren, R., Muirhead, W., and Dunnett, C.W. (1978). Measurement of carcinoembryonic antigen in patients with bronchogenic carcinoma. *Cancer* **42** 1484 - 1491.

Johnson, A.L., Taylor, D.W., Sackett, D.L., Dunnett, C.W., and Shimizu, A.G. (1978). Self-recording of blood pressure in the management of hypertension, *Can. Med. Assoc.* **119** 1034 - 1039.

Juniper, E., Frith, P., Dunnett, C.W., Cockcroft, D., and Hargreave, F. (1978). Reproducibility and comparison of responses to inhaled histamine and methacoline, *Thorax* **33** 705 - 710.

Dunnett, C.W. (1979). A Monte Carlo study to determine the validity of a multiple comparisons procedure. In *12th European Meeting of Statisticians*, Varna, Bulgaria, September 3–7, Abstract no. 87.

Dent, P., Louis, J., McCulloch, P., Dunnett, C.W. and Cerottini, J-C. (1980). Correlation of elevated C1 binding activity and carcinoembryonic antigen levels with clinical features and prognosis in bronchogenic carcinoma. *Cancer* **45** 130 - 136.

Dunnett, C.W. (1980). Pairwise multiple comparisons in the homogeneous variance, unequal sample size case. *Jour. Amer. Statist. Assoc.* **75** 789 - 795.

Dunnett, C.W. (1980). Pairwise multiple comparisons in the unequal variance case. *Jour. Amer. Statist. Assoc.* **75** 796 - 800.

Dunnett, C.W. and Goldsmith, C.H. (1981). When and how to do multiple
comparisons. In *Statistics in the Pharmaceutical Industry*, Marcel
Dekker, Inc., 397 - 433.

Dunnett, C.W. (1981). Multiple comparison methods, In *3rd Hungarian
Conference on Statistical Analysis of Observational Data*, Budapest,
Hungary,165 - 169.

Bechhofer, R.E. and Dunnett, C.W. (1982). Multiple comparisons for
orthogonal contrasts: Examples and tables. *Technometrics* **24** 213 -
232.

Dunnett, C.W. (1982). Robust multiple comparisons. *Commun. Statist.
- Theory Meth.* **11** 2611 - 2629.

Alavi, M., Dunnett, C.W., and Moore, M. (1983). Lipid composition
of rabbit aortic wall following removal of endothelium by balloon
catheter. *Arteriosclerosis* **3** 413 - 419.

Gilbert, J.R., Feldman, W., Siegel, L., Mills D.-A., Dunnett, C.W., and
Stoddart, G. (1984). How many well-baby visits are necessary in the
first two years of life? *Can. Med. Assoc. Jour.* **130** 857 - 861.

Dunnett, C.W. (1984). Selection of the best treatment in comparison
to a control, with an application to a medical trial. In *Design of
Experiments, Ranking and Selection, Essays in Honour of Robert E.
Bechhofer*, Marcel Dekker, 47 - 66.

Dunnett, C.W. (1985). Multiple comparisons between several treatments
and a specified treatment. In *Lecture Notes in Statistics No. 35,
Linear Statistical Inference*, Springer-Verlag, 39 - 47.

Bechhofer, R.E. and Dunnett, C.W. (1986). Two-stage selection of the
best factor-level combination in multifactor experiments: Common
unknown variance. In *Statistical Design: Theory and Practice, Proceedings of a Conference in Honor of Walter T. Federer*.

Bechhofer, R.E. and Dunnett, C.W. (1987). Subset selection for normal
means in multi-factor experiments. *Commun. Statist.- Theory Meth.*
16 2277 - 2286.

Dunnett, C.W. (1987). Multivariate Student's t: Applications and numerical computations, Statistical Society of Canada, Annual Meeting, June 1.

Bechhofer, R.E. and Dunnett, C.W. (1988). Percentage points of multivariate Student t-distributions. *Selected Tables in Mathematical Statistics* **11** 1 - 315.

Dunnett C.W (1988). Statistical design in drug screening. Presented at *Symposium on Statistics in the Pharmaceutical Industry (in honor of Joseph L. Ciminera)*, Philadelphia, June 15.

Mohide, E.A., Tugwell, P.X., Caulfield, P.A., Chambers, L.W., Dunnett, C.W., Baptiste, S., Bayne, J.R., Patterson, C., Rudnick, K.V., and Pill, M. (1988). A randomized controlled trial of quality assurance in nursing homes. *Medical Care* **26** 554 - 565.

Bechhofer, R.E., Dunnett, C.W., and Tamhane, A.C. (1989). Two-stage procedures for comparing treatments with a control: Elimination at the first stage and estimation at the second stage. *Biomet. Jour.* **31** 5, 545 - 561.

Dunnett, C.W. (1989). Multivariate normal probability integrals with product correlation structure. *Appl. Statist.* **38** (3) 564 - 579.

Holford, T.R., Walter, S.D., and Dunnett, C.W. (1989). Simultaneous interval estimates of the odds ratio in studies with two or more comparisons. *Jour. Clin. Epidem.* **42** 427 - 434.

Bechhofer, R.E., Dunnett, C.W., Goldsman, D.M., and Hartmann, M. (1990). A comparison of the performances of procedures for selecting the normal population having the largest mean when the populations have a common unknown variance. *Commun. Statist. - Simul.* **19**(3) 971 - 1006.

Dunnett, C.W. and Tamhane, A.C. (1991). Step-down multiple tests for comparing treatments with a control in unbalanced one-way layouts. *Statist. Med.* **10** 939 - 947.

Dunnett, C.W. and Tamhane, A.C. (1992). A step-up multiple test procedure. *Jour. Amer. Statist. Assoc.* **87** 162 - 170.

Dunnett, C.W. and Tamhane, A. C. (1992). Comparisons between a new drug and active and placebo controls in an efficacy clinical trial. *Statist. Med.* **11** 1057 - 1063.

Index